2-19
NLS

D1315319

WHERE TO FIND BIRDS IN NEW YORK STATE

A York State Book

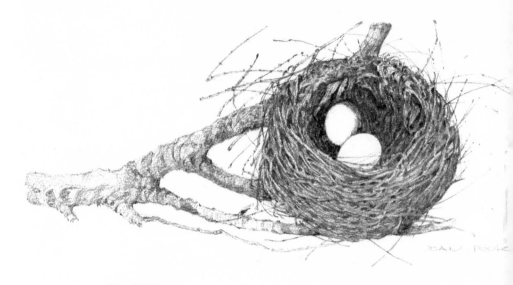

Myrtle Warbler nest with eggs. By permission of Joan Poole.

WHERE TO FIND BIRDS IN NEW YORK STATE

The Top 500 Sites

SUSAN RONEY DRENNAN

SYRACUSE UNIVERSITY PRESS
1981

Library of Congress Cataloging in Publication Data

Drennan, Susan Roney.
    Where to find birds in New York State.

    (A York State book)
    Bibliography: p.
    Includes index.
    1. Bird watching—New York (State) I. Title.
QL684.N7D73   598'.07'234747   81-16744
ISBN 0-8156-2250-3   AACR2
ISBN 0-8156-0173-5 (pbk.)

Manufactured in the United States of America

For the sharp-eyed trio —
Matthew, Matthew Paul, and Maureen Roney.
Anyone who knows them will know why.

Susan Roney Drennan is Associate Editor of *American Birds,* the ornithological field journal of the National Audubon Society. A recognized authority on birds of the Americas, she has written articles for both popular magazines and scientific journals and has had a hand in the production of important books on birds, among which are *The Birder's Field Notebook* and *Audubon's Birds of America.*

Roger Tory Peterson is a world-honored author of field guides, a renowned naturalist, and an artist and photographer.

CONTENTS

FOREWORD

ALTHOUGH PHILADELPHIA can be called the cradle of American ornithology, because both Alexander Wilson and John James Audubon once lived near that city, it was Cambridge, Massachusetts, that nurtured the infant science. William Brewster and other New England ornithophiles formed the Nuttall Club, the first bird club in America, which in turn fathered the American Ornithologists' Union (the AOU), the top scientific bird organization in the New World.

Field birding as we know it today, however, had its roots in New York. At the turn of the century Dr. Frank M. Chapman of the American Museum of Natural History initiated the Christmas Bird Count (then called the "Christmas Census"), as a substitute for the traditional Christmas bird-shooting parties. But serious field ornithology continued to be largely a matter of the collecting gun until Ludlow Griscom, a junior staff member of the Department of Ornithology at the American Museum, proved to his colleagues that a skilled observer with a binocular could identify most birds correctly even at a distance by their "field marks." There was no need to shoot.

I recall those days back in the late 1920s when Ludlow Griscom held court in the museum at the Tuesday night meetings of the Linnaean Society, and the grilling he gave the young members of the Bronx County Bird Club when they reported their unusual finds. Although I came from the other end of the state—Jamestown, in Chautauqua County, I was admitted as the first non-Bronx member of this exclusive group of teenagers which included Joseph Hickey, Allan Cruickshank, and later, John Bull and others who were to become well known in ornithological circles.

Griscom and his cohorts eventually spread the gospel to the far corners of the land. My own *Field Guide to the Birds,* first published in 1934, presented in visual form the lessons I had learned from the master and from his young disciples. Bird-watching was rapidly becoming a super-sport, and it was inevitable that a different sort of guide was needed to inform the burgeoning army of birders where to go. In 1947, when I updated the *Field Guide* for the second time, I wrote in the Preface, "The author is working on a book that he hopes will be a useful companion to this guide, a sort of Baedeker that will tell you *where* to look for birds. . . . I would welcome ideas about this." Almost immediately I heard from

Dr. Olin Sewall Pettingill who informed me that he was already deeply into such a book, having conceived the idea in 1945, so I dropped my own plans which had not progressed very far.

It is appropriate that New York, where modern field techniques really started, now has, in *Where to Find Birds in New York State,* the most detailed and sophisticated state directory yet produced, thanks to the extraordinary dedication and skill of Susan Roney Drennan, associate editor of *American Birds,* a publication that no serious birder should be without. It is the only national magazine that monitors the ebb and flow of bird populations of our continent on a seasonal basis and reports the rarities through a network of regional editors.

During the last half century, and especially in recent years, there have been some remarkable changes in the pace and the techniques of field birding in New York State. I recall the pronouncement of Ludlow Griscom in 1927 that it was physically impossible to rack up a list of 250 species in the New York City region in the course of a single year. On the very last day of that same year, Frank Watson, a staff entomologist at the museum, ticked off his 250th bird, a razor-billed auk. Griscom soon duplicated the feat himself, and today yearly lists of 270 or more are not infrequent in the New York region thanks to the Rare Bird Alerts, regularly scheduled pelagic trips, hawk lookouts, and the countless feeders where avian goodies may turn up.

And now we have this very detailed handbook telling us where to go, when to go, and what we should see. During Ludlow Griscom's reign in New York it would have been impossible to prepare a book such as this. Many, indeed most, of the hot-spots were unknown. Birders did not enjoy the mobility or the means of transportation now available; few of them had cars. A visit to Montauk Point was almost a major expedition. Long Beach and Jones Beach were wild barrier islands reached only by ferry. What changes will we witness by the year 2000?

During the past fifty years more than two million acres of state parks have been made available to birders by Robert Moses and other far-sighted conservationists. Jamaica Bay and Tobay (formerly Guggenheim) Pond at Jones Beach are well known to New York City's field glass fraternity and Allegheny State Park is a mecca for thousands of birders living in the western part of the state. The Adirondacks and the Catskills are massive mountain preserves that must be vigilantly safeguarded by all of us if we would continue to enjoy good birding. We must never take them for granted.

It was not until the 1970s when trips were scheduled regularly by the Federation of New York State Bird Clubs that we began to realize the full potential of pelagic birding off the continental shelf. Such marine localities as Hudson Canyon, Block Canyon, Cox Ledge, and Georges Bank became familiar names to the field glass fraternity. Fulmars, Manx shearwaters, and skuas were found to be normal visitors to our offshore waters, not casual strays as we had thought, and even albatrosses have been sighted.

Cape May in New Jersey and Hawk Mountain in Pennsylvania have long been known as concentration points for birds of prey, particularly during the fall migration, but in recent years at least twenty-five vantage points or "lookouts" for migrating raptors have been discovered in New York State alone, one of which—

Derby Hill on Lake Ontario—is regarded as the top spring hawk-watching site in the Northeast. In general, the largest spring flights occur in the vicinity of the Great Lakes; coastal sites are best in the fall.

In addition to a network of pyramiding telephone calls, the tape recorded Rare Bird Alerts in New York, Albany, and Buffalo pinpoint stray birds, and hundreds of birders may flock in to watch one poor hungry waif far from home as it seeks to survive at somebody's feeder.

Stocking feeders has enhanced the winter survival of such non-migratory "southern" species as cardinals and titmice which have extended their ranges and are now resident throughout the greater part of New York State and New England. Evening grosbeaks, formerly birds of the western Canadian forest, have expanded explosively because of the largesse of sunflower seeds until they now nest east and south to upper New York State and New England.

Mockingbirds, seldom seen north of southern New Jersey during the 1930s and 1940s, are now commonplace as far north as the Great Lakes and New England, due perhaps to the widespread planting of *multiflora* roses. Several of the southern herons and egrets as well as the glossy ibis have recently colonized as far north as New York and New England for reasons that are not clear. Two European gulls, formerly accidental strays—the little gull and black-headed gull—have established nesting beachheads on our continent.

It is this dynamic side of bird-watching that fascinates the author of this book. Read carefully her descriptions of the geology, botany, and land forms of the places she directs you to.

The 106 skillfully executed maps and the more than 500 sites that Susan Drennan has researched, many with personal tours, is an unparalleled tour-de-force that will set a standard for state or regional guides for a long time to come. She gives us more than directions and a listing of the birds to be looked for; she also sketches the ecological components that determine why the birds are where they are. This is especially evident when she discusses the marine factors that dictate the occurrence of seabirds and the meteorological and physiographic conditions that influence hawk flights.

It is inevitable that the thoughtful person who watches birds becomes an environmentalist, although a few birders—a very few with tunnel vision—ignore the butterflies, the flowers, and the other components of the ecosystem. To them "the list" is the thing. Listing is a perfectly valid sport or recreation, and the best listers usually make good naturalists and conservationists. However, there are many other facets to bird-watching. It can be not only a skillful game, but also a science, an art, an esthetic experience, a healthful recreation, a philosophical pursuit—or just a bore—depending on the observer.

The author's aim is to help the birder move on from listing and identification to the level where he or she actually inquires into the whys and wherefores. It is only in that way that the birder can keep his or her interest meaningful and alive, contributing to the body of ornithological knowledge as well as to the environmental movement. The road signs, highway numbers, and landmarks are there, too, but the author's intent is to stress the ecological interplay.

Perhaps the most rewarding byproduct of birding is a sharpening of the senses—the eyes and the ears. It develops *awareness* and a regard for life—*all* life.

Old Lyme, Connecticut   Roger Tory Peterson
Fall 1981

PREFACE

IN THE MID-1970s, the Federation of New York State Bird Clubs, Inc., realized
that New York stood as one of an ever-decreasing number of states that had
not published a birding Baedeker. The Executive Committee of the Federation decided that preparation of such a publication was long overdue, and I was invited to
write the proposed book. So began what has been for me a long and exciting adventure culminating in these pages. These sites are not the only trophies, however,
for along the way I discovered some of the finest, most delightful, and inspiring
people I have had the good luck to know.

The scope of this book includes the entire state of New York. Its purpose is
to provide practical information on many more than 500 sites at which birding is
rewarding. The state has been divided into ten regions based on the presently existing reporting regions as defined by the Federation and adopted in its quarterly
publication, *The Kingbird.* Each region has an overall map showing the location
and name of each of its sites. Detailed accounts of each site, organized by county,
include directions for reaching the site, some of its botanical, geological, and
avian merits, its best seasons, and, of course, the exceptional and many common
birds which may or should be there. There is a generous sprinkling of very detailed
site maps throughout the text to aid the birder. Revealing the exact nesting whereabouts of rare or scarce species is an indiscretion I have often avoided by being deliberately vague.

I have included a discussion of the state's primary and secondary physiographic regions and have located it before the site guide section. This is rather
lengthy but its function is to enhance and deepen the appreciation of the traveling
birder for the sites he or she is visiting.

The first section of this volume contains a discussion of the New York State
Avian Records Committee and suggestions on how to take thorough field notes in
order to submit records of unusual or rare birds or birds out of season to the Committee. I have tried to emphasize the value of the aggregate body of such sightings
and urge all birders afield to carry a notebook in which to record their sightings.

Because field birding has undergone a metamorphosis in the last 10–15
years, today's student of birds wishes to employ a host of techniques geared to
sharpen and refine his or her skills. With this in mind I have included a separate

section devoted to the best places throughout the state where people can study preserved skins and mounts and where there are bird-book libraries open for study. I hope that this section is well used by birders in every corner of New York.

Following the sections detailing sites in each of the *Kingbird* reporting regions is a separate section on seabirds and pelagic birding whose aim is to orient the birder to oceanographic properties affecting the local abundance of seabirds. I designed the accompanying map for this section to familiarize the birder with his offshore surroundings in any direction from the state coastline.

The final section of the book details the most productive sites in New York at which to watch hawks in migration.

If the book is successful it is owing to a host of talented observers — some of whom rank among the best in the state — and a number of other people, including virtually all of the friends who have aided me by reading various chapters and giving me the benefit of suggestions for them. These readers are specifically acknowledged following those chapters with which they helped.

This book could not have become a reality without the help and inspiration of Roger Payne, who saw the manuscript in all its stages and gave invaluable assistance and warm encouragement when there were very many other calls on his time. He read each instalment of the manuscript, and his imaginative and perceptive commentary and incorporated suggestions have added infinitely to its quality; without his direction it would have been a far less useful guide.

In addition, I owe a special vote of thanks to Gordon Meade, who encouraged me to undertake the book, and to Robert Arbib, whose involvement in the project as Chairman of the Federation's Publications and Research Committee has been unflagging and who repeatedly and thoughtfully discussed many aspects of the book with me. They both generously and critically read the manuscript, and the book has benefitted from their contributions.

Throughout the evolution of the project I have relied heavily on Matthew Paul and Matt Drennan. They always went far out of their way to make my task less difficult and consistently buoyed me with their caring camaraderie. That sort of friendship goes beyond thanks. To Maureen Roney Drennan, whose imperishable influence and boundless energy permeate the pages of this book, I extend thanks.

Most of the maps in the book have been executed by myself and Sam Schoenbaum. The contribution of Mr. Schoenbaum can scarcely be overemphasized and will be fully appreciated only by those who regularly take the book afield.

Lee Roversi typed and proofread the manuscript and its many drafts and assisted in preparing the index, all with particular care, patience, and skill. Her spontaneous good humor lightened many work-weary days. Also vital was the help and support of Sara Casmer and Lois Heilbrun who cheerfully read galleys, and to them for this and much more I am indebted. Ron Zisman originally designed the book's cover using Kenneth W. Gardiner's stunning phalarope photograph.

Despite the help of all of these people and the care of the publisher, there will be some errors that have slipped through; these are, of course, my responsibility.

New York, New York   Susan Roney Drennan
July 1981

Part I

DOCUMENTATION AND RESOURCES

New York State Avian Records Committee

B IRDING IS NOT A MILD BUT AN ACUTE ADDICTION. An inveterate birder's entire consciousness is preoccupied with—even dominated by—the desire to be out looking at birds. The birding addict is really happiest out of doors giving free reign to the urge to observe, reflect, and discover. If, by some marvelous twist of fate, you happen to be such a person, relax about it. Do not contemplate psychoanalysis. It is a far better strategy to be out walking a beach hunting up the last remaining Eskimo Curlew than to try to reprogram your psyche. Just admit that you too are curious about nature and consider that there are less meritorious preoccupations. Grab your binoculars and head for the door. But wait a minute! You forgot your notebook.

Field birding by avocational ornithologists has consistently and indisputably amplified the science of ornithology. Some of the most brilliant and pioneering work in behavior studies has been the contribution of nonprofessional ornithologists. The thousands of observations published yearly in the national and state ornithological journals constitute an irreplaceable record of mass movements of birds and fluctuations in their numbers. Hundreds of studies are based upon these data. Scores of studies regarding the geographical and seasonal distribution of birds would have been impossible without the notable contributions of amateurs.

The scientific value of these studies would not exist without the collective mass of supporting details that distinguish the individual observation record. The necessity to document sight records should never be considered an affront. It is simply the scientifically sound way to perpetuate a record. But if good field records of unusual sightings remain only with the observer, then their scientific value is nil.

In order for field records to add to the body of ornithological knowledge, they must be collected, reviewed and disseminated. Recognizing this, the Federation of New York State Bird Clubs, Inc., established in 1977 a committee to review significant New York State records. This committee is known as the New York State Avian Records Committee (hereafter, NYSARC). The serious amateur birder in New York State should send observations of rare or out-of-season occurrences to this committee for review along with any substantiating material, *e.g.,* photographs or tape recordings of the voice of the bird, or specimens. Material of this type should be considered supplementary to and not the basis of submitted reports. Every extraordinary observation, irrespective of the observer, must be accompanied by a detailed description *written in the field,* that confirms the identity of the bird(s) beyond a reasonable doubt. Written descriptions should accompany sightings of rare species or common species out-of-season. Include supporting details about the observer's distance from the bird or birds, the exact location of the observation, and include the duration of the bird's stay, as well as the date and the time of day. Mention whether the bird was associating with other species. Include the names and addresses of others who saw the bird and the type of optical equipment as well as the power and conditions of light at the time of the observation. Describe the habitat in which the bird was observed. It is most important to include a detailed description of the bird's appearance with emphasis on the color and pattern of plumage, size and shape, behavior, voice, and any other pertinent diagnostic data. Describe the bird both in flight and at rest if possible. After all of that is done, write your reasons for the identification you arrived at and how similarly appearing species were eliminated. Add a note on your previous familiarity with the reported species and those species similar to it. A drawing or sketch made in the field, while observing the bird would be good supporting evidence for your conclusion as to what species you were observing. Excellent evidence would include photographs of the bird and tape recordings of the song or chip notes of the bird. Send all observations to NYSARC, Laboratory of Ornithology, Cornell University, Ithaca, New York, 14853.

There are today a number of excellent illustrated guides designed for ease of use in the field and organized to aid the birder in quick recognition. Among the most notable of the field guides useful for identifying birds that occur in New York State are: *A Field Guide to the Birds,* the new fourth edition, by Roger Tory Peterson; *A Guide to Field Identification to Birds of North America,* by Chandler S. Robbins, Bertel Bruun, and Herbert S. Zim; *The Audubon Society Field Guide to North American Birds (Eastern Region),* by John Bull and John Farrand, Jr.; and *Audubon Bird Guide: Small Land Birds,* and *Audubon Waterbird Guide: Water, Game and Large Land Birds,* both by Richard H. Pough.

Use of all of these, in conjunction with John Bull's *Birds of New York State* and its supplements published by the Federation of New York State Bird Clubs, Inc., should be the basic tools of those birding within the state. In addition to submitting field observations to NYSARC, the binocular birder can contribute to the detailed knowledge of bird distribution of the state by paying careful attention to when and where species are and even are not found and by carefully and conscientiously submitting detailed field notes on rarities, either temporal or locational, to the regional editors of the state journal, *The Kingbird.*

NYSARC has a suggested procedure which it encourages those submitting written reports to follow; however, it is not mandatory that reports be sent in on this format. In its short history many excellent reports have been submitted and accepted by the committee in entirely different formats than that suggested by NYSARC. The suggested guidelines set forth by the committee are:

1. Reports should be submitted on a form such as the one located at the end of this chapter.

2. Details of reports should be submitted on one or more such forms with no more than one report per page.

3. If available, photographs along with as much technical photographic data as possible, should be submitted with the report.

4. If a report is substantiated by tape recordings or specimen, the current location of such evidence should be indicated.

In addition to rare or out-of-season reports of birds occurring in New York State, NYSARC will review reports of observations of any species new to New York, any addition to the list of species proven to nest within New York State, and the following species and identifiable forms from any locality within the state:

Yellow-billed Loon	Corncrake
Arctic Loon	Lapwing
Western Grebe	Wilson's Plover
Yellow-nosed Albatross	Long-billed Curlew
Audubon's Shearwater	Eurasian Curlew
Black-capped Petrel	Whimbrel
Scaled Petrel	(white-rumped races only)
South Trinidad Petrel	Eskimo Curlew
Leach's Storm-Petrel	Bar-tailed Godwit
White-faced Storm-Petrel	Wood Sandpiper
Red-billed Tropicbird	Great Skua
White-tailed Tropicbird	South Polar Skua
White Pelican	Long-tailed Jaeger
Brown Pelican	Ivory Gull
Brown Booby	Thayer's Gull
Magnificent Frigatebird	California Gull
Wood Stork	Arctic Tern
White-faced Ibis	Sooty Tern
White Ibis	Bridled Tern
(Black) Brant	Common Murre
Barnacle Goose	Common Puffin
Fulvous Whistling Duck	White-winged Dove
Cinnamon Teal	Hawk Owl
Smew	Burrowing Owl
Black Vulture	Great Gray Owl
Swallow-tailed Kite	Boreal Owl
Mississippi Kite	Gray Kingbird
Swainson's Hawk	Scissor-tailed Flycatcher
Gyrfalcon	Ash-throated Flycatcher
Sandhill Crane	Say's Phoebe
Yellow Rail	Black-billed Magpie

Brown-headed Nuthatch
Bewick's Wren
Sage Thrasher
Fieldfare
Redwing
Mountain Bluebird
Townsend's Solitaire
Wheatear
Bell's Vireo
Swainson's Warbler
Yellow-rumped (Audubon's)
    Warbler
Black-throated Gray Warbler
Townsend's Warbler
Painted Redstart

Northern (Bullock's) Oriole
Brewer's Blackbird
Western Tanager
Black-headed Grosbeak
Painted Bunting
Hoary Redpoll
Rufous-sided (Spotted) Towhee
Green-tailed Towhee
Lark Bunting
Le Conte's Sparrow
Baird's Sparrow
Bachman's Sparrow
Harris' Sparrow
Smith's Longspur
Chestnut-collared Longspur

The committee will also review all observation records submitted to it of the following species occurring away from downstate New York (that is, away from Long Island and offshore waters, the New York City area; and established nesting areas on the lower Hudson River):

Any species of shearwater
    petrel or
    storm-petrel
Gannet
Great Cormorant
Louisiana Heron
Yellow-crowned Night Heron
Tufted Duck
Common Eider
Black Rail
Purple Gallinule
American Oystercatcher
Back-necked Stilt
American Avocet
Piping Plover
Marbled Godwit

Curlew Sandpiper
Gull-billed Tern
Roseate Tern
Sandwich Tern
Royal Tern
Black Skimmer
Any alcid
Chuck-will's-widow
Western Kingbird
Yellow-throated Warbler
Boat-tailed Grackle
Summer Tanager
Blue Grosbeak
Savannah (Ipswich) Sparrow
Sharp-tailed Sparrow
Lark Sparrow

NYSARC will review reports of the following species occurring at locations away from the Adirondacks:

Spruce Grouse
Gray Jay

Black-backed Three-toed Woodpecker
Northern Three-toed Woodpecker

REPORT NO._____ STATUS_____

For use of NYSARC

- -

NEW YORK STATE AVIAN RECORDS COMMITTEE REPORTING FORM

Observers should include only one report with this form. Obtain as complete a description as possible during the observation and before consulting a field guide; copies of original notes are welcomed.

NAME OF BIRD_____ DATE_____

Locality (give county):

Time and length of observation:

Light conditions, distance from bird, and optics used:

Description of bird

  a.) Number, size and shape:

  b.) Color and pattern, including soft parts:

  c.) Behavior, including vocalizations:

  d.) Habitat, including associated species:

Additional comments, including comparisons with other species and observer's experience:

Does any other substantial evidence document this report? _____
If yes, what is it and where is it deposited?

Time and date of writing this report:

Observer(s) (when multiple, each is encouraged to submit an independent report):

submitted by _____; and _____.

Address of person submitting this report:

Return completed report to: The New York State Avian Records Committee, Laboratory of Ornithology, Cornell University, Ithaca, N.Y. 14853.

New York Rare Bird Alerts

NEW YORK STATE IS BLESSED with several highly efficient schemes for communicating current avian sightings. Many of these information networks rest on a broad base of pyramiding telephone calls. At least one other involves subscription to a commercial calling and answering telephone service. Additionally there are (in 1981) three well-established rare bird alerts that employ tape-recorded messages to disseminate the news of current bird rarities. With this system the curious birder can avail him or herself of the latest happenings on a specific avian scene at any time during the day or night without being concerned about waking, interrupting, or in other ways disturbing others in the birding fraternity. Furthermore, the risk of not finding one or another friends at home is eliminated. Each of the recorded services listed below gives current rare or unusual birds sighted, date and time of sightings, location and usually down-to-the-fencepost instructions on how to reach the area. Often these services give plumages, color, and other variable markings noted on the birds highlighted. Sometimes a behavioral note or two is included. The editors of these messages occasionally expand their coverage to include news of scheduled field trips, open bird club or other public meetings, and upcoming legislation or public hearings of particular concern to birders. These messages vary in length from two to six minutes and usually conclude with information on where watchers can report their observations.

When telephoning these services long distance, considerable expense can be avoided by (a) tape recording the message (thereby eliminating repeated calls) and (b) calling between 11:00 P.M. and 8:00 A.M. and Direct Dialing (no operator assistance).

The Buffalo Museum of Science sponsors the state's most western *Dial-A-Bird* and reports sightings from northwestern New York, the Niagara River, the eastern end of Lake Erie, and the western end of Lake Ontario (both U.S. and Canada sides). Call (716) 896-1271.

The Hudson-Mohawk Bird Club in Albany sponsors a rare bird alert that covers sightings in the Capitol District, the Adirondacks, and often includes the highlights of similar services in Massachusetts, Vermont, Buffalo, and New York City. Call (518) 377-9600.

The New York Rare Bird Alert is sponsored jointly by the Linnaean Society of New York and the National Audubon Society. Coverage includes New York City, all of Long Island, northern New Jersey, lower Connecticut, and Westchester, Putnam, and Rockland counties. Call (212) 832-6523.

For those birders in the Cayuga Lake Basin, the Cornell Laboratory of Ornithology keeps abreast of recent rare or unusual sightings and will inform callers during business hours. Call (607) 256-5056.

Ornithological Collections and Libraries
in New York State

M OST SPECIES OF BIRDS AND THEIR EGGS are protected by laws and indiscriminate collecting of them is rightfully forbidden. It is permissible to collect them only for scientific or educational purposes, and only after necessary federal and state permits have been secured from the proper agencies. Today, unlike years ago, private bird and egg collections are not only frowned upon but are also governed by very strict laws forbidding even temporary possession of dead birds by persons without the proper credentials. Permits, when available at all, are usually restricted to one or two species. Permission to collect a broader spectrum is generally granted only to researchers working on a sanctioned project.

Ornithological collections usually consist of mounts, study skins, "alcoholics," skeletons, eggs, and nests. A *mount* is the bird's skin filled with cotton or another suitable material, and arranged in a life-like posture—feeding, preening, perching—on some sort of stand. It is a sort of manikin of a bird. Mounts are usually displayed in exhibition halls of universities or museums, along with nests, eggs, and habitat groupings, to show the bird in its natural surroundings. They are used primarily to impart information about birds to the public visually. Unfortunately, mounted birds' skins, when exposed to light and dust over a long period of time, lose much of their natural color. Moreover, they are often mounted in poses that make thorough examination and standardized measurements impossible, so they are of limited scientific value.

Collections of study skins, however, are among the most important tools of the ornithologist. *Study skins* are specimens from which the axial skeleton and associated musculature have been removed. The skin is then filled with cotton and laid out on its back in a standardized position for ready comparison. They are kept in light-tight cabinets, which are also protected against moths and dermestids (carrion-feeding beetle larvae). Because they can be handled and minutely examined, they are of the utmost importance in systematic ornithological research. Each skin carries a small tag on which are entered the date and the place where the specimen was collected, as well as the species, name, sex, weight, collector, and any other pertinent data.

The term *alcoholic* does not refer to the drinking habits of a bird, but to its perishable remains, preserved intact in alcohol or formalin for the study of the

11

bird's soft anatomy. *Eggs, nests,* and *skeletons* are also important additions to a well-rounded museum or university ornithological collection.

Collections are acquired by purchase, exchange, gift or through expeditions. Expeditions often gather many duplicate specimens, which can be exchanged with other museums to improve the collection. In this way, an extensive and representative worldwide collection can be built up even by museums that do not sponsor worldwide expeditions.

Many people are not aware that most museums welcome serious amateurs. Access to a study skin collection is of incalculable use in ornithology, because study skins demonstrate the whole plumage spectrum of a species, a subspecies, or hybrids. Examination of a large series, for example, can help in determining the breeding or wintering range of species and subspecies. By examining a large series of study skins of the same species, the ornithologist can determine the relative age of a bird and its sequence of plumages and molts. Only a large series can reveal the various subtle juvenile plumages of some species. Sometimes study skin collections contain useful information bearing on migration and habitat selection, as well.

These are only a few uses to which ornithological collections are put. The following listing enumerates the major collections within the state of New York. Many of the institutions allow use of their collections by visitors. The nonprofessional can usually make arrangements for access to a good bird skin collection for purposes of study and identification. Prospective visitors should write the person overseeing the collection well ahead of time, giving alternative dates for the proposed visit, and stating as specifically as possible which part of the collection they wish to study. They should also remember that during the summer months, university collections may be closed.

New York State enjoys a well-developed public library service as well as numerous university, college, museum, and privately endowed reference libraries. Many of these have strong resources for avian research. Numerous libraries throughout the state contain large and complete collections of avian publications, so that the combined knowledge of the highly specialized field of ornithology is within easy reach of both the amateur and the professional scholar. Access to a major natural history library should provide countless opportunities for those interested in avian evolution, systematics, biogeography, behavior, ecology, physiology, ethology, and allied subjects to expand their knowledge. The following table lists the approximate number of ornithological volumes owned by the institutions that house preserved specimens. The first fifteen contain working libraries that include most, if not all, of the standard ornithological reference works on North American birds; complete sets or extensive runs of all of the major journals published in the United States and Canada; a selection of state, province, or regional works; general ornithological textbooks and other books on specialized subjects or groups of birds; and, of course, most of the field guides to birds. Some of these libraries own various rare and valuable books and archives containing letters, personal journals, notebooks, and memorabilia concerned with ornithology and famous ornithologists. One contains considerable collections of avian sound recordings, films, and slide transparencies. A very few provide a wealth of ornithological data on microfilm or microfiche. All of the institutions well repay the effort of dipping into their resources.

The National Audubon Society, a citizens' conservation organization for over three-quarters of a century, has its headquarters in mid-town Manhattan. In 1981 it housed a reference library of more than 15,000 books, including about 5000 bird books and some 8000 natural history volumes. In addition to this well-integrated collection, it subscribed to more than 260 periodicals in the fields of ornithology, natural history, ecology, and related fields. The primary purpose of the library since its founding in 1905 has been to assist the Society's staff, both within and outside New York City, with its informational needs. It is, however, open to the public by appointment Monday through Friday from 9:00 A.M. to 4:00 P.M., for reference only. This library has a systematic acquisitions policy, an exemplary catalog, and an organized administration. Over the years, it has acquired small private collections and serves as the repository of several considerable endowments. Among its volumes, pamphlets, back issues and current periodicals, technical reports, monographs, and environmental indices are abundant resources for the amateur and professional. The library mirrors the diversity of the society's interests, both of which are extensive. Request the librarian for an appointment as far in advance as possible. Address all requests directly to the library at the National Audubon Society, 950 Third Avenue, New York, New York 10022.

Finally, the collections of the American Museum of Natural History in New York City, including the library, are in a class by themselves. They serve the interests of scholarship in a panoramic style. The museum's specimen collections are unequalled in number and species representation in the Western Hemisphere. In the world, only the British Museum can compete. Its ornithological library is one of the largest, most comprehensive, and cosmopolitan. Many of us feel it is the juiciest bite in the Big Apple.

Thanks to Richard C. Banks, Mary H. Clench, and Jon C. Barlow for permission to use the results of their survey of bird collections, conducted during the mid-1960s, the final report of which can be found in *The Auk* 90:136–170 (January 1973), and its supplement, *The Auk* 93:126–129 (January 1976). The data presented below update those presented by Banks *et al.*

I should like to extend my gratitude to the following persons who provided information about their institutions' collections. Clearly, without their cooperation this section of the book could not have been written: Robert F. Andrle (Buffalo Museum of Science), Robert C. Beason (SUNY at Geneseo), Allen H. Benton (SUNY at Fredonia), Jerry H. Czech (Rochester Museum and Science Center), Stephen W. Eaton (Saint Bonaventure University), Michelle Epstein (National Audubon Society), John Farrand, Jr. (American Museum of Natural History), Ronald Giegerich (State University College of Environmental Science and Forestry at Syracuse), Robert E. Goodwin (Colgate University), Barbara Linton (National Audubon Society), Heinz Meng (State University College at New Paltz), Neil S. Moon (Rochester Museum), John G. New (State University College at Oneonta), James W. Parker (SUNY at Fredonia), Edgar M. Reilly (New York State Museum at Albany), James M. Ryan (Hobart and William Smith Colleges), Lester L. Short (American Museum of Natural History), Charles R. Smith (Cornell Laboratory of Ornithology), Gerald A. Smith (SUNY at Oswego), James E. White (Keuka College). Also I am grateful to Roger S. Payne and Kenneth C. Parkes for their careful reading and fine suggestions on this chapter.

The institutions in Table 1 are listed in order from highest number of study specimen skins to lowest number. Institutions with fewer than 200 study skins are listed alphabetically by institution after the main body of the table.

TABLE 1

MAJOR ORNITHOLOGICAL COLLECTIONS AND LIBRARIES IN NEW YORK STATE

Institution	Study Skins	Mounts	Alcoholics	Skeletons	Egg Sets	Nests	Ornithological Volumes*
AMERICAN MUSEUM OF NATURAL HISTORY Central Park West at 79th Street New York, New York 10024 (212) 873-1300	909,500	2000	7700	9250	20,000	5100	335,000† (5000)
CORNELL UNIVERSITY Bird and Mammal Collections Section of Ecology and Systematics Ithaca, New York 14853	37,000	1000	1000	4000	1800	350	
LABORATORY OF ORNITHOLOGY Ithaca, New York 14850 (607) 256-5056							3000 (350) plus 10,000 reprints
BUFFALO MUSEUM OF SCIENCE Buffalo, New York (716) 896-5200	7000	2500	—	1000	2000	175	700 (18)
NEW YORK STATE MUSEUM State Education Building Albany, New York (518) 474-5877	5600	1000	—	600	6500	‡	1000 ± (250)
STATE UNIVERSITY COLLEGE AT ONEONTA Biology Department Oneonta, New York 13820 (607) 431-3500	2015	700	—	—	—	—	N/A
STATE UNIVERSITY COLLEGE OF ENVIRONMENTAL SCIENCE FORESTRY AT SYRACUSE UNIVERSITY Department of Forest Zoology Syracuse, New York 13210 (315) 473-8611	1700	370	16	12	590	40	N/A

TABLE 1 (continued)

MAJOR ORNITHOLOGICAL COLLECTIONS AND LIBRARIES IN NEW YORK STATE

Institution	Study Skins	Mounts	Alcoholics	Skeletons	Egg Sets	Nests	Ornithological Volumes*
STATE UNIVERSITY COLLEGE AT OSWEGO Rice Creek Biostation Oswego, New York 13126 (315) 341-2243	1460	45	—	5	50	20±	100± (0)
ST. BONAVENTURE UNIVERSITY Biology Department St. Bonaventure, New York 14778 (716) 375-2000	1200	50	50	935	50	—	N/A
ROCHESTER MUSEUM Rochester, New York 14607 (716) 271-4320	1100	600	—	—	1000+ sets,	some in nests	N/A
HOBART & WILLIAM SMITH COLLEGES Museum Geneva, New York 14456 (315) 789-5500	1000±	260±	—	—	200±	small local collection	150± (3)
STATE UNIVERSITY COLLEGE AT NEW PALTZ Department of Biology New Paltz, New York 12561 (914) 257-2121	880	10	—	10	100	50	150± (3)
STATE UNIVERSITY COLLEGE AT GENESEO Department of Biology Geneseo, New York 14454 (716) 245-5211	400	40	—	—	—	10	500± (4)
STATE UNIVERSITY COLLEGE AT FREDONIA Department of Biology Fredonia, New York 14063 (716) 673-3111	375+	250	—	15	20	20	3000 (3)

TABLE 1 (continued)

MAJOR ORNITHOLOGICAL COLLECTIONS AND LIBRARIES IN NEW YORK STATE

Institution	Study Skins	Mounts	Alcoholics	Skeletons	Egg Sets	Nests	Ornithological Volumes*
STATE UNIVERSITY COLLEGE AT POTSDAM Potsdam, New York 13676	290 ±	–	–	–	25 ±	30 ±	25 ± (7)
COLGATE UNIVERSITY Department of Biology Hamilton, New York 13346 (315) 824-1000	200 +	1180	–	–	300	–	500 ± (1)

*Number of ornithological journals subscribed to shown in parentheses.
†Natural history volumes including every major and most minor ornithological works.
‡Good collection representing most state breeders.

COLLECTIONS WITH FEWER THAN 200 RESEARCH SPECIMENS (alphabetically ordered by institution)

1. Bayard Cutting Arboretum
   Oakdale, New York 11769
2. Hartwick College
   Oneonta, New York 13820
3. Houghton College
   Houghton, New York 14744
4. Keuka College
   Keuka Park, New York 14478
5. Museum of Natural History
   Pawling, New York 12564
6. Niagara University
   Niagara, New York 14109
7. Rogers Conservation Education Center
   Sherburne, New York 13045
8. George H. Lesser Collection
   State University College at Buffalo
   Buffalo, New York 14222
9. State University College at Cortland
   Cortland, New York 13045
10. Vassar College
    Poughkeepsie, New York 12601

Physiographic Regions

MUCH OF THE EARLY WORK of writing this book was concerned with developing a format that would be consistent in approach, offer distinct advantages over others of its type, and provide an even statewide coverage of wildlands, parks, nature centers, and sanctuaries. This was to be done in a manner that would be useful to the birder throughout the year. After some frustration and soul-searching over the advantages of arranging the material according to avian biotic districts instead of the traditional human cultural districts, I chose in favor of the cultural divisions. This book is essentially a cooperative project sponsored by the Federation of New York State Bird Clubs.

In late 1954, the Federation of New York State Bird Clubs, Inc., divided the state into ten regions (See map of *Kingbird* Reporting Regions.) The purpose of these regions was to organize distinct units from which details of the avifauna of each could be seasonally collected and published in the federation's quarterly, *The Kingbird.* Many of the boundaries and configurations of the reporting regions are based on cultural features (major highways, county and township lines, economic and administrative sectors, etc.), rather than on natural communities defined by distribution of plant and animal life. From a pragmatic point of view, the intricate machinery of *The Kingbird*'s system is well-oiled and effective. To have organized this guide's reporting regions around biotypes probably would have diminished its usefulness.

In this guide, each of the ten *Kingbird* reporting regions is preceded by a general map indicating approximately where particular sites can be found within the region. Detailed maps to those sites are interspersed throughout the chapter describing that region.

At the same time, because *The Kingbird* reporting districts have little to do with biological factors, it would be misleading not to discuss the multitude and variety of physiographic regions throughout the state. The modern army of birders is knowledgeable and curious. A familiarity with the state's land forms and their elevation, vegetation, geology, climate, and hydrographic features will provide the birder with a deeper appreciation of the interdependencies and interconnectedness that characterize biological assemblages. Such knowledge links subtle and often unexpected phenomena, which might otherwise be considered in isolation. Rela-

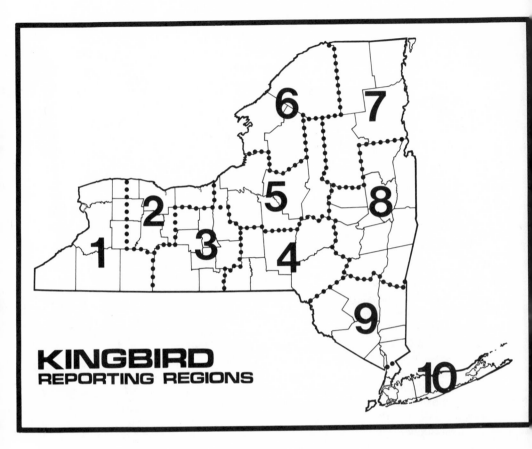

KINGBIRD
REPORTING REGIONS

tionships that separate birds geographically, by altitude, or by habitat preferences (including separation by between-habitat preferences, or within-habitat preferences or guilds), should be made more readily visible. Without such knowledge, the bird observer could overlook many facets of the biology of the state's avifauna and the biological significance of ecological niches.

Even a brief glance at the accompanying map of New York's physiographic regions, each with its own distinguishing resources, will show that the state (which encompasses an area of some 48,000 square miles), is blessed with a wide range of land forms. These set the scene upon which the state's avifauna evolved and depend. This map should be used in conjunction with others throughout the book.

The following discussion presents an overview of the physical condition of the state by examining its physiographic regions. Those disposed to study New York's avifauna in all of its aspects will profit by acquainting themselves with the concise account of each region's geography.

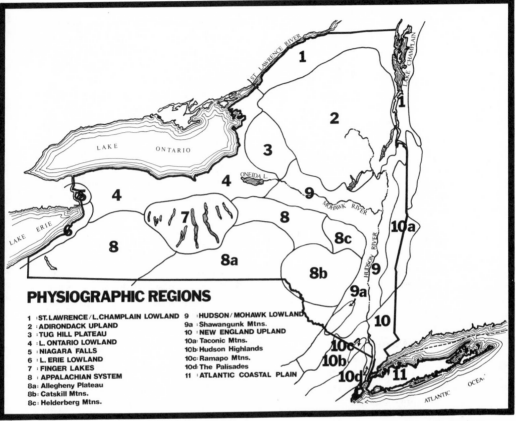

Physiographic Regions

PHYSIOGRAPHIC REGIONS

1 : ST. LAWRENCE / L. CHAMPLAIN LOWLAND	9 : HUDSON / MOHAWK LOWLAND
2 : ADIRONDACK UPLAND	9a : Shawangunk Mtns.
3 : TUG HILL PLATEAU	10 : NEW ENGLAND UPLAND
4 : L. ONTARIO LOWLAND	10a: Taconic Mtns.
5 : NIAGARA FALLS	10b: Hudson Highlands
6 : L. ERIE LOWLAND	10c: Ramapo Mtns.
7 : FINGER LAKES	10d: The Palisades
8 : APPALACHIAN SYSTEM	11 : ATLANTIC COASTAL PLAIN
8a: Allegheny Plateau	
8b: Catskill Mtns.	
8c: Helderberg Mtns.	

1: ST. LAWRENCE–LAKE CHAMPLAIN LOWLAND

Lying east of the St. Lawrence River and west of Lake Champlain, the St. Lawrence–Lake Champlain Lowland is part of a larger land form that extends north and east of the Great Lakes and that terminates, after passing through Canada, at the sea. Close to the waterways, the bedrock is covered by limestones; farther inland, by sandstones. Considerable deposits of pure sand or sandy soil have accumulated in some places.

Within this major physiographic land form there are several subforms. The *St. Lawrence Marine Plain* is a gently rolling, almost flat strip that follows the land contours along the St. Lawrence River to the Canadian border. It consists partly of sandy soils but mainly of poorly drained deposits of marine clays. The *St. Lawrence Hills* are located between the Marine Plain and the Adirondack Mountains. They extend east to within a few miles of Lake Champlain and north to and beyond the Canadian border. This subform features more rugged terrain, with its sandstone bedrock largely covered with glacial drift. The *Champlain Lake Plain* is underlain with limestone and covered mostly with marine clays. It is the low, flat, narrow belt of relatively level farm land west of Lake Champlain and east of the Adirondacks. That narrow belt ends near the southern end of Lake Champlain, where the mountains reach the shore of the lake. The area within the St. Lawrence–Lake Champlain Lowland is a mixture of rolling plains with almost no relief, where elevation changes are measured only in tens of feet.

This region receives between 30 and 39 inches of rain annually, which is relatively low for the state but, approximately equal to other areas within the Lake Ontario and Lake Erie lowlands. The St. Lawrence River Valley and the Lake Champlain Lowland often suffer drought conditions in summer months, and because the soil retains moisture poorly, irrigation has been necessary to grow crops. However, substantial amounts of precipitation fall in the form of snow (60 to 100 inches annually), which remains for some time.

It is a region of very cold winters and sunny summers. Mean January temperatures below 20°F. are found almost everywhere, and July mean temperatures are approximately 69°F. The region's economy is based on dairy farms, which are widespread and successful. In addition, fruit and vegetable farms thrive near Lake Champlain, where the soil is sufficiently dry to work the fields in early spring.

The St. Lawrence–Lake Champlain Lowland is forested with frequent stands of American elm and red maple, especially in those areas with poor drainage and numerous abandoned fields. Where farming still thrives, both oak and northern hardwoods are present, but in reduced numbers, due to the large areas that have been cleared and drained for crops and pastures.

The natural vegetation in this physiographic region then, is largely explained by three factors: the surface configuration; human disturbance; and to a lesser extent, climatic conditions.

2: ADIRONDACK UPLAND

The Adirondack Upland consists of ancient domed igneous and metamorphic rock, with erosional remnants forming the very high, more rugged peaks. Extreme glacial scouring has smoothed its surface and has removed most of the soil. Eroded material has often rerouted stream patterns and created new lakes.

Within this major land form there are three distinct subforms. The *Adirondack Mountain Peaks,* approximately in the center of the region, comprise the very highest and most rugged section. Mount Marcy (elev. 5344 feet) and Algonquin Peak (elev. 5114 feet) are the highest. The summit elevations within this subform exceed 3000 feet and local relief surpasses 2000 feet. Most of the major peaks are composed of resistent, dark intrusive rock and younger, lighter igneous rock.

The *Adirondack Low Mountains* encircle the Adirondack Mountain Peaks with some summits in excess of 2000 feet and local relief usually under 1000 feet. The area is characterized by hundreds of lovely glacial lakes, Saranac, Raquette, and the Fulton Lakes Chain among the most prominent.

The *Western Adirondack Hills* subform is a broad zone of sand-covered foothills, located between the westernmost Adirondack Low Hills and the Tug Hill Plateau (see physiographic region 3). Within this area are some of the state's richest iron, zinc, talc, and lead deposits.

This region has extremely cold, snowy winters and very cool, wet summers. Throughout, it averages more than 35 inches of rain annually, and in the southwestern portion the average exceeds 50 inches. Annual snowfall ranges between 90 and 165 inches throughout the Adirondacks, and snow usually remains on the ground longer than in other parts of the state. Winters are extremely cold, with mean January temperatures approximately 15°F. July mean temperatures of 65°F. are almost the coolest in the state. There is never any real water deficit in this region, with numerous mountain streams carrying large water surpluses at the end of winter and through spring. There is usually an overall water surplus of between 16 and 32 inches annually in the Adirondack Upland.

Owing to the rugged terrain, general inaccessibility, and extreme climate, this region is unsuitable for agriculture or industry, and no dense population centers have developed. The potential for recreational use, however, is boundless. It is well-suited to recreational activities that are consistent with good conservation practices.

The natural vegetation of the Adirondack Upland consists of spruce-fir-northern hardwood forests. Dense stands of spruce and fir prosper in the coolest parts of the mountains, mixed with beech and sugar maple hardwoods and smaller stands of ash, basswood, cherry, hemlock, birch, and white pine. Red maples flourish where drainage is poor, but oaks are generally difficult to find. Of course, above the tree line (approximately 4500 feet), small areas of alpine tundra and bare rock are common.

3: TUG HILL PLATEAU

The Tug Hill Plateau lies north and east of the Lake Ontario Lowlands and west of the Western Adirondack Hills. The Black River and the Black River Valley form the boundary between the Tug Hill Plateau and the Adirondacks. At one time the site of a glacial lake, the Black River Valley's broad, fertile alluvial flats are underlain with granitic rock. This lowland is a vital access route between the Mohawk Valley and the North Country. The Tug Hill Plateau rises 1800 to 2000 feet above the lowlands that surround it in all directions, and is underlain by sandstones, limestones and shales. The plateau is tilted slightly toward the west. Very poor drainage, poor soils, and excessive snowfall have severely limited agriculture on the plateau and visitors passing through the area are often impressed with the number of abandoned farms. It is, in fact, the least populated section of the state (even the Adirondacks are populated more densely).

The Tug Hill Plateau has extremely cold, snowy winters and very cool, wet summers. Winters are nearly unbearable, with mean January temperatures of about 14°F. Minimum temperatures between −40°F. and −50°F. have been recorded on the plateau. Mean average temperatures in July cluster around 64°F.

Rainfall averages near 50 inches annually. Tug Hill lies on the lee side of Lake Ontario and gets the brunt of snow squalls moving east from Lake Ontario; it is infamous for annual snowfall averages between 130 and 225 inches. The area never suffers for lack of water.

In its highest elevations, the soil lends itself to cultivation of the natural spruce-fir zone vegetation, interspersed with larch, white pine, hemlock, and white cedar, which thrive in the poorly drained areas. Bogs containing typical bog vegetation are scattered throughout. Most of the softwood tree species have been reduced by heavy lumbering. Density of natural vegetation varies considerably on the plateau, but efforts at reforestation are obvious in some areas. In the steep uplands thick forests of spruce and fir abound. Natural forest fires have also denuded whole expanses, and these areas take a relatively long time to regenerate; in the interim, they provide excellent habitat for many bird species.

4: LAKE ONTARIO LOWLAND

Just south and east of Lake Ontario lies the Lake Ontario Lowland. It is underlain with sandstones, limestones and shale. Much of the lowland is rather featureless, but there are some rolling hills formed by moraine deposits. Within this major land form are six subforms.

The *Eastern Ontario Hills* stretch from the eastern shore of the lake to the Tug Hill Plateau and to the western limits of the Adirondack Hills. Prominent sand dunes occur along the lakeshore in this area of low glacial drift.

The *Oneida Lake Plain* is the almost flat expanse south of Oneida Lake. Extensive swamps are a major feature. The *Ontario Ridge and Swampland* extends east from the Lake Ontario shore from the mouth of the Oswego River to the Oneida Lake Plain to the south. To the east is the Tug Hill Plateau, the Eastern Ontario Hills lie to the north, and the Ontario Drumlins border it on the west. An area of many swamps and some ridged moraine, poor drainage is its major characteristic.

The next land subform to the west is the *Ontario Drumlins,* which extend southward from the lake shore to the Finger Lakes. The Ontario Drumlins stretch nearly from Syracuse to Rochester. The drumlins are elongated, usually oval, ridges resembling an egg halved lengthwise. Although they seem to be glacial in origin, the drumlins' geological history is a matter of controversy. It is probable, however, that they are fragments of ground moraine, compressed and molded into their characteristic shape by ice movements. Wherever they occur, drumlins are found in large numbers. Here, there are literally thousands, spaced so closely that they give the area a distinctively hilly appearance.

Adjoining the Ontario Drumlins to the west is the *Ontario Lake Plain,* which extends from the lake shore south to the Niagara limestone escarpment. It is bounded by the Niagara River on the west. This lake plain is especially suited to the growing of specialty fruit crops.

South of the Ontario Lake Plain to approximately the Cattaraugus Hills (see physiographic region 8) is a rolling landscape covered with glacial drift known as the *Southern Ontario Plain*. A good deal of its western portion is now occupied by the expanded Buffalo urban system.

In general, the Lake Ontario Lowland has cold, snowy winters and warm, dry summers. Mean January temperatures of 25°F., are characteristic, and July mean-averages of around 70°F. are typical. This is one of the driest areas of the state, with annual rainfall varying from 25 to 40 inches. Summer brings not only its maximum precipitation, but also its maximum need for water, and so, in most years, small water deficits occur. Snowfall also varies from 40 to 70 inches in the Finger Lakes Hills, to more than 80 inches on the Ontario Plain, which catches the impact of squalls from the lake. The state's finest fruit belt lies along the lake shore, and wheat and other grains prosper in the Genesee country south of Rochester; elsewhere in the region, dairying is the primary form of agriculture.

The natural vegetation throughout the Lake Ontario Lowland is predominantly that of the American elm–red maple–northern hardwood association. The occurrence of these tree species is primarily due to the poor drainage of the region and the nearly complete removal of the original forests by lumbering interests many years ago. The many abandoned fields contribute to the growth of American elm and red maple.

5: NIAGARA FALLS

Unquestionably, Niagara Falls is the most famous physical feature in New York State. It is certainly most spectacular to see millions of gallons of water rushing over the precipice in a thunderous roar. The falls are 167 feet high on the American side, and nine feet lower on the Canadian side. The American Falls are 1000 feet wide and are separated from the 2500 foot-wide Canadian Falls by Goat Island. The American Falls are nearly at right angles to the river flow, but the uneven erosion of the Canadian Falls has produced an arc shape, prompting the common name "Horseshoe Falls." Because of their much greater width, more water flows over the Canadian Falls. The gorge of the Niagara River extends 6.5 miles downstream from the falls. Between the north end of the gorge and Fort Niagara, the river flows calmly through a wide valley with no hint that the turbulent falls, gorge, and rapids lie upstream. The erosion rate of the Canadian Falls averages four to five feet per year, while that of the American Falls is only about one foot per year. Geologists have calculated that approximately 8800 years have elapsed since the river first began to carve out its channel, and that it will be another 75,000 years before the Niagara River cuts all the way back into Lake Erie and drains it.

6: LAKE ERIE LOWLAND

The Lake Erie Lowland is a narrow strip of somewhat featureless land immediately south and east of Lake Erie. It rises to about 100 feet above the lake and stretches from the lake to the Cattaraugus Hills (see physiographic region 8). Along the lake shore are numerous beaches and several excellent harbors.

The Lake Erie Lowland experiences cold, snowy winters and warm, dry summers. Average January and July temperatures are 25°F., and 70°F., respectively. Annual rainfall hovers about 35 inches, and annual snowfall can register as high as 70 inches.

The region is especially well adapted to cultivation of plants with deep root systems and is famous for its fine and productive grape vineyards.

7: FINGER LAKES

The Finger Lakes are a group of long, narrow lakes in the west-central part of the state. They lie in the Oswego River watershed of the Lake Ontario drainage. Popular terminology recognizes seven Finger Lakes; from east-to-west they are Otisco, Skaneateles, Owasco, Cayuga, Seneca, Keuka, and Canandaigua lakes. Four additional lakes west of this group are considered by some authorities to be part of the Finger Lakes: Honeoye, Canadice, Hemlock, and Conesus lakes. In pre-colonial times this area was the center of the very powerful Iroquois Indian Confederation; thus the names Cayuga, Seneca, Keuka, Canandaigua, Owasco, and Skaneateles.

The lakes are of glacial origin; the receding ice was powerful enough to scour some of the lakes to depths of 180 feet below sea-level but left others relatively shallow. The maximum depth of Seneca Lake, for instance, is 618 feet, while Otisco Lake is only 66 feet at its deepest point.

Scenic shadowy glens, hidden grottoes, and shimmering waterfalls line the lakes' 400 miles of shoreline. They have played a major role in the cultural and agricultural history of the state. The southern end of Keuka Lake is the center of major and famous vineyards, and wineries have flourished there since 1860. Orchards annually yield bumper crops of apples. Farms, ranging in size from small family affairs to corporate businesses, produce abundant fresh fruit and vegetables.

This region, like the Lake Ontario Lowland, experiences cold, snowy winters and warm, dry summers. Mean January and July temperatures are 25.3°F. and 71.4°F., respectively, and the average annual mean rainfall is a little more than 32 inches. The snowfall ranges from 40 to 70 inches, averaging 54 inches annually.

The natural vegetation is largely of the oak–northern hardwood association, and forests alternate with farmlands and scattered woodlots. Interspersed throughout the Finger Lakes urban centers flourish where once almost continuous forests of beech, sugar maple, basswood, ash, and cherry once stood. However, large and beautiful tracts of varied hardwood forests remain, with hemlock lining moist, shady slopes and ravines and white cedar crowding abandoned fields and poorly drained areas. Alder and some larch appear, primarily in wet areas, and white pine is also common in the region.

8: APPALACHIAN SYSTEM

The Appalachian System is certainly the largest of the various physiographic regions of New York State. It occupies nearly half of the state, when all of its subdivisions and subforms are considered. It is almost totally underlain with Paleozoic sedimentary rock, which dips slightly south and west. Almost all parts of the system have been glaciated, especially in areas of pronounced escarpments.

The area is roughly bounded by the Lake Erie Lowland on the west, the Southern Ontario Plain on the north, the Finger Lakes on the northeast and the Pennsylvania border on the south. This section of the Appalachian System contains the *Cattaraugus Hills,* relatively flat-topped uplands with deep valleys. Drainage here flows in three directions: to the Ohio River system (and ultimately the Gulf of Mexico); to the Susquehanna River system (and ultimately the Atlantic Ocean); and to Lake Ontario by way of the Genesee River. The upland surface is not too rugged to support successful farms, but the best farm land is in the valleys.

In the south-central section, straddling the Pennsylvania border, lie the *Allegheny Hills,* the only part of the Appalachian System that was not glaciated. Almost the entirety of the Allegheny Hills are contained within the Allegany State Park, where the terrain is slightly more angular and the valleys are more regular in direction. In this subdivision the bedrock is markedly more exposed than in most other parts of the state.

Owing to its size, the Appalachian System is divided into three subdivisions. The *Allegheny Plateau (8a* on the map), and the area designated as 8, northeast of it, contain the *Susquehanna Hills,* which are drained by the upper Susquehanna River and its tributaries. The bedrocks are primarily shales, but sandstones and limestones are found throughout. The surface of the plateau appears to be fairly even when viewed from a distance, but closer examination shows little level land with numerous deep, narrow valleys. Divides are found at heights of 1700 to 2100 feet.

South of the Mohawk River Valley and west of the Hudson River Valley lie the *Catskill Mountains (8b).* These are low mountains with summit elevations ranging from about 2000 feet to 4000 feet above sea level. (The highest peak in the Catskills is Slide Mountain, whose summit is 4204 feet.) Glacial erosion has combined with stream erosion to produce deep dissection, and the topography appears coarsely textured. There are few valleys in this region, so it seems additionally "bulky." The underlying bedrocks here are very resistant sandstones deposited by a Paleozoic sea. Because the sandstones are highly permeable, few small tributaries traverse the region. The water effectively soaks in, rather than runs off.

Just south of the Catskills lie the *Delaware Hills,* which are drained by the Delaware River. The local relief is merely a few hundred feet, but there is little flat land. Often spoken of as the Catskill foothills, the Delaware Hills seem destined to be developed for tourism.

Southwest of Albany, in the northeastern section of the Appalachian System, lie the *Helderberg Mountains (8c),* which rise approximately 2000 feet above the Hudson and Mohawk Valleys. This series of sharp northeast-facing escarpments survived the glacial onslaught because of very resistant bedrock capped with limestones. From the escarpments the glacial lake plain can be seen stretching below and northward to the Adirondack foothills. Throughout the Helderberg Mountains there are many sinkholes (the largest of which forms Thompson's Lake) and many limestone caves. The sheer cliffs and talus slopes of the Helderberg Mountains are classic study areas for students of geology, due to the abundant rock formations and well-preserved fossils.

Throughout the Appalachian System the winters are cold and snowy and the summers are cool and wet. January mean temperatures range between 20°F. and 25°F., while July means are a cool 67°F. Minimums of −25°F. to −40°F. have been recorded but extremes in summer seldom reach 100°F. Rainfall is heavy, nearly throughout, with very few weather stations recording under 40 inches annually. The upper southern slopes of the Catskills receive as much as 60 inches of rainfall yearly. Snowfall is substantial and stays on the ground longer at higher elevations than in the valleys. The western part of the Appalachian System lies in the path of storms crossing Lake Erie, and therefore receives more than 100 inches of snowfall per year.

Throughout the Appalachian System are some of the most extensive concentrations of forests in the state. Allegany State Park, along the Pennsylvania border in the region's western section, is famous for its limited stands of old-growth trees. Lands generally were denuded of any valuable or large trees before being acquired by the state, but, large state preserves are now covered with second-growth forests. There is a very large forested state park in the Catskill Mountains. The alluvial soils in valley bottoms should be suited for cultivation, but farming the steep terrain is not currently profitable, and the eastern section of physiographic region 8 has farming only on locally favorable sites. Where cultivation is not feasible, forests have been allowed to flourish.

All throughout the Allegheny Plateau and farther north in the Susquehanna Hills, the progressive abandonment of marginal farm land has been slowly followed by invasion of trees. The resulting forests are being managed to cater to the growing army of recreationalists. The highest and coolest parts of the Catskills have spruce, fir, and larch, as well as white pine. Stands of northern hardwoods extend south across the Pennsylvania border and into the higher reaches of the Appalachian System. This section is mostly dominated by stands of beech and sugar maple but basswood, white ash, and black cherry are also found in the warmer areas. Hemlock, white pine, white cedar, and evergreens are abundant throughout, but are distributed unevenly. Hemlock is characteristic in ravines and on shady slopes and in some places forms pure groves.

At low and intermediate elevations across the system some northern hardwoods can be found mixed with moderate amounts of oak, or oak and hickory, and other tree species. These mixed stands are often found on slopes facing south or southwest, where they receive more sun, and where consequently more stands of oak grow. North and northeast facing slopes receive the least sun and tend to resemble more northerly vegetational zones. Everywhere in these mixed forests there is evidence of earlier logging, with both large, older trees and smaller, younger ones. The typical ground cover consists of dense ferns, low brush, and smaller trees.

9: HUDSON-MOHAWK LOWLAND

The Hudson-Mohawk Lowland includes the Hudson and Mohawk rivers and their surrounding valleys. The *Mohawk Valley,* an east-west lowland drained by the

Mohawk River, lies between the Adirondacks on the north and the Appalachian System on the south. The width of the valley varies between ten and thirty miles. The bedrock is a composite of soft shales eroded by the harder glacial till washing over them, to a depth of about 100 feet below the higher country to the north and south. At Little Falls southeast of Utica, the Mohawk River narrows considerably into a deep gorge. This has long been a major artery of commerce, containing first the Erie Canal, then the railroad, and then the New York State Thruway. The *Hudson Valley,* a north-south lowland drained by the Hudson River, lies between the Taconic Mountains (10a), on the east, and the Helderberg Mountains (8c), and Catskill Mountains (8b), on the west. It varies from ten to twenty miles in width. North of Albany it is wide and flat and is covered with glacial lake deposits. To the south, the valley narrows and is broken by low-lying hills and terraces underlain with sedimentary rock.

The *Shawangunk Mountains (9a)* are steep-sided ridges rising 1000 feet, in the southwest section of the Hudson-Mohawk Lowland. These mountains are formed of sharply folded sedimentary rock capped with a hard sandstone conglomerate that has provided protection against erosion.

Climatologically, the Hudson-Mohawk Lowland is a transition zone between the uplands and the warmer New York–Long Island area. The valleys are considerably drier than the uplands surrounding them; however, there are sufficient, if not generous, amounts of precipitation to the south. Mean January temperatures range from 15°F. to 30°F., and July means range from the high sixties to 74° F. Annual rainfall varies from 35 to more than 45 inches. Snowfall is usually no more than 40 inches in the south but is double that amount in the north and west of the region.

Outside of urban and suburban areas, the natural vegetation alternates between numerous scattered woodlots, a few forest preserves, and farms.

10: NEW ENGLAND UPLAND

The New England Upland is an area of ancient crystalline rock, stretching from the eastern border of the state, east of the Hudson River, all the way south and across the *Lower Hudson Valley* into New Jersey and Pennsylvania. Manhattan Island is in the very southernmost part of this region. The *Taconic Mountains (10a)* run in a north-south direction, reaching their maximum ruggedness in Vermont. Local relief runs to about 2000 feet; valleys are narrow and slopes are very steep.

The *Hudson Highlands (10b),* sometimes called the Hudson Hills, are also composed of hard crystalline rock, here carved by running water. The summits of the highlands are some 1000 feet above the Hudson River. The river flows through the area in a steep-sided gorge that extends from Storm King, past West Point, to a point just north of Peekskill. The *Ramapo Mountains (10c)* are rolling hills of low relief lying on the west side of the Hudson River just northwest of the Hudson Highlands. This range is, in fact, simply a continuation into New York State of the New Jersey Highlands.

The *Palisades (10d)* line the west side of the Hudson River north from Hoboken, New Jersey. Two types of rock border the river here: the eastern (Manhattan) side is a combination of igneous and metamorphic rock of Pre-Cambrian date, while the western (New Jersey) side is basalt rock of Triassic age. The Triassic basalt forced its way between previously formed sedimentary beds. The Hudson River carved through the sedimentary rock until it reached the harder basalt, which now forms the sheer cliffs we call the Palisades. Complicated jointing, which occurred during the cooling and solidification of the basalt, caused the columnar appearance of the Palisades. Because the Hudson bore huge quantities of glacial melt water and sediment, its erosive power was tremendous, and it deepened the gorge without leaving much sediment behind. Most of this glacial material was deposited along the Continental Shelf, which extends approximately fifty miles from the shore. The Hudson River is, in fact, an estuary in this area. Its channel reaches a depth of 700 feet near West Point, the northernmost point of tidal influence on the Hudson.

The climate of the New England Upland is similar to that of the lower Hudson River Valley. The highlands overlooking the Hudson River have been the scene of much residential building and commercial development, but several scenic parks have been set aside on the west side of the river. It is hoped that these preserves, amounting to some 60,000 acres, will remain green islands along the Palisades.

11: ATLANTIC COASTAL PLAIN

The Atlantic Coastal Plain occurs only on Long Island and on Staten Island. These islands are the terminal moraines of the great ice sheet; their elevation rises to over 300 feet above sea level. Southward from the crest of the terminal moraines on Long Island, and sloping toward the sea, is a broad outwash plain. Some of the hills and glacial outwash plains are covered with rich soil, which, combined with the relatively mild climate, makes Long Island one of the most important truck-gardening centers of the country. The proximity of the New York City market is certainly instrumental in keeping it that way.

Long Island extends 118 miles east-northeast from the mouth of the Hudson River to Montauk Point, varying in width from twelve to twenty-three miles, with an area of about 1680 square miles. It is the largest island adjoining the continental United States, and is bounded on the north by *Long Island Sound,* which separates it from the south shore of Connecticut. The Atlantic Ocean bounds it on the east and south, while the Narrows, New York Bay, and the East River connect the ocean with the sound and complete the boundaries on the west and northwest. Several small islands around its coast are included in its political boundaries, for example, Coney, Rikers, Fire, Shelter, Gardiner's, Plum, and Fisher's.

Long Island is composed of low hilly plateaus on its north side, longitudinal ridges of glacial moraines through its central parts, and gently sloping plains to the south of the moraines. The shore lines have steep bluffs, which are constantly being eroded by the action of the sea and weather. Several parts of the island are covered by ground moraine as well as kettle ponds and eskers, of which, the kettle ponds of Lake Ronkonkoma and Kellis Pond (near Bridgehampton), and the esker near Smithtown Branch are typical. Some of the lower areas on the island have been slowly filled in with vegetation and are potential bogs.

Natural ponds and lakes are very common on the island, and there are many bays and inlets along its 280 miles of indented coast line. In addition, about 120 ± square miles of salt marsh are scattered throughout Long Island. The eastern portion is especially well-wooded and noted for its pine forests. Several of Long Island's most beautiful beaches, especially on the south shore, are the result of wave action and long shore currents, which have developed barrier beaches, lagoons, spits, and sand bars out of materials deposited by early glacial melt.

The Atlantic Coastal Plain experiences mild, wet winters and warm, humid summers. The warm temperatures are frequently lowered by cool winds from the ocean. The January mean temperature of 31°F. is at least fifteen degrees higher than in the Adirondacks. July mean temperatures are approximately 72°F. This physiographic region receives about 43 inches of rainfall annually and usually has a water surplus of nearly 20 inches. Owing to the relatively warm winter temperatures, no more than 40 inches of snow falls here, and many areas, including a good deal of Long Island, receive under 25 inches annually.

So much of the Atlantic Coastal Plain has become urbanized that a discussion of the natural vegetation and agriculture is almost academic. Farms here have continued to reap profits from vegetable, fruit, and potato production, however, even in the face of encroaching urbanism.

Part II

BIRDING SITES BY *KINGBIRD* REGIONS

Niagara Frontier — Region 1

THE BROAD PHYSIOGRAPHIC ASPECTS OF THE NIAGARA FRONTIER REGION include Lake Ontario, the Niagara River, and Lake Erie, which form its northern and western borders; the wide valleys and rounded peaks of the Allegheny Plateau on the south; and the rich rolling farmland with its fertile vineyards and the productive orchards of the Ontario and Erie plains on the north and west. The western boundary plain stretches up from the Pennsylvania state line northeastward along the south shore of Lake Erie to Buffalo, where the Niagara River forms the border as it flows northward through a series of lowlands to its mouth at Lake Ontario. Allegany, Cattaraugus, Chautauqua, Erie, and Niagara counties are wholly contained in this region, along with the western halves of Orleans, Genesee, and Wyoming counties. The natural vegetational associations found within the region are those of the Southern hardwood forest — largely oak and other hardwoods, covering much of the lowland area bordering the lakes; and interiorly, those of the Northeastern hardwood belt — largely birch, beech, maple, and hemlock, with white pine occasionally in the river bottoms. The elevation varies from approximately 250 feet at the Lake Ontario shoreline to 2500 feet at the crest of the Allegheny Plateau. (See Map 1.1.)

The Lake Erie–Niagara River Basin and the Lake Ontario Basin are two of the largest and most critical drainage basins in the state. The quality of these waters to a large extent determines the quantity and diversity of the avifauna of the region. Nowhere in the state can more extensive or more beautiful gravel river terraces be found than in the southwestern part of the Niagara Frontier. They can be seen all along the upper Allegheny River and its tributaries.

The abundant bird life in this region should be viewed year-round and not merely in migration seasons. A series of three northward-facing escarpments situated in the western and northern parts of the Niagara Frontier affect migration. The Portage Escarpment along the Erie Plain, inland from Lake Erie and the Niagara, for instance, occasionally tends to channel migration movements of raptors, particularly in the spring. But the area has a high proportion of resident breeders, and it also has the nation's premier gull-viewing site and New York's largest state park.

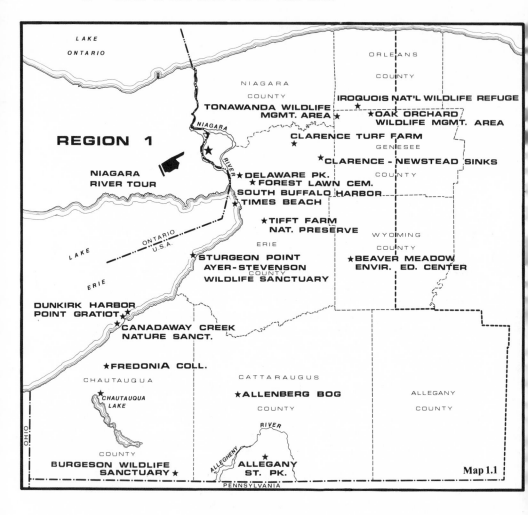

CHAUTAUQUA COUNTY
Burgeson Wildlife Sanctuary (Map 1.2)

Rating: Spring****, Summer**, Autumn****, Winter**

The Burgeson Sanctuary is approximately 6 miles southeast of the city of James-town. Surrounding the sanctuary are rolling and hilly lands dissected by small stream valleys feeding into the area's principal drainage way, Conewango Creek. The 189-acre preserve contains an enormous diversity of plant species (approximately 392) and a variety of habitats, including old and cultivated fields, hardwood and conifer plantings, swamp forest and upland forest.

A system of dikes has created two large wetland areas and several small ponds, and a network of nine trails, totaling about five miles, has been carefully developed, featuring observation towers and photographic blinds overlooking two of the larger ponds. The sanctuary is owned and managed by the Jamestown Audubon Society, which provided the sanctuary with a forward-looking facility for its already very active educational program: the Roger Tory Peterson Nature Interpretive Building. The building contains an auditorium, a reference library, a permanent exhibit of art works (many by Roger Tory Peterson), a collection of more than 200 mounted bird specimens, and administration offices. It also serves as chapter headquarters for the Jamestown Audubon Society. Exhibits housed in the center emphasize the ecology of the region and are designed to stimulate interest in the trail walks. The large windows create an atmosphere of openness and intimacy with the outdoors; one window is equipped with a high-powered telescope, which visitors may use for distant bird observation out on the ponds or for closer views of birds at the nearby feeding stations.

Roger Tory Peterson was born in the city of Jamestown, attended school there, and there first became devoted to birds and art. His subsequent career as artist, ornithologist, teacher, naturalist, author, lecturer, photographer, conservationist, and birder *extraordinaire* reveal what may be accomplished through unwearied industry and development of one's talents. He left Jamestown at the age of nineteen, after an adolescence spent in the cultivation of his vivid artistic imagination and keen field skills.

To reach the Burgeson Sanctuary, leave I-90 (the New York State Thruway) at Exit 60. Proceed southeast on NY 17 and NY 394 to the Southern Tier Expressway (NY 17). Get on the Southern Tier Expressway and proceed east to Exit 12. Exit here and get immediately on NY 60 heading south. Follow NY 60 through and out of the city of Jamestown. At the junction of NY 62 turn right (south) and proceed to Riverside Road. Turn left on Riverside Road and proceed to the entrance gate of Burgeson Wildlife Sanctuary. Park in the interpretive building parking lot. From there, all of the existing trails are easily accessible.

Walk east and north around the edge of Big Pond to the observation tower on its southwest side. Common Loon, Horned and Pied-billed grebes, four heron species, Cattle and Great egrets, Least and American bitterns, some ten species of surface-feeding, marsh, and pond dabblers, and a minimum of six diving duck species have been recorded on the grounds and should be searched out. After birding the upland forest all along the Tower and Sparrow trails, walk around the swamp forest habitat of Warbler Trail. Be vigilant for Soras and Virginia Rails. You'll come out on the northern edge of Spatterdock Pond. Walk along the dike birding the small shadowy recesses of Spatterdock. Walk to its south end and turn right onto Red Pine Trail. Ramble through the conifers. Barn, Screech, Great Horned, Snowy, Barred, Long-eared, Short-eared, and Saw-whet owls have been recorded in the preserve, so be on the alert while roaming through this area.

In spring, scan the sky occasionally for the transient accipiters or buteos. Turkey Vultures and Ospreys occur frequently here. Pass through the mowed fields and pastures bordering Maple Trail. You will now be walking east. Again bird the planted hardwoods and conifers on your right. In spring these woods harbor such migrant passerines as Least Flycatchers, Eastern Bluebirds, and Blue-gray Gnatcatchers. White-eyed, Yellow-throated, Solitary, Red-eyed, Philadelphia, and Warbling vireos all occur here. More than thirty species of warblers have been observed on the sanctuary grounds, including Prothonotary (and amazingly, Swainson's once), Worm-eating, Blue-winged and Golden-winged, Orange-crowned, Nashville, Cape May, Cerulean, Blackburnian, Chestnut-sided, Bay-breasted, and Blackpoll. In spring look for Indigo Buntings singing from prominent perches and American Goldfinches flitting among the tops of the flowering dogwood.

During an average year more than seventy passerine species are banded on the sanctuary grounds. More than ten sparrow species have been logged, including Grasshopper, Henslow's, Vesper, and Lincoln's; in the appropriate seasons be on the lookout for them.

Chautauqua Lake, a Tour (Map 1.2)

Rating: Spring***, Summer*, Autumn***, Winter***

Chautauqua Lake has enjoyed a rich heritage as an important water route between the Trans-Appalachian area and western New York in the 1700s and 1800s, as a route for rafting timber when the lumbering industry was working in the lake's surrounding watershed. Today it serves as a beautiful recreational resource and a long-established cultural center. The Seneca Indians thought of it as a bag tied in the middle, and indeed it is almost cut in two at narrow Bemus Point. Water quality and depths in the lake's two basins differ markedly. The southern section of the lake is shallow, with a maximum depth of twenty feet, has a high concentration of dissolved nutrients, dense algal growths, and seasonally suffers oxygen deficiencies. The northern section is much deeper, reaching more than sixty feet at several places and seventy-seven feet southwest of Dewittville. The water quality and depth account in part for the difference in avian life found on the lake.

Leave I-90 (the New York State Thruway) at Exit 60. Proceed southeast 6 miles on NY 394 (NY 17) through Westfield. In Mayville cross NY 430 (East and West Chautauqua Streets). Turn left at Whallon Street. Proceed a short distance to excellent overlooks from which you will be able to bird Hartfield Bay and the north end of the lake. In October and through late November concentrations of several Common Loons, 600 to 800 Whistling Swans, Gadwalls, Canvasbacks, 100 to 200 Redheads, 150 to 200 Hooded Mergansers, Ruddy Ducks, and 300 to 400 American Coots can be viewed at the north end of the lake. By the first week of December the Whistling Swans peak will have reached 1000± birds. In early March through late April look at this location for small flocks of Common Loons, 100 to 150 Whistling Swans, and White-winged Scoters.

After birding the north end go back to South Erie Street (NY 394) and turn left. Proceed south along the west side of Chautauqua Lake. Continue approximately 6.3 miles, passing through Chautauqua (Chautauqua Institution). Turn left at the signs indicating Prendergast Point Boat Launching Site. Go to the point and

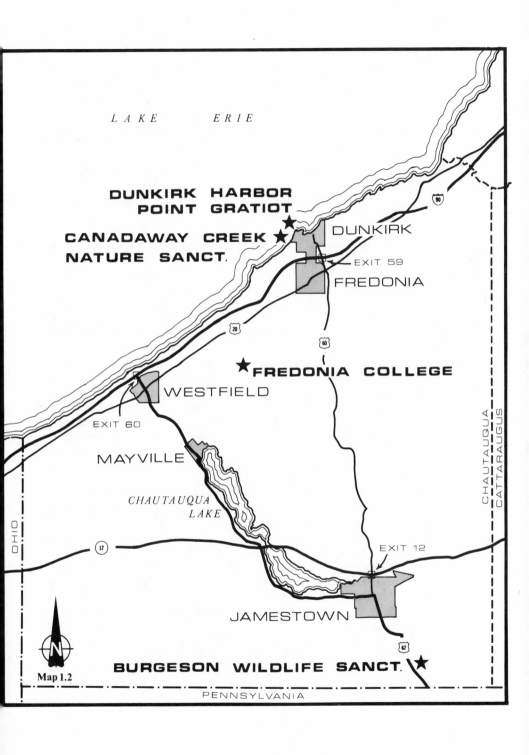

LAKE   ERIE

DUNKIRK HARBOR
POINT GRATIOT

CANADAWAY CREEK
NATURE SANCT.

DUNKIRK

90

EXIT 59

FREDONIA

20

60

★ **FREDONIA COLLEGE**

WESTFIELD

EXIT 60

MAYVILLE

CHAUTAUQUA
LAKE

17

EXIT 12

CHAUTAUQUA
CATTARAUGUS

OHIO

JAMESTOWN

62

N

BURGESON WILDLIFE SANCT. ★

Map 1.2

PENNSYLVANIA

bird the lake from there. The very locally rare Brant has been recorded here in late November along with other water species; but, in general this is a location that is more productively birded in summer. Among other shorebird species, Solitary Sandpiper, Short-billed Dowitcher, and Stilt Sandpiper can be well seen on their southward migrations in late July and early August.

Continue south on NY 394 through Stow and cross the Southern Tier Expressway. Proceed another 2.0 miles to Cheneys Point. At Cheneys Road turn right and bird the marshy areas along the road. Here in the breeding season (late June–early July) the diligent birder can hear and see several pairs of Long-billed Marsh Wrens. This is perhaps the only breeding location for the species in southwestern New York. A left turn off NY 394 will terminate at Cheneys Point, from which the lakefront can be birded.

Continue south along the lake turning off at any convenient spot just outside of Lakewood. Park and walk down to Shermans Bay and scan the open water here for various migrants. Proceed through Lakewood on NY 394. At its eastern end turn left on Southwestern Drive (Sessions Avenue) and proceed across the railroad tracks. Take the next right and continue along the lake shore to the park at the foot of Jackson Avenue in Celoron. Bird along the lake to the outlet of the Chadakoin River. From mid-December through March search here for Common Loons, Red-necked, Horned, and Pied-billed grebes, Great Blue Herons, Whistling Swan concentrations, Gadwalls, rare but recorded Barrow's Goldeneyes among the flocks of 150 to 200 Common Goldeneyes, and other waterbird species.

Follow Jones and Gifford Avenue southeast across the Chadakoin River to the junction with NY 394. Here turn left and proceed to NY 60. Turn left and proceed to Fluvanna Avenue. Turn left onto it and continue along the east side of Chautauqua Lake, heading northwest. There are several good vantage points along this side of the lake. From east to west they are Fluvanna at the outlet, Dutch Hollow Creek and Stockholm Point just west of Greenhurst, Martha's Vineyard between Greenhurst and Bemus Point, and Fluvanna Avenue overlooking the lake between Griffith and Driftwood.

On the northern section of the lake, proceed northwest on NY 430 (NY 17). After crossing Bemus Creek, with the village of Bemus Point on the south (left), turn left on Center Street. Proceed to the ferry crossing at Bemus Point and scan the bay. Retrace the Lakeside Drive, staying on it around Bemus Bay and stopping wherever there is an unobstructed view. Continue on Lakeside Drive to Long Point on Chautauqua Lake State Park. Park and walk around in the park searching for migrating passerines in spring and autumn and the lakeshore in every season. Continue the approximately 8.5 miles circuit of the lake stopping at convenient vantage points from the park to Hartfield Bay at the north end, where the lake trip began.

The entire tour around Chautauqua Lake is approximately 40.0 miles and should take several hours to cover thoroughly. In some cold years, nearly all of the southern half of the lake is frozen by December 15. The much deeper waters of the northern half take considerably longer to freeze over and drive off hardy lingering waterfowl.

Fredonia College Camp (Map 1.2)

Rating: Spring***, Summer**, Autumn**, Winter*

This 200-acre plot (otherwise known as College Camp of the State University College at Fredonia) is situated in the hilly upland above Lake Erie. Essentially, it has not been disturbed for more than forty years. Before then it was farmed, and several pine plantations were started. It now comprises several habitats, including more than fifty acres, which are locally unique in that they are beginning to resemble a climax beech-maple eastern deciduous forest; mature pine plantations; several acres of relict hemlock; one permanent and one intermittent pond; and considerable edge habitat.

The varied habitat seems to attract a wide assortment of bird species during both migrations, but spring yields more migrants, as well as breeding species especially dependent on large patches of mature, undisturbed woodlands for nesting habitat. This generally rugged topography southeast of the lake plain contains several steep ridges (most of which are on state forest lands), and conditions are typical of extensive plots in deep woods, with several moist depressions and tiny ponds. There are no particular hazards or discouragements to birding, and doing so could be rated as easy, if one were to restrict walking to flat ridge tops, or difficult if one traverses the numerous steep hills at a fairly brisk pace. Trails at the Fredonia College Camp take both routes, so the choice is entirely up to the birder.

To reach the area, leave I-90 (the New York State Thruway) at Exit 59. Proceed southeast on NY 60 to NY 20. Turn south (right) on NY 20 and follow it into Brocton. Here pick up NY 380 heading south. At 3.0 miles look for the sign indicating the camp on your right.

Sharp-shinned and Cooper's hawks are found here and may nest on the grounds. Red-shouldered and Broad-winged hawks are confirmed breeders. Yellow-billed and Black-billed cuckoos can be seen as spring and autumn migrants. Screech, Great Horned, and Barred owls are common here, with the last two and the Long-eared Owl classified as possible breeders. Whip-poor-wills and Chimney Swifts are easily seen in spring. Five species of picids are regularly seen at the camp with four as breeders and Yellow-bellied Sapsucker as a possible breeder. Five flycatcher species have been recorded, and some breed in the woodlands. A full complement of chickadees, nuthatches, creepers, wrens, catbirds, thrashers, and thrushes can be found in almost every season, with the Wood Thrush recorded as a breeder and the Veery as a possible breeder. Gnatcatchers, kinglets, waxwings, and Yellow-throated, Solitary, Red-eyed, and Warbling vireos commonly occur in spring and autumn. Of the twenty-five warbler species recorded on the College Camp bird list, all occur in both spring and autumn; Yellow, Black-throated Blue, Blackburnian, Hooded, and American Redstart are probable breeders and Ovenbird and Common Yellowthroat are confirmed breeders. Both Orchard (rarely) and Northern orioles can be seen on northbound and southbound migrations, and Scarlet Tanagers. Rose-breasted Grosbeaks, and Indigo Buntings nest on the grounds. The Evening Grosbeak may be seen as a migrant. Eight sparrow species have been recorded to date at the Fredonia Camp, with Chipping and Song sparrows occurring as nesters and the remainder passing through on both spring and autumn migrations.

This relatively unbirded area holds a good deal of potential for the visiting birder, especially during the height of the spring migration.

Dunkirk Harbor and Point Gratiot (Map 1.2)

Rating: Spring****, Summer*, Autumn***, Winter****

Two main factors make these sites especially good for birding: all winter, the Niagara-Mohawk Power Corporation's power plant on Dunkirk Harbor releases warm water into the harbor, making it icefree and attractive to many duck and gull species; and Point Gratiot, surrounded on three sides by Lake Erie, acts as a funnel, concentrating birds into a wooded area on the point, where totals sometimes reach into the several hundreds during heavy spring migrational waves. The two areas together encompass approximately 100 acres and are situated in the town of Dunkirk and near its western edge. In winter the harbor pier, which lacks safety barriers, is often especially icy, and the birder should exercise care when walking it.

Leave I-90 (the New York State Thruway) at Exit 59 (Dunkirk-Fredonia). Turn north on NY 60 and follow it to NY 5 (Lakeshore Drive West). Turn left (west) here and at the intersection of Central Avenue (0.3 mile) turn right onto harbor pier.

The best vantage points on Dunkirk Harbor are the drive along the harbor edge in the north, the driveway on the city fishing pier, and the marina dock area in the southwest corner of the harbor. In recent winters, area rarities have included Red-throated Loons, Eared Grebes, Harlequin Ducks, Red Phalaropes, Glaucous, Iceland, Black-headed, Laughing, Franklin's, and Little gulls, Black-legged Kittiwakes, and Snowy Owls out on the breakwall. Red-necked Grebes in small numbers can be seen in February. Commonly, large numbers of Common Mergansers and Greater Scaups raft here in January. Small flocks of Snow Buntings can usually be seen here from November through winter. Caspian Tern flocks can be found here in late April. If the birder finds himself in the general vicinity from November through April, a special trip to Dunkirk Harbor is recommended.

From the intersection of Central Avenue and NY 5 (Lakeshore Drive West) continue driving west to Point Drive North. Turn right here and proceed to a "One Way, Do Not Enter" sign. Here park and walk into the wooded area at the tip of Point Gratiot. This is an excellent site for many rare and uncommon migrants in spring. Merlins have been recorded perched and in flight in the open woods here in late April. Red-headed Woodpeckers breed here. Great Crested Flycatchers, Winter Wrens, Orange-crowned Warblers, and literally hundreds of tanagers, grosbeaks, thrushes, and flycatchers can be viewed here on a good May day. Loggerhead Shrikes, White-eyed Vireos, Yellow-throated Warblers and Orchard Orioles are rarities that have occurred here. These brushy woods are usually closed in winter.

Canadaway Creek Nature Sanctuary (Map 1.2)

Rating: Spring****, Summer**, Autumn***, Winter*

This nature preserve of approximately thirty acres is just southwest of the city of Dunkirk. The first shots in the War of 1812 were fired on this site. A good deal before that the area hosted first an Erie Indian fishing village and later an Iroquois Indian village. The Indian term *Canadaway* translates "stream among the hemlocks," and the lush and varied vegetation here makes it easy to imagine why it would be attractive to both humans and birds. The sanctuary is now owned by the Nature Conservancy and it is composed of two sections separated by NY 5. The two sections lie on the Canadaway Creek floodplain, so after periods of heavy rains or ice melt, waterproofed footwear is suggested. It has long been recognized that this unspoiled tract lies squarely on a major migratory flyway. The rampant growth of vines, herbs, shrubs, and trees provides food and shelter for migrants stopping over on their spring and autumn passages. In some places the thickets and tangles are nearly impenetrable. The red osier and panicled dogwoods, speckled alders, sycamores, willows, maples, blackberries, and aspens are glorious in spring bloom. Ostrich fern, violets, trout lilies, wild cucumbers and grapes, and various jewelweed species abound on the sanctuary. More than 140 species of birds have been recorded on the sanctuary grounds, and as field work and general birding continues here the list will certainly grow.

Leave I-90 (the New York State Thruway) at Exit 59 (Dunkirk-Fredonia). Turn north on NY 60 and follow it to NY 5 (Lakeshore Drive West). Turn left (west) and proceed approximately 2 miles, where Temple Road comes in on the left and a small bridge on NY 5 crosses Canadaway Creek. The sanctuary is both west of NY 5, at the mouth of the Creek on the south shore of Lake Erie and east of NY 5, upstream. Large signs on both NY 5 and on Temple Road designate the entrances. There are parking spaces on the west side of Temple Road.

Birding the area is relatively easy. The path from NY 5 to the lake follows the creek. At the creek mouth there are mud flats on which Spotted, Solitary, and Western sandpipers, Willets, Red Phalaropes, and Caspian Terns have all been recorded. Bird along the path, at the mouth of the creek and along the shore of the lake. Penetrate the woods along the route wherever possible and search the area for flocks of migrant passerines. On the east side of the bridge in the Temple Road segment, follow the obvious path which roughly circles the tract. Carolinian species nest here in summer and the natural food supply sustains 25 to 30 species here in most winters. In spring Great Crested Flycatchers, thrushes, Veeries, Blue-gray Gnatcatchers, at least ten warbler species, White-eyed Vireos, Rusty Blackbirds, and several sparrow species can be observed here.

During the height of the spring migration it is advisable to bird both the brushy woods of Point Gratiot and Canadaway Creek Nature Sanctuary for a truly representative sampling of the species passing through.

CATTARAUGUS COUNTY
Allegany State Park (Map 1.1)

Rating: Spring***, Summer****, Autumn**, Winter*

After all that has been written about natural beauty, it seems rather absurd to begin this description by stating that "this is undoubtedly one of the most splendid pieces of property in the Northeast." That is my lasting impression, and I doubt anyone could feel otherwise. One cannot walk about for very long in the forests of beech, sugar maple, hemlock, and yellow birch without being genuinely moved by their wild grandeur. Abundant streams and valleys are found throughout. Rugged hills and broad reaches of gently rolling countryside provide considerable variation to the landscape. The massive, unaligned geologic formations, embellished by the forces of wind and water provide visual interest. Upland pastures, meadow edges, orchards, stream margins, thickets, lush forest floors, and dense forest canopies provide infinite variety for resident bird species. This marvelous area boasts an extensive bird list primarily because of its large tracts of diverse habitats. Size and variety combine to make this one of the most interesting and rewarding places in which not only to identify birds but also to understand their habits, ecological plant and insect associations, physical and geographical ranges and limitations. Although migrant birds can be found throughout spring and fall, and winter finches occur usually in season, it is the resident species, especially the less common ones, which are of greatest interest here.

Allegany State Park is located in western New York, in the southern part of Cattaraugus County, just south of the city of Salamanca. It is roughly circumscribed on the west, north and east by a great arc of the Allegheny Reservoir and River and its northward flowing tributary, Tunungwant Creek. The Allegany Indian Reservation includes the land for approximately 0.5 mile on either side of the river. On the south it is bounded by the Pennsylvania state line, across which lies the extensive wilderness tract of the Allegheny National Forest. This forest of over 700,000 acres is more than ten times the size of the 65,000-acre Allegany State Park. The road and trail systems of the state park and the national forest give access to one of the largest continuous forests in the east.

The following description includes trails and places only within New York State. Within the park the mountains rise abruptly 800 to 1000 feet above the broad valley floors, and the elevation ranges from 1280 feet to 2400 feet above sea level.

Birding in the park can be as physically easy or taxing as one wishes to make it. The season, ruggedness of terrain traversed, and extent of walking will determine the energy one must expend here, although a good deal of birding could be accomplished from a car. During the summer season the park visitor must pay a minimal admission charge.

Leave I-90 (the New York State Thruway), at Exit 59 (Dunkirk-Fredonia exit). Turn south on NY 60 and follow it south to NY 17 (Southern Tier Expressway). Just north of Jamestown, get on NY 17 heading east. Follow it approximately 35 miles to Salamanca. Alternatively Salamanca can be reached on NY 219 from either the north or south. Proceed through town on NY 17 to its eastern end, where there will be signs designating Allegany State Park. Turn left (south) on Parkway Drive, which becomes Scenic Drive and then Park Road or ASP 1. This is the road indicated to Red House Lake and the Park Headquarters. Stop along the way at the two marked scenic vista overlooks; the first lies just inside the park boundary and the second is at the junction of ASP 1 and a gravel road on the right, which leads to an observation tower. From both of these vantage points, look for hawks soaring above the adjacent ridges. Proceed south on ASP 1 to the headquarters building, where a map of the hiking trails and a check list of the park's birds can be obtained.

Around the administration buildings look for Chimney Swifts, Eastern Phoebes, Barn and Cliff swallows, and listen for the short, rapid notes of the House Wren's song. While in the vicinity of Red House Lake, take time to bird it well. Red House Lake is an artificial lake occupying the lower Red House valley above the dam. It is rimmed with slopes covered principally with sugar maple and beech on the north and mostly aspen and red maple on the south. Toward the Allegheny River, the forest is largely second-growth mixed types. The nearly 100-acre lake receives the drainage from upper Red House, Bova, Bee Hunter's, Stoddard, and McIntosh creeks. Pine groves and other small stands of planted evergreens are set around the lake's perimeter. Red House Lake occasionally hosts Common Loons in early June, grebes, Great Blue and Green herons, bitterns, and ducks. Look overhead for Goshawks, Sharp-shinned, Red-shouldered, Broad-winged, and Red-tailed hawks — all of which nest within the park. Be watchful for Killdeer, Common Snipe, and Spotted Sandpiper while birding around the lake. On the little island at the southwest side of the lake, a few years ago Andrle *et al.,* confirmed the first known breeding of Ruby-crowned Kinglets for western New York state. Golden-crowned Kinglets nest within the park also. Near the right-angle turn on the maintenance road close to the Red House Lake dam, both Yellow-throated Warblers and Northern Parulas have been observed and heard singing at the beginning of June.

South of the lake, and south of the administration buildings, is the Big Basin trailhead. The Big Basin Forest is by any standards impressive. Its large tract of mature trees, principally hemlock, beech, yellow birch, and sugar maple would take at least 300 years to replace. In some places within the area there is a dense understory of young trees, fallen and decaying logs, bracken, and goldenrods. In some places wintergreen and other shade-enduring species form the groundcover. The interior of the forest hosts a rich flora of woodland plants; where the humus is deep, the ground is covered with an unusual display of mosses and ferns. More than twenty fern species can be found here as well as wood sorrel, trilliums, violets, and a wealth of similar plants. In addition to a variety of animal life living in the area, Sharp-shinned Hawks, Ruffed Grouse, Turkeys, American Woodcocks, Whip-poor-wills and at least four woodpecker species have been easily recorded here. By walking the trails through the Big Basin in early morning the birder should be able to see and hear Eastern Wood Pewees, Least Flycatchers, chickadees, nuthatches, Brown Creepers, Winter Wrens, Wood, Hermit and Swainson's thrushes, as well as Veeries. A few of the notable and some expected breeding warblers found not only in the Big Basin Forest, but also at additional sites throughout the park are Black-and-white, Northern Parula, Magnolia, Black-throated Blue, Yellow-rumped, Black-throated Green, and Blackburnian. Listen along the paths for Ovenbirds and waterthrushes. If the birder really knows the songs of breeding male warblers, he will be able to discern those of Mourning, Hooded, and Canada warblers.

ASP 1 provides a short route from the administration and camping areas of Red House Area to the Quaker Run camping and cabin area. The road passes through the center of the Big Basin and divides the mature forest area almost exactly in half. It allows the visitor access to areas and trails which he could not reach otherwise. The birder can also walk through the magnificent forest on the Big Basin Trail and discover a representative sampling of the avian life present throughout the forest.

In the vicinity of Quaker Run one can find the bird species associated with aspen forest and open or cultivated lands. Working northwest up Quaker Run one will encounter several swampy areas and very small, nearly remnant areas of mature timber and forest plantings.

In the more moist areas of the Quaker Run yellow birch predominates. The sugar maple-beech association is widely distributed throughout this part of the park, and large-toothed aspen and quaking aspen, though less common, prevail in many more open sections. Bird all around Quaker Run, walking down two or three of the very productive trails, say, Firetower, Three Sisters, or Bear Caves.

There is a wide variety of habitat here. Be sure to go west on ASP 3 to a road coming in on the right opposite the eastern end of Quaker Lake, Cain Hollow Road. This road leads to Cain Hollow camping area. Follow along it, birding all the way. In this general vicinity there is a pasture and stream margin, tilled fields and orchards and unbroken forest, some of it oak, on the western slopes. Listen for the guttural note of the Red-headed Woodpecker. You should be able to spot Eastern Kingbirds snapping up insects and hear the harsh rasps of Great Crested Flycatchers. Try to locate singing Olive-sided (migrating) and Least flycatchers. Be sure to bird the old fields in and around the periphery of the Quaker Run vicinity. These broad, flat meadowlands are principally in successional stage, reverting now to forest. This environment is hospitable to Brown Thrashers, Bobolinks, Eastern Meadowlarks, Indigo Buntings, and of course, towhees. It is in the area of Quaker Run, during years of cankerworm infestation, that adult Evening Grosbeaks with recently fledged young have been recorded, providing rare records for the Niagara Frontier region. In the tall grass fields near the lower end of Quaker Run look for Savannah, Grasshopper and Vesper sparrows. Along the margins of old pasture hillsides where thickets have sprung up look diligently for Field Sparrows.

Be sure to check out Quaker Lake for migrating waterfowl in spring and fall.

Proceed west on ASP 3 up the west side of Quaker Lake to NY 280. Bird the Allegheny Reservoir up and down this road, looking for Great Egret and Black-crowned Night Heron north of Quaker Lake as well as migrating waterfowl in early May. Keeping in mind that Bald Eagles have been occasionally recorded here in October, January, and June. Look especially along Allegheny Reservoir Road for Snowy Owls in early winter and Eastern Bluebirds, Blue-gray Gnatcatchers,

Yellow-throated and Red-eyed vireos, Pine Warblers, orioles, and tanagers in warmer seasons.

Outside the park, cross the reservoir on NY 17, entering at Exit 18 and leaving at Exit 17. Proceed south on Pierce Run Road, stopping along it to bird the reservoir on its west side. Flocks of Common Loons, grebes, small flocks of twenty to thirty Oldsquaws and five to ten White-winged Scoters can be found here mid-April to mid-May. Explore along Onoville Road and Bone Run, Sawmill Run, and Brown Run roads. Snowy Owls have been seen regularly along Bone Run Road in winter. Goshawks nest between the village of Ivory and the Allegheny Reservoir. Turkeys can also be seen here, and early spring migrant passerines can often be seen near the intersection of Bone Run Road and the West Perimeter Road in mid-April and early May. Acadian Flycatchers breed along the Sawmill Run Road between Frewsburg and Onoville. The oak-hickory forest between Steamburg, just north of Exit 17, and Onoville, just north of the Pennsylvania state line, holds a dense concentration of thirty to thirty-five pairs of breeding Cerulean Warblers. In this same area migrant Yellow-breasted Chats and Hooded Warblers can often be heard singing in the first week in June.

There are other entrances to the park in addition to the one noted above, and depending upon which one the birder uses, he may choose other sites. It is reportedly quite easy to see Turkeys and Ruffed Grouse in the park's interior, and the birder should ask at the park headquarters building where they are most likely to be found at the time of his visit.

Allenberg Bog (Map 1.1)

Rating: Spring****, Summer****, Autumn**, Winter*

Allenberg Bog (sometimes spelled Allenburg), also known as Congdon's Pond, Waterman Swamp, and Owlenburg Bog, is on the border of the towns of Napoli and New Albion. It is a unique and fascinating wildlife refuge of more than 325 acres, owned and maintained by the Buffalo Audubon Society. Even before the Buffalo Audubon Society took possession of the property in 1957, this area was famous among botanists for its rare wild orchids, more than 30 species of liverworts, nearly 60 species of mosses, and approximately 285 species of vascular plants. The Buffalo Audubon Society emphasizes that collecting of any plants in the refuge is strictly prohibited. The unique vegetation and flora includes dense, beautiful, but absolutely impenetrable thickets of rhododendrons, numerous species of wildflowers, ferns, huge hemlocks, pines, black cherry, black spruce, and tamarack stands. This beautiful wildlife refuge sports, in fact, two sphagnum bogs surrounded by tall trees, fragrant blossoming wildflowers, deciduous shrubs, clumps of ferns, and shoulder-high bracken. It is unquestionably one of the most glorious areas in southwestern New York and its wealth of natural beauty should simply not be missed.

Allenberg Bog is situated in Napoli and New Albion townships in Cattaraugus County. The elevation ranges approximately 80 feet (1720 feet to 1800 feet above sea level), and its springs and ponds drain south and east into the Allegheny

River. Once inside the refuge, it is so natural and wild that it is fairly easy to get lost or turned around. Therefore, it is advisable to carry along a compass, at least on the first trip.

To reach the bog, leave I-90 (the New York State Thruway), at Exit 59 (Dunkirk-Fredonia). Turn south on NY 60 and proceed south to the village of Laona, where you will turn left on NY 83 heading east. Follow NY 83 and its extension, NY 322, to NY 62. Here turn right (south) and follow NY 62 to the village of Leon. Here turn left on Leon Road and proceed east for approximately 6.5 miles. Turn right on Farm Market Road and proceed south approximately 0.75 mile to the right-of-way entrance to the trails on the west side of the refuge. This roadway is on the east (left) side of Farm Market Road. Park your car off the road and proceed from here on foot. Walk down the farm roadway to a path cut across the meadow, cut through the grass and underbrush, along a barbed wire fence. Note the flowering dogbane along the path and the black-eyed susans, daisies and pink mallows in the meadow. Where the path runs between the woods and the fence you'll note bracken. Coming through the meadow watch and listen for trilling House Wrens, Common Yellowthroats, Bobolinks, Eastern Meadowlarks, Indigo Buntings, American Goldfinches, Rufous-sided Towhees, and Savannah, Chipping, Field, and Song sparrows.

Continue on the path into the cool shade of the undisturbed woods. Wherever shafts of sunlight break through, yellow hawkweed, wood sorrel, and jewelweed bloom. Underneath the trees can be found the sweet-smelling flowers of the partridge berry. Notice little stands of Indian pipes as you walk along. To go to the north bog, follow the trail as it veers toward the left. The path going to the south bog branches to the right. Walking toward the north bog, the path is ordinarily fairly damp, listen for the call of the Red-shouldered Hawk, which nests in the woods. At first ferns are relatively scarce but soon huge clumps of cinnamon and interrupted fern appear along with fewer royal ferns. The abundant club mosses are delightful. Along the path be attentive for the irrepressible song of the Winter Wren. To me, it has always seemed quite just that so tiny a creature should have a song that is so beautiful it defies description. The liquid notes of the Wood Thrush will be heard along this path along with those of Hermit and Swainson's thrushes and Veery. The various species of viburnum are fun to try to identify and note the several small cucumber trees along the way.

Soon you'll reach the enchanting north sphagnum bog. The marvelous green of the black spruce and tamarack contrasts with the rhododendrons in full bloom. No photographer could pass by the exquisite rose pogonias or calopogon (grass pink). Everything seems to have sprung up out of the thick leatherleaf. The tiny sundew, pitcher plants, Labrador tea, and bog cranberry are especially delightful to anyone who has never visited a bog. This bog, smaller than the south bog, is nearly half the size of a football field and about fourteen feet deep.

Flycatchers flit about, catbirds call and dart about the willows. Cedar Waxwings whisper from the tall spruces. Solitary and Red-eyed vireos sing. This is an excellent place to learn the song of the breeding Black-throated Blue and Blackburnian warblers. Brilliant Scarlet Tanagers also sing from dead branch perches. Towhees appear and disappear among the vegetation. Listen for the staccato trilling of the Swamp Sparrow.

If the birder chooses to hike to the south bog, he turns to the right where the original path forks. This pond, larger than the north bog, lies in drier habitat and is surrounded by deep woods and gigantic trees. As the birder walks along this path to the southern "Puls pond" he will be amazed by the profusion of wildflowers, deciduous shrubs and ferns. The bog rosemary and tiny berries of the bog cranberry in the immediate vicinity of the bog are no less than a revelation. Along the way listen and watch for the wild and wily Turkeys, which are the pride of the refuge.

Prepare yourself for the beauty of the native rhododendrons as you near the bog. The blossoms of the grass pinks—white delicately tinged on the outer petals with pink, are breath-taking. It is a real experience to just stand on the edge of the quaking bog mat and become aware that yes indeed, it *is* quaking underfoot.

In both this area and a little bit farther southeast look carefully for Goshawks and Red-tailed Hawks, both of which have been documented as nesting in the general vicinity. Eastern Kingbirds and White-breasted Nuthatches can be heard scolding and attracting attention in their own way. Winter Wrens nest in this section of the preserve as well as in the northern section. Along the trail you will hear the clear refrains of the Wood and Hermit thrushes. Quite close to the pond at the edge of the woods, look for nesting Yellow-throated Vireos. Black-throated Blue, Black-throated Green, Cerulean, and Chestnut-sided warblers can all be located along the trail, and it would be wise to brush up on the songs of these warblers before visiting the bog in the spring. Walk carefully along the paths watching for the occasional sudden eruption of the nesting Dark-eyed Junco.

It is possible to hike to the east side of the wildlife refuge, but one member of the group should volunteer to go back to the west side entrance and drive the car over to pick up the hikers. The walk to Pigeon Valley Road, named after the enormous nesting colony of Passenger Pigeons that occupied this area in the 1820s, passes through a fine evergreen plantation of considerable size, along a path that skirts a ravine, outlined in part by a row of ancient sugar maples, and along a pond edge, finally emerging in an open meadow on the other side. The peerless beauty of this refuge is always compelling.

WYOMING COUNTY
Beaver Meadow Environmental Education Center (Map 1.1)

Rating: Spring***, Summer***, Autumn***, Winter**

In 1931 the New York State government brought beavers from Idaho and released them in this area, which is some thirty-five miles southeast of Buffalo. They dammed a stream and created the marsh, which is the dominant topographical feature of this 267-acre nature center owned by the Buffalo Audubon Society. The preserve is named after the beavers that established themselves so successfully here. It is about two miles west of the village of North Java. This carefully maintained preserve has well-cleared trails on which to both bird and study the delicately balanced ecosystems of the woodland and pond areas. Within the center there is an education building with top-quality wildlife displays highlighting the typical life forms found throughout Beaver Meadow, and a reference library with books on all subjects of natural history. Birders can stop at the center to get an area map and a checklist of local birds before walking the trails. Birding the area is rated as "easy." In some winters, however, snowfalls are even heavier than in Buffalo proper, so dress appropriately.

The dominant types of vegetation at Beaver Meadow are luxuriant wet woodlands of hemlock, birch, white spruce, and jack pine. The latter two species are not native to western New York but have been introduced at the refuge. On the property there are large open meadows, upland forests, and the Jenny Glen Trail, a marvelous boardwalk trail for the handicapped constructed across a beautiful swamp which was previously inaccessible and had gone almost unnoticed. Although this boardwalk is called a "trail for the handicapped," use is by no means restricted.

Leave I-90 (the New York State Thruway) at Exit 54 and immediately go to NY 400 (the East Aurora Expressway). Follow it to NY 20A. There turn left and proceed east on NY 20A about 11 miles to its junction with NY 77. Turn right at Persons Corners and proceed south on NY 77 for 7 miles, watching for the green highway marker indicating "Beaver Meadow Audubon Center" at the junction of Welch Road. Here turn left onto Welch Road and continue less than a mile to the Beaver Meadow parking lot on the right.

After visiting the Center, walk the trails looking for the specialties of the refuge. Twenty-six Wood Duck boxes have been erected throughout the marsh. In addition to housing Wood Duck families, Hooded Mergansers are occasionally fledged from them. Red-shouldered and Broad-winged hawks nest fairly regularly in the woods. Listen for the drumming of the nesting Ruffed Grouse. Great Horned (commonly) and Long-eared (much less frequently) owls nest in their appropriate habitat here. Watch for Ruby-throated Hummingbirds and Red-headed Woodpeckers, which nest regularly near the main buildings. Brown Creepers, Veeries, Blackburnian Warblers and Northern Waterthrushes can be found nesting in wet woodland habitat dominated by hemlock. Warbling Vireos, Mourning, Hooded and Canada warblers can be sighted if the birder listens for the notes of their distinctive territorial songs. Look in the tree tops for the flashy coloring of the Scarlet Tanager. Walk through the field areas where the Henslow's Sparrow nests.

To walk the boardwalk of Jenny Glen Trail, cross Welch Road opposite the main portion of Beaver Meadow Center and proceed a short distance west (toward NY 77), where you will see a sign indicating the path to the beginning of the elevated swamp boardwalk.

ERIE COUNTY
Sturgeon Point and the Ayer-Stevenson Wildlife Sanctuary (Map 1.1)

Rating: Spring***, Summer**, Autumn***, Winter*

If the birder has only a short time to devote to birding on his way to Buffalo and is approaching from the southwest in spring, there is a place to stop to get a taste of gulls migrating northward and the warblers and other passerines heading for the breeding grounds.

Leave I-90 (the New York State Thruway) at Exit 57A and immediately obtain Versailles Road. Turn right on it headed north. Follow it a short distance to Sturgeon Point Road. Turn left (west) on Sturgeon Point and follow it to its terminus at Lake Erie. Park and scan the lake. If you are lucky, you might hit it on a spring day of strong northwest winds and be fortunate enough to witness as many as a thousand Bonaparte's Gulls migrating southwestward. Up to and through early May, there should still be grebes, Gadwalls, wigeons and scaups in the vicinity. Look for Semipalmated and Black-bellied plovers and Ruddy Turnstones. For the best view of the lake at this point, take along a telescope.

To get to the Ayer-Stevenson Sanctuary, begin back at Sturgeon Point Road. Follow it a very short distance to Dennis Road. Here turn right and proceed south for about a mile. Look for the fifty-foot-wide access path from Dennis Road back to the refuge area. It will be on the east (left) side of the road. The Ayer-Stevenson Sanctuary is a mere two acres but is in every way delightful. It is owned and maintained by the Buffalo Audubon Society.

Walk back on this path, which borders Little Sister Creek. The path may appear cramped, but there is a more commodious trail inside the preserve under the beeches, maples, hemlocks, and large pines. The birder will be amazed at the Solomon's seals, starflowers, gold threads, devil's paintbrushes, and rhododendrons in bloom; at least ten fern species also flourish. Look carefully for sulphur mushrooms and wood anemones. Tiger swallow-tail and sulphur butterflies flit in and among the patches of wild strawberries. Listen for the rattle of the Belted Kingfisher. Look for the Red-headed Woodpecker, which is rarely recorded locally. You should easily see Common Flickers, Downy Woodpeckers, Great Crested Flycatchers, Eastern Wood Pewees, Barn Swallows, chickadees, White-breasted Nuthatches, Brown Creepers, House Wrens, catbirds, thrashers, robins, Wood Thrushes, kinglets, Cedar Waxwings, Red-eyed Vireos, Ovenbirds, American Redstarts, Field and Song sparrows. All of these species nest within the boundaries of the preserve. If you also manage to sight some of the many migrants that pass through, it should be rewarding birding.

Tifft Farm Nature Preserve (Map 1.3)

Rating: Spring****, Summer**, Autumn***, Winter*

This unusual nature preserve of 264 acres occupies land owned by the City of Buffalo and leased and managed by Tifft Farm, Inc., a not-for-profit corporation whose current address is 1200 Fuhrmann Boulevard, Buffalo, New York 14203. It lies close to Buffalo Harbor on the eastern shore of Lake Erie, approximately 2.5 miles south-southeast of Niagara Square in downtown Buffalo. Of the total acreage, nearly 75 acres is freshwater cattail marsh, another 50 acres is covered by four 45-foot-high grassy mounds of contoured, soil covered, decomposed refuse. There is a large pond in the northwestern part of the preserve, and the remaining

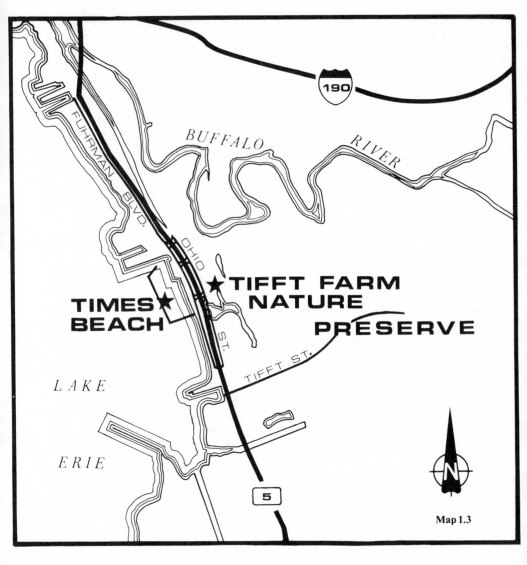

Map 1.3

acres are in varied habitats of large, wet black willows, scattered grasslands and shrubs, an aspen stand, a cottonwood forest, and scattered dense thickets.

The Tifft Farm is situated on a primary migration route and provides a feeding and resting stopover for migrant passerines and waterbirds. The preserve boasts a list of more than 200 species and is rated as moderately easy to bird. All travel within the preserve is on foot. Be careful while walking, as partially buried and surface debris requires caution. Mosquitoes are often numerous, especially around the marsh and adjacent wooded sections, from late May into July.

From downtown Buffalo, proceed to NY 5 (the Fuhrmann Boulevard Expressway). Get onto it heading south. Leave NY 5 at the Tifft Street Exit. Exit onto the feeder road *but just before* Tifft Street turn left underneath the expressway and then left again heading north on the feeder road. Travel on this feeder road for 0.3 mile to the preserve nature center parking lot designated by a sign, located on the right hand side of the road, adjacent to a large pond. If coming from south of the city of Buffalo, proceed north on NY 5. Take the Ohio Street Exit onto the feeder road from the Father Baker Bridge, and proceed approximately 0.3 mile to the preserve nature center parking lot.

Search the pond at the nature center for herons, waterfowl, gulls, and terns in season. The locally rare Snowy Egret and Glossy Ibis have been recorded here. In spring both before and following ice-out, look over migrating flocks of gulls for the "white-winged" species, the Glaucous and Iceland gulls. Walk to the north pond where any of several waterfowl species will probably be seen. Walk along the perimeter of the 75-acre freshwater cattail marsh, birding at any or all of its numerous short "in" paths. Herons, bitterns, ducks, Virginia Rails, Soras, Common Gallinules, coots, and Killdeers nest in the immediate area of the marsh. The rarely recorded King Rail has been seen and heard here. Black Terns usually nest in the marsh and its adjacent ponds, and can be easily seen in the breeding season. The hilly mounds can have field and migrant landbirds on them, during a particularly good spring season, or a Snowy Owl in winter. Walk to the southeast quadrant of the preserve to the impressive stand of black willows. Search all the trees and adjacent thickets for migrant flycatchers (Acadian has been recorded here), thrushes, vireos, and warblers. Prothonotary Warblers have been recorded in this section, and one never knows what rarities will appear among the migrants and resident species. Leave the black willows sector and proceed northwest along the edge of the cottonwood forest and then due west through the grasslands, shrubs, and thickets. Along these paths watch for rare migrants as well as nesting Ring-necked Pheasants, Downy Woodpeckers, chickadees, creepers, marsh wrens, migrant kinglets, vireos, blackbirds, and nesting Swamp and Song sparrows.

Times Beach and South Buffalo Harbor (Map 1.3)

Rating: Spring***, Summer***, Autumn***, Winter*

Times Beach in Buffalo Harbor and Tifft Farm Nature Preserve can easily be covered on the same outing, for they are near each other and are both productive during spring and autumn. Times Beach is a diked former disposal area for harbor sediment dredgings. It is located about one mile (or three miles by road) due south-southwest of Niagara Square in downtown Buffalo. It is easy to bird its 57 acres partly by car and the rest on foot, but, because it was a sediment disposal area, it is nonetheless advisable not to walk into the shallow water.

The site consists of open water with shallows and mud flats as well as marshy sections of higher land with stands of trees and thickets. The water depth is lake-controlled and varies with wind direction and velocity. Its range is sufficient to make the area suitable for gulls, terns, shorebirds and both dabbling and diving waterfowl. Of course, marsh birds also occur here, and the treed and shrubbed sections attract migrant land birds in good numbers and excellent variety (30 species of warblers and 24 fringillids recorded) during spring and fall migrations. To date, 219 species have been recorded here, including such rarities as Yellow-crowned Night Herons, Cinnamon Teals, Golden Eagles, Marbled and Hudsonian godwits, Ruffs, American Avocets, Laughing, Black-headed, and Little gulls, Acadian Flycatchers, and Prairie and Connecticut warblers.

From downtown Buffalo: Take NY 5 (Main Street) southwest toward the Fuhrmann Boulevard Expressway. Pass over the Skyway and take the first exit at the base of the Skyway. Proceed along the feeder road to its first underpass. Make a U-turn under NY 5 and continue north on the feeder road. Cross under the Skyway. Proceed about one mile and park along the lakeside adjacent to Times Beach. From Tifft Farm Nature Preserve and other points south, proceed north on NY 5. Take the exit just before the Skyway and continue north for about one mile on the feeder road. Park along the lakeside adjacent to Times Beach.

Delaware Park and Forest Lawn Cemetery (Map 1.4)

Rating: Spring****, Summer**, Autumn***, Winter*

If you happen to be in Buffalo during the spring or autumn migrations and have only a little while to devote to birding, your best bet is to either take public transportation or walk the two miles north from Niagara Square on NY 384 (Delaware Avenue) to these two sites. Together they cover approximately 3020 acres and are landscaped and manicured in the manner one might expect of such places. They have somewhat varied topography and the natural and planted vegetation make them isolated pockets of variety in the continuous band of urbanization of Buffalo. During migrations all of the more common birds turn up, with some really remarkable waves and noteworthy migrants thrown in. For example, in Delaware Park one is apt to see Ruddy Ducks, Least Flycatchers, Veeries, Eastern Bluebirds, Nashville, Northern Parula, and Chestnut-sided warblers, as well as Scarlet Tanagers, Rose-breasted Grosbeaks, and various sparrow species. In Forest Lawn Cemetery one could see Turkey Vultures, and very likely Ruby-throated Hummingbirds, Belted Kingfishers, Red-headed Woodpeckers, Yellow-bellied Flycatchers, Blue-gray Gnatcatchers, Chestnut-sided Warblers, and Rusty Blackbirds. Of course, this is a mere sampling of the sizable list of these two areas, and they are quick alternatives to the other Buffalo "hot-spots."

Clarence-Newstead Sinks (Map 1.4)

Rating: Spring****, Summer*, Autumn*, Winter*

A sink hole, swallow hole, or what is popularly called a pot hole, is a saucer-shaped depression usually found in limestone regions. Rainwater collects in these depres-

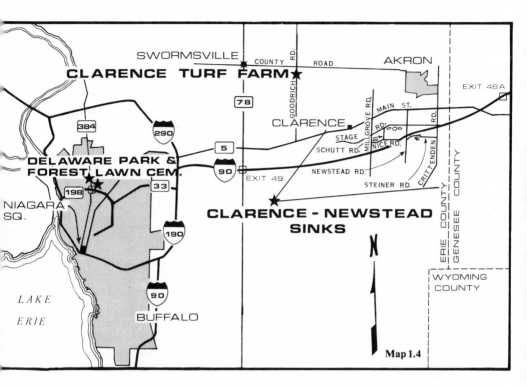

Map 1.4

sions and slowly passes into the ground. Waterfowl concentrations, particularly of Whistling Swans and Canada Geese, reach exceptional proportions in the flooded bottomlands and sinks in Erie County, located about two miles southeast of Clarence. The area takes in some 2000 acres, and many of the sinks are on private farmlands. Birders should stay off the owners' property and avoid disturbing these very understanding people. Confine birding to the public roads, from which all of the important viewing areas can easily be seen. Carry a scope to ensure good views.

Spring, primarily late March and April, is the best time to bird the sinks. Glossy Ibises, White-fronted Geese, and European Wigeons are rarities reported from this area. All water-associated migrants—plovers, yellowlegs, Pectoral Sandpipers, and various swallow species—can be expected here in season. Water Pipits are sometimes abundant at the height of migration.

Leave I-90 (the New York State Thruway) at Exit 49. Drive north on NY 78 for 1.5 miles to its junction with NY 5. Turn right (east) on NY 5 (Main Street) and proceed east 6 miles to Clarence. Stay on NY 5 for about 2 miles to Millgrove Road. Here turn right (south) and proceed 1 mile to Stage Road. Now drive slowly south on Millgrove Road for less than a mile, watching the fields on the left. Turn left (east) at Nice Road, which is the next road on the left. Just beyond the turn, the first of the sinks can be seen near Nice Road on the left. Proceed east a little farther to Schutt Road, which is the next road on the left. Here turn left and follow it north a short distance back to Stage Road. Here turn right (east) and bird the three sinks close to the road on its south (right) side. Proceed to the next road, which is Newstead Road. Turn right (south) and proceed across the New York State Thruway. Take the first available left onto Steiner Road. Follow it nearly 0.7 mile, where another sink will be visible on the right. At the next road, Ayers Road, turn right and then immediately left onto the continuation of Steiner Road. Follow it to Crittenden Road. Turn left and proceed north to the New York State Thruway (I-90) or to NY 5. Remember that Stage, Nice, and Steiner Roads pass all of the best viewing sinks; and because they are not heavily travelled, birders can get out of the car and bird from the roads, thereby avoiding trespassing.

Clarence Turf Farm (Map 1.4)

Rating: Spring**, Summer**, Autumn***, Winter*

This area encompasses approximately 1200 acres of private land, which can be partly covered by car and partly on foot, about three miles east of the village of Swormsville. Although it is private land, the owners of Lakeside Sod Farm are very understanding of birders who take care to not disturb new grass plantings. Birding the area is easy, and although some areas are muddy and wet, normal footwear is sufficient.

Open country and shorebirds are the specialties here. Migrant shorebirds appear in greatest numbers and variety when rain-spawned puddles dot the fields. American Golden Plovers are numerous from late August through early November. Upland Sandpipers breed here and are present from late April through August. Buff-breasted Sandpipers are regularly noted, generally one week either side of Labor Day. Water Pipits are plentiful in both spring and fall, and Lapland Longspurs and Snow Buntings are occasional in winter.

Leave I-90 (the New York State Thruway) at Exit 49. Drive north on NY 78 for 6.5 miles to the Swormsville traffic light. Here turn right (east) on County Road and proceed east about three miles to Goodrich Road. The turf farm is located here on the south (right) side of County Road, both east and west of Goodrich Road. The fields on the north (left) side of County Road are often good for sparrows and open field species.

THE NIAGARA RIVER, A Tour (Map 1.5)

Rating: Spring***, Summer*, Autumn****, Winter****

The Niagara River, covered in this section from its source to its mouth, is international in character, for its opposite banks are in different countries. Since birds seem not to care a whit for international boundaries, we will consider the entire river a single birding area, and discuss the best vantage points on both sides of the river—in Ontario, Canada, as well as New York State.

There are superb points along both shores from which to view cormorants, herons, waterfowl, marsh and shorebirds, occasionally jaegers, and gulls and terns attracted to the river and its immediate shorelines and its few marshes. The Niagara is an excellent migration feeding and resting area for waterfowl and gulls, which can best be seen from roads along both banks of the river. Viewing is most satisfying in autumn, winter, and early spring, for in summer, a plethora of boats dominate much of the river. Niagara Falls, including the river gorge, whirlpool, rapids, and upstream islands, are a mecca for birds, the best places on the river for birding—not to mention that they are fantastically interesting physiographically and geologically.

Although, birds can be found in season along large stretches of the Niagara River, the following numbered locations, which correspond to sites on the accompanying map, are most of the better points from which to bird. Remember that although birding along the river is considered easy, in winter wear extra warm clothing to Niagara Falls, for rushing water and ice tend to refrigerate the area.

The 36-mile Niagara River is more precisely a strait connecting Lake Erie and Lake Ontario. It flows north and slightly west, is almost never so much as one mile wide (usually much less), and on its passage between the two lakes descends 326 feet. Most of that plunge is made in the spectacular sound and fury of Niagara Falls. Queen Victoria Park, on the western shore of the Niagara River in Canada, commands the best views of the falls, extending above and below the falls for 2.5 miles. The birder should, at the very minimum, sample the observation points along the Canadian shoreline. The river is rich in small fish, swarms of wormlike aquatic larvae, and myriad small invertebrates, all of which attract and sustain the rich avian life found in its vicinity. In winter the Niagara is relatively ice-free and therefore serves as a source of food for many species of gulls.

1. The River Source, the United States Side

From I-90 (the New York State Thruway) take Exit 53. Proceed west on I-190 to the Peace Bridge Exit. Exit and take the first right, making a sharp U-turn onto Busti Avenue (one-way south). Proceed south on Busti Avenue four blocks to Porter Avenue. Turn right (west) on Porter, cross over I-190 and proceed to Black Rock Canal. Make a left turn and drive south up Amvet Drive into La Salle Park. Here stop to scan the inner and outer breakwalls, Black Rock Canal, the river and harbor for ducks, shorebirds in spring and fall, migrating and wintering gulls, nesting Common Terns in spring, and the possible Snowy Owl in winter.

Drive back (east) on Porter from Canal and turn left (north) before the I-190 overpass on the side road at the Great Lakes Laboratory sign beyond the laboratory building. Here there is an overlook above the canal, river, and Peace Bridge.

Next, return to Porter Avenue and cross over I-190. Go east six blocks to Niagara Street. Turn left (north) up NY 265 (Niagara Street), which runs roughly

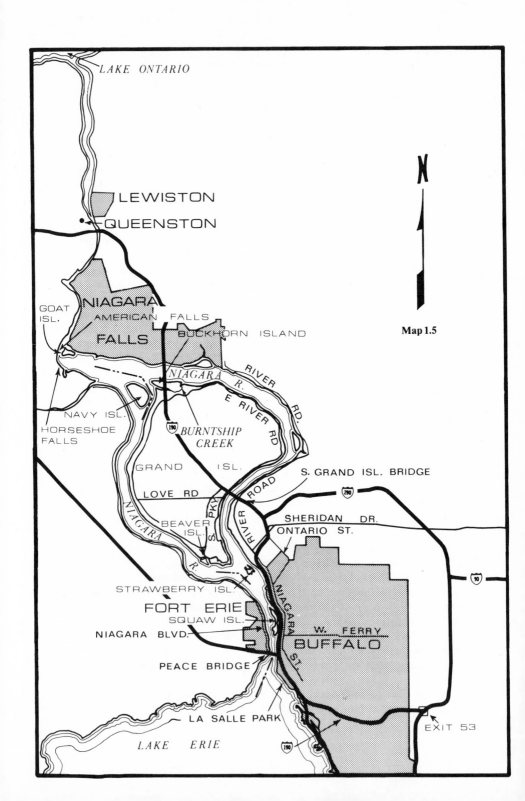

Map 1.5

parallel to I-190 and just east of it, for approximately 1.2 miles to West Ferry Street. Turn left. Cross the bridge over the Black Rock Canal. Turn left onto Squaw (or Bird) Island into Broderick Park. Scan the river and the Bird Island Pier (actually a long breakwater extending south from the island). Look for water-fowl, shorebirds in season, and gulls and terns.

Retrace the route and regain Niagara Street. Turn left heading north. Drive about one mile on Niagara Street to where it curves left. Stay left. Continue approximately 1.3 miles to just before Ontario Street, which comes in from the right. Turn left, crossing under I-190 to a parking area overlooking both the river and Strawberry Island. This is again a good place to scan, preferably with a scope.

Regain Niagara Street. Turn left. Proceed north approximately 1.7 miles, crossing I-190 to the intersection of Sheridan Drive. Here Niagara Street becomes River Road. At the intersection of River Road and Sheridan Drive turn left and proceed to the boat launching site to gain views of the river upstream from Strawberry Island.

Not all of these sites, of course, must be visited. This route simply connects several good vantage points. On any given day, owing to variable weather, traffic conditions, and gull movements, one or another of these vantages may be far superior.

2. The River Source, the Canadian Side

Cross the Peace Bridge into Canada. Take the first immediate right, which U-turns onto a street leading to Niagara Boulevard. Turn right into Mather Park and drive above the Peace Bridge. Then use various roads and parking areas from that point downstream to view the river. Continue north along Niagara Boulevard for approximately 1.4 miles, crossing over the railroad tracks and stopping wherever observation of the river is easy. Many waterfowl and gulls in season frequent this stretch of river, especially off the marsh in the bird sanctuary.

Continue north to about one mile beyond the International Railroad Bridge. Opposite is the Strawberry Island sector described above. If time permits, continue north on Niagara Boulevard to North Fort Erie. Drive out past the village limits to any convenient overlooks where parking presents no problem and scoping the gulls at leisure is possible. Viewing the river here should be both productive and panoramic.

3. Navy Island, the Canadian Side

Continue driving north along the Niagara River Parkway stopping to overlook the West River wherever convenient and wherever the river is ice-free. Drive south thru Chippawa. Park at the eastern ends of Weaver, Mc Creedie, Edgeworth or Willoughby roads to obtain good views of the northern and southern ends of Navy Island. Large flocks of waterfowl concentrate at both ends of the island in fall, spring, and winter.

4. & 5. Grand Island, Beaver Island State Park and Buckhorn Island State Park

From Buffalo take I-190, or I-290 to the South Grand Island Bridge. Cross over and take the first exit. Take the South Parkway south to the parking areas and the road loop overlooking Beaver Island, the creek and the East River. In winter look for wigeon, teal, Canvasback, and both scaup species. Inside Beaver Island State Park, walk along the roads looking for owls and shrikes that may overwinter in that area of the park. The park is usually an excellent place to explore for warblers and other migrants in spring.

After birding Beaver Island State Park, regain the South Parkway headed north, and follow it to Love Road. Turn left. Follow Love Road to the entrance to a wildlife refuge owned and maintained by the Buffalo Ornithological Society. This is a pleasant preserve of less than 50 acres and is far more productive to bird in spring than winter. Turn around and travel east the length of Love Road to its intersection with the loop road on the east side of the island. Take the loop road around the island viewing the river from obvious observation points along the way. It is usual for the West River shore to be choked with ice in winter, which keeps grebes, swans, waterfowl, and even gulls beyond decent viewing range. Travel up the East River side on East River Road to Ransom Road, which comes in on the left. Turn left and follow the road across Gun Creek to the Ransom High School, which you will see on the left. All along this road you should look for Sharp-shinned and Red-tailed hawks and American Kestrels overhead. Search the fields for Common Snipe. Scan the wires for Loggerhead Shrikes. Proceed west on Ransom Road to the first through-street, Stony Point Road. Turn right and drive to East River Road. Turn left. While driving be sure to be alert for Snowy Owls. Proceed to Baseline Road and scan the East Branch of the Niagara (sunken islands) from this point.

Continue into Buckhorn Island State Park along the shore road which dead ends at Burntship Creek, where it enters the East Branch. Walk through the woods to the point where the viewing is excellent. Another good vantage point overlooking Burntship Creek is where I-190 crosses it, south of the North Grand Island Bridge. The parking is restricted here, but one can usually scan outside the car for a few minutes.

If your tour of Grand Island lasts until almost dusk, go back to Baseline Road and proceed down it, watching the roadside meadows for the Short-eared Owls that often overwinter. They usually come out around dusk and are active enough to see well. If they cannot be found here, try Ransom Road, from which they are also often visible. The owls are usually present in good numbers throughout winter but may be gone as early as the beginning of March.

6. Niagara Falls, the U.S. Side

The birder should use all of the available parking areas, although fees are charged in many of them. Goat Island divides the Niagara River at the height of the falls and offers an exceptional view from the top. It can be reached from the mainland by automobile. From there walk out to Three Sisters Islands to obtain better views of the rocky areas, the rapids, the stunning curtain of water, and the boulder-strewn river above the Canadian Falls. Promontories over the gorge are not so rewarding from the U.S. side as from the Canadian side, but sometimes viewing from the observation tower is excellent. The New York State Observation Tower elevator rises 100 feet above the falls and then descends 280 feet to the river level, from which one can look up from the base of the falls. The elevator is located in Prospect Park, just downriver from the American Falls and upstream from the Rainbow Bridge. Prospect Point within the park can often be a good place to look for gulls.

7. Niagara Falls, the Canadian Side

From the west bank of the Niagara Gorge the views of the wider, more deeply indented Horseshoe (or Canadian) Falls offer breathtaking vistas of this great natural wonder — and, moreover, spectacular views of the thousands of gulls and various waterfowl in the gorge and especially in the rapids above the Horseshoe Falls. Go to the area above the falls and use all the available parking sections, particularly the one just below the old stranded barge and the one above the water control structure, where many gulls often roost on the concrete pier. Beyond the pier there should be Horned Grebes and several hundred Canvasbacks, Redheads, and scaups, and a few pairs of Hooded Mergansers. Make use of all the available parking from the water control structure downstream to opposite the greenhouse to view the rapids and rocky islands above the Canadian Falls. Search the rapids and power pools above the falls for occasional Harlequin Ducks and Barrow's Goldeneyes, whose elegant plumage should be easy to study. Overhead in fall and early winter especially there will be gulls by the hundreds, and an astute birder may be able to pick out Glaucous, Iceland, Herring, Great Black-backed, possibly Lesser Black-backed, and perhaps even Thayer's Gull. Look among the Ring-billed Gulls for the possible Mew Gull. This region has been the site where Black-headed Gulls have been recorded among the flocks of Bonaparte's Gulls sure to be present in season. Even Ivory Gull, Black-legged Kittiwakes, and Sabine's Gull have been logged within the general area of the gorge. Every birder should be keenly alert for the rare possibilities that this area presents.

There is parallel parking opposite the Dufferin Islands. Walk through the islands, where you will probably see Belted Kingfishers, chickadees, nuthatches, Brown Creepers, and kinglets. Go to the river's edge for Common Loons, Red-breasted Mergansers, and all three scoter species. Again, look for every gull species common or even infrequent to the area.

Go to Queen Victoria Park and park. Walk to the gorge overlook opposite the restaurant and along the low wall in the park to scan the rocky areas downstream toward the Rainbow Bridge. The Maid-of-the-Mist ferry boat landing is also a fine vantage point. This is the best place for Northern and Red phalaropes in fall. It has also proven very good for King Eider. Scrutinize bare rock ledges anywhere in the rapids above the falls for Purple Sandpipers. Check for Little Gulls among the Bonaparte's.

8. Robert Moses Power Plant, New York, and Sir Adam Beck Generating Station, Canada

Some five miles downstream from the falls, stop at the gorge edge on either the American or Canadian side of the river. These power station outlets are known for their large gull assemblages. Many large gulls congregate in late fall and early winter, including "white-winged," Lesser Black-backed, and Thayer's gulls. Parking spots on both sides of the river provide views, but perhaps sighting is best from the Canadian side.

9. Lewiston, New York, and Queenston, Canada

In Lewiston, travel west downhill on NY 104 to a parking area near a green water tower at the bottom. View the river from here. In Queenston, below the village there are several level areas where one can park, and the road past the sand piles downstream from the village also provide good views. This is a good place for Ring-billed, Black-headed, Bonaparte's, and Little gulls.

10. The River Mouth, the United States Side

From Fort Niagara, views can be obtained either near the parking area or by walking from there to the Coast Guard Station to look out over the river mouth. The water is rougher here, where the river flows into Lake Ontario, and it is a favored place for waterfowl and gulls.

11. The River Mouth, the Canadian Side

Between Queenston and the mouth of the river, there are several convenient areas where the birder can park and scan; however, because the river tends to be wider here, distances from the birds are usually greater. From Niagara-on-the-Lake excellent observation points are located by driving shore roads downstream from the marina to the river mouth and parking wherever possible. Walk downstream to obtain good views wherever it is safe and practical.

ORLEANS AND GENESEE COUNTIES
Iroquois National Wildlife Refuge, Oak Orchard Wildlife Management Area, and Tonawanda Wildlife Management Area (Map 1.6)

Rating: Spring****, Summer***, Autumn***, Winter**

Iroquois National Wildlife Refuge comprises approximately 10,800 acres and Oak Orchard and Tonawanda Wildlife Management areas together total approximately 6500 acres more. The habitat of the three is a happy combination of ponds,

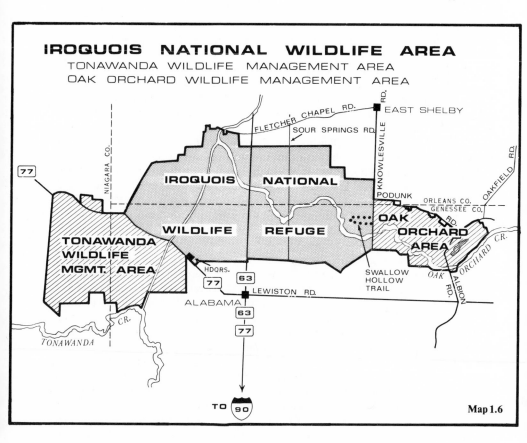

marshes and other flooded places, fields, thickets, and woodlands capable of supporting a large number and variety of migrating and breeding birds. In addition to being a major stopover area for Canada Geese on spring migration, the refuge hosts Snow Geese (including the blue form), Wood Ducks, wigeons, Green-winged Teals, Mallards, Black Ducks, Pintails, Blue-winged Teals, scaups, Buffleheads, goldeneyes, and a few Ruddy Ducks and shovelers. In addition to these waterfowl, the refuge has recorded a long list of water-associated species that should be of considerable interest to birders far and near.

Leave I-90 (the New York State Thruway) at Exit 48A. Travel six miles north on NY 77 to the village of Alabama. About one mile north of town go left (west) on Casey Road for about one mile to the Headquarters building, where you can pick up a checklist of local birds recorded and also an area map.

Consult the accompanying map. As you can see, Oak Orchard Wildlife Management Area borders the Iroquois National Wildlife Refuge on the southeast and is divided from it by Knowlesville Road. Tonawanda Wildlife Management

Area borders the Iroquois refuge on the southwest and is divided from it by NY 77. The three areas together make up a rich and varied birding venture.

There is a checklist of well over 200 species recorded in the area. In March and April thousands of Canada Geese and waterfowl species are present. In addition, herons and shorebirds frequent the wetlands. Good views can be obtained from the roads and the entire area can be easily covered by automobile; but, of course, birding will be enhanced if some of the trails and paths are birded on foot. Go to Cayuga Pool and Ringneck Pond for close-up views of herons, swans, and ducks. There is a Great Blue heronry of over 400 nests in Iroquois. Consequently, many of these birds will be fishing in the ponds and sluiceways from late spring to fall. Drive along Knowlesville Road. In the area where it divides Iroquois and Oak Orchard, wet woods and marsh lands line the road. Here look for Pied-billed Grebes; Great Blue and Green herons; American and Least bitterns; King and Virginia rails, Soras, gallinules; and coots. Go north of Knowlesville Road where it crosses Oak Orchard Creek. Look here for nesting Wood Ducks, and listen for nesting Ruffed Grouse drumming. Also, go to the flooded woods where Oak Orchard Creek crosses Sour Springs Road, a special Wood Duck haunt. Be aware that Rink-necked Pheasants nest in the slightly drier areas as you walk through. In the woodlands bordering the marshes watch for Great Crested Flycatchers, Eastern Phoebes, Alder Flycatchers, Yellow-throated and Warbling vireos, Golden-winged and Cerulean warblers. Listen for tanagers and grosbeaks. The beginning of Swallow Hollow Trail comes in on the west side of Knowlesville Road (left), and is marked with an obvious sign. This is an excellent trail to bird. Be aware that Barred Owls and Prothonotary Warblers nest along Oak Orchard Creek, and sometimes the warbler can be seen and heard south of the trail. Continue north to the junction of Knowlesville and Podunk Road. Turn right here and follow Podunk Road to its end. Park in the lot on the left. From here you can observe Stafford Pond. This is a great spot to view the congregating waterfowl in late March through April. Some of the rare species recorded in these refuges are White Pelicans, White-fronted Geese, Cinnamon Teals, King and Yellow rails, and Ruffs.

The lowlands west of the village of Alabama in the Tonawanda Wildlife Management Area are flooded in spring. These lowlands also serve as a gathering place for swans, geese, and ducks. There Glossy Ibises have occasionally been seen. Remember to check any mud flats for migrating shorebirds. The lowlands of all three areas are good places to spot Turkey Vultures, Marsh Hawks, Red-shouldered, Red-tailed, and Rough-legged hawks. Look all along the wires for American Kestrels. Occasionally, birders spot an Osprey or Bald or Golden eagles migrating through the area.

In July 1981, the New York Department of Environmental Conservation launched the second five-year phase in its program to restore the endangered Bald Eagle as a breeding species within the state. "Project Mega-hack" has as its goal the successful hacking, or rearing in the wild, of 21 Bald Eagles in 1981 and by 1985 a total of 130 eagles. Hacking is to be conducted on a giant tower of double tiers and multiple bird-holding chambers. This second phase of the eagle release program has been moved from Montezuma N.W.R., in Seneca County to Oak Orchard W.M.A. owing to its larger size, abundant fish food resources, and adjacent management lands. The release tower can be observed by the public from the Goose Pond overlook at Oak Orchard W.M.A.

ACKNOWLEDGMENTS

The author would like to express her sincere thanks to the following people who contributed invaluable information on the Niagara Frontier chapter: Robert F. Andrle, Allen H. Benton, David B. Freeland, Dave Junkin, James W. Parker, Vivian Mills Pitzrick, Frances M. Rew. Special thanks to Robert Andrle of the Buffalo Museum of Science, whose reading of the Niagara Frontier chapter and comments on it considerably improved it.

Genesee—Region 2

THE GENESEE REGION (Map 2.1) encompasses more than 2883 square miles and includes all of Monroe, Wayne, and Livingston counties; the eastern halves of Orleans, Genesee, and Wyoming counties; and the western portion of Ontario County. The principal urban center is Rochester, which has a population of around 293,000. This city produces photographic and optical equipment used all over the world, and the University of Rochester enjoys a reputation as one of the state's leading educational institutions. The entire region is rich in natural and cultural history.

The region's northern boundary lies along the southern shore of Lake Ontario, along the low-lying Lake Ontario Plain. This northwestern area of New York State immediately south of the Lake contains large deposits of Silurian limestone, shale, salt, and gypsum. Much of it is swampy; other parts, where glacial debris was deposited, are hilly. This area is drained by the Genesee River and its tributaries. The Genesee flows north from the Pennsylvania state line, emptying into Lake Ontario above Rochester. Elevations range from 248 feet above sea level along the lake to 2256 feet on the Allegheny Plateau.

The northern reaches of Region 2 are very attentively birded. Spectacular passages of northbound raptors can be seen from sites on the west-east flight paths along the lake. Not a few of the areas described below witness memorable hoards of warblers and shorebirds during both spring and autumn migrations. From some, the birder can view an outstanding number and great variety of migrating and wintering waterfowl species.

ORLEANS COUNTY
The Borrow Pits, Lakeshore Road, Pt. Breeze State Pier (Map 2.2)

Rating: Spring****, Summer****, Autumn****, Winter***

The Borrow Pits lie along a section of the Lake Ontario State Parkway from twenty-five to thirty-two miles west of Rochester and ten miles north of Albion. When the Parkway was under construction, sand and dirt were taken (borrowed)

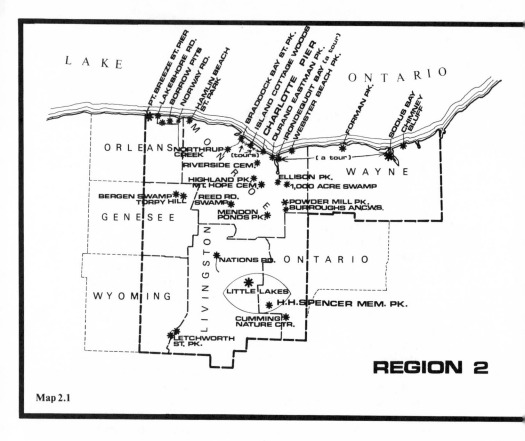

Map 2.1

from this area which left excavations (pits) that subsequently filled with water. These depressions eventually became known as "the Borrow Pits."

Obviously, because the area is on state-held land, permission to bird the area is not required. The pits can be birded from a car or on foot. If from a car, be sure to pull well off onto the wide shoulder, because it is illegal to stop on the Parkway, and doing so strains the patience of even the most understanding and sympathetic of the Parkway police. However, if you pull off well onto the wide shoulder the Parkway police are usually understanding with birders. Covering the area is rated moderate to easy, depending upon how it is done. Fortunately, the area is insect-free; *however,* fields of ragweed and goldenrod abound. Those subject to hay fever attacks are forewarned. The discovery of the Borrow Pits as a birding hot-spot has been relatively recent, but already its potential is manifest.

The Lake Ontario State Parkway parallels Lake Ontario from above Rochester to Lakeside Beach State Park, north of Kuckville. It will be designated on any current road map. The Borrow Pits extend over a seven-mile stretch from the Kendall-Holley exit to the Point Breeze-Route 98 exit. To reach there, enter the Parkway from Hamlin Beach and travel west 11.7 miles to the Point Breeze exit.

Immediately turn left and then re-enter the Parkway headed east toward Rochester. The first pit is 1.1 miles from starting onto the Parkway entrance ramp (Pit D), and is visible from the Parkway. The second pit (Pit E) requires getting off at the Lakeshore Road exit, which is 1.2 miles farther. Then make a right turn 0.4 mile beyond on Sawyer Road; advance 0.4 mile farther and park. The pit is on private land, approximately 100 yards inland across an open field. The owner has welcomed birders, except during duck hunting and skeet-shooting season, when the area is very well posted.

Re-enter the Parkway at Lakeshore Road and proceed to three other borrow pits at 1.9 miles, 3.0 miles, and 3.6 miles from the Lakeshore Road entrance. With the exception of Pit E, all of the pits are visible from the Parkway. However, to maximize coverage, bird the pits on foot from under Parkway bridges, reached *via* roads perpendicular to the Parkway.

Although winter is not the prime time to bird the area, various gull and waterfowl species, including an occasional Harlequin Duck can be viewed from the Pt. Breeze State Pier (Map 2.2). Purple Sandpipers can be found at the pier from November on. Also in winter, Marsh and Rough-legged hawks, Short-eared Owls, Northern Shrikes, Lapland Longspurs, and Snow Buntings can be seen in the fields adjacent to Lakeshore Road. Spring birding begins as soon as the pits are ice-free. Hawk flights can be seen from here in both April and May with an occasional Bald Eagle putting in an appearance. Southern heron species and breeding-plumaged shorebirds arrive in spring and, as that season ripens into summer, puddle ducks breed in the pits. Wilson's Phalaropes have, in some years, spent the major part of June in or about Pit C; Orchard Orioles breed regularly in the area, especially along the north side of Lakeshore Road. This is also one of the last strongholds for lakeshore-breeding Loggerhead Shrikes, and Upland Sandpipers often breed along the Parkway edge. Be sure to listen for the infrequently occurring Western Meadowlark along the Parkway median, especially around Pit A.

Autumn at the Borrow Pits is called the "super season." Great Egrets traveling north and west on their postnuptial wanderings arrive in mid-to-late August. Whimbrels, Willets, Buff-breasted Sandpipers and Northern Phalaropes appear regularly in the fields near the intersection of Center and Lakeshore roads from late August through September. American Golden and Black-bellied plovers, along with Baird's Sandpipers, frequent these fields. Baird's and Western sandpipers are rather commonly found in any of the Borrow Pits. Black-headed and Little gulls and Forster's Terns have occurred at Pit B during August and September. Beginning in September, the lake should be carefully birded from the Pt. Breeze Pier. This is an excellent place to observe spectacular late-fall waterfowl flights, and as an added bonus, an occasional Pomarine or Parasitic jaeger or Black-legged Kittiwake may be found there.

Norway Road (Map 2.2)

Rating: Spring****, Summer****, Autumn***, Winter**

Norway Road is a very special local birding area seven miles northwest of Brockport. The area encompasses about 1900 acres, if the area contained in a 300-foot

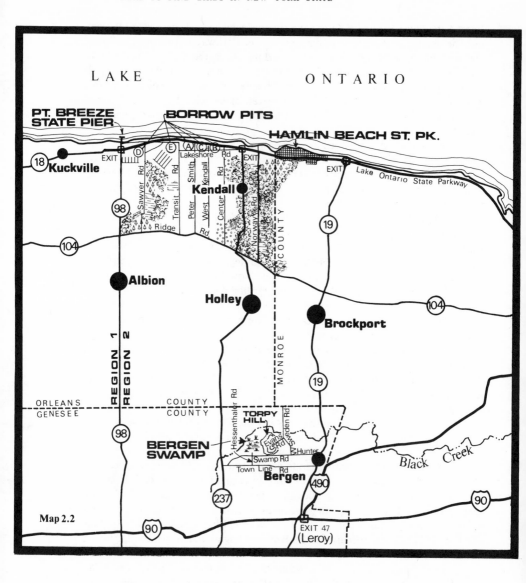

Map 2.2

swath on both sides of the road is included. Birding Norway is usually quite easy; however, venturing into the woods to observe the interesting habitat or some of the many varieties of field and woodland wildflowers, or to locate a particularly wary bird, can become moderately strenuous. Along the road the habitat is a mixture of second-growth oak, sycamore, birch, maple, hemlock, locust, heavy scrub undergrowth, grassy open fields, swamps and hedgerows teeming with frogs and turtles, wood margins and hayfields. This variety in habitat is, of course, what accounts for the corresponding diversity in the avifauna. Be alerted that poison ivy flourishes in the wood's interior and that mosquitoes and other irritating insects are plentiful.

To reach Norway Road from the New York State Thruway, take Exit 47 at LeRoy. Head north on NY 19 through Brockport to NY 104. Turn west (left) on NY 104 and drive about three miles across the Monroe-Orleans County Line Road. Proceed for another 1.5 to 2 miles to Norway Road, which joins on the right. *Note:* Norway does not cross NY 104, it simply meets it. Norway is a dirt, unpaved road.

Norway Road is not only easy to bird, but also extremely rewarding. It is more than nine miles in length from NY 104 (Ridge Road) to its terminus at Lake Ontario. The area that is most heavily birded is the 2.5 miles from NY 104 north. Among the birds nesting along the road and deeper in the woods are Ruffed Grouse, Yellow-billed and Black-billed cuckoos, Pileated Woodpeckers, Winter Wrens, Golden-winged, Blue-winged and Cerulean warblers, Yellow-breasted Chats, Orchard Orioles, and many of the more common species. Because the heterogeneity of the habitat is so attractive to migrant passerines, those who regularly bird the Rochester area wouldn't miss Norway Road in spring, especially on a Big Day!

GENESEE COUNTY
Bergen Swamp and Torpy Hill (Map 2.2)

Rating: Spring***, Summer****, Autumn**, Winter**

This swamp lies between a ridge of Lockport dolomite on the north and a ridge of Onondaga limestone on the south. These two types of sedimentary rock are very resistant to weathering. Less resistant sedimentation was eroded from between the two ridges and the low-lying region ultimately filled with water. After the last great ice cap receded, the area was doubtlessly the site of a glacial lake.

Bergen Swamp is a wetland that developed in the bed of that former glacial lake. It lies 22 miles west of Rochester and about three miles west of the village of Bergen. Its 2000 acres are on private land. The swamp can be reached by car but must be birded on foot. Most of the swamp is now owned by the Bergen Swamp Preservation Society. Groups wishing access to the swamp must obtain permission from the Bergen Swamp Preservation Society, P.O. Box 18043, Twelve Corners Branch Office, Rochester, New York 14618. A bird checklist and area map and various other helpful publications may also be obtained from the Society for a small fee. Birders should write well ahead of a planned visit. The visitor is urged to read two other publications of the Society, "Guide to Visitors Trail to Bergen-Byron Swamp" and "Trips through Bergen-Byron Swamp."

Covering the area is rated as moderately easy if the observer stays on the established trails, which he is strongly requested to do. The hiker may encounter Massasauga rattlesnakes during the warm months, so watch your footing. They are protected, as is all wildlife in the swamp. Collecting of any kind is not permitted.

The village of Bergen is reached by leaving the New York Thruway at Exit 47. Get on NY 19, head north, and take it to the center of Bergen. From there head north three more blocks and turn left (west) on Hunter Street, which is on the north side of the village. Hunter becomes Swamp Road. Follow this road for four miles to the second crossroad, Hessenthaler Road. Turn right and park on the right side about 100 yards before the end of the road, along the woods' edge. The trail veers to the right at the end of the road.

The trail passes through several vegetative zones; old field, hardwood, pine-hemlock, white cedar, and open marl. During the nesting season the bird species encountered will be characteristic of the Canadian Zone: Red-breasted Nuthatch; Brown Creeper; Winter Wren; Black-and-white, Nashville (in open marl area), Black-throated Green, Blackburnian, and Canada warblers; Northern Water-thrush; and White-throated Sparrows.

Bergen Swamp is a botanist's paradise – a 1961 catalog of the plants of Bergen Swamp done by the botany department of Cornell University listed 2392 species! It is a storehouse of New York's native orchids.

Another interesting area is Torpy Hill at the east end of the swamp. From Hessenthaler Road, turn left on Swamp Road toward the village of Bergen and proceed two miles to the first crossroad. Turn left (north) on West Sweden Road. Just before reaching Black Creek, turn left on Evans Road and drive to the end. In addition to providing an excellent overlook of the swamp, Grasshopper and Henslow's sparrows can usually be found in the open fields here. Birds typical of fields and swamps can be seen by following along the field and swamp edges. Turkey Vultures are often seen perched in nearby dead trees or soaring overhead.

WYOMING AND LIVINGSTON COUNTIES
Letchworth State Park (Maps 2.3, 2.4 and 2.5)

Rating: Spring****, Summer****, Autumn***, Winter**

Letchworth is a very beautiful state park located partially in Wyoming and partially in Livingston counties. It encompasses 14,340 acres due north of Portageville and about half a mile west of the town of Mt. Morris. Birding can be done on foot, bicycle, or from a car. Most of the birding is easy to moderate; driving from one area to another facilitates covering widely separated areas, and parking is ample throughout the park. A small daily admission fee is charged from May through September. Poison ivy is abundant, but it can be avoided by staying on designated walking trails. Letchworth contains one of the region's – and the state's – most outstanding geologic and scenic resources, the Genesee River Gorge. It is also rich in native wildlife and plant communities. Miscellaneous publications elucidating some of the park's geology and history are available at the park headquarters. Birders can also stop by for an area map and bird checklist.

Convenient park entrances are located at Mt. Morris, off NY 36 on the north; Portageville, off NY 436 on the south; Perry, off NY 39 on the park's

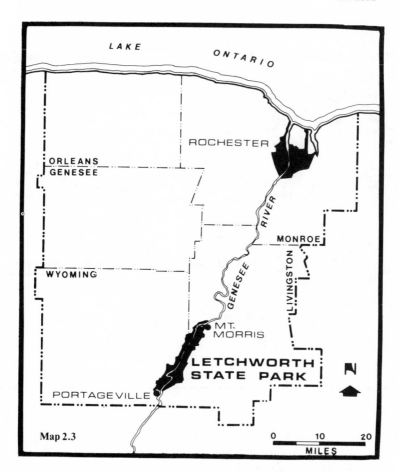

LAKE   ONTARIO

ROCHESTER

ORLEANS
GENESEE

MONROE

WYOMING

GENESEE RIVER

LIVINGSTON

MT.
MORRIS

**LETCHWORTH
STATE PARK**

PORTAGEVILLE

Map 2.3

0   10   20
MILES

northwest; and Castile, one mile north of the junction of NY 19A and NY 39 on the park's southwest side. *Note:* The Castile entrance is currently the only entrance kept open year-round. All of the entrances are very well marked; gaining access to the park should offer no problem.

Within the park (Map 2.4), birders will find a wide variety of natural resources. It contains a number of natural and man-altered habitats, including plantations of both exotic and native trees, open fields in varying stages of succession, a natural pond and two that are man-made, an old orchard, and several streams and ravines. Land adjacent to the park boundaries provides a suitable buffer zone for the park. The trail system layout (see Map 2.4) is somewhat restricted by the Erie-Lackawanna rail line with its associated hazards and two steep-walled ravines. On the east side of the Genesee River lie the Genesee River floodplain forest and a small swamp east of the main parking road. However, the east side of the

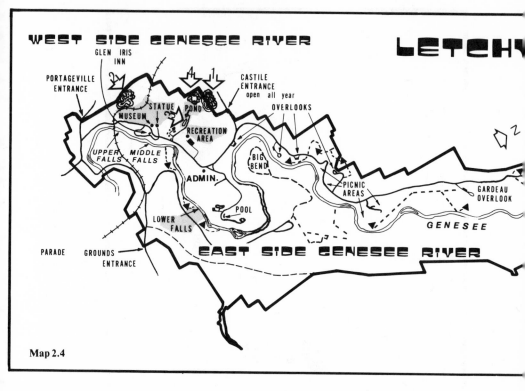

Map 2.4

river cannot compete with the west side for variety of habitats. The park's spectacular Genesee River Gorge is widely known as the "Grand Canyon of the East," and its three resplendent waterfalls seem to be the prime attraction for most park visitors. The nesting warblers however, are definitely the main attraction to the park birders.

With the exception of the Deh-Go-Ya-Sah Creek ravine, the park on the west side of the river is composed of gently rolling hills which become more precipitous the closer one gets to the main park road. The steep-walled ravine is heavily-forested and lies nearly 200 feet below the relatively flat surrounding ground level. Elevations in the park range from a high of 1380 feet in the western and northwestern portions, to a low of 1100 feet at the bottom of the Deh-Go-Ya-Sah Creek bed, near the main road.

Letchworth is well known to both the amateur and professional geologist. About one million years of rocks — primarily limestone, sandstone and shale — lie exposed in the river gorge. Shaped and sculptured by ice, water, and wind working over immense periods of time, the Genesee riverbed history is over 425 million years long.

Of the four park ponds, only Willow Pond (1) is natural. This quarter-acre

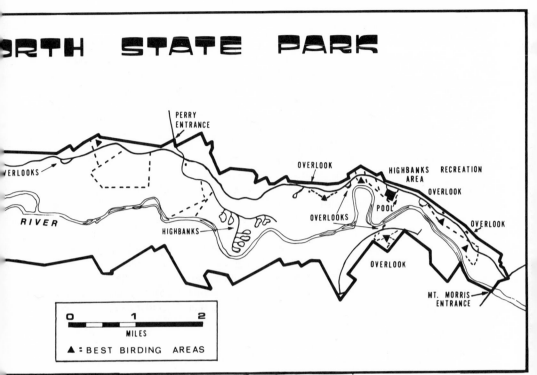

RTH STATE PARK

PERRY
ENTRANCE

OVERLOOK

HIGHBANKS RECREATION
AREA

ERLOOKS

OVERLOOK

POOL

RIVER

OVERLOOKS

HIGHBANKS

OVERLOOK

OVERLOOK

OVERLOOK

MT. MORRIS
ENTRANCE

0 1 2
MILES

▲ = BEST BIRDING AREAS

pond lies just north of the present park boundary, and is surrounded by cattails, rushes and willows. It is attractive and an excellent birding site. Railroad Pond (2), which lies inside the park boundary west of the present Erie-Lackawanna railroad tracks, was originally constructed to provide a source of water for the railroad's steam locomotives. Its sides are somewhat steep, but it also provides an excellent birding site. Trout Pond, (3), located relatively close to the park maintenance center, was originally built as a fire-protection reservoir and is now stocked for fishing. Pine Pond (4), is in a delightful setting. Its one and one-half acres are completely surrounded by a red pine plantation and native deciduous forest.

The dominant forest association found in the park is the Northern hardwood type, consisting of sugar maple, beech, and yellow birch. Other common species observed include basswood, black cherry, cucumber tree (magnolia), hemlock on north-facing slopes, white ash, white pine, and yellow poplar. A secondary association is the oak-hickory type which occurs on the drier sites in the park. Some of the most important understory species are blue beech, flowering dogwood, spice bush, striped maple, and witchhazel. During the middle 1930s, the Civilian Conservation Corps planted more than 25,000 trees and shrubs, representing sixty species, along the west side of the river between the railroad bridge and the Lower Falls road. A recent study identified 262 species of trees, shrubs, vines, ferns, and flowers within the park.

A list of over 140 bird species has been casually compiled over a ten-year period. Black Ducks, Red-tailed Hawks, Ruffed Grouse, and Pileated Woodpeckers nest in the park. It is presumed that Turkey Vultures nest here for they are spotted in the gorge throughout the summer. Listen also for the distinctive gobble of the male Turkeys, which nest along the edge of the gorge. Rock Doves nest along the cliffs in the gorge. The long list of park breeders also includes Solitary Vireos, Golden-winged, Blue-winged, Nashville, Pine warblers, and Northern Parula. Ovenbirds, Louisiana Waterthrushes, Yellow-breasted Chats, Hooded Warblers, and Dark-eyed (Slate-colored) Juncos are also breeders.

In winter there is usually some open water in the gorge where a few duck species and some mergansers can usually be found. In some winters, finch incursions have been noteworthy in the park.

The trail system as depicted on the site map is designed to expose the visitor to the major natural features of the area and give adequate access to a particular area or point of interest. The site map also indicates the very best birding sites. Not all of these points may be accessible during the winter, but a great many of them will be. The trails usually follow a loop design and are interconnected where feasible, so walks can be extended by including additional loops. In general, the trails wind across the natural contours of the land, so that the birder can pass through several types of habitat on his quest for birds.

LIVINGSTON COUNTY
Nations Road (Map 2.5)

Rating: Spring***, Summer***, Autumn**, Winter****

Nations Road is about four miles southwest of Avon and is exceptionally easy to bird. Leave the New York State Thruway (I-90) at Exit 46 and drive south on NY 15 for about ten miles to NY 5 and 20 (East Avon). Turn right (west) and proceed to Avon. From the traffic circle in Avon on NY 20 go west 0.2 mile. Turn left (south) on NY 39 toward Geneseo. Continue 1.7 miles to Fowlerville Road. Turn right and go west one mile. At this point, stop and look north across fields to a wood lot approximately 0.25 mile away. This is the site of a Great Blue heronry of about thirty nests. Continue another 1.4 miles to Nations Road, which comes in on the left. Turn left. Continue 0.5 mile to a small pond on the left. It often has Pied-billed Grebes, Wood Ducks, Green-winged and Blue-winged teals, and Virginia Rails in the summer. Continue south, watching for Upland Sandpipers, Loggerhead Shrikes, Bobolinks, and Savannah, Grasshopper, Henslow's and Vesper sparrows. In winter watch for Horned Larks, Tree Sparrows, Lapland Longspurs and Snow Buntings. Along this stretch of road in winter there can be large numbers of Marsh, Red-tailed, and Rough-legged hawks, and American Kestrels, and an occasional Northern Shrike have been seen. Most noteworthy are the remarkable concentrations of Short-eared Owls from December through March. They

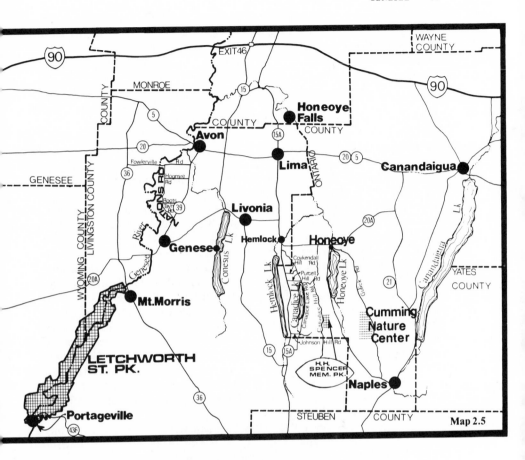

Map 2.5

can be seen at almost any point coursing over the fields, sitting on fence posts or in hedgerow trees. In particular, look carefully in the locust tree grove at the northeast corner of Nations Road and Hogmire Road (0.7 mile south of the pond). As many as fifty owls have been seen here at once. Although they may be roosting out of sight during the daylight hours, they can usually be found in the late afternoon, when they become more active.

Continue 2.2 miles south to Root's Tavern Road. Here, on the northeast corner of the intersection is an oak woods, where Red-headed Woodpeckers are common both summer and winter. In summer the now uncommon Eastern Bluebird may be found from here south to the end of the road. South from the intersection 0.7 mile is a creek, where Tree, Bank, Rough-winged and Barn swallows are often sighted. Look for Red-bellied Woodpeckers, especially in the woods on the left. In summer Red-eyed and Warbling vireos, Cerulean Warblers, and Scarlet

Tanagers are consistently found here. Pileated Woodpeckers have been known to nest in this area and should be watched for. Continue driving another 1.2 miles to NY 39 (Avon-Geneseo Road). At NY 39 turn left to return to Avon or turn right to head toward Geneseo.

LIVINGSTON AND ONTARIO COUNTIES
The Little Lakes Area: Conesus Lake, Hemlock Lake, Canadice Lake, Honeoye Lake, Harriet Hollister Spencer Memorial Park, Cumming Nature Center (Map 2.5)

Rating: Spring**, Summer****, Autumn**, Winter**

The following areas are more than twenty miles south of Rochester and are scattered across 200 square miles. They include all the roads and adjacent territory around the Little Lakes, Harriet Hollister Spencer Memorial Park, and Cumming Nature Center. Birding these areas ranges from easy to moderately difficult. To properly explore some of the glens which lead back into the hills from the lakes in the spring and early summer may require boots, and birders should carry insect repellent. This area is beautiful at all seasons, but especially in the fall. Deer and smaller mammals are common. The hills, which range up to 2000 feet, are largely forested with mixed deciduous and coniferous trees. The lakes, except for Conesus, freeze over in most years from January to March. The rich diversity of the flora of the area lends itself to botanizing. *Nota Bene:* This is a huge, choice birding area with numerous specialties. Because of the size of the area and its variation, it is probably best to seek the guidance of local Rochester-area birders familiar with the region. Many of the most noteworthy species have a limited distribution and could easily be missed without guidance.

From Rochester, take NY 15A (East Henrietta Road) south for about twenty miles. The best birding areas are south of NY 20. Stay on NY 15A past the town of Hemlock. About three miles farther, turn left on Coykendall Hill Road. Cliff Swallows nest on a large red barn on the north side of the road. Proceed to the bottom of the hill and turn left onto a dirt road. In this brushy area one can usually find nesting Golden-winged and Blue-winged warblers and Yellow-breasted Chats. Follow this road a little farther to Canadice Lake Road. Here turn right and head south. Bird this stretch of road intensely to the next road coming in on the right, which is Purcell Hill Road. Both "Lawrence's" and "Brewster's" warbler hybrids have been recorded within this stretch. Clay-colored Sparrows have been recorded approximately half-way up Purcell Hill Road at a prominent pine plantation. Return. Back on Canadice Lake Road, proceed south, stopping every quarter mile to check for nesting and migrant warblers. Careful birding should turn up Black-and-white, Black-throated Green, Blackburnian, Mourning and Canada warblers. Quite frequently Nashville, Magnolia, and Hooded warblers are spotted along this road. Pileated Woodpeckers, Winter Wrens, and Louisiana Waterthrushes breed in the glens. Be watchful for Broad-winged Hawks and the occasional accipiter darting across the road at tree-top height. The foot trail at the very south end of Canadice Lake is an excellent place to see Chestnut-sided Warblers and to hear Ruffed Grouse and Turkeys (less frequent). Farther south on Canadice Lake Road, Johnson Hill Road crosses. This is a good place to investigate the fields on either side. Eastern Bluebirds are still not uncommon, and Grasshopper and Henslow's sparrows nest here.

At the south end of Hemlock Lake a natural active Bald Eagle nest is probably unique in the state (see Oak Orchard W.M.A. in region 1 and Montezuma N.W.R. in region 3). Be sure to scan the sky repeatedly, for eagles from this site are frequently seen soaring over the surrounding countryside.

One exceptionally fine area that should not be missed is the Cumming Nature Center, whose entrance is located on Gulick Road, southeast of the village of Honeoye. This is actually part of the Rochester Museum and Science Center. Here there is a resident naturalist, and a small entrance fee is charged. Excellent trails lead through the woods to a beaver pond, past boggy areas and through woodland floor that is a delight to birders and botanists.

The very well-marked entrance to Harriet Hollister Spencer Memorial Park is on the east side of Canadice Hill Road, seven miles south of its beginning at NY 20A, and about one-half mile west of the village of Honeoye. Along the winding way, especially if the day is clear, several panoramic and spectacular views will present themselves. This park is owned and managed by the Finger Lakes Park Commission. The most interesting feature of the avifauna here is the number of breeding birds. Owing to the 2200-foot elevation and the relative remoteness of the park, the following breeding species can usually be found: Ruffed Grouse; Black-billed Cuckoos; Screech Owls; Pileated, Red-bellied, Hairy and Downy woodpeckers; Hermit Thrushes; Eastern Bluebirds, Golden-crowned Kinglets, Yellow-throated, Solitary, and Red-eyed vireos, Black-and-white and Black-throated Green warblers, Ovenbirds, and Yellow-breasted Chats. Barred Owls, relatively rare in western New York State, breed within the park.

Drive into the recreation area and park along the circle at the end of the road. Begin walking into the woods on the trail that begins behind an old barn basement foundation on the right. Follow the car-width trail through the woods, taking any of several foot trails branching off it, all indicated by obvious yellow markers. These trails generally circle around and return to the main parking area.

MONROE COUNTY
Reed Road Swamp (Map 2.6)

Rating: Spring****, Summer**, Autumn**, Winter*

In 1949 the Genesee Ornithological Society purchased one of its favorite birding spots to save it from being lumbered: Reed Road Swamp. It is seven miles southwest of Rochester, and it covers 100 acres. On both sides of Reed Road the land is a low wooded swamp, with red maple and red ash predominating, but with smaller amounts of willow, aspen, dogwood, basswood, white pine, and arbor vitae. Dense undergrowth is formed by spice bush, raspberry, gooseberry, and fox grape. Beware the poison ivy!! On the forest floor are cinnamon, ostrich and maidenhair fern, wild ginger, meadowrue and baneberry. Comfortable birding is largely confined to spring. During the breeding season, the swamp is insect-infested. In winter it is difficult to penetrate the snow-covered brush, and the birds are few. In spring, however, it is a site for sharp eyes.

To reach Reed Road Swamp from the Rochester area, begin at the junction of NY 252 (Ballantyne Road) and NY 383 (Scottsville Road), on the west side of the Genesee River, about two miles south of the Monroe County Airport. Proceed south on NY 383 for 1.8 miles to Brook Road. Turn right (west) and proceed 1.2 miles to the first intersection (Ballantyne Road). Turn left on Ballantyne, and after 1.2 miles Reed Road comes in from the left. Turn left. After 0.3 mile more, the wooded swamp begins and extends on both sides of the road for 0.3 mile. An additional 0.8 mile beyond the south edge of the swamp is a small marsh at the corner of Morgan Road, which is worth checking in the early morning for rails, bitterns, ducks, and marsh wrens.

Birding within the sanctuary is most easily accomplished by walking along paved Reed Road, which transverses the swamp; but, some breeding species, especially Ruffed Grouse, Brown Creeper, and Winter Wren, can usually only be found back from the road. An effort is made to keep open some unobtrusive trails into the interior. Other nesting species are, or have included recently, Red-tailed Hawks, American Woodcocks, Yellow-billed and Black-billed cuckoos, Great Horned, Barred and Saw-whet owls, Blue-Gray Gnatcatchers, Black-and-white and Cerulean warblers, Northern Waterthrushes, American Redstarts, Scarlet Tanagers, Rose-breasted Grosbeaks, and Indigo Buntings. In the past Hooded Warblers were common here, but they are now rare. Mourning Warblers appear to be quite common at this time.

Powder Mill Park and Burroughs-Audubon Nature Club Wildlife Sanctuary (Map 2.6)

Rating: Spring****, Summer***, Autumn**, Winter**

During the Civil War, in what was then a remote and hidden spot among the hills along Irondequoit Creek, the Rand family operated a powder mill. Today parts of the mill race can still be seen and the water wheel still runs in Powder Mill Park, which is ten miles southeast of Rochester and three miles southeast of Pittsford. The area encompasses 575 acres of county park land. It can be both reached and birded by bicycle, car, or on foot. There is no admission charge or parking fee. The birding here is easy, and there are no special hazards. Maps of all the county parks may be obtained free at the County Parks Office, 375 Westfall Road, Rochester, New York 14620.

To reach the park, leave the New York Thruway at Exit 45, and just beyond the toll booth take NY 96. Proceed along it about four miles to the park's marked entrance on the left. If approaching from Pittsford, proceed from the main downtown intersection (four corners) south on South Main Street to the first cross street, which is NY 96. At the stoplight turn left. Proceed on NY 96 thru Bushnell's Basin for approximately four miles, and look for the park entrance on the right.

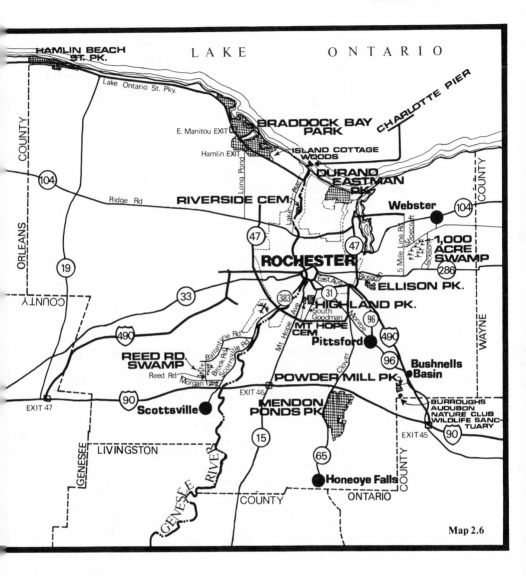

Map 2.6

Warbler study during migration is ideal from the road that runs south from the fish hatchery past the maintenance building to Woolston road, especially in the early morning sun. Here Cerulean Warblers are common breeders and the song of the Mourning Warbler can be almost guaranteed in the low, swampy areas on the east side of the road. In the damp areas of the trillium swamp, Golden-winged and Blue-winged warblers regularly occur and the Brewster's Warbler hybrid has been recorded. Seven picids nest in this park: Common Flicker, Pileated, Red-bellied, and Red-headed woodpeckers, Yellow-bellied Sapsuckers, Hairy and Downy woodpeckers. The sand bank across from the ski jump merits investigation. An occasional Rough-winged Swallow is seen anticing with members of the very large Bank Swallow colony located there. Three vireo species, tanagers, Rose-breasted Grosbeaks, and Indigo Buntings are common breeders in the park. Winter birding among the mixed evergreen plantings can yield chickadees, kinglets, Pine Siskins, Red and White-winged crossbills, and juncos. In the most southeasterly section of the park is Rabbit Valley Farm, otherwise known as the "sparrow stop." Savannah, Grasshopper, Henslow's, Vesper, Chipping, Field, Swamp, and Song sparrows can all be observed here.

Adjacent to Powder Mill Park is the thirty-acre Burroughs-Audubon Nature Club Wildlife Sanctuary. A leisurely walk along its nature trails will pique the curiosity of the observant birder, and he will be rewarded with the sights and sounds of nesting and migrating specialties. By following the foot path at the base of the drumlin up the hill from the clubhouse, one can reach the area of the predictably occurring Hooded Warbler.

Mendon Ponds Park (Maps 2.6 and 2.7)

Rating: Spring****, Summer***, Autumn**, Winter**

This unique area of glacier-formed ridges and post-glacier ponds and potholes nine miles southeast of Rochester, is a birder's delight, with 2443 acres of conifer-planted upland, deciduous woods, brushy fields, and seven ponds with surrounding marshes. A wise administration has kept the park free of overdevelopment, commercialism, and overuse.

From Monroe Avenue (NY 31) and Clover Street (NY 65) in the township of Brighton, take Clover Street south for about 5.5 miles to the Canfield Road entrance on the left. Farther along NY 65 there are two other entrances at Hopkins Point Road and Pond Road. A rare aspect of this park is the year-round nature center and a full-time naturalist who administers the small interpretive museum and educational programs. The nature center is on the right just inside the Pond Road entrance. A network of trails leads from it into some of the best birding terrain. Watch out for the aggressive chickadees along the Birdsong Trail — they beg to be fed and will take handheld seeds, if offered.

During spring and fall migrations the park serves as a resting and feeding place for waterfowl and passerines. One of the best places for ducks is Quaker Pond. Walk down the service road along its east side. From March through April, wigeons, Ring-necked, and Ruddy ducks may be found. Look also for Common Loons, Green Herons, night herons, American Bitterns, and rails. In spring there are often large concentrations of Canada Geese on Quaker and Hundred Acre ponds — and occasionally Snow Geese. In this park, in May 1940, Meade *et al.,* provided the second New York State record of Bachman's Sparrow. Grasshopper and Henslow's sparrows can be found in the grassy fields east of Quaker Pond.

Hundred Acre Pond is annually host to a wide variety of waterfowl, and

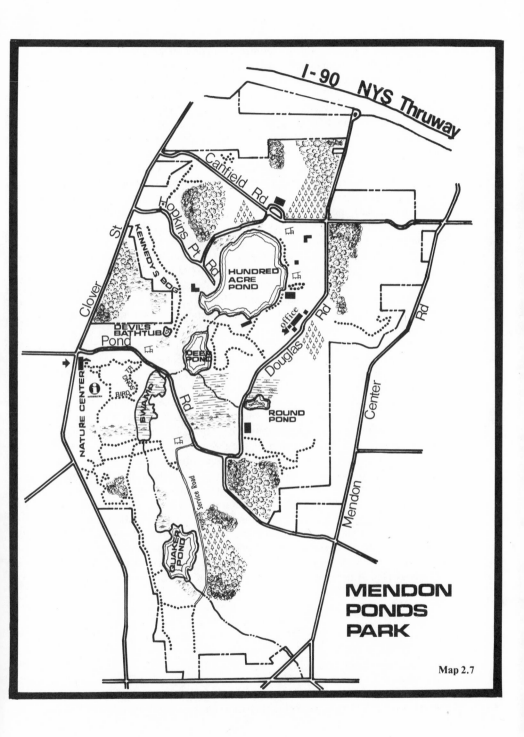

I-90 NYS Thruway

Canfield Rd

Hopkins Pt Rd

Clover St

KENNEDY'S BOG

HUNDRED ACRE POND

DEVIL'S BATHTUB Pond

DEEP POND

office

Douglas Rd

NATURE CENTER

INFORMATION

TERSONG

BIRD

SWAMP

Rd

ROUND POND

Center Rd

Mendon Rd

Service Road

QUAKER POND

MENDON PONDS PARK

Map 2.7

Deep Pond, just south of it, sometimes has diving ducks, especially Buffleheads and Hooded Mergansers. South of Deep Pond is a marshy area that now and then has water of some depth, and it apparently attracts herons, bitterns, and rails, but seldom ducks. West of Deep Pond is a very deeply recessed glacial pothole known as the Devil's Bathtub, which is a good place to find Wood Ducks. Round Pond should be investigated for mergansers and duck species.

Mendon Ponds offers hospitable habitat for a wide variety of nesting birds: common woodland and brushland species, including cuckoos, Pileated Wood-peckers, Great Crested Flycatchers, Eastern Wood Pewees, Yellow-throated Vireos, and tanagers. During the height of spring migration more than 100 species may be found in the park in a single day. Some of the less common warblers that breed in the park include Golden-winged, Blue-winged, and Cerulean warblers, and the Yellow-breasted Chat. Screech, Great Horned, and Long-eared owls can all be found in the park.

Toward the northwest corner, south of Hopkins Point Road, is Kennedy's Bog, which is of special interest to the botanist; Rose Pogonia has been found here.

The park is visited in winter by good numbers of northern finches, especially in incursion years. Near the nature center there is a flourishing stand of bayberry, which has proven a great attraction some winters to Yellow-rumped Warblers.

Thousand Acre Swamp (Map 2.6)

A potentially good area, which has only recently become available to birders, is the Thousand Acre Swamp east of Rochester. It is tucked away in the far eastern corner of Monroe County in the town of Penfield. It has many different vegeta-tion successional zones, typically southern, without boreal components. The Na-ture Conservancy owns a section of the swamp, and there is every hope that through cooperation of the various owners and with conservation easements, the swamp will be preserved for present and future generations. Because so many of the secrets of the area have yet to be uncovered, anyone visiting it is requested to keep a record of any bird, mammal, or plant species encountered there and to then send this information to The Nature Conservancy or the *Kingbird* Region 2 editor.

Access to Thousand Acre Swamp is best gained by a cable right-of-way be-hind the church parking lot at Five Mile Road and Embery Road. The path leads across the southern section of the swamp and terminates behind the Penfield Highway Department building on Jackson Road near Shoecraft Road.

Ellison Park (Map 2.6)

Rating: Spring***

Approximately one mile east of the Rochester city limits lies Ellison Park, whose entrance is on Blossom Road at the bottom of the long hill just east of Landing Road. In May a birder can profitably spend time viewing Winter Wrens, thrushes, warblers, Purple Finches, towhees, and several species of sparrows. Wood Ducks, Mallards, and Blue-winged Teals can be found in wet grassy areas here also in spring. In the marsh, Long-billed Marsh Wrens and Swamp Sparrows are summer residents. Screech and Great Horned owls are also residents, and now and then one catches sight of the irregular flight of a Pileated Woodpecker quickly vanishing out of view.

Highland Park and Mt. Hope Cemetery (Maps 2.6 and 2.8)

Rating: Spring****, Summer**, Autumn**, Winter***

Highland Park, located in south-central Rochester and wholly contained within the city limits, has probably been a favorite birding spot longer than any other Rochester-area location. Richard Horsey and William Edson, park botanists, began observing birds and keeping meticulous records in this 100-acre park in about 1910 and continued until 1948. For years, their office was a center for local ornithologists. *Nota Bene:* Planted along the gently sloping hillsides of Highland Park are over 1600 lilac bushes of more than 550 varieties. At the peak of their glory, these and the other magnificent plantings are visited by crowds estimated in excess of 100,000 people during what is known as "Lilac Time". The festival is always held in May and lasts for one week. During this period the park can be birded in the early morning. It is, of course, impossible later as the crowds increase, except in the areas north of the east-west through road. With this exception, the park is an uncrowded and beautiful arboretum featuring one of the world's largest collections of lilacs, conifers from all over, hundreds of varieties of azaleas, hawthorns, and flowering and fruiting shrubs, and trees providing an abundance of food attractive to birds.

Highland Park and Mt. Hope Cemetery lie on portions of a ridge of hills that extends across the southeastern border of the city and that is made up of glacial debris. Highland Park is bordered on the east by South Goodman Street, on the south by Elmwood Avenue, and on the west by Mt. Hope Avenue. Mt. Hope Cemetery is west of Highland Park just across Mt. Hope Avenue.

During the year one can find well over 100 species in Highland Park. Notable are the "warbler days" during spring migration, and the winter-finch populations in the conifers during that season. In spring, after a good night of passerine flight, the park can swarm with birds. If there is a good small-land-bird movement it will be worth going over to Mt. Hope Cemetery where the trees and shrubs can be "dripping" with flycatchers, vireos, warblers, etc. This is especially true after a warm May night with southerly winds. At other times of the year the cemetery is not worth birding.

Evening and Pine grosbeaks, Pine Siskins, Red and White-winged crossbills are common in the conifers and hawthorns of the park from October through April. Try the conifers lining both sides of the entry road from Goodman Street and the hawthorns around the lily pond. Do not neglect to look at the reservoir for occasional waterfowl and gulls. Grebes and Oldsquaws have been seen here.

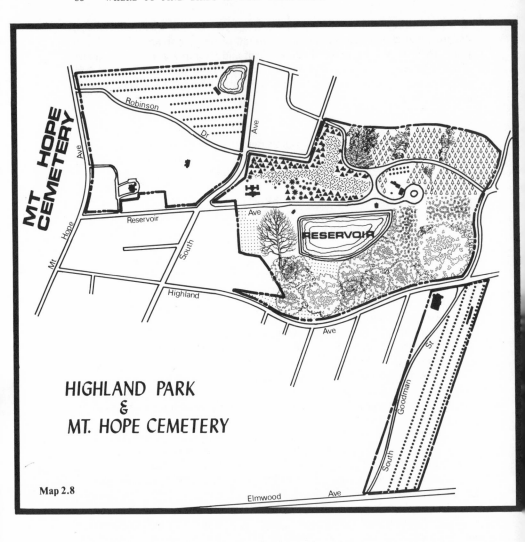

Map 2.8

Riverside Cemetery (Map 2.6)

Rating: Spring***, Summer**, Autumn***, Winter**

Riverside Cemetery occupies 200 acres within the city limits of Rochester and offers wonderful birding in both spring and autumn migrations; it also harbors poison ivy. Another *caveat* is due its gorge, which is approximately 100 feet deep and surrounded by an unprotected bank. Riverside is rated as easy to bird unless one chooses to go down to the river levels — which would then be termed "very difficult."

To get to Riverside from Rochester's interior, begin at the intersection of Ridge Road (NY 104) and Lake Avenue. Proceed north 1.6 miles on Lake Avenue to Riverside Cemetery (at 1.4 miles there is an entrance to Holy Sepulchre Cemetery — do not enter, but go 0.2 mile farther to Riverside). Enter through the gate on the right (east side of Lake Avenue). Parking is allowed about 200 yards inside the cemetery, beside the pond.

The area within the cemetery has open water, except in midwinter, and very large hardwoods, which make it a good place for flycatchers, vireos, and warblers during migration seasons. In the fall and winter, the river bank gives a good vantage for waterfowl, particularly goldeneyes, mergansers, and an occasional Ruddy Duck. The riverbank is easily accessible for about a mile, and to the north is a hardwood stand which is usually productive for woodpeckers. At the north end of this stand is a turning basin for lake ships, which may be surveyed from the south bank. This is a good area for gulls and shorebirds in late fall and early winter. Here also herons and kingfishers are regularly seen during the winter months.

Hamlin Beach State Park (Maps 2.2 and 2.6)

Rating: Spring***, Summer*, Autumn****, Winter****

Hamlin Beach, three miles north of the town of Hamlin, encompasses 1500 acres. It is not so much a place for rare specialties as one for concentration of a wide variety of migrants — and it fulfills its potential for occasional surprises. The best birding areas can be covered largely from an automobile; of course, walking enhances the whole venture. From Memorial Day through September this park can be and often is very crowded. Birding here during the tourist season is really not recommended. However, although birding is rated "easy" during the best fall periods, it is sometimes cold and windy, and the birder should dress appropriately. Although there is no parking fee, there is a daily admission charge during the summer and early autumn months.

Leave the New York State Thruway (I-90) at Exit 47 (LeRoy). Immediately get on NY 19 heading north. Proceed north sixteen miles to its intersection with NY 104. Continue on NY 19 3.25 miles past the intersection to a fork in the road. Stay left until the road ends at Moscow Road. Turn left on Moscow Road and follow it to a sign that indicates the entrance to Hamlin Beach State Park.

Enter the park on the east end. Just past the entry booth swing around the traffic circle and note a roadway to the right (east) to a boat launching area marked with the sign "Car Top Boat Launching Only." In the off-season one can park here and hike along the shore and eventually come to an overlook on a marsh where Yanty Creek enters Lake Ontario. This area can be good for waterfowl from autumn through spring, particularly for concentrations of geese and scoters; Brant and even Harlequin Ducks are possible sightings. A guided trail and also a Braille Trail are in various stages of construction.

Heading west into the park from the traffic circle, there are five numbered parking lots between the main road and the lake. A cement pier off Parking Lot

No. 1 extends into Lake Ontario; it can be a good site from which to view late fall and winter diving ducks, Purple Sandpipers, and Glaucous and Iceland gulls. Snow Buntings will probably be scattered along the beach.

The area around Parking Lots No. 2 and No. 3 has rarely been very interesting; but, check them out anyway. The camping area across the main road (both pairs of lanes) away from the lake has been a good place for visiting winter finches, Evening and Pine grosbeaks, redpolls, and crossbills in early winter. Some of the densely overgrown swales between stands of pines harbor very late lingering migrants. Large concentrations of Long-eared Owls have been found in the pines.

The high stretch of shore along Parking Lot No. 4 is the center of activity for several local expeditions to Hamlin Beach in the autumn. One anticipates seeing Common and Red-throated loons, Red-necked and Horned grebes, Brants, scoters, Glaucous, Iceland, and Little gulls. Those born on the sunny side of the hedge might even see Pomarine or Parasitic jaegers. Look for Purple Sandpipers, Water Pipits, Snow Buntings, and assorted shorebirds between waterfowl flights. All of the rock jetties should be checked.

In the forested areas of the park Boreal Chickadee and Bohemian Waxwing have been recorded and in various other sections Great Horned and Snowy owls have occurred.

Braddock Bay Park and Vicinity (Maps 2.6 and 2.9)

Rating: Spring****, Summer**, Autumn***, Winter*

If one is birding the Rochester area in spring and has limited time, this would be the one area to visit. The hawk flights can be spectacular from late February into early June (see separate section on hawk watches). The marshes are attractive to herons, ducks, rails, and shorebirds, and the woods provide a first shelter for small migrant passerines crossing from the bay and marshes. Most days, especially weekends, the area is manned by hawk watchers, who can also provide directions to other interesting birding areas. This area may be birded by car or bicycle, although to reach it on bicycle, use one of the routes outlined for the Charlotte-Island Cottage Woods tour, because the Lake Ontario State Parkway is closed to bicycles. Although it can be covered mostly by car or bicycle, birders should hike throughout the area to cover it thoroughly.

Leave the New York State Thruway (I-90) at Exit 47 (LeRoy). Take I-490 east to NY 47 (I-590 east) north. Follow it to US 104 (Ridge Road). Exit west and go 1.1 miles to Long Pond Road. Turn right (north) and proceed 4.9 miles to the Lake Ontario State Parkway entrance, designated Hamlin Beach Park. Proceed 1.8 miles west to the East Manitou exit. The entrance to Braddock Bay Park is just to the right of the exit ramp with the parking lot one-half mile into the park. For purposes of orientation: looking across the bay one sees a large white hotel and dance hall; this is north. The woods behind the observer then are southeast.

The birds to be seen in this area include virtually the entire area checklist, including most of the rare visitors. The best way to cover it is by a brief tour stopping at the indicated points to observe the specialties. The starting point is certainly the

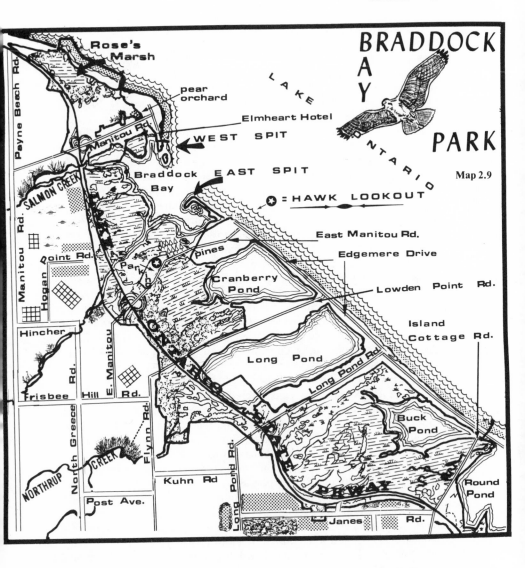

Hawk Lookout. In addition to hawks, late winter and early spring bring large concentrations of waterfowl. April and May see the advent of waders, sometimes including a good representation of the more southerly herons. Shorebirds are generally spotty in this area. American Woodcocks can be heard at dawn and dusk and
can sometimes be flushed. Common Snipes will certainly be winnowing overhead
from March to June. From autumn through spring, gulls often concentrate on the
bay and may include Glaucous, Iceland, and Little gulls. Forster's, Common, Caspian, and Black terns can be seen here, with the first three species peaking in late
July and August.

Owls are definitely specialties here. Winter finds one or more Snowies in the marsh; spring brings Short-eareds, and from the last week of March to the third week of April Long-eareds and Saw-whets can be found. The pines stretching along East Manitou Road at the east end of the parking lot provide a favored roosting place for owls.

Small passerines pour through this area. Anything may be found in the woods. The edge of the woods, again along East Manitou Road, is an excellent place to study warblers in the dawn light. The warbler parade usually starts with Pine and Yellow-rumpeds in mid-April and peaks with various specialties in mid-May. This is generally not a place where one looks for particular species; it is rather a place with great variety and an occasional memorable surprise.

After leaving the Hawk Lookout, go to the East Spit. This is reached by walking northeast from the parking lot through the state marina to the lakeshore (0.6 mile), then left onto the spit. Alternatively, drive from the entrance to Braddock Bay State Park, turn left on East Manitou Road and proceed one mile to Edgemere Drive; turn left, advance 0.15 mile to a wide shoulder at the end of the road. This is the entrance to the East Spit. A No Parking regulation is sometimes enforced. The options are to chance a ticket or to drive to one of the nearby side roads and find a parking spot—not an insurmountable problem. The spit extends about a quarter-mile toward the bay outlet and then about the same distance at right angles into the bay. Specialties of the area include loons, diving ducks, grebes —including Common and Red-throated loons and Red-necked Grebes, King Eiders, and Harlequin Ducks. Of major interest, however, are shorebirds in May and also from July through October. In the spring this is a good vantage point to try to see the Whimbrel flight, which usually occurs sometime during May 18 to 25. On a good August day over twenty shorebird species may be tallied. Among the more interesting regular visitors are Black-bellied Plovers; Ruddy Turnstones; Red Knot, Baird's, and Western sandpipers; and Hudsonian Godwit. Occasionally, Willets and Ruffs have been reported. For additional interest in spring, large flights of Sharp-shinned Hawks, American Kestrels, and Blue Jays can be observed from this point moving east parallel to the shoreline. In autumn and spring an interesting variety of migrants pass along the spit, stopping briefly in the willows. In late summer the gatherings of swallows are incredibly spectacular.

From the East Spit travel to the West Spit *via* several other interesting stops. Return to the park entrance by retracing the route followed to the East Spit. Continue on East Manitou Road until it tees with Frisbee Hill Road (one mile). The large hayfield on the left has had Short-billed Marsh Wrens, Bobolinks, and Savannah and Henslow's sparrows from early May until the first mowing. Turn right on Frisbee Hill Road, proceed 0.6 mile to North Greece Road. Turn right and go 0.7 mile to Hincher Road. Take a left and proceed 0.35 mile to Hogan Point Road and take a right onto it. Along the route scan the fields for Cattle Egrets, and Bobolinks from late April through May. The fields on either side of Hogan Point Road for the next half-mile are very good in May for grassland sparrows: Savannah, Grasshopper, Henslow's and Vesper. Listen and look from the roadside only. The fields are on private property and are posted. All of these birds sing from high points. Scan the taller weed stalks. After a half-mile the paved road turns 90° to the right. At this point, the fields straight ahead on either side of a gravel road are often partly flooded in the spring, and Cattle and Snowy egrets and Glossy Ibises have been observed there. The gravel road may be walked, or even driven, if it is not excessively wet. Continue to the end of Hogan Point Road (0.4 mile). There are fields on the left which when plowed, spring or fall, often attract American Golden or Black-bellied plovers. Return now to Hincher Road (0.9 mile).

Turn right on Hincher Road. Advance 0.6 mile. Turn right on Manitou Road, go one mile to a forked intersection. Be sure to bear right on NY 261 and go 0.5 mile to a spot where it is possible to pull off on the right side of the road and overlook Salmon Creek and the fields across the creek at the end of Hogan Point Road. In early spring Salmon Creek harbors a wide variety of waterfowl. In May egrets missed from the other side might be seen from this height. In the fields along the creek often large gatherings of gulls claim sanctuary; rarely, a Franklin's is among them. Farther on, about 1.1 miles, an abandoned pear orchard runs straight back from the road on the left. This is part of the park, so do not be fooled by the two houses on either side of the orchard. Stop along the shoulder. To cover this area boots may be necessary—it is often wet and muddy in the spring. Specialties to be found here are Long-eared and Saw-whet owls in late March and early April, all of the eastern Empidonax flycatchers and also Olive-sided Flycatchers in the latter half of May, and warblers and sparrows throughout May. At the crest of a good mid-May migrant passage, a few hours here should yield nearly 100 species in the orchard or overhead.

Back to the car. Advance on Manitou Road follwing the right-angle bend to the right and forward to end at the parking lot of the Elmheart Hotel (0.4 mile). Ahead is the West Spit, an area of sandy ground and large cottonwoods. The spit tapers to a narrow point on which are scattered willows. This area is now private property and permission to bird it can be obtained at the hotel. There is usually a small charge; but, often local birding groups are allowed free access. *Nota Bene:* Birders are not welcomed during hunting season.

The best periods to bird here are during spring and fall migration, as the spit acts as a funnel, concentrating the birds in the willows toward its end. The large willow on the right at the entrance to the parking area seems to be a first stop for many migrants coming in from the lake before they advance to the tree-trap at the end of the spit. The concentrations of birds can be superb on a peak May day with all of the eastern swallow species lined up on the utility wires near the old dance hall next to the hotel—with more than twenty species of warblers passing through; with several species of hawks, including large numbers of Sharp-shinneds moving east over the spit; with loons, grebes, cormorants, ducks, gulls, and terns on and over the lake.

Northrup Creek to Island Cottage Woods, a Tour (Maps 2.6 and 2.9)

Rating: Autumn****

This is a short tour that offers profitable shorebirding under optimum conditions in the autumn.

Return to the intersection of East Manitou Road and Frisbee Hill Road. Continue east on Frisbee Hill for 0.3 mile to its intersection with Flynn Road. Turn right and park along Flynn Road. Less than 100 feet south of the intersection a trail through the cattails leads to the bank of Northrup Creek and to one of the best shorebird spots in the area, if the water levels are right. The trail may be muddy, so be prepared with boots, etc. Between fall 1975 and spring 1976, thirty-one species of shorebirds were observed here. Among the more interesting regular visitors are all the "peeps" seen in the east, both dowitcher species, Stilt Sandpipers, Hudsonian Godwits, Wilson's and Northern phalaropes. Rare visitors are Whimbrels, Willets, Buff-breasted Sandpipers, Marbled Godwits, Ruffs, and American Avocets. It is quite definitely worth a look.

Continue south on Flynn Road a half-mile; turn left on Kuhn Road; after one mile more, turn right on Long Pond Road, and proceed nearly a mile to Jane's Road. Turn left here. Near the beginning of Jane's Road the multiflora rose hedges flanking the first house on the right have harbored Mockingbirds for several years. Cover the entire length of the road (1.5 mile), birding the plowed fields for American Golden and Black-bellied plovers, to Island Cottage Road. There turn left and proceed one mile to Island Cottage Woods parking lot, which will be on the right beside Scarlata's Party House.

Charlotte to Island Cottage Woods, a Tour (Map 2.6)

Rating: Spring****, Summer**, Autumn***, Winter*

This area is in the town of Greece and encompasses approximately 90 acres of private land; but permission to bird is not now required.

Leave the New York State Thruway (I-90) at Exit 45. Take I-490 west for 11 miles to the intersection of NY 47 (I-590). Take it north to US 104. Take US 104 (Keeler Street Expressway) west for 4.2 miles to Lake Avenue. Turn right (north) on Lake Avenue. For the first 0.5 mile the Eastman Kodak factories will be on your left. Drive north 4.5 miles to its end where it tees with Beach Avenue. To the right is the parking lot for Ontario Beach Park. Park near the far (east) end at the retaining wall, and walk out along the Genesee River pier. This area is best birded from fall to early spring. Avoid the area altogether in summer; the crush of people is lamentable, to say the least. In winter the pier is ice-covered and dangerous or impassable.

While walking along the pier, scan the Genesee River for ducks, both diving and puddle varieties. Usually good numbers of Greater Scaups with fewer Lessers are present in late autumn and winter. The inverse is true in spring. Ruddy Ducks in small numbers are found here. The gulls present around the piers and surrounding areas should be studied for the occasional Glaucous, Iceland, and even Black-headed or Little gulls among the Great Black-backeds, Herrings, Ring-billeds and Bonaparte's. Common and Caspian terns usually appear in April and are seen periodically throughout the summer.

A careful investigation of the beach and the wave-washed windrows of weeds to the left of the pier usually produces Semipalmated and Black-bellied plovers, Ruddy Turnstones, Dunlins, Sanderlings, and assorted peeps. Ducks should be passing nearby constantly, and an occasional American Golden Plover or Hudsonian Godwit might be sighted out over the lake, or off the beach or pier. Oldsquaws, scoters, Buffleheads, Common Goldeneyes, and mergansers are usually to be found in numbers varying according to the season. In late October and early November Brants should be present.

During migration, flocks of passerines pass along the lakeshore; in autumn they often drop exhausted into the first trees or even onto the pier or beach. Investigate the pier for flocks of Purple Sandpipers, an occasional Snowy Owl, and flocks of Snow Buntings.

Departing this area, move west along Beach Avenue for 1.2 miles to the Russell Station outflow. Notice an open area with a wide gravel shoulder on the right where cars may pull off. This is a profitable birding site from autumn to early spring. It is, in fact, one of the most likely places to locate King Eiders and Harlequin Ducks. Glaucous, Iceland, and Little gulls may be feeding over the outflow.

From Russell Station west, Beach Avenue becomes Edgemere Drive. Proceeding along the lakeshore, stop at any of several obvious turnouts on the right, designed to provide views of either Lake Ontario or Round Pond and its marshes on the left.

At 1.5 miles from Russell Station, just past Scarlata's Party House (the old Island Cottage Hotel), turn left into the gravel parking lot. At the far southwestern corner of the lot (300 yards from Island Cottage Road and Edgemere Drive), an old overgrown road leads past an abandoned dance hall (thought by some to be an old clambake house) which appears now as no more than a dilapidated white shelter, into the Island Cottage woods. The trails from there are obvious, although overgrown and not distinctly defined, and the woods are small. The prime time to bird this marvelous migrant trap is during the peak of migration in May, especially after a southwest wind when temperatures are between 65° and 70° F.

Hiking is necessary to achieve really adequate coverage. In the spring the trails may be wet and muddy and then boots are a boon if one is not indifferent to wet feet. Mosquitoes and deer flies are quite common in late spring and early summer. In autumn the dense undergrowth is difficult to penetrate; otherwise, birding this area is rated as easy to moderately difficult.

The woods contain a mixture of mature and secondary growth of willow, poplar, beech, maple, sassafras, ironwood, ash, and dogwood, along with an understory of saplings and brush. In the area of dense groundcover adjacent to the old dance hall is an excellent place from which to watch waves of migrants filter through the woods. Gray-cheeked Thrushes, Fox and Lincoln's sparrows are commonly seen here; however, the area is probably best known for the variety of its flycatchers, vireos, and warblers — Yellow-bellied, Acadian, and Olive-sided flycatchers; Yellow-throated and Philadelphia vireos; and Worm-eating, Kentucky, Connecticut and Mourning warblers are some of the more uncommon species that occasionally are found here.

Usually, most of the best birds are spotted at the north end of the woods within a quarter-mile of the old dancehall shelter. However, to conclude that this is

the optimum section of the woods may be an ungrounded assumption. Its treasures may well be fortuitous; obviously, this portion is trailed and is consequently
more thoroughly covered. Perhaps the more interior, less penetrable reaches
would be equally or more productive. When activity is slow at the north end, particularly after a warm spring night, a check at the south end of the woods, adjacent to the Lake Plains Waterfowl Association Headquarters will probably be
worthwhile. This is approximately 0.5 mile south on Island Cottage Road. The
higher land and characteristic associated vegetation there will yield different bird
species — Yellow-breasted Chats, for instance.

A walk to the marsh adjacent to the north end of the woods in spring will
usually produce bitterns, Wood Ducks, American Wigeons, Blue-winged Teals,
Virginia Rails, Soras and coots. This is probably the best place in the county to observe Least Bitterns, which are fairly common breeders in some years here. They
are best seen along the marsh-field borders.

In some autumns, depending entirely upon the water level, shorebirds may
be found on mudflats in the marsh. Island Cottage woods provides only good-to-
fair birding in September and October; owing to its very dense foliage canopy, it is
most difficult to observe birds during this season, although Red-headed Woodpeckers and Mourning Warblers are included on the list of the wood's breeders. To
date, birding here in winter has been exceedingly unrewarding.

Durand Eastman Park (Maps 2.6 and 2.10)

Rating: Spring****, Summer***, Autumn***, Winter****

Durand Eastman is a park along the shore of Lake Ontario, entirely located within
the city limits of Rochester, about seven miles north of the center of the city. It encompasses 943 acres and can be reached and birded in a car, on bicycle, or on foot;
however, in winter some of the roads are not plowed, and are closed to automobile
traffic. Birding here is generally rated as easy, but one can climb up and down the
hills as well as walk on level ground. The degree of physical effort is largely a matter of choice. These hills are actually drumlins formed of erosion-resistant bedrock that has been abrasion-polished into streamlined hills by glacial ice. The ac-
cumulations of surface till are rich enough to support varied flora. As a result,
Durand Eastman has a fine, extensive horticultural collection (much of it labelled).
An outstanding collection of evergreens — many of them introduced — thrives in
what is called the Pinetum. In early spring the blooming Sargent cherry trees along
the east end of Lakeshore Boulevard (Sweet Fern Road) present a marvelous tree-
scape, and early *Magnolia stellata* (Star) and *Magnolia soulangeana* can be found
on the Zoo Road, just south of the flowering crabapples.

Nota Bene: This is a city park, and like so many located near large urban
areas, suffers the misfortune of sometimes harboring persons of dubious intent.
By all means bird the park; but take care. Neither male nor female should bird the
park singly, especially in the early morning hours. A pair, or preferably a trio, of
birders, staying close together would probably ensure an outing untainted by an

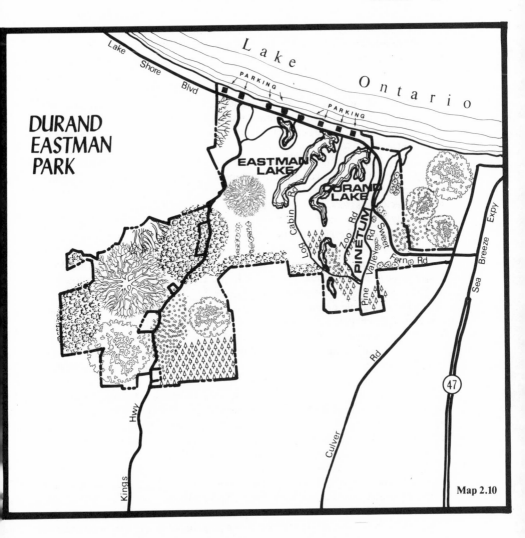

Map 2.10

unpleasant confrontation. Avoid vandalism by locking your car, and don't leave valuables in it. Strong and concerted efforts are being made locally to keep this park safe.

To reach the park from south and east of Rochester, leave the New York State Thruway (I-90) at exit 45. Get on I-490 north, proceed to NY 47 (I-590) north (on the *east* side of Rochester). About one-half mile from Lake Ontario, which can be seen ahead, exit left at the Durand Park sign. Follow the road west a short block to the park entrance straight ahead across Culver Road.

If approaching from west of Rochester along the Lake Ontario State Parkway, go to its eastern terminus, which is Lake Avenue. Turn left on Lake Avenue, then turn right abruptly at the first traffic light, onto Stutson Street. Cross the Genesee River on the Stutson Street Bridge, and continue east to St. Paul Boulevard. Turn right on St. Paul Boulevard. After one block, turn left at the traffic signal, which will put you on Lake Shore Boulevard. Continue east for approximately one mile until a sign becomes visible that reads, "Rochester Welcomes You." That sign marks the beginning of the park. Proceed east along the lakeshore through the park.

Lake Ontario acts as a giant barrier to northbound birds in spring migration. There seems to be a general unwillingness of the smaller passerines to venture out over the water. Durand Eastman offers an outstanding collection of evergreens, deciduous trees, berry-bearing shrubs: an ever-valued safety zone for these apparently timid venturers.

The park's interior network of roads consists of the main road, Lake Shore Boulevard (runs east-west between the lake and the park), and various roads that branch south from it into the park. Most important of these minor roads are: Pine Valley Road, Zoo Road, and Log Cabin Road. If you are so inclined, follow several of the numerous foot trails — with friends.

The crabapple orchard on Zoo Road attracts robins; and, in winter, both Cedar and Bohemian waxwings, grosbeaks, and finches. In November 1977, a Varied Thrush appeared and was viewed for months by many observers as it fed on frozen crabapples. Rochester's meticulous hard-core birding devotees were well rewarded for their diligence. In spring, especially April and May, the orchard is also an excellent place to find warblers.

The Pinetum, on the west side of Pine Valley Road, is where the Boreal Owl was found in February, 1978. Again, an extralimital gift, but this area is always a good winter birding spot, nonetheless. Owls, woodpeckers, jays, nuthatches, redpolls, siskins, and crossbills are constant winter residents. Several ponds, although frozen in winter, are productive in spring and autumn.

Parking space all along the lake side of Lake Shore Boulevard provides ample opportunity for overlooking the lake. Stop where you will and scan for gulls, terns, and waterfowl. Parasitic Jaegers have been sighted in October and November from this bluff.

The park is fortunately situated on the pathway of northward bound hawks in spring. When there is a southwesterly wind look for them traveling west to east along the lakeshore.

Irondequoit Bay, a Tour (Map. 2.11)

Rating: Spring***, Summer**, Autumn****, Winter****

One of the translations given for the Indian name "Irondequoit" is place where the waters meet. The waters of Irondequoit Bay empty into Lake Ontario seven miles northeast of Rochester. The bay lies in the ancient bed of the Genesee River and was a major route taken by the Seneca Indians, as they traveled to their main village, near what is presently Geneseo. Rather than portage around the high falls of the Genesee River gorge, they traveled up Irondequoit Bay and Irondequoit Creek to a place still called Indian Landing, and then portaged overland to the placid waters of the Genesee, above the falls.

The tour described here passes over both private and public property and encompasses approximately twelve square miles. It can be done by car, or if the birder possesses an enormous amount of energy, by bicycle. Any amount of walking can be incorporated into the tour. The time required to do this route varies according to the time spent scrutinizing the various stops. It could take as few as three hours or as many as six. Irondequoit Bay is approximately five miles in length and the actual distance involved in this tour is twenty-one miles.

In winter, north winds coming off the lake can be especially strong; dress warmly, taking into consideration the wind-chill factor. The tour around the bay and through Irondequoit, Penfield, and Webster is most productive in late fall and early winter before the bay is frozen, and then again in early spring after ice-out. In summer the area is too densely populated to bird fruitfully.

To reach the start of the tour, leave the New York State Thruway (I-90) at Exit 45 and get on I-490 north. Proceed north to NY 47 (I-590) and follow it north for approximately 5.7 miles, looking for Pt. Pleasant Road; it now has no street sign, but it is at the first stoplight past Seneca Road and a tall blue watertower. Turn right on Pt. Pleasant Road, which is the beginning of the tour. Once on Pt. Pleasant Road proceed one mile and turn right onto Pleasant Avenue, which shortly bears left and down a rather steep hill leading to the bay at Pt. Pleasant. In winter and spring Pt. Pleasant is an excellent lookout for gulls and waterfowl. A low knoll provides a decent vantage point.

Because nearly all roads dead-end at the bay, it becomes necessary to retrace the route to the traffic light and then turn left on NY 47 (I-590). Proceed 0.6 mile to the traffic light at Seneca Road. A landmark is a watertower on the left. Turn right and follow Seneca Road as it curves a half-mile downhill to the Newport Yacht Club parking lot. This area, called "Birds and Worms" locally, provides another fine vantage point from which to view the bay. Irondequoit Bay Bridge is less than a half-mile south. The water around its pillars is often open when the rest of the bay is frozen. Scope the opposite shores for ducks huddling close to the eastern side. Little Massaug Cove (right), may shelter Pied-billed Grebes or Hooded Mergansers. Once back in the car, retrace the route back to NY 47 (I-590) and turn left there. Proceed 0.8 mile to the exit ramp leading to Ridge Road. Turn left onto Ridge Road. Advance 0.4 mile to Newport Road. Here turn left and proceed to the parking lot at Newport House. The Bay Bridge is one-quarter mile north. Directly across the bay is Inspiration Point. There the bay widens to form Devil's Cove. Scan the area well here for wintering gulls and ducks, because they are extremely difficult to see from the Webster side, opposite. Retrace the route back up the hill, stopping at the Town of Irondequoit's landfill, where you may find hawks, larks, sparrows, and buntings.

At Ridge Road turn left and follow it as Ridge Road becomes Bayshore Boulevard. At 0.5 mile, just past Point Lookout, the road extends along the top of an embankment beneath which is a delta marsh. This soft, low-lying tract has been

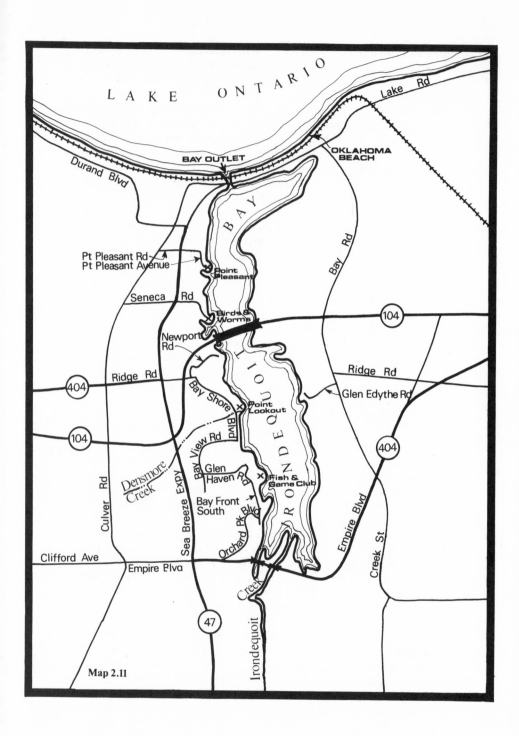

Map 2.11

formed by the runoff from Densmore Creek, which runs underneath the road and which itself carries the runoff from a small sewage plant on the right. The nature of its composition hinders the marsh water from completely freezing in winter. Virginia Rails, Killdeers, and Purple Sandpipers have all been seen here in December and January. On the opposite side of the road wintering sparrows can regularly be found. Proceed on Bayshore 0.4 mile to its end. Turn right onto Bay View Road. On the left there is a condominium development with a private access road to the bay, down which one may walk. Take Bay View Road away from the bay and up its hillside 0.7 mile to Glen Haven Road, which comes in on the left. Follow Glen Haven Road as it winds and twists back to the bay. This may appear unnecessarily circuitous but it is not — there is no direct access along the shore. Where Glen Haven Road turns sharply, bird for wintering landbirds and spring migrants. Proceed to level ground near the bay and go 0.2 mile farther to the intersection of Bay Front South. Stops should be made anywhere along here to "pish" for whatever isn't visible, to call a Screech Owl, or to scan the feeders.

Turn left at Bay Front South, passing Irondequoit Bay Fish & Game Club. The club is relatively abandoned in winter and birders are welcome to look over the bay from the parking lot. Past the club the road parallels the bay. Rely on instinct and vigilance to dictate where to stop for coots and dabbling ducks feeding in the shallows or secreting themselves under the marina piers. At 0.5 mile there is a stop sign restricting access to the south. Turn right at Orchard Park Boulevard, which leads away from the bay rising in sweeping curves to a residential area. At Empire Boulevard (NY 404) make a sharp left; descend a hill with Irondequoit Bay on the left and Irondequoit Creek marshes on the right. Bird anywhere along here. Expect wintering Great Blue Herons, Red-winged Blackbirds, and sparrows in the cattails.

Cross the bridge over Irondequoit Creek and enter the town of Penfield. This area is known as "Float Bridge," a name which dates from an era when a floating bridge spanned the creek. Proceed to the Colonial Inn; turn right into its parking lot. At the rear of it a gravel road can be found which extends 0.6 mile into the wetlands bordering Irondequoit Creek. This road is usually passable, but exercise discretion in deep snow. While walking or driving this road, stop to investigate feeders.

After returning to the parking lot, exit by turning right onto Empire Boulevard. After 0.2 mile there is a parking lot on the left from which the bay can again be scoped. Crossing over two lanes of oncoming traffic may be perilous, however. Proceed east. The small pond opposite the Buccaneer Restaurant (0.1 mile) has often been a site where herons, dabbling ducks, and late or lingering shorebirds have been seen. Survey the open sky here for Red-tailed Hawks.

Follow Empire Boulevard north and away from the shore toward Webster. Proceed 0.9 mile to Smith Road, opposite the Denonville Inn. Turn left and scan the fields on the right for wintering Horned Larks. Snow Buntings are a remote possibility here. After another 0.4 mile one encounters a circle. Park and walk 0.3 mile down to the shore by an obvious service road, past the treatment plant. *Nota Bene:* This road is precipitous, and driving it is certainly ill advised. Return then to Empire Boulevard and turn left.

Cover another 0.2 mile to an intersection with a traffic-light; Creek Street will be on the right and Bay Road on the left. Turn left. Pass Eastway Plaza and go through an area that is primarily residential to Glen Edythe Road. Turn left. Follow the curving drive down to the bay. Be cautious, for this road, although lovely to bird, narrows to one lane at some points and has sharp drop-offs. Proceed to the Glen Edith Restaurant parking lot. From the adjacent lawn the mile-wide bay can be clearly and easily scoped.

Return up the glen by the route just traveled and turn left at Bay Road. Bird the wires in this predominantly residential and farm area. Follow it to the traffic light at the intersection of Lake and Bay Roads. Turn left on Lake Road, which runs alongside the bay for a short distance. Oklahoma Beach (discussed later) is on the right. Stop only on the north (right) side of the road to bird here; afterwards, go 0.9 mile farther to the Irondequoit Bay Outlet.

Cross the bridge. Park on the west side of the bay outlet on the bay side of the road. The "No Trespassing" regulations are not enforced. Bird both the lake and bay sides of the outlet. This is one of the best places in early winter to view gulls, providing the outlet is not ice-clogged. Eight species have been recorded here with white-wingeds occurring fairly regularly. Little Gulls may be present when Bonaparte's Gulls congregate in large flocks during November and December. Phalaropes also occasionally occur here. When ice-out is in progress large concentrations of mergansers are observable along with Canvasbacks, Redheads, Old-squaws, scoters, Buffleheads, and goldeneyes. Infrequently, an eider is reported from this spot.

After birding the outlet return to NY 47 (I-590). Advancing south on it completes the circuit and points the driver toward metropolitan Rochester.

Irondequoit Bay to Sodus Bay, a Tour (Map 2.12)

Rating: Spring****, Summer**, Autumn****, Winter***

The Lake Ontario shoreline from Irondequoit Bay east to Sodus Bay is rich in geological, historical, biological, and ecological significance. Recreational development here has been very substantial and has gone unmonitored — as have its attendant detrimental aspects of refuse removal, slough dredging, effluents in the lake, disruption of plant and animal communities, beach overuse, etc. However, the overall character of this region remains notably undisturbed and it contains several small ecosystems, which help to offset the negative effect produced by the sites degraded through human exploitation.

Note that there are a number of cobblestone houses along the route. These unusual and attractive dwellings are peculiar to this part of the country and were built by resident English masons during their years of work on the Erie Canal. Most of the houses have cobblestones of a uniform size. These cobblestones are naturally very well cemented and extremely durable. They were rounded by the wave action of Lake Ontario and were gathered along the shore. Gathering and sizing them for use as the primary building material was back-breaking work at best. The formula for the mortar used has been lost. The endurance of the standing houses is mute testimony to the craftsmen who designed and erected them.

Thorough coverage of this tour requires the better part of a day. A shorter version of the route is possible by stopping at the Irondequoit Bay Outlet, Oklahoma Beach, Pellett Road, Pultneyville Harbor, and Sodus. This means that the side routes should be ignored. This tour is designed to be done in late fall to early winter, before the bays are frozen. It would also be profitable in early spring. Places that are best covered in spring at the height of migration are specified. In order to view the lake adequately, a telescope is almost a necessity. Begin this trip with a full tank of fuel because stations are scarce. Public restrooms may be found in the parks.

To reach the beginning point of this tour, leave the New York State Thruway (I-90) at Exit 45. Take I-490 north to NY 47 (I-590). Take NY 47 (I-590), the Sea Breeze Expressway, north to its end at the Irondequoit Bay outlet, and park on the right before crossing the outlet bridge. When the outlet is open in autumn, winter, and spring, large concentrations of gulls and waterfowl can be observed here. Look for Glaucous and Iceland gulls all winter, and for Little and Bonaparte's gulls through early January. Phalaropes are sighted here in late fall. Most of the rare gulls recorded in the Rochester area have been seen here.

Proceed one mile along Lake Road (town of Webster) to the end of Irondequoit Bay. The railroad underpass on the left leads to Oklahoma Beach. Park on the north side of Lake Road. The south side is marked with "No Parking" signs, and the Webster police enforce the regulation. Follow the steep path up the railroad embankment to the right of the underpass. Proceed northeast along the tracks for half to three-quarters of a mile. Lake Ontario is on the left, and birders can view it from many points. This is essentially a residential area, but because the railroad right-of-way is fifty feet wide, there is little chance of accidentally trespassing here. This area can be quite wonderful in mid-May for passerine migration. In 1961 this was the site of a spectacular chickadee migration. Eared and Western grebes have been seen here, and in autumn all three scoter species have been recorded.

Return to the car. Proceed 2.4 miles along Lake Road to Pellett Road, after passing through an extensive beech woods. Turn right at Pellett Road, which is bordered on the right by beech woods and on the left by orchards. Stop and listen at various intervals. At 0.3 mile from the junction of Lake and Pellett Roads there is an open meadow on the left. Opposite it are woods in which Pileated Woodpeckers nest and Hooded Warblers have nested. This area is fine to bird in spring, winter, and even in summer. Pellett Road ends at Herman Road, at which intersection the driver should turn around and retrace the route back to Lake Road. One mile east along Lake Road is the camping entrance to Webster Park. If the gate is open, a side trip here could be very worthwhile.

The main parking lot of Webster Park is 0.6 mile farther along Lake Road at the intersection of Holt Road. Park and cross the road to the high bluffs which overlook the lake and scan it with a scope. In spring it is possible to witness a very interesting migrational phenomenon here: cross-migration. One can see birds at one altitude flying east, while others at a different altitude are flying west on a

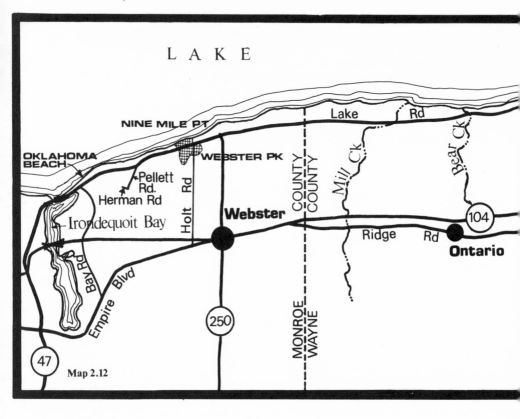

Map 2.12

southwest wind. During April and May hawks can be seen moving east along the lakeshore.

Drive 0.1 mile south on Holt Road to another parking area on the left. This affords access to the Nature Trail along Mill Creek. Mourning Warblers have nested here and it provides good spring warbler watching. The pine stand along the creek on the opposite side of Holt Road is another good warbler spot. Return to Lake Road and proceed along it into Wayne County. All of this area is farmland with extensive orchards. Hawks are frequently seen hunting over the area in both winter and summer.

Proceeding forward 2.6 miles, one reaches the Ontario Town Boat Landing bordering Bear Creek. Birders will find a decent view of the lake here. Lake Road is quite a distance from the lake along this stretch, so fine views of it are few. Continue 4.7 miles to a stop sign in the village of Pultneyville. Turn left immediately onto Hamilton Street, a dirt road. Proceed past the Yacht Club basin. After another 0.3 mile the road turns right. Park here. Climb the hill to a very fine vantage point overlooking the lake and the Salmon Creek outlet. This spot seems to be ex-

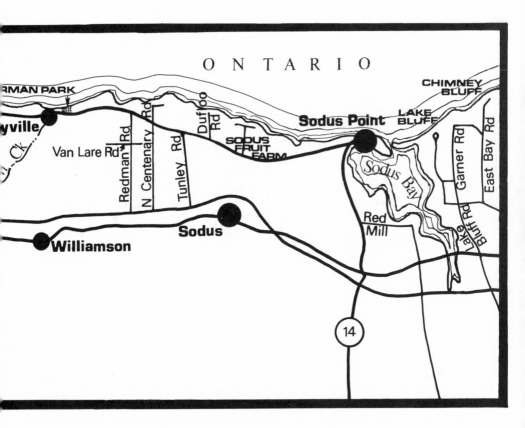

tensively used, and access has never been questioned. Arctic Loons have been seen from this hill in April. Common and Red-throated loons, diving ducks, and scoters frequently overwinter here. After scrutinizing the lake, return to the intersection and make a hard left turn onto Lake Road. There is a gas station and small grocery store at this point, and the owners allow use of the comfort facilities. By the way, there will be no other gas stations before Sodus Point.

Follow Lake Road to another overlook at 0.2 mile; 0.7 mile farther is the entrance to Forman Park. Public access to the lake is available here, but due to winter conditions, it may be closed to traffic. Proceed east to Redman Road, 2.2 miles farther. Turn right on Redman, which is a gravel road. Enter a woods after 0.3 miles. In spring this can be excellent for warblers. Ruffed Grouse and Screech Owls nest here. Pileated Woodpeckers are quite common. Proceed to the first crossroad, which is Van Lare Road.

Turn around and return to Lake Road. Go along Lake Road, heading east, another 0.8 mile to Centenary Road. A fieldstone building is on one corner. Here turn left and drive 0.7 mile to the lake. This spot is a regular stop on the mid-winter

Waterfowl Census route, so scan for any that may be present. Return again to Lake Road and head east (left) 1.8 miles to gravel-surfaced Dufloo Road. Turn left onto it and drive north 0.5 mile to a cobblestone house with a rail fence on the right. The residents usually maintain a very active feeding station all winter. Check it out. Continue 0.1 mile to a sign reading "Private Road." This road is usually snowbound in winter, so the remaining 0.1 mile to the lake and a small cove must be walked. Birders are welcome on all of these roads.

Again return to Lake Road and proceed east 1.2 miles to Sodus Fruit Farm. Turn left on an oiled gravel road. Advance 0.1 mile to a large ivy-covered white stucco house. Feeders near the house in winter host a representative sampling of winter resident birds. Evergreens at the front of the building provide excellent cover. This is a private road, but the manager, Mr. Putnam, has given permission to birders to enter. Follow the S-curve drive past the barns and outbuildings, and proceed north through the apple orchard 0.7 mile to a creek crossing. Ruffed Grouse, Saw-whet Owls, Mockingbirds, and Hermit Thrushes have all been found here. Approximately 0.3 mile farther, the road ends in a tee intersection. Turn left, and check the lake at various points along the way. If the road is open, drive 0.4 mile to the cement block pump house on the right that affords an excellent view of the lake and a vantage point to the west. Look for Gadwalls and other waterfowl. Then retrace the route to Lake Road, proceeding straight ahead past the barns.

Go east on Lake Road to the village of Sodus Point. Lake Road ends at a cross-street at a stop sign 3.6 miles farther. Proceed straight ahead on Bay Street, and bird the loop around Sodus Point on Gregg Street, a total distance of 1.1 miles. In winter and early spring, stop to look for ducks at any available vantage. Many of the residences are occupied in the summer only, so in the off-season, it is possible to bird the periphery of the yards.

Retrace the route to Bay Street Extension, where there is a Gulf fuel station on the northwest corner. Turn right, and at 0.1 mile turn right again onto Wickham Boulevard, following bay shore 0.3 mile to Sodus Point Park parking lot, Coast Guard Station, and lighthouse. Check the bay, the lake, and the channel between the piers for waterfowl and gulls. The rocks across the channel should be scanned for Purple Sandpipers even late into December. There is a drafty portable toilet here whose main virtue is its availability.

Retrace the route to the Gulf station and turn right at Bay Street. Drive 0.4 mile to the intersection where NY 14 begins. Turn left (south) and follow NY 14. At the Trestle Landing Marina, 0.4 mile farther, there may be open water. The road borders the bay for a short distance with a marsh on the right. Check any likely spots. Follow NY 14 for 2.8 miles to Red Mill Road (left). Turn left. Stop at 0.3 mile to check out the ravine of a small stream; go 0.8 mile farther where Mary Drive offers access to the southwest shore of Sodus Bay. An additional 0.2 mile beyond Mary Drive, Sodus Bay is on the left and a marsh is on the right. This is a great place to stop and reconnoiter. Proceed up the hill on Red Mill Road, a nice stretch for observing the hawk flights in spring. Hunters Point Road veers off to the left 0.4 mile farther and follows the bayshore, offering several good bay views. Return to the intersection. Red Mill Road becomes Shaker Tract Road. Turn left and proceed south. In April, when southwest winds are blowing, any point along the 1.4 miles stretch of Shaker Tract Road provides first-rate views of migrating hawks. Look carefully to determine the line of flight. A count of sixty-three Ospreys, two Golden Eagles, and hundreds of Broad-winged Hawks was made on one May 2 from this area. Once the site of a Shaker community, Shaker Heights is easily identifiable by its water tower.

Reaching Ridge Road turn left (east); go 0.9 mile to the bridge crossing the south end of the bay. Stop at access points on the west side of the bay. Advance another 1.1 miles to Lake Bluff Road. Turn north and go through an attractive woods 1.7 miles to the intersection with Lummisville Road. Turn right (east), and proceed 0.8 mile to East Bay Road. Turn left (north), and pass through orchards. After another 1.4 miles find a pond on the left, which is productive for ducks and shorebirds in spring and fall. Only 0.1 mile and 0.3 mile farther the Department of Environmental Conservation has provided lookouts. Approximately 0.5 mile farther Chimney Bluff Road leads right (north). Check the marshes here in spring and fall.

Advance 1.0 mile, where the road ends at the lake and the Chimney Bluffs. These headlands, formed by the natural processes of erosion and deposition, are scenic projections with bold, almost perpendicular fronts. Retrace the route 1.0 mile to Garner Road. Turn right (west), following it as it makes a sharp curve southward. Lake Bluff Circle veers right after 2.8 miles. Follow it. Make another sharp right, and return to Sodus Bay (on the left).

At 1.3 miles from the turn onto Lake Bluff Circle, the LeRoy Island Circle Road goes off to the left. Explore this before continuing on Lake Bluff Circle. Ignore the "Dead End" sign, and follow the road to the Lake Bluff. Erosion has caused large portions of the bluff to fall into the lake. Some of the cottages are not occupied because they are also in danger of plummeting into the lake. Park, being careful not to block the access. Walk to the edge of the bluff, a superb place in autumn to watch for Brants, scoters, and jaegers. It can be awfully cold, so seek some protection from the wind while birding.

Retrace the route 1.8 miles. Take the first right, which is Sloop Landing Road. After 0.7 mile this road dead ends at Lummisville Road. Turn left. Drive 0.3 mile back to Lake Bluff Road, where the tour began. Having completed a generally circular route, turn right and return to Ridge Road. Turn right onto Ridge Road (NY 104) to return to Rochester.

ACKNOWLEDGMENTS

The author wishes to acknowledge and thank the following people for the detailed and well organized information they supplied for the writing of the Genesee chapter: Gordon M. Meade, Robert G. Spahn, Warren Lloyd, Chip Perrigo, Mary Ann Sunderlin, Michael E. Carlson, Roger H. Clark, John W. Foster, Robert G. McKinney, Mr. and Mrs. Joseph E. McNett, Barb Molyneaux, Laura Wyand Moon, Richard T. O'Hara, Jeanne Skelly, William D. Symonds, Leo Tanghe, Stephen F. Taylor, Thomas E. Tetlow.

I would especially like to thank Gorden Meade, to whom I am particularly indebted for his unflagging patience and generous assistance in every aspect of the preparation of the manuscript, for his critiques of early drafts,and especially for his wise counsel and cheerful encouragement throughout the writing of this book.

Finger Lakes — Region 3

THE FINGER LAKES REGION comprises all of Chemung, Schuyler, Seneca, Steuben, Tompkins, and Yates counties, the southern two-thirds of Cayuga County, and the northeastern section of Ontario County. Of the eleven Finger Lakes gouged out during the last ice age, five lie within Region 3. These are, in order of largest to smallest, Cayuga Lake (38 miles long, 1 to 3.5 miles wide), Seneca Lake (35 miles long, 1 to 3 miles wide), Keuka Lake (21 miles long, 1 to 2 miles wide), Canandaigua Lake (16 miles long, 1 to 1.5 miles wide), and Owasco Lake (11 miles long, 1 to 1.5 miles wide). The maximum depth of these lakes ranges from Owasco Lake's 177 feet to Cayuga and Seneca lakes' 435 feet and 618 feet, respectively). Lakes of these depths are slow to freeze over, and therefore afford open water for large numbers of waterfowl in most winters (Map 3.1).

Stream-pattern evidence reveals that the retreat of the last glacial ice cap was arrested at the valleys south of the lakes for an unusually long time. The glacial moraine deposited then now functions as a barrier to the southward draining of the lakes. This moraine constitutes the divide between waters draining into Lake Ontario to the north, and waters draining into the Susquehanna system to the south. The famous Taughannock Falls, located several miles northwest of Ithaca, is the highest waterfall east of the Rocky Mountains. A true hanging-valley waterfall, it should not be missed, both for its beauty and for its birdlife. The landscape throughout the region is certainly some of the most picturesque in the country, with high rolling hills, wild and lush swamps and mucklands, cultivated farm lands and grape vineyards, and large forested sections of oak, beech, butternut, ash, maple, elm, poplar, basswood, white pine, and hemlock. The basic rock of the region is secondary slate overlayed with gypsum and limestone. Gravel hills scattered over the plain contain large segments of limestone, so extensive quarries have been developed throughout the region.

An important feature of the Finger Lakes country is the many stream-cut glens (ravines), incised into the hillsides rising up on the sides of the lakes. The cool microclimates are conducive to the growth of hemlock, walking fern, etc., and such breeding birds as Louisiana Waterthrushes. These glens are great birding spots, and their predominantly limestone walls are, incidentally, a fossil hunter's paradise.

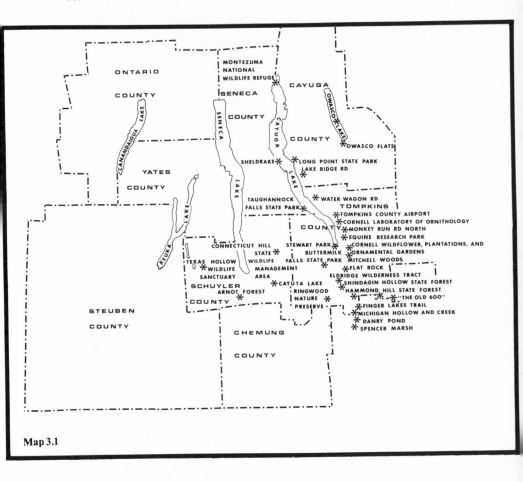

Map 3.1

Most of the birding sites described here lie in Tompkins, Cayuga, Schuyler, and Seneca counties and represent a wealth of diverse habitats; the birder who visits these sites will find they offer an opportunity to explore systematically the area's representative species and its specialties.

SENECA COUNTY
Montezuma National Wildlife Refuge (Maps 3.2 and 3.3)

Rating: Spring****, Summer****, Autumn****, Winter*

Too often the pressure to drain marshlands for agricultural or real estate development, has led to large expenditures of money, from both public and private sectors, and created areas that are effectively wastelands. Drainage has too frequently yielded land infertile for agriculture and unusable by wildlife. In some areas, such drainage has lowered the water table, dried up springs, and increased the danger of forest fires. In short, because for too long we acted on the erroneous idea that all drainage was beneficial, today a good marsh is hard to find. Few and far between are the great green profusions of wind-blown cattails, blue irises, pickerel weed, and swamp loosestrife with herons and bitterns frogging on the creek edge and Virginia Rails calling in the marsh. Truly, the natural beauty of an extensive, wild, undisturbed marsh has few parallels.

With the disappearance of vast marsh areas, the wealth of wildlife resident in the Montezuma National Wildlife Refuge is a precious boon to the naturalist. The birder will find it a bountiful storehouse of avian lures. The refuge of 6334 acres lies wholly within Seneca County at the northern tip of Cayuga Lake. It is directly on the Atlantic Flyway, a major waterfowl migration route. Due to the considerable range in the depth and vegetative covering of the refuge marshes, a wide variety of birdlife is represented. *Chara* species, pondweeds, and duckweeds grow abundantly in the wetlands and provide an excellent food source for migrating birds. In addition, several hundred acres of the refuge are planted in winter wheat and corn, and the large amounts of waste grain accommodate thousands of feeding geese and ducks.

At the peak migration periods during mid-March and mid-October more than 100,000 ducks and geese of some 25 species put down inside the refuge to rest and refuel. Of course, many of those species find the security and habitat they require and remain to nest. The refuge bird list boasts some 240 species, and about 110 of those breed there. Montezuma is carefully managed for maximum waterfowl production, which does not always mean maximum shorebird habitat; however, every year in late summer or early fall, one of the pools is drained. The resulting mud flats attract varying numbers of migrating shorebirds of at least 30 species. Herons, hawks, rails, terns, owls, woodpeckers, and a minimum of 60 passerine species nest on the refuge grounds. Many more are seen in transit.

The area is also rich in history. From the northern boundary of the refuge traveling east to west, one is able to cross the enlarged Erie Canal, the original Clinton's Ditch, and the present Barge Canal all within one mile.

The refuge itself contributes seasonally to the ornithological history of New York State—it is one of the sites where young Bald Eagles have been re-introduced ("hacked-back") into the wild. The ultimate goal of the program, conducted under the joint auspices of the U.S. Fish and Wildlife Service, the New York State Department of Environmental Conservation, and the Cornell University Laboratory of Ornithology, is to re-establish a breeding population of Bald Eagles in north-central New York State (see Oak Orchard WMA, region 1).

More than ten miles of driving and hiking trails allow the birder and photographer excellent views. Two fine observation towers overlook the various marshes. Several of the hiking trails are closed during breeding season to prevent disturbance of parents and young. Mosquitoes breed in the marshes, just as they do in ordinary woodlands, so the birder should remember to carry along insect repellent. A telescope will also greatly enhance the birder's viewing pleasure.

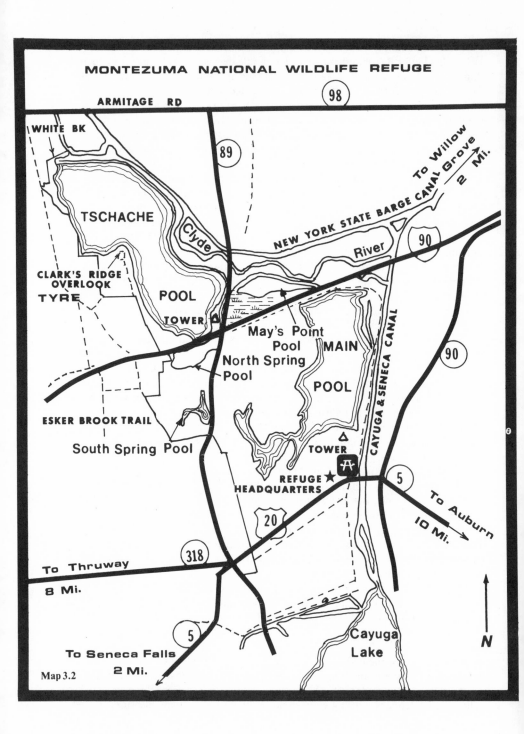

MONTEZUMA NATIONAL WILDLIFE REFUGE

ARMITAGE RD   98

WHITE BK

89

To Willow Grove 2 Mi.

TSCHACHE

Clyde

NEW YORK STATE BARGE CANAL

River

90

CLARK'S RIDGE OVERLOOK

POOL

TYRE

TOWER

May's Point Pool

North Spring Pool

MAIN

POOL

90

CAYUGA & SENECA CANAL

ESKER BROOK TRAIL

South Spring Pool

TOWER

REFUGE HEADQUARTERS

5

To Auburn

10 Mi.

20

To Thruway

318

8 Mi.

5

To Seneca Falls

2 Mi.

Cayuga Lake

N

Map 3.2

Remember while birding to scrutinize closely any bird that appears "different." Montezuma has been the site of such rare occurrences as White Pelican (in 1961 and 1979), Cinnamon Teal (in 1974), Sandhill Crane (in 1948, 1968, 1978), Ruff (in 1973, 1975, 1979), Marbled Godwit (in 1966, 1970, 1979), and American Avocet (in 1953, 1955, 1966).

To reach the refuge, leave I-90 (the New York State Thruway) at Exit 41. Turn right on NY 414 and proceed south a very short distance to NY 318. Here turn left (east) and proceed to the junction of NY 318 and NY 5 and 20. Turn left (northeast) and proceed approximately 1.6 miles to the entrance, which will be on the left (north).

After stopping at the Display Pool on the right to check for herons, bitterns, waterfowl, rails, and shorebirds, proceed immediately to the visitor's information booth at the picnic area (see accompanying map), for an auto tour route map, interpretive brochure, refuge bird checklist, and other helpful materials. *Be sure* to consult the log of birds seen daily for any rarities that may be currently present.

After leaving the information booth, check out the cottonwood trees and the shrubs in the picnic area; this small area often harbors raptors and small passerines. Just beyond the picnic area, climb the observation tower, which affords a terrific overview of the Main Pool. Herons, geese, and ducks will be visible in impressive numbers. Marsh Hawks will be coursing low out over the marsh. If the water level is down, scope carefully for shorebirds. In summer Black Terns can be spotted everywhere.

Follow the tour route along the gravel road. This road follows alongside the Main Pool Dike, and terminates at the Tschache Pool Tower. Driving along this dike, which parallels the Cayuga and Seneca Canal, look for herons, bitterns, rails, gallinules, and coots on the left. If the water level is low, the section of the dike drive paralleling the Thruway may also reveal rails and shorebirds. On the right, in the wet, grassy sections, listen for Long-billed Marsh Wrens and in the cattails, for various sparrow species. Overhead you should see several swallow species in good numbers. In autumn, Cliff Swallows, which no longer nest in the Cayuga Lake Basin, pass through.

Approaching NY 89, be sure to look for ducks and shorebirds in the open shallow pond area, in the cattail marsh to your left. If the water is low, you may see rails. At the macadam road (NY 89), turn right, cross the bridge over the New York State Thruway, and proceed to the Tschache (formerly Storage) Pool, on the left. May's Point Pool is on the right. The Tschache Pool Tower (left) is a good place to view both of the pools. The Tschache Pool Dike and May's Point Pool Dike are closed to traffic from April 1 through July 1. After July 1, hiking is allowed on these dikes, which are easy, level, gravelled roads. Both of the pedestrian dikes are dead-end trails; the May's Point Pool Dike is about two miles long and the Tschache Pool Dike is about four miles long—and it is just as far (or it may seem a little farther) coming back.

Leaving the tower area, turn around and proceed south across the Thruway bridge again. On your right will be North Spring Pool. In 1953 this area was flooded with the goal of developing more marsh. In the process trees were killed, which has provided excellent nesting habitat for Wood Ducks, American Kestrels, Common Flickers, Pileated, Red-bellied, and Red-headed woodpeckers, and probably numerous chickadees, nuthatches, and wrens.

Proceed approximately 0.5 mile to an overlook on your left. This is located on top of a knoll across the road from the northern tip of the South Spring Pool. From this point large numbers of waterfowl can be seen on the marsh. Just to the south of the overlook is a stand of conifers. In winter Short-eared Owls roost in these conifers. Walk across the road and check out the foliage on the west side of NY 89 for small land birds.

Proceed a quarter-mile south and turn right onto a dirt road. Park and walk the short trail skirting the southern shore of South Spring Pool. This hike can be rewarding at the right season and has often been the site of an unusual find.

Proceed about a half-mile to East Tyre Road. Turn right. Proceed another half-mile to a small parking area and the entrance to Esker Brook Trail on the right. This trail, a narrow path through the woods along Esker Brook on the west side of the refuge, is 1.4 miles long. It can be excellent for viewing flycatchers, thrushes, kinglets, vireos, warblers, and sparrows during both spring and fall migrations.

CAYUGA COUNTY
Owasco Lake and Owasco Flats (Map 3.3)

Rating: Spring***, Summer**, Autumn***, Winter**

Here, about fourteen miles south of Auburn, is an area of some 300 acres which holds all the promise of virgin territory. To date, although the area has been (in 1981) relatively little birded, the reports received from those die-hards who bushwhack through its rougher spots and carefully bird its easier ones have been glowing. This spot is included as a potential hit.

Leave I-90 (the New York State Thruway) at Exit 40. Proceed south on NY 34 to the center of Auburn. In Auburn, NY 34 becomes South Street. From the intersection of NY 34 and NY 20 (Genesee Street), proceed south to Swift Street. Turn left. Proceed east four blocks. At Lake Avenue (NY 38) turn right. Proceed south on NY 38 (Lake Street) to Owasco Lake.

At the north end of the lake, go around the traffic circle, exiting on the third street. Cross the outlet bridge, and take the first right immediately past the bridge. This road parallels the outlet and extends south toward the lake. Park and walk out on the breakwater. This is an excellent place for waterfowl in fall, spring, and early winter. Gulls can be viewed here before the lake freezes in winter. Be sure to check the outlet itself for close views of bitterns and ducks. Return to the traffic circle, again exiting at the third street, which is NY 38. Regain it and head south. For the first few miles the road parallels the lake very closely with a few wide pull-offs, from which the birder can get good vantages of the lake. At the south end of the lake, after about twelve miles (clocked from the traffic circle) turn left toward the lake in the village of Cascade. Park at the marina and scan the lake again for

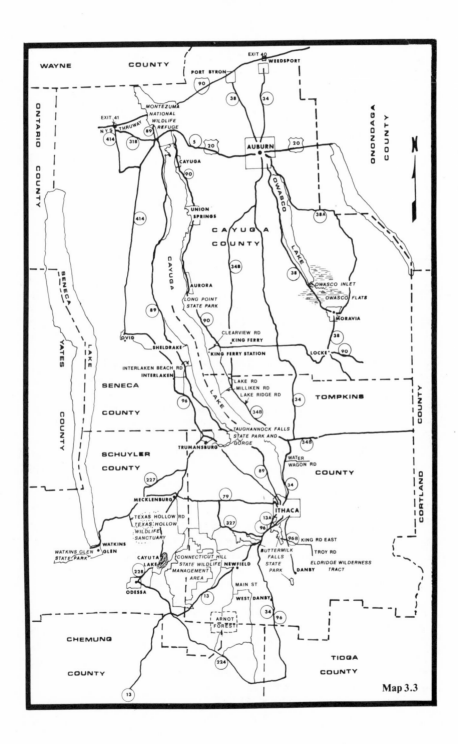

Map 3.3

water birds. From here one may walk (recommended) or drive (entirely at your own risk), south on the dirt road before you. This road leads a small distance through a wooded swamp to the Owasco Inlet. This wooded area has been very productive in spring for migrating passerines. Wood Ducks breed in the flooded areas of the woods. From here you can, if you are in the mood for some bushwhacking, make your way south along the inlet bank and on and along the abandoned Lehigh Valley railroad tracks. In the spring this area is wonderful for water and land birds.

Approximately 1.7 miles south of Cascade there is an excellent overlook of the Owasco Flats and adjacent wetlands from NY 38. This overlook is just before an old barn bearing the sign "Trudy's Ranch." From here American Wigeons, Pintails, and Northern Shovelers, as well as rails, gallinules, and several shorebird species can be seen on the flats.

Another two miles south of the overlook (approximately 3.7 miles south of Cascade) brings the birder to the south end of the Owasco Flats, through which the Owasco Inlet flows. To walk through the flats, a fun-filled but not-so-easy feat, locate the slaughterhouse on NY 38. Ask permission to park in the lot and walk to the railroad tracks east of the slaughterhouse. Once on the tracks walk north on them through the cattails of the fields, stream, and wooded swamp habitat on both sides of the tracks. Walking up the tracks the birder will be passing through property mainly owned by the city of Moravia or the city of Auburn or over no longer used railroad right-of-way property. Farther back from the tracks, along some stretches, there is private property which should not be trespassed upon. The birder should find the two-and-a-half to three-mile hike rich in birds. Listen for the distinctive calls really only commonly heard in wooded swamps with marshy borders. Herons in numbers and Least and American bitterns should be readily sighted. Wood Ducks, Green-winged Teals, Mallards, and Black Ducks should breed along the inlet. This indeed should be a breeding and stopping-over place for rails, gallinules, coots, and hosts of shorebirds on migration. The *Kingbird* Region 3 editor welcomes reports of birds sighted in the Owasco Lake and Owasco Flats areas along with accompanying photographs, if possible.

TOMPKINS COUNTY
The Laboratory of Ornithology (Map 3.4)
(Sapsucker Woods Sanctuary and Lyman K. Stuart Observatory)

Rating: Spring****, Summer***, Autumn***, Winter**

In 1927, at a memorial service for Louis Agassiz Fuertes, Frank M. Chapman said, "If the birds of the world had met to select a human being who could best express to mankind the beauty and charm of their forms, their songs, their rhythmic flight, their manners for the heart's delight, they would unquestionably have chosen Louis Fuertes." For anyone familiar with Fuertes' work, even this warm sentiment and honest appraisal understates the genius of the great artist.

The renowned ornithologist Arthur A. Allen was professor of ornithology at Cornell University for forty-two years. In the more than fifty-five years of his active career, Allen taught several thousand students the essence of avian biology and pioneered significant programs researching the living birds. Allen and Fuertes were together when the latter gave the name Sapsucker Woods to the inviting woodland they so often birded together.

In the early 1950s, Lyman K. Stuart, a friend and informal student of Dr. Allen, purchased 130 acres of the tract and donated it to Cornell to express his appreciation for Allen's unstinting instruction and advice. Very soon thereafter, another fifty acres were purchased and donated. Arthur Lane later gave the laboratory five acres and his home, where he and his wife Edna had observed and fed birds for many years.

At last, Allen and Peter Paul Kellogg, an ornithologist who was Allen's colleague and distinguished in the fields of electronics and acoustics, could realize the dream of developing this enticing area into a bird sanctuary and research center. A 1,500-foot dike was constructed, flooding a wet meadow and small woodland and creating a ten-acre pond on the west side of the sanctuary. The Lyman K. Stuart Observatory, the first research laboratory for the study of living birds, was built in 1957. In it are housed the administrative offices of the Laboratory of Ornithology; a very good ornithological library containing the major ornithological journals and essential reference works in the field, a small book and gift shop, which is well-stocked with books, prints, cassette and disc recordings, and other nature-oriented products; the headquarters of the long-established Nest Record Card Program and the newer Colonial Bird Register; the Library of Natural Sounds; an observatory room; photographic rooms; and the Brewster Wing.

The laboratory's famous Library of Natural Sounds originated through the energy, imagination, and perseverance of Peter Paul Kellogg. In 1981, as a cooperative and on-going major project of the laboratory, more than 30,000 individual bird vocalizations of more than 3,000 species have been edited, cataloged, and credited. This impressive repository is available to scholars worldwide and is the source for the well-known Cornell University series of bird records.

To reach the Laboratory of Ornithology, leave I-90 (the New York State Thruway) at Exit 36 (Syracuse). Take I-81 south to Exit 11 (Cortland). Here leave I-81 and gain NY 13 headed south and west. Proceed along NY 13 through Dryden. After approximately 5.7 miles more, NY 13 turns right and crosses Fall Creek. At the next junction, NY 13 intersects Lower Creek Road. Here, turn left (west) and follow Lower Creek Road, continuing as it turns and becomes Hanshaw Road, for about 1.2 miles to Sapsucker Woods Road, which comes in on the right. Turn right and proceed 0.9 mile to the parking lot of the observatory.

Sapsucker Woods Road divides the sanctuary into east and west sections of approximately equal size. Across the road from the Stuart Observatory is a large facility known as "the hawk barn." In fact, the building is a research facility. All of the work done here is devoted to the breeding and raising of captive Peregrine Falcons and other raptorial birds, with the hope of eventually restocking Peregrines in the natural areas from which the bird has disappeared as a breeder. This has been one of the laboratory's major scientific research programs since 1970; the work seems to be progressing steadily and has even experienced some success. In

1980, two pairs of Peregrines released under this program nested in the coastal marshes of New Jersey and successfully fledged young. This plant may not be visited without making prior arrangements through the Laboratory of Ornithology.

Visit the Lyman K. Stuart Observatory, the north and south corridors and the Brewster Wing before walking the trails of Sapsucker Woods Sanctuary. In the entrance hall note the poignant exhibit of the probably extinct Ivory-billed Woodpecker. The huge picture windows along the west wall of the observatory overlook a ten-acre pond, a source of food and shelter for several waterfowl species during spring and autumn migrations and a permanent residence for others. As you study the scene outside the windows, you will be treated to a range of natural sounds, amplified by a stereophonic address system, either being picked up by microphones just outside the building under its eaves, or being played from one of the records produced by the laboratory. Through the windows you should be able to spot Great Blue and Green herons, geese, Wood Ducks, Mallards, Black Ducks, transient waterfowl, Killdeers, Spotted Sandpipers and other shorebirds during migration, Ring-billed Gulls, Belted Kingfishers, and numerous Tree Swallows hawking insects or perching on the wires around the pond. Look out the south window toward the feeding stations. Here one might expect to see small land birds in all seasons as well as Mourning Doves, Hairy and Downy woodpeckers, jays, chickadees, titmice, nuthatches, blackbirds, finches, and sparrows.

In the north and south corridors, rotating displays of the representative works of successful artists and photographers and beautifully illustrated monographs are exhibited. Be sure to walk down the corridor leading to the Brewster Wing. This corridor serves as a gallery for some of the L. A. Fuertes artworks the laboratory owns, as well as a bit of Fuertes memorabilia. This revered artist was an Ithacan and Cornellian. Go into the Fuertes Room, which is both an auditorium and a gallery displaying some of Fuertes' finest paintings as well as a special collection of books he illustrated.

After all of this indoor observation go outdoors into Sapsucker Woods Sanctuary. Louis Fuertes christened Sapsucker Woods one morning early in June, 1909, when he, A. A. Allen and two students discovered there the nest of a Yellow-bellied Sapsucker, thereby establishing the first breeding record for Ithaca and the Cayuga Lake Basin. Walk along some of the four miles of trails, noting the rich and varied flora and watching especially for those natural species having boreal affinities breeding alongside those having southern affinities. For example, note painted trilliums and hepaticas thriving next to graceful white trilliums. These will probably be interspersed with Ostrich, Cinnamon, Interrupted, Christmas, or Royal ferns. Walk carefully along looking for some of the more than 200 bird species recorded within Sapsucker Woods. Red-shouldered and Red-tailed hawks, Ruffed Grouse, American Woodcocks, Great Horned and Barred owls, Pileated Woodpeckers, Yellow-bellied Sapsuckers, Brown Creepers, Wood and Hermit thrushes, Veeries, and Northern Waterthrushes are but a few of the species breeding on the sanctuary grounds. Throughout this moist woodland are stands of beech, maples, hemlocks, and birches. Along the Dike Trail, on the west side of the sanctuary, there is a rich, wet bog that invites close observation. Numerous boardwalks bridge the wettest places along the trails, and several roofed shelters keep birders from being drenched by one of the many surprise showers so frequent in the Ithaca area.

The membership-supported Laboratory of Ornithology enjoys an honored place in the ornithological world, owing to its continuing contributions of educational programs, scientific research, and public relations. Every New York State birder should take time to visit and learn firsthand why the Laboratory of Ornithology is so affectionately regarded throughout the nation and world. You will be welcomed as a visitor, but better yet, join and support the laboratory's worthy programs as a member.

Tompkins County Airport Vicinity (Map 3.4)

Rating: Spring****, Summer***, Autumn***, Winter**

The Tompkins County Airport properties, which cover about one square mile, offer productive birding. There are several shallow ponds, brushy marshes, and short-grass upland and overgrown fields, providing diversified habitat and correspondingly diversified birdlife. *Please note:* because some areas of the airport are posted for security, birders *must* telephone the airport manager before they visit for permission to bird these areas. Every birder is obligated to respect this ruling or risk losing for all birders the present good will and indulgence of the airport management and personnel.

The ponds in the vicinity harbor Pied-billed Grebes, Green Herons, dabbling ducks, rails, and shorebirds in spring and fall. The single possible but undocumented record for the Black Rail in the Cayuga Lake Basin was obtained at the airport pond. American Woodcocks and Common Snipes can be viewed in their courtship rituals near dusk over the fields and ponds every early spring (April-May). Marsh Hawks may be seen coursing over the fields hunting prey, and in winter this is an excellent site for Rough-legged Hawks, Northern Shrikes, Pine Grosbeaks, and Common Redpolls. Search out all of the short-grass and overgrown fields in every season for various sparrow species; in spring and summer, Savannah, Grasshopper, Henslow's, Vesper (on migration only), Field, Swamp, and Song sparrows are present. In addition to the attractions of the fields skirting the airport, the airport grounds usually host a small population of Henslow's Sparrows. Many birders have had most success locating them in the early morning hours during late May and early June. In fall, migrant sparrows come through the area, and in winter Tree Sparrows can be found in the company of Snow Buntings.

To reach the airport, follow the directions to the Laboratory of Ornithology (above) through Dryden and to the point where NY 13 turns right and crosses Fall Creek. Stay on NY 13 as it follows a northwest course. At the first traffic light, look for Warren Road, which is a through road. At Warren Road, turn right (north) and continue 1.1 miles to Snyder Road. Here turn right and park near the junction of Warren Road and Snyder Road. Walk west across Warren Road and bird the ponds and brushy areas surrounding them on that side. This is a good place for Willow and Alder flycatchers and Yellow-breasted Chats in late spring and summer in addition to those species previously mentioned. After leaving this section drive along Snyder Road, being careful to bird the fields and their edges. Check for Henslow's and Grasshopper sparrows.

Only if you have obtained permission, turn right on Mohawk Road and proceed south to Etna Road, which comes in on the left. Park and walk south about 200 yards, where you note the entrance to a dirt path on the right just before a hedgerow. This path leads right to a shallow pond and provides relatively easy access into an otherwise dense area. Sometimes marsh dwellers can be viewed from the road. This has proven to be a highly productive pond, so give it a little time. Along the road and path rails are present in the marshy edges of the pond, and Willow and Alder flycatchers are fairly common. When through here, walk back (north) along Mohawk Road to your car. Here turn right (east) and bird along the road near the several small overgrown ponds and marshes to the left and right off Etna Road. Yellow-breasted Chats may be present.

From here it is a simple matter to get to the next birding site, and it should be visited—so budget enough time.

Monkey Run Road, North End (Map 3.4)

Rating: Spring****, Summer****, Autumn***, Winter***

This area is one of the popular birding sites around Ithaca. Birding can only be done here on foot; the habitats include wooded deciduous slopes, brushy and cultivated fields, spruce and pine plantations, and creekside.

To reach Monkey Run North, proceed southeast on Etna Road from the junction of Etna and Mohawk Roads (above) to its junction with Hanshaw Road. Turn right and proceed south. Cross NY 13 and continue south to the junction of Hanshaw Road and Lower Creek Road. Park here off the road and walk south on the abandoned road before you, which is Monkey Run Road. Spring migrants can be seen from the steep banks along Fall Creek. You should be able to locate nesting Ruffed Grouse, Pileated Woodpeckers, Red-breasted Nuthatches, Brown Creepers, Golden-crowned Kinglets, and Blue-winged Warblers. Search the mature spruce plantation and all of the brushy areas at the bottom of the hill.

Equine Research Park (Map 3.4)

Rating: Spring****, Summer**, Autumn****, Winter**

In recent years, this area has acquired a new name, but the pond it contains gained fame as Bull Pasture Pond for its breeding migrations of salamanders. The habitats birded here are pond, pasture, brush, golf course lawns, fields, and woods.

To reach the area, follow directions for the Laboratory of Ornithology (above) to the junction of Hanshaw Road and Sapsucker Woods Road. Here do not turn right, but proceed west on Hanshaw Road to its junction with Warren Road. Here turn left (south) and advance to the Cornell golf course road that

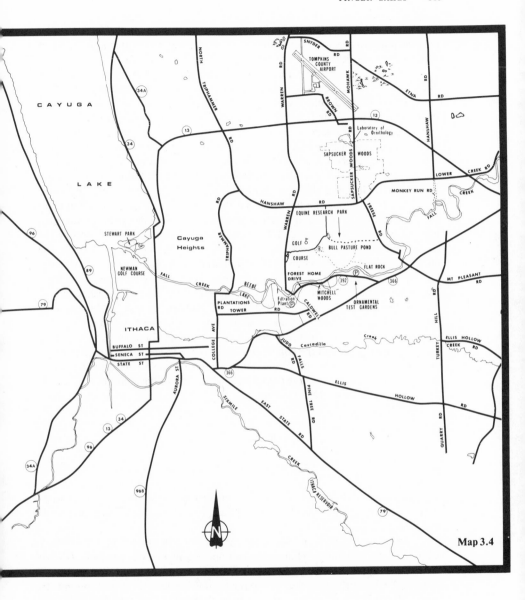

Map 3.4

comes in on the left. Turn left and proceed east through the golf course to the Equine Research Park.

Along the way, note the small pond on the right side of the road just before the road curves northeast. This is the old Bull Pasture Pond, and in addition to its salamanders, rails have been observed. There is a long swampy depression opposite the pond. It marks the course of a former stream. Park on the road shoulder near the pine grove just before reaching the pond. The parking area near the barns may be used, but it is frequently filled, especially on weekends.

Walk northwest along the edge of the field, west of the swampy area. The thickets here often contain late-lingering migrants in early winter and attract various sparrow species during migration. Cedar Waxwings and House Finches feed in the fruiting shrubs in the fall. Walk up the road and check out the pond. Continue up the road to a farm lane on the left just before the barn. Follow this lane along the woods' edge along the northeast side of the swampy depression. Look for Red-bellied Woodpeckers near the barns and Eastern Bluebirds in the open areas, migrating thrushes, vireos, warblers, and sparrows along and in the woods in spring and fall, and Horned Larks and Snow Buntings frequenting the fields in winter.

Walk east along the lane between the paddocks. In spring and fall there should be Horned Larks and Water Pipits here. Continue beyond the oval track looking for such field birds as Bobolinks and Savannah Sparrows. Opposite the east end of the oval track, walk south to the edge of the woods. Duck under the barbed wire fence at the yellow-blazed tree and turn right on the ski trail, which is marked all along its length with yellow blazes. Follow it back to the golf course. Along the way check for resident Pileated Woodpeckers, migrant thrushes, vireos, and warblers, and in winter for Red-breasted Nuthatches and Golden-crowned Kinglets. At the golf course, walk diagonally across the fairway and through the woods and pines to the road. Eastern Wood Pewees are abundant here. One may wish to walk left around the edge of the golf course to check for migrants in the area. A circular path goes down the slope to Fall Creek. Near the creek a path on the left follows the creek to the suspension bridge, providing access to the Cornell Test Gardens and Flat Rock (below).

Cornell Wildflower Garden (Map 3.4)

Rating: Spring****, Summer*, Autumn*, Winter*

This area, developed in the early 1970s, has extensive plantings of native wildflowers. The edge of Fall Creek, a small stream, and a wooded slope provide excellent birding.

To reach the area follow directions for the Laboratory of Ornithology (above) to the junction of Hanshaw Road and Sapsucker Woods Road. *Do not* turn left, but proceed west on Hanshaw Road to its junction with Warren Road. Here turn left (south), pass the road to the Equine Research Park and continue to Forest Home Drive at the tee intersection at the foot of the steep hill. Turn left. Just across the bridge over Fall Creek there is a parking lot on the right near the Cornell filtration plant. Park here.

Walk the path west to the garden. Explore all the paths. In spring this is a stopping place for thrushes, vireos, and warblers. In 1976 a Kentucky Warbler spent a week here. For several winters a Screech Owl roosted in a hole in a sycamore tree near Fall Creek.

Cornell Plantations and Ornamental Test Gardens and Mitchell Woods (Map 3.4)

Rating: Spring****, Summer**, Autumn**, Winter**

One may walk to this area from the Cornell Wildflower Garden by turning right on Caldwell Road at the parking lot entrance, then taking the first road left, which goes east through a nut-tree collection and downhill to the Test Gardens.

By car, when leaving the parking lot for the Cornell Wildflower Garden, proceed straight ahead along Fall Creek on the continuation of Fall Creek Drive about a half-mile and park on the left, opposite the building in the Test Gardens.

In this part of the Cornell University grounds there is an extensive variety of blooming plants and shrubs, along with a collection of hedges. It is worth spending time in the labyrinth of criss-crossing trails and paths to look for migrants and the few more common summer residents. Walk around the periphery of the thickets, and be alert for the rare warbler. One path climbs up the slope on the east side and circles around to the sculpture garden. On the west side two paths go through Mitchell Woods. The steep wooded slopes attract woodland migrants and small passerines. A road through the middle goes uphill and then west around the top of Mitchell Woods. An open field and the brushy edge of the slope attract migrant sparrows in spring and autumn. A notable nut-tree collection lines this road.

Flat Rock and Cornell Plantations (Map 3.4)

Rating: Spring****, Summer***, Autumn****, Winter*

From the parking place for the Test Gardens, walk east on the road and turn left onto the path at the gate. Follow the path toward the creek. A path to the right goes along Fall Creek to the Hemlock Forest, a likely spot to find good numbers of migrant vireos and warblers at the height of the spring migration and kinglets and finches in winter. After birding the forest, retrace your route and turn right at the path to the suspension bridge across Fall Creek. Spotted Sandpipers teeter along the creek edge, and occasionally a migrant Solitary Sandpiper stops along it. Swallows course the creek, and migrating warblers may be found in the trees.

At the north end of the bridge, paths go east and west. To the left the path follows the creek, then circles up the slope to the Cornell Golf Course near Bull Pasture Pond. To the right, trails go along the creek as far as Freese Road and up the hill to a lookout. This area has rich mixed deciduous bottomland forest interspersed with brushy openings and a relatively thick brush understory, with open fields and hedgerows at the eastern end.

In the forest, Ruffed Grouse nest. Be watchful for flycatchers, wrens, thrushes (including migrant Swainson's Thrushes), kinglets, vireos, migrant warblers, Scarlet Tanagers, Indigo Buntings, and sparrows.

This is an excellent birding site, and the fortunate birder could catch the Flat Rock area when big warbler waves are coming through — which is always a treat.

Ferguson Road, Ringwood Nature Preserve, and Ellis Hollow Creek Road (Map 3.5)

Rating: Spring****, Summer***, Autumn**, Winter*

The best time to bird this tour is in spring. Begin early in the morning and spend several hours enjoying the open fields and pastures, the climax forest of northern hardwoods, and finally the brushy fields and marshy glades. The habitat types are varied enough on this tour to encounter a whole range of species.

Leave I-90 (the New York State Thruway) at Exit 36 (Syracuse). Take I-81 south to Exit 11 (Cortland). Here leave I-81 and take NY 13, headed southwest. Proceed along NY 13 to the center of Dryden. Pass through the four-way intersection (at the traffic light), where NY 13 (North Street) turns right (west), continue south on NY 38 (South Street). Cross Virgil Creek. Ferguson Road comes in on the right. Turn right (west) onto it.

Proceed west on Ferguson Road for about three-quarters of a mile. Now begin looking on both sides of the road, as you drive slowly along, for Upland Sandpipers, which nest in the open fields adjacent to both sides of the road. Check out the fence posts and field edges. Look along the woods edge to the south (left) for Turkeys, which also nest in the area. Drive west at a leisurely pace, making sure to stop and carefully bird where the road crosses Willow Glen Creek to the junction of Irish Settlement Road. This is an excellent site for Bobolinks in spring and Horned Larks and Snow Buntings in winter. Continue another 0.8 mile to Yellow Barn Road, birding all along the way. At Yellow Barn Road, turn left (south). At approximately 0.6 mile the road curves right. Here get out and check the deep gorge on the left (east) for nesting Louisiana Waterthrushes. Listen carefully for the familiar gobbling of the polygamous tom Turkey, calling and strutting as part of his elaborate display to attract a harem of females.

Return north up Yellow Barn Road about one mile to NY 13. Turn left (west) and proceed approximately 0.6 mile to Ringwood Road, which comes in on the left (south). Turn left here and travel south on Ringwood Road 2.4 miles.

The Ringwood Nature Preserve is owned and maintained by Cornell University. The forest preserve contains beech, oak, birch, and hemlock. Its several swampy areas are surrounded by large stands of oak and beech. One of the typical hardwood forests near Ithaca, it also harbors bird species of that habitat. During spring migration, this area can be very productive. In summer many species nest within the preserve. Look especially for Ruffed Grouse, thrushes, including Hermit Thrushes, several vireo species, and of course, northern warblers.

There is a sign designating the preserve and birders may park on Ringwood Road near the sign. Walk along the trails on both sides of the road to bird the forest preserve thoroughly.

Proceed south along Ringwood Road to its junction with Ellis Hollow Creek Road. Here turn right (northwest). Bird both sides of this road as you drive west for the next half-mile. Although much of Ellis Hollow Creek Road has been developed within the last several years, this end is still wild and usually full of birds in spring and summer. After another half-mile, just before the road takes a definite left turn, stop to bird the area where a seasonal creek crosses the road. Bird along the next half-mile, where you will pass through open and brushy fields and woods. Look in the brushy areas for Golden-winged and Blue-winged warblers. The old Deer Ridge Farm, east of the pumping station, and its accompanying buildings is kept in a wild condition and is one of the finest places along this stretch to bird. This is private property, so bird only along the road.

Hammond Hill Road, Hammond Hill State Forest and "The Old 600", a Tour (Map 3.5)

Rating: Spring****, Summer***, Autumn***, Winter**

This area is again a favorite haunt of Ithaca birders, who visit it most often in spring and summer. The topography of these areas is a mixture of hills and valleys interspersed with ravines. The habitat is varied, with representative sections of pastures and open fields, transitional brush, hardwood hemlocks, and some conifer plantations.

One of the most delightful aspects of birding this area, especially "The Old 600," is the diversity of ferns on these tracts. Two of the more unusual species are Long Beech Fern, *Dryopteris phegopteris,* and Walking Fern, *Camptosorus rhizophyllus,* and one can sometimes find various wood-fern species hybrids (*Dryopteris* spp.).

On my last visit, I discovered that many of the native plants growing beneath the tall tree canopy have very interesting root systems, which are most easily observed as soon as the first full warmth of the spring season arrives. The fantastic shapes of the roots and tubers — tentacled, hairy, lacy, and marvelously intricate — seem straight out of some science fiction fantasy.

To reach this area, follow the directions for Ferguson Road as far as Ferguson Road. Proceed west on Ferguson Road to the first road coming in on the left (from the south). This will be Irish Settlement Road. Turn left onto it. Proceed slowly south for 1.9 miles, noting Beam Hill Road, which will have joined Irish Settlement Road on the left. At the next road, turn left. This dirt road is named Hammond Hill Road. It runs east a short distance, veers southeast, then straightens out, headed due east. Birders unfamiliar with this area will be amazed by its wildness and quiet serenity. Cross a tiny creek, which is an offshoot of Six Mile Creek, and find a place to park where you will not be blocking the road. You'll know you are in the right place if you are able to locate several foundations from an old farm on both sides of the road. From here to the stream, along all the spring-fed boggy runs, but primarily in the open areas, look for American Woodcocks performing courtship displays toward dusk, when they become much more active. The woodcock feeds on earthworms or angleworms, which it ravenously devours, mainly at night. Walk down to the ravine of the creek offshoot north of the road. The forest association here is northern — hemlock, birch, beech, and maple woods of various ages and heights. This, of course, attracts birds with a northern habitat preference; here the birder should find Yellow-bellied Sapsuckers, Brown Creepers, Winter Wrens, Hermit and Swainson's thrushes, and Veeries. Look also for the frequently occurring Black-throated Green, Blackburnian, and

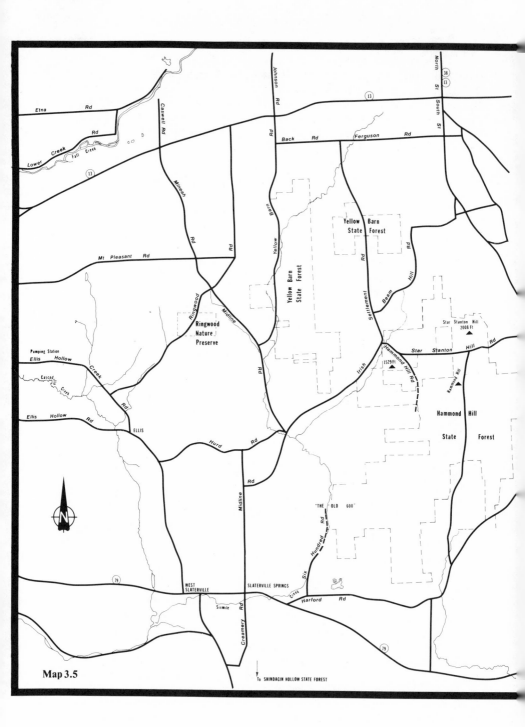

Map 3.5

Mourning warblers. Scarlet Tanagers, Rose-breasted Grosbeaks and Dark-eyed Juncos nest throughout the area.

As you walk through this heavily forested region, you will notice a yellow-marked boundary line angling off into the woods. This defines the border between private property and the Hammond Hill State Forest. You may hike on both sides of this property line. Once back in the car, proceed to the junction of Hammond Hill Road and Star Stanton Hill Road. Star Stanton Hill Road heads due east here, approximately 0.3 mile from the junction of Irish Settlement Road. The right fork, Hammond Hill Road, is a relatively unimproved road, and for its first 0.2 mile you will still be in the Hammond Hill State Forest. As the road proceeds south, it passes alternately over private and public land, in and out of the forest. The elevation at the junction of Hammond Hill Road and Star Stanton Hill Road is 1529 feet, and it rises to approximately 1670 feet before beginning to descend. If the weather is fair and dry, Hammond Hill Road may be passable for quite a distance before one must turn back to Irish Settlement Road. The high hill on your left and approximately 0.6 mile due east, is Hammond Hill, which lies within the state forest. After advancing a short distance, get out and do a little birding. Look for Red-shouldered Hawks, which nest in this area, and for Ruffed Grouse, Turkey, and migrating picids, flycatchers, vireos, and warblers. Hooded Warblers may be found along this road.

Turn around, return to the junction with Star Stanton Hill Road and take this road (Star Stanton Hill Road) to the right (east). There are several side trails as one proceeds uphill, any of which may be productive, if explored. Just over the crest of the hill there is a small parking area at a four corners. From here, walk in any direction. Thrushes, Golden-crowned Kinglets, Solitary Vireos, Magnolia, Black-throated Blue, Yellow-rumped, Black-throated Green, and Mourning warblers nest. The track to the left goes slightly upgrade, crosses an east-west track and then starts down a steep wooded hill. To the left, Hooded Warblers have been present on this hillside for several years.

Turn around and proceed back to Irish Settlement Road. When you reach it, turn left and proceed south for approximately 2.7 miles, where it joins Midline Road (NY 162). Continue on it south for approximately 1.4 miles to the hamlet of Slaterville Springs. Here turn left at the junction of Midline Road and NY 79. Continue on NY 79 for approximately 0.6 mile and keep left at the first fork where NY 79 goes right and Harford Road goes left. Almost immediately after gaining Harford Road, a road named Six Hundred Road comes in on the left. Watch very carefully; It can be easily missed. Turn left onto Six Hundred Road. To the left of the road, in a high bank of one of the creek crossings is a large colony of Bank Swallows. Follow the road to its end and park the car well off the road, so that passage is not blocked. You will now be approximately 2.2 miles northeast of Slaterville Springs and in the general vicinity of the 435 acres known as "Old 600."

This area is also known as the Slaterville Preserve and the Lloyd-Cornell Wildflower Preserve. The preserve mostly lies along and east of the headwaters of Six-Mile Creek and is joined on the east by Hammond Hill State Forest. The "Old 600" was a part of a land gift to Cornell University in the mid-1930s, and it is currently administered by the university. The original conditions of the gift required that the property be allowed to revert to its "natural and primeval conditions," and that no dead trees or undergrowth were to be cut or removed, regardless of their commercial value. It is owing to these carefully worded stipulations and the university's willingness to respect them, that the "Old 600" is today a botanical and avian gem. Ferns and orchids as well as a profusion of wildflowers can be viewed there at the peak of the blooming season. Plants may not be collected within the preserve.

Once parked, begin walking north. Follow the course of Six Mile Creek along its banks, until you have had your fill of this enchanting place. It has indeed nearly reverted to its natural state, and a great many bird species can easily be seen here, especially during spring migration. Additionally, this carefully neglected preserve provides a combination of nesting requirements for many species.

Species that one might expect to see include Goshawks and American Kestrels, which nest in the area; several picid species, including Yellow-bellied Sapsuckers; Yellow-bellied and Least flycatchers; House and Winter wrens; Swainson's and Gray-cheeked thrushes; kinglets and vireos; Black-and-white, Prothonotary, Worm-eating, Orange-crowned, Nashville, Yellow, Magnolia, Yellow-rumped, Cerulean, and Pine warblers, Northern and Louisiana waterthrushes, which nest along the stream in the preserve, Mourning and Canada warblers. Additionally, one can usually expect to see Northern Orioles and very occasionally, an Orchard Oriole; Scarlet Tanagers; Rose-breasted Grosbeaks; Pine Siskins, which might nest in the area; White-winged Crossbills, which have been recorded once singing along the creek in June and July; migrant Fox and Lincoln's sparrows and Vesper Sparrows in the fields along Harford Road.

Shindagin Hollow State Forest along Shindagin Hollow, Gulf Creek, and Downey Roads
(Maps 3.1 and 3.5)

Rating: Spring****, Summer***, Autumn***, Winter*

In the most southeastern section of the region and of Tompkins County lies the Shindagin Hollow State Forest, which is an excellent place to bird, especially at the height of spring migration and early in the summer. Shindagin Hollow Road cuts through some of the most dramatic and beautiful parts of the forest, and it is along this road that the birder should travel.

From downtown Ithaca, at the junction of NY 13 and NY 79 (West State Street), proceed east and southeast on NY 79 to the hamlet of Slaterville Springs. There, at the junction of NY 79 and Midline Road on the north (left) and Creamery Road on the south (right), turn right (south) on Creamery Road. Turn left (again south) on Central Chapel Road (NY 330), and continue south at the crossroad for approximately 1.4 miles, past a small cemetery on the left, to the fork of Shindagin Hollow Road and Braley Hill Road. Follow the Shindagin Hollow Road (left fork) southeast and into the state forest. All along this road there are convenient pull-offs where the car can be parked without blocking traffic. Take the opportunity to stop at several of these to get out and bird. Make your way slowly to the junction of Shindagin Hollow Road and Gulf Creek Road. Park at this corner and bird south along the road. The creek gorge has many spring wild-flowers and nesting Louisiana Waterthrushes. Here make a very sharp left turn and follow Gulf Creek Road north and east to South Road, again birding all along the way. At South Road turn left and proceed to the next through road, Downey Road. Here again turn left and bird your way back to Chestnut Road. This entire circuit could take as many as three or four hours with time for birding away from the car at several junctures. Leave plenty of time to do this wilderness area justice.

Several of the nesting species you might expect to see are: Red-breasted Nut-hatches; thrushes; Solitary Vireos, Black-and-white, Magnolia, Black-throated Green, Blackburnian, Canada, and Chestnut-sided warblers; Ovenbirds; and American Redstarts; plus other migrant warblers in May.

In addition to all areas mentioned, east and slightly northeast of Ithaca, there are some wonderful areas along the eastern and western shores of Cayuga Lake that really should be visited, if the birder has time. Reserve some time, for the lake's shores are among the most productive birding sites in the region. Cayuga Lake is the longest of the six major lakes of the Finger Lakes — nearly forty miles. Because of its great depth, it maintains a very brisk temperature year-round. Fol-lowing are several descriptions of sites along the shores of Cayuga Lake, taken in south-to-north order. Naturally, not every good site is included; however, the most productive sites are listed, and birders should picnic on the tree-shaded shores or swim in the bracing waters of the lake, as well as enjoy its avian life. Directions to the following sites all assume that the birder has already reached downtown Ithaca.

CAYUGA LAKE, THE EASTERN SHORE IN TOMPKINS AND CAYUGA COUNTIES
Water Wagon Road (Map 3.3)

Rating: Spring*, Summer*, Autumn**, Winter****

Approximately 4.8 miles north of Ithaca, a short drive along upland cultivated fields is probably the sure place left near the city to find Horned Larks, Lapland Longspurs and Snow Buntings. It takes a minimal amount of time to cover the road and should be a part of an excursion geared at sampling the winter bird-populations around the Ithaca area.

From the junction of State Street (NY 79) and NY 13, in downtown Ithaca, proceed north on NY 13 a short distance to the NY 34 Exit. Take NY 34 north and proceed along the lake to Water Wagon Road, which is about four miles from Ithaca. Turn right (east) on Water Wagon Road and proceed to Triphammer Road, birding the fields on the right and left sides of the road. In spring and autumn this is a good place to see Horned Larks, Water Pipits, and various field-loving species. In the winter you will undoubtedly be able to see wintering hawks, Horned Larks, American Goldfinches, sparrows, Snow Buntings, and one or two Lapland Long-spurs, among the larks.

TOMPKINS AND CAYUGA COUNTIES
Lake Ridge Road, Lake Road, and points west (Map 3.3)

Rating: Spring***, Summer*, Autumn*, Winter***

From the previous site continue north on NY 34 to its junction with NY 34B. Look for this point in the village of South Lansing. Turn left and follow NY 34B approximately northwest along the lake. Pass Milliken Road, which comes in on the left (where you will see a large Gas and Electric Company sign). Approximately 100 yards north of it Lake Ridge Road and NY 34B fork. You will be able to identify this point even if you've missed Milliken Road by the small cemetery on the right side of NY 34B at the fork. Keep left and follow Lake Ridge Road. If time permits, before veering left on Lake Ridge Road, proceed west on Milliken Road to the power plant and railroad tracks at its terminus on the lakeshore. The power plant, of course, has a warm water outlet and often when nearby water areas are frozen, waterfowl tend to congregate at or near the outlet. In spring walk south along the tracks past the cottages and check the lake for ducks. Proceed north on Lake Ridge Road to the lake, remembering to check out the upland fields and wooded hedgerows bordering the road and its side roads. After it crosses the county line, Lake Ridge Road becomes Lake Road. Follow it north, checking the hedges west on Nut Ridge Road for sparrows and east on Center Road for hawks, pipits, and various sparrow species, including Vesper. At the junction of Lake Road and NY 90 (on the right, or east) and Clearview Road (on the left, or west) turn left, and proceed down the steep incline to the lake. There is a summer cottage community at the foot of Clearview Road, but in the late autumn and winter it will be fairly deserted. Check the lake here for the enormous rafts of geese and ducks that usually assemble during migrations. On the way to the lake do not be surprised to see Red-tailed and Rough-legged hawks soaring overhead, and be alert for small passerines in the shrubs and fields all along the way.

CAYUGA COUNTY
Long Point State Park (Map 3.3)

Rating: Spring***, Summer*, Autumn*, Winter****

This small state park offers the birder excellent vantage points for viewing loons, grebes, diving and dabbling ducks, raptors, gulls, and small passerines in spring and especially in winter. In addition to covering the state park thoroughly, it would also be wise to bird the fields and woods east of NY 90 to NY 34B. It is in those sheltered fields and cedar stands that picids (including Red-headed Woodpeckers), kinglets, pipits, half-hardy winter species, sparrows, buntings, and longspurs may occur in feeding groups or even singly.

From the intersection of Lake Road and Clearview Road and NY 90, follow NY 90 north for about four miles. Turn left on Lake Road and follow it to Long Point State Park, checking for waterfowl along the way. There is no admission fee.

Spend sufficient time birding the park and its environs during spring and winter. Search the duck and goose flocks for the occasional straggler or accidental. You should spot a "blue" goose or two among the dense flocks of Canada Geese. Be alert for hawks, Snowy Owls, or Northern Shrikes. Concentrate on sheltered areas, the park's adjacent cornfields, open water, and berry-bearing bushes. Check out all areas with unfrozen ground and older pine plantations. Proceed west and north, birding all along Lake Road. Remember especially to bird well the habitat edges, where one can expect to encounter a greater diversity of species than in large tracts of woodland or open country. In the smaller areas along habitat edges birds have less opportunity to disperse and are thus more likely to be seen.

CAYUGA LAKE, THE WESTERN SHORE IN TOMPKINS AND SENECA COUNTIES
Taughannock Falls State Park (Map 3.3)

Rating: Spring***, Summer****, Autumn**, Winter*

The unique, rugged beauty that abounds throughout this region is typified by the spectacular Taughannock Gorge. Taughannock Falls, with its 215-foot straight drop, is the highest waterfall east of the Rockies. Within this and other deep, steep-walled valleys near Ithaca, one can find enough potholes, overhangs, prows, pillars, and types of rock to fascinate even the more experienced geologist. The Taughannock Gorge, quite near the falls, was the site of a famous Peregrine Falcon eyrie memorialized in a well-loved painting by Louis Agassiz Fuertes. In 1913, the late Dr. Arthur A. Allen wrote a long descriptive article in *Bird-Lore* (forerunner of *American Birds*) recounting the thrill of watching the parent Peregrines bring prey back to the nest for the young birds and the memorable moment when he witnessed the youngest bird's first flight. Allen described what prey items the adults brought back to the nest and enumerated the Rough-winged Swallows of the colony in the gorge below, Eastern Meadowlarks, American Goldfinches, Common Flickers, Eastern Bluebirds, and pigeons (Rock Doves), as well as ". . . species of the more open country."

Because so much of the southern talus slope is shady and moist, that side is the one often chosen by bird species with definite northern affinities for nests along the gorge walls. Species with southern affinities also breed along the gorge, and they should be watched for. Finally, during spring migration visit this 790-acre state park to view migrants passing through this area. An especially fine site from which to watch the migration is the overlook on Taughannock Park Road.

Taughannock Falls State Park is about ten miles north of Ithaca, on NY 89. From the intersection of NY 13 and NY 79 (State Street) in downtown Ithaca, proceed west on NY 79 across Cayuga Lake Inlet. Take the first right onto NY 89. Continue north following the lake's shoreline until you see signs for the state park. Proceed to the designated parking areas. If your time and energy permit, take the pleasant walk to the Falls. This is a 20 minute hike. By driving west up Taughannock Park Road, on the north side of Taughannock Creek, one reaches an overlook with a parking area where the birder is afforded marvelous views of the 200- to 300-yards-wide wooded talus slopes and the cascading waterfall.

When walking up the gorge, look for flycatchers, Rough-winged Swallows, Brown Creepers, Winter Wrens, Eastern Bluebirds, Cerulean Warblers, Northern and Louisiana waterthrushes, Canada Warblers, Rose-breasted Grosbeaks, American Goldfinches, possibly crossbills, White-throated Sparrows, and Dark-eyed Juncos.

Sheldrake, a Tour (Map 3.3)

Rating: Spring***, Summer*, Autumn**, Winter***

A drive along Lake Cayuga in the vicinity of Sheldrake often yields the best all-around winter birding within Region 3, as both water birds and late-lingering land-birds concentrate on it or move along it here. This area is about twenty-three miles north of Ithaca and can be a rewarding birding area in spring. While approaching Sheldrake be sure to check all of the open country, manured fields, and sunny sheltered areas close to the ground along the route. Check the lake shoreline and bays. The sheltered waters attract species that consistently avoid the open lake.

From Downtown Ithaca follow directions for Taughannock Falls State Park to get to NY 89. Proceed north on NY 89 for about twenty-two miles, then begin looking for the names and numbers of roads going east (right) from NY 89. Pass by Bergen Beach Road and Interlaken Beach Road. The next road is County Road 141, the northeastern extension of Powell Road. Turn right here and proceed to Sheldrake Park Road (County Road 153). Follow it left around a curve and bird all along Sheldrake Park Road, proceeding north.

This drive along the lakeshore is certainly one of the most profitable water-fowl viewing sites in the entire Cayuga Lake Basin. Proceed slowly checking out large flocks of wintering waterfowl that are sure to be present. Scan and rescan the lake's surface. Remember that often wave motion tends to obscure smaller birds on the water; a scope will prove helpful. Loons, grebes, geese, and ducks raft in considerable numbers along this western, more sheltered, side of the lake. Along the drive north the birder should see loons, possibly Red-necked and certainly Horned grebes, Canada Geese, maybe Brants, wigeons, (European Wigeons have been infrequently reported), large flocks of Gadwalls, Mallards, and Black Ducks, rafts of Redheads and both scaup species, a few Oldsquaws, certainly Buffleheads and Common Goldeneyes and perhaps a rarely sighted Barrow's Goldeneye, one or two Ruddy Ducks, small numbers of Hooded, and, of course, larger numbers of Red-breasted and Common mergansers. Be sure to check all of the birds in flight and check all of the trees along the shoreline for Goshawks, Sharp-shinned and Cooper's hawks and any of the smaller, more hardy, winter finches—Evening and Pine grosbeaks, Common Redpolls, Pine Siskins and American Goldfinches. In addition to the enumerated diverse duck species, the birder should also be alert for Snowy and Short-eared owls and woodpecker species.

Birding on the lakeshore can be very cold in winter, however; it is possible on this tour to cover the area thoroughly from the car, if the weather is too frigid. Excellent views of the lake can be had all along Sheldrake Park Road. A short distance after passing Morgan Road on the left, the road crosses Sheldrake Creek.

Take time to check out the creek carefully on both sides of the road. It sometimes harbors surprises. Continue to the center of the village of Sheldrake. Turn right and go to the park at the end of the road to get additional fine views of the lake. Return to the center of the village. Turn right. You will now be on Weyer's Point Road. Continue birding north and follow the road where it turns left. Follow it back to NY 89, the end of the tour.

TOMPKINS COUNTY – ITHACA AND POINTS SOUTH
Stewart Park (Map 3.4)

Rating: Spring****, Summer*, Autumn**, Winter*

This park, maintained by the City of Ithaca, is located at the southeastern corner of Cayuga Lake where Cayuga Street meets the lake. For many years it has been a favored birding site of local birders, especially in spring. In May the Cayuga Bird Club and the Cornell Laboratory of Ornithology sponsor bird walks at 6 a.m. on Saturday mornings (when leaders are available), to which the public is cordially invited.

From downtown Ithaca at the junction of NY 34 and NY 13, proceed north on NY 34 under the overpass and turn left to the well marked entrance to Stewart Park. Turn right into the park.

Walk the trails, bordered with willows, maples, sycamores, and in some places by thick undergrowth. Follow the footbridge over Fall Creek to the Newman Golf Course. Proceed through the woods to the pier and lighthouse. Here you will be afforded excellent overviews of the Cayuga Inlet and the mouth of Fall Creek. In winter, small rafts of surface-feeding ducks can be seen from here as well as Canvasbacks, Redheads, Ring-necked Ducks and scaups, Buffleheads, and mergansers. In August, when the water level of the lake is low, small numbers of shorebirds can be seen on the mud flats and sand bars relatively close to shore. In most seasons, along the woodland margins, Great Horned Owls and Pileated Woodpeckers are usually found. In spring search out passerines in the cedar clumps and thick undergrowth and within the woods proper. Yellow-billed and Black-billed cuckoos, several flycatcher species, swallows, wrens, thrushes, Blue-gray Gnatcatchers, Cedar Waxwings, vireos, Yellow, Black-throated Blue, Cerulean, and Canada warblers are likely finds. Yellow-throated Warbler is strictly accidental here. Tanagers, Cardinals, grosbeaks, buntings, and sparrows will be found within the park, either as residents or migrants.

Upper Buttermilk Falls State Park (Maps 3.3 and 3.6)

Rating: Spring****, Summer*, Autumn*, Winter**

Within the park is a four-mile trail system, many stretches of which are lined by fine deciduous woodlands. The banks of Buttermilk Creek are mostly bordered by hemlocks, and throughout the park there are tangles and thickets. In many areas of the park's 675 acres there is decent birding, but the more rewarding areas are found in Upper Buttermilk Falls. To get to the upper falls, follow the directions below.

From downtown Ithaca, at the junction of NY 13 and NY 79 (West State Street), proceed south on NY 13 to the first traffic light at Clinton Street, which is NY 96B. Turn left here and drive east on NY 96B across Six Mile Creek to South Aurora Street. Here 96B turns south. Turn right and follow it south to King Road West. Here, turn right and proceed into the park.

Once parked, walk the trails on the east side of King Road West. Expect to see typical migrants on their journey to the boreal forests and passerines that nest within the park. Trails to the falls are west of the road. Remember to pack a picnic lunch. Eating it on the slope-side in full view of the falls is always a treat. Swimming is allowed in the natural pool at the base of the waterfall. A maintained trail connects the upper and lower parts of the park.

Eldridge Wilderness Tract (Map 3.3)

Rating: Spring****, Summer*, Autumn*, Winter*

The Eldridge Wilderness Tract is included because it is primarily an upland birding area—relatively rare within this region. A small stream and pond on the tract round out the site and provide habitat for a range of birds associated with upland deciduous woods, fields and ponds, hemlocks, thickets, and glen. The Nature Conservancy owns the tract, and it is managed by the Biology Department of Ithaca College. Approximately 115 bird species have been identified on the property, including about 25 warbler species. This is a great place to find cuckoos, owls, and caprimulgids. An evening visit or two in spring should be extremely rewarding.

From downtown Ithaca and the junction of NY 13 and NY 96B (West Clinton Street), turn left on NY 96B and proceed east across Six Mile Creek to South Aurora Street. Turn right and follow NY 96B south to King Road East. Here turn left and proceed to Troy Road. Turn right and proceed a quarter-mile. Look for Eastern Bluebirds along Troy Road; in some years, however, the nest boxes are not maintained. Just over the crest of the hill is the entrance to the tract, on the left side of the road. At the entrance there is a small parking area.

Follow the main trail into the tract and wander its side trails searching out elusive migrants and nesters.

TOMPKINS AND TIOGA COUNTIES
The South Cayuga Lake and North Susquehanna River Basins, a Tour (Map 3.6)
(including Michigan Hollow, Michigan Creek, Finger Lakes Trail, Spencer Marsh, Center Schoolhouse Road, Bald Hill Road, and Jennings Pond in Danby)

Rating: Spring****, Summer**, Autumn**, Winter*

The following tour passes through some of the best habitat in *Kingbird* Region 3 for migrant flycatchers, thrushes, vireos, and warblers. It includes a marsh that boasts nearly every species common to central New York marshes. A wilderness trail, beaver ponds, sand and gravel pits, and several spectacular vistas are also part of the tour. To cover the entire tour, count on spending several hours. Take along waterproofed boots, insect repellent, and a picnic lunch. Do not attempt any of the many side dirt roads mentioned here after very heavy rains or in winter.

From downtown Ithaca and the junction of NY 13 and NY 96B (West Clinton Street), turn left on NY 96B and proceed east to South Aurora Street, crossing Six Mile Creek. Turn right here and follow NY 96B south to Danby. Just south of Danby, turn right (south) off NY 96B onto Michigan Hollow Road. The road is marked, but also look for a sign directing travelers to the Cayuga Lake Beagle Club down Michigan Hollow Road.

All along Michigan Hollow Road from the turn-off at NY 96B to the intersection with Bald Hill Road to the south avail yourself of numerous opportunities to stop and bird. Follow the road south. Stop frequently and bird off, as well as from, the road, taking appropriate care for automobile traffic. During spring migration this is an excellent site for a variety of flycatchers, thrushes, vireos, and warblers; during summer this area contains a high proportion of breeding warblers. Proceed south approximately 2.2 miles, to Smiley Hill Road (on the left). Pull off the road and park. Scan the beaver ponds on the right (west) side of the road for nesting Wood Ducks. These ponds form the headwaters of Michigan Creek, which flows south toward the Susquehanna River. Just south of the beaver ponds, opposite and a little bit north of Smiley Hill Road, you will notice a hiking trail. This is part of the Finger Lakes Trail, which is planned to run from Allegany State Park to the Catskills. This segment runs from Michigan Hollow Road west, intersecting Bald Hill Road on the other end. Walk west on the trail, exploring this eastern section. Later on in the tour there will be an opportunity to bird the other end of the trail from Bald Hill Road. The woods, beaver ponds, creek, and abundant thickets in this vicinity will undoubtedly harbor many passerines, and they should be carefully and patiently birded.

After thoroughly exploring the Finger Lakes Trail, continue south on Michigan Hollow Road, birding as you go. South of the beaver ponds, the hollow becomes more dramatically scenic. The road passes through the Danby State Forest. Along this stretch of Michigan Hollow the habitat is excellent for nesting deciduous-forest and northern conifer-forest species. Look and listen for Yellow-bellied Sapsuckers, Brown Creepers; Winter Wrens; Hermit Thrushes; Solitary Vireos; Black-and-white, Blue-winged, Nashville, Magnolia, Black-throated Blue, Black-throated Green, Blackburnian, Chestnut-sided, and Prairie warblers. It will be quite easy to log Ovenbirds as well as Northern and Louisiana waterthrushes. Other nesting warblers here include Mourning Warblers and American Redstarts. Dark-eyed Juncos nest along the hollow slopes.

Some 2.5 miles south of Smiley Hill Road, Bald Hill Road meets Michigan Hollow Road. Look for it on the right (west) side of the road. Here turn right and proceed 0.3 mile. At the first small lane on the right, again make a sharp right turn and drive a short distance north to a sand and gravel pit. Here in the banks, you should be able to see nesting colonies of Bank and Rough-winged swallows. After making that detour, retrace your route back out to Michigan Hollow Road. Turn right and proceed south. Cross the Tompkins-Tioga county line. Continue another

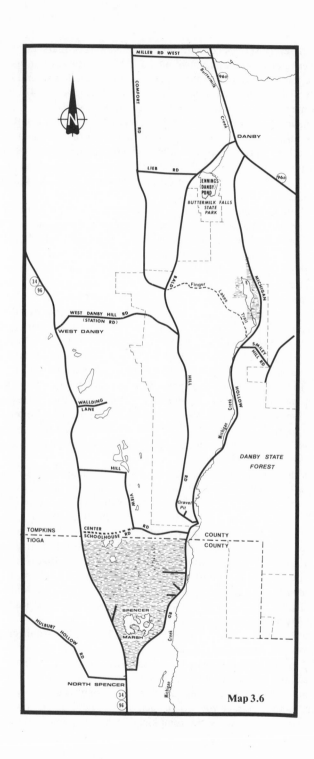

Map 3.6

1.2 miles and begin looking for the access to Spencer Marsh on the west side of Michigan Hollow Road. Along this, the east side of the marsh, access is relatively difficult, as a good deal of the surrounding land is privately held and close overlooks are rare. However, some birding can be done from the road in spite of the limited vistas. Investigate the few dead-end lanes on the right side of the road. Continue south to the junction of Michigan Hollow Road and NY 96 (NY 34). Here turn right (north) and proceed north on NY 96 a short distance to a dirt road coming in from the right, immediately south of the railroad overpass on NY 96. Turn sharply right and drive uphill to the old railroad watertank. From here one is afforded good views of Spencer Marsh. Be persistent and you should hear and see herons, bitterns, dabbling ducks, rails, gallinules, shorebirds, woodpeckers, and marsh wrens. At this site and on Center Schoolhouse Road (coming up), the birder is most advantageously situated to view the enormous waves of migrants traversing the valley between the Cayuga Lake Basin on the north and the Susquehanna River Basin to the south. To thoroughly bird Spencer Marsh, however, a scope is indispensable.

After leaving the old watertank overlook, proceed to NY 96 (NY 34) and turn right, headed north. Drive north and again cross the Tompkins-Tioga county line. After approximately 2.2 miles turn right (east) on Hill View Road. Follow it east and south 1.3 miles. Be alert for an abandoned dirt road coming in on the right at the sharp left turn. It is very easy to miss, for there is no sign. Hill View Road and this dilapidated road were once together called Center Schoolhouse Road, and this deserted section is still locally referred to as Center Schoolhouse Road. Park off Hill View Road, without blocking oncoming traffic, and begin walking west on the abandoned road. You will note that the property on either side of the road is posted. Do not be distressed; excellent birding is to be had without ever leaving the roadway. This site is on the divide between two watersheds, and its deciduous woods, fields, ponds, brushy tangles, and marshy areas provide some of the most productive spring birding in the region. Continue walking westward on the old thoroughfare until you are afforded superb views of a large marshy spot and a very large beaver complex to the south (left). This pond deserves careful scrutiny, for in most seasons it harbors herons, ducks, and rails. After thoroughly covering old Center Schoolhouse Road, return to the car.

Continue traveling east on Hill View Road to its terminus at Michigan Hollow Road. Turn left and proceed to Bald Hill Road, which is only about 0.2 mile away, the first available left turn off Michigan Hollow Road. Turn left onto Bald Hill Road, and if you skipped the swallow colonies and gravel pit mentioned above, check them out now. Bald Hill Road is more than seven miles (long) and passes through overgrown fields, spruce plantations, and rich deciduous woodlands. All along its route you will find convenient places to pull off and get out of the car to bird. Do just that.

Along this lower 2.7 miles of Bald Hill Road, between Michigan Hollow Road and West Danby Hill Road (formerly known as Station Road), a number of small, infrequently used primitive roads should provide fruitful birding. Investigate these lanes and bird the public hunting area through which Bald Hill Road passes. Stop and bird the overgrown fields near the small cemetery south of West

Danby Hill Road. Approximately 0.6 mile north of West Danby Hill Road, which comes in on the left, another road comes in on the left, which may or may not be labelled Comfort Road. Park here and walk east on what appears to be a primitive extension of Comfort Road. This is in fact, the other, western end of the portion of the Finger Lakes Trail mentioned earlier. After birding this end of the trail and west along Comfort Road, continue north on Bald Hill Road approximately 1.6 miles, past Lieb Road on the left, to the next road coming in from the right. Turn right and proceed to the parking area. You are now at Jennings Pond (also called Danby Pond), a disconnected section of Buttermilk Falls State Park.

Jennings Pond is maintained by the Finger Lakes State Park system. The source of Butternut Creek, it is a clear, natural pool, where loons, grebes, and ducks rest during migrations. Fine views of the entire pond may be had at the dike near the parking lot. Watch for Ospreys fishing in the pond and Marsh Hawks coursing along the moist edges of the pond. From the east side of the parking lot, there is a trail that passes through a heavily wooded section. Follow the path east and south around the pond where you may well see migrant flycatchers, thrushes, vireos, and warblers. Return to Bald Hill Road. Turn right and proceed one-half mile to NY 96B. Here, turn left and return to Ithaca, having completed the tour.

TOMPKINS AND SCHUYLER COUNTIES
Connecticut Hill State Wildlife Management Area and Cayuta Lake, a Tour (Map 3.3)

Rating: Spring****, Summer***, Autumn**, Winter***

The following tour covers the 11,000-acre Connecticut Hill State Wildlife Management Area and the small but very rewarding Cayuta Lake west of it. The management area is essentially a conglomerate of habitats, ranging from abandoned farmland to dense stands of mixed coniferous-deciduous trees. The highest elevation on the tract is Connecticut Hill, which rises to 2100 feet, and is crossed by a short section of the Finger Lakes Trail. The tract is managed for Turkey, grouse, and deer, and a good deal of the land that once supported farms is being reforested with spruce, oak, maple, and pine.

To appreciate Connecticut Hill fully, the birder should walk several of its roads and sample the unusual variety it offers. For example, Goshawks nest here; Ruffed Grouse can be heard drumming in spring; and there is a year-round Turkey population, some of which can be seen with little difficulty, feeding in open fields or near the razed headquarters building on Carter Creek Road. Birds with northern affinities nest in the higher elevations of the tract — Red-breasted Nuthatches; Wood and Hermit thrushes; Black-throated Blue, Yellow-rumped, Black-throated Green, and Chestnut-sided warblers. The nesting Prairie Warbler in several regenerating clearcuts along Ridge Road is a local specialty. Louisiana Waterthrushes can usually be found in spring and summer along Carter Creek. Canada Warblers, Grasshopper Sparrows, and Dark-eyed Juncos are among the more unusual species that breed in the management area. Among the more common breeding species look for American Woodcocks soaring and diving in nuptial displays during spring evenings at dusk, Brown Thrashers along scrubby woodland borders, Blue-winged Warblers in second-growth fields, and Yellow-breasted Chats and Rufous-sided Towhees in brushy thickets and overgrown tangles. The coniferous forest at elevations above 2000 feet attracts numbers of winter finches — Purple Finches, Pine and Evening grosbeaks, Common Redpolls, Pine Siskins, and Red and White-winged crossbills.

The patient and persevering birder will find an abundance of birdlife. In season, deer hunting is allowed here, so birders should be especially cautious then.

The following directions are for spring, summer, and fall. For winter directions, go to the next paragraph. From downtown Ithaca and the junction of NY 13 and NY 79 (West State Street), proceed south on NY 13. Continue traveling southwest toward Newfield for about nine miles. Pass through Newfield and continue another four miles to Carter Creek Road. Just beyond this corner, Turkeys may be observed feeding in the fields. Turn right on Carter Creek Road and continue north to Cabin Road, which comes in on the left. Here there is a barn and the foundation of the former headquarters. Bird the management area, being sure to pay special attention to those areas where two different habitats meet. Travel on all open roads, and if time and inclination allow, take a walk along the Finger Lakes Trail and climb Connecticut Hill.

Not all of the roads traversing Connecticut Hill State Wildlife Management Area are open in winter. Entrance to the area is restricted to a route that begins about three miles northeast of the village of Newfield. From downtown Ithaca, drive southwest on NY 13. After five miles, watch for NY 327, which meets NY 13 on the right. Continue driving, being sure to take note of the junction of NY 13 and NY 96. Just one mile southwest of that junction, Millard Hill Road meets NY 13 on the right. Turn right (west) onto Millard Hill Road. Bear right at Douglas Road. Turn left at Connecticut Hill Road, and investigate the road, which is sure to be plowed as far as the water tower; in fact, explore any open roads on the tract. Bird the entire area. During this season, it will be especially rewarding to hike or cross-country ski to the coniferous forest near the top of Connecticut Hill, where you are most likely to find winter finches.

In any season, after birding the management lands proper, proceed to the junction of Connecticut Hill Road and Vanloon Road. Proceed west on Vanloon Road to Willamee Road. Here turn left and proceed south along the eastern shore of Cayuta Lake. Circle the lake, stopping at the Cayuta Creek outlet on its southeast corner. Continue on Willamee Road to its intersection with NY 228. Turn right and proceed north up the west shore of the lake. At Cayutaville Road turn right and right again at the second road, which is Swamp Road. At Willamee Road, again make a right turn and proceed to where you began to circle the lake. Along the lakeshore circuit you will find many opportunities to stop and bird it to excellent advantage. Since its relatively recent inclusion on many birders' jaunts, the lake and its environs have turned out to host a surprising variety of birds.

Arnot Forest (Map 3.3)

Rating: Spring***, Summer**, Autumn**, Winter (closed December 1–May 1)

The Department of Conservation of the New York State College of Agriculture manages the 4025-acre tract known as Arnot Forest, about twenty miles southwest of Ithaca. It lies between the towns of Newfield, Cayuta, and Van Etten. The forest is available to hikers, birders, picnickers, cross-country skiers, and all interested in nature study during daylight hours from about May 1 through the end of November. A pleasant and minimally used area, it has proven in the past several years to be an excellent birding site. Throughout the managed forest there are some five miles of unimproved dirt (or jeep) trails, and about nine miles of hiking trails.

From downtown Ithaca and the junction of NY 13 and NY 79 (West State Street), proceed south on NY 13 about nine miles to Newfield. Turn left from NY 13 at the fire station onto Main Street, and follow it a short distance to its end at Van Kirk Road (County Road 132). Turn right onto Van Kirk Road and drive south to its junction with Irish Hill Road. Turn left here and continue south to the sign designating the entrance to Arnot Forest.

Probably the best method of birding the forest is to travel some distance along the main dirt road that winds through it, stopping along the way at the gravelled pull-offs to bird on foot one of the many side trails. These trails extend into a wide variety of habitats suited to a corresponding variety of breeding birds. During late spring, Arnot is an excellent place to view large waves of migrating thrushes, vireos, and warblers. Within the forest there are ponds, a creek, demonstration plantations, hilly, forested slopes, woodlots, and farmland. Among the breeding birds are Broad-winged Hawks, Ruffed Grouse, Barred Owls, Hermit Thrushes, various vireo and warbler species (Hooded Warblers nested in 1980), and Dark-eyed Juncos.

One final note as an aid in planning ahead; although there are several picnic tables within the grounds, neither water nor restrooms are provided.

SCHUYLER COUNTY
Texas Hollow Wildlife Sanctuary (Map 3.3)

Rating: Spring***, Summer***, Autumn**, Winter (inaccessible)

Texas Hollow Wildlife Sanctuary is about fifteen miles west of Ithaca and about six miles northeast of Watkins Glen in Schuyler County. Its nearly fifty acres lie on private land, but permission is not required to gain access. The sanctuary is inaccessible during winter and the trails may be moist in early spring. In summer, it is advisable to carry insect repellent, to ward off deer flies and mosquitoes, and to avoid the patches of poison ivy along some stretches of trail.

The preserve has a rich mixture of conifers and hardwoods as well as several thick brushy areas and a fine marsh. Its unique feature is an old sphagnum bog — the only bog in Schuyler County. Surrounding the bog is a rich and varied plant life with bog cranberries, sundew, grass pinks (*Calapogon*), and rose pogonia. Throughout the woods there are abundant pink moccasin (*Cypripedium acaule*), rattlesnake plantain, grape fern, maidenhair fern, and spectacular cinnamon ferns close to the bog.

From downtown Ithaca and the junction of NY 13 and NY 79 (West State

Street), proceed west on NY 79 across the Cayuga Lake Inlet. Follow NY 79 north up the hill, then due west for about fifteen miles. You will pass through Mecklenburg, and join NY 227. After passing Mark Smith Road on the right, watch the left hand side of the road for Texas Hollow Road, which comes in just east of Bennettsburg. At Texas Hollow Road turn left and proceed south one mile to the sanctuary entrance, which is well marked. Park along the road and walk to the entrance.

Walk along the obvious trail that winds through the conifers and hardwoods and circles the bog. It skirts the brushy growth then leads to the marsh. Be watchful for all of the regular nesters — Green Herons; Wood Ducks; Cooper's, Broad-winged, and Red-tailed hawks; Great Horned Owls, Solitary Vireos; Blue-winged and Blackburnian warblers. Abundant migrants also pass through the area in spring, and many of the more common nesters are resident on the preserve.

ACKNOWLEDGMENTS

To the following people go the author's sincere thanks for the information they supplied for the Finger Lakes chapter: The Cornell Laboratory of Ornithology, Walter E. Benning, M. C. Comar, Elaine Grandjean, the late Lawrence J. Jackmin, Douglas P. Kibbe, Dorothy W. McIlroy, Thomas M. Riley, Margaret Shepherd, Charles R. Smith, Betty J. Strath, Sam Weeks. An extra vote of thanks is extended to Dorothy McIlroy, Margaret Shepherd, and Doug Kibbe for their readings of and helpful suggestions on the Finger Lakes chapter.

Susquehanna—Region 4

THE SUSQUEHANNA REGION consists of all of Broome, Chenango, Cortland, Delaware, Otsego, and Tioga counties—about 5100 square miles. It includes the lands drained by the upper Susquehanna River and its tributaries, and the hills in the upper reaches of the Delaware River drainage basin. The major water bodies within the region are the southernmost tip of Skaneateles Lake, Otsego Lake, Pepacton and Cannonsville reservoirs, and the much smaller Canadarago Lake. Numerous smaller lakes and ponds are scattered throughout the region. The area is generally rolling, hilly uplands cut by many beautiful ravines, with valleys narrow enough to make farming difficult. In the hills drained by the Delaware River, the local relief reaches from several hundred feet to over 3000 feet. Flat bottomlands along the major rivers and tributaries respond well to modern agricultural methods (Map 4.1).

Sycamore, basswood, elm, willow, and ash are the most common trees along these waterways, with some silver maple. The tree-clothed slopes of the uplands are wooded with gray and black birch, poplar, red maple, and some stands of pitch pine. The higher slopes of the Catskills are lined with birch, beech, and hard maple; the more southerly slopes support oak, hickory, and some balsam and spruce. Throughout the region one comes across former farmland, now deserted and reverting to a relatively wild state. Farms have had to compete with an ever-growing demand for highway right-of-ways, as well as commercial and residential interests. All of these factors, added to the increasingly marginal return on many once successful farms, has left the Susquehanna Region a mosaic of agricultural successes and failures. The successes, of course, are situated in alluvial valley bottoms or well-drained uplands, and the failures are largely in the steep uplands, where soils are thin, acid, rocky, and poorly drained. In many such "failure" areas, the land has been left to grow back to forest, and some of it has even been replanted with conifers. This rough wooded country has definite recreational value, particularly for birders, who treasure the dense woods, hedgerows, thickets, abandoned orchards, and former pastures, which now house and nourish populations of wild birds.

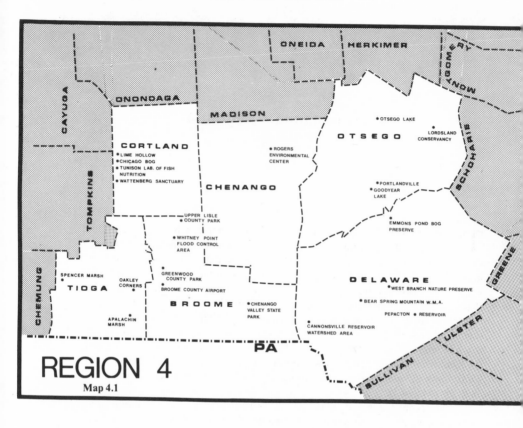

REGION 4
Map 4.1

TIOGA COUNTY
Spencer Marsh (Map 4.1 and see Map 3.6)

Rating: Spring****, Summer**, Autumn**, Winter*

Spencer Marsh lies in the northwestern section of Tioga County one mile from Spencer, immediately north of the village of North Spencer. It drains south, and is therefore part of the Susquehanna River drainage basin. Spencer Marsh is an excellent birding site, especially in spring, when the keen birder can view most forms of marsh wildlife under unrestricted natural conditions. This marsh and its surrounding roads and observation points have been treated in the preceding chapter, because birders from Ithaca and environs have traditionally monitored this area more closely than birders from the Susquehanna Region. Except for a few select points, Spencer Marsh is surrounded by privately held land and is not easily approachable. It is however, a site at which there is a rich marsh flora and an equally rich birdlife. Almost all of the birds found in marshes in central New York State can be found here. It is worth the time and effort to bird the marsh well.

TIOGA AND BROOME COUNTIES
Oakley Corners and Greenwood County Park (Map 4.2)

Rating: Spring****, Summer***, Autumn****, Winter***

Oakley Corners, some twenty miles northwest of Binghamton, was formerly sometimes called Newark Valley State Land. It encompasses roughly a thousand acres of public land about five miles west of the village of Union Center in Tioga County. The area is heavily birded by members of the Naturalist's Club of Broome County, who cover it in spring, especially at the height of the migration, and in winter to find finches, Pine Grosbeaks, redpolls, siskins, and crossbills. The tract contains a very productive lake and large stands of deciduous and coniferous trees accessible along well-maintained paths.

To reach Oakley Corners, proceed west on NY 17 from Binghamton for about nine miles. Leave NY 17 at Exit 67 and take NY 26 north. At the junction of NY 26 and NY 17C turn left; shortly thereafter, turn right and follow NY 26 north to Union Center. There, at the junction of NY 26 and NY 38B, turn left and proceed west on NY 38B for 2.8 miles to the junction of Dutchtown Road. Here, turn left, and continue along Dutchtown Road for 2.3 miles to Oakley Corners. This name is misleading, for there is no longer a settlement here; however, it is easily recognized by the state forest sign marking it. Turn left and continue one-half mile to the parking lot, which is essentially a small pull-off area.

From this point, one can choose one of two trails to follow. The one on the right leads immediately downhill to the vicinity of the lake. The other leads off to the left and eventually meets the lake, but by a circuitous route. Take the trail on the left if you have time to bird the area leisurely. The trail passes through areas reforested with deciduous and coniferous trees and passes around the south edge of the water. The lake is filled with dead tree stumps and water-loving vegetation — leatherleaf, for instance, which has proven especially attractive as cover for grebes, herons, and waterfowl.

Continue along this trail past the outlet to an area of large overgrown hedgerows, scrub, and conifers, where thrushes and warblers congregate in spring on their northward journey. This wetland border is productive in autumn, but, less so than in spring. Bird this same area in winter for finch specialties, particularly during "invasion" years. The heavy thickets are also worth investigating. The general character of the forest is changing, owing to natural processes as it matures, as well as to management practices. The heavily wooded terrain deserves thorough coverage.

After birding Oakley Corners there may be plenty of time to bird another site; the most easily accessible from Oakley Corners is Greenwood County Park. There is a nominal admission and parking fee here during the summer season.

From Union Center and the junction of NY 26 and NY 38B, continue north on NY 26 through the village of Maine. Just north of Maine, NY 26 turns right. *Do not* turn right with it. Continue north on Nanticoke Road to the hamlet of Nanticoke. Here Nanticoke Road becomes Caldwell Hill Road. Proceed north on it to Greenwood Road and the well-marked entrance to the park. From the village of Maine to the park entrance is around five miles.

Birding in the park is rated as moderate-to-easy, and birding is good along several marked trails. None of these trails will present any difficulties to the birder, but some are longer than others. Migrating warblers abound during May, and finches, redpolls, and crossbills are specialties during winter. Take one of several trails through the woods and upland fields, and around the large lake. On the lake, migrating Canada Geese and many waterfowl species congregate during late autumn. Sharp-shinned, Cooper's, Red-shouldered, Broad-winged, and Red-tailed hawks and American Kestrels are all seen regularly within the park's several hundred acres. Screech, Great Horned and Barred owls are resident within the park. The trails through the mixed woodlands and open grasslands, and those skirting the camping section of the park, are less heavily used than those of more well-known parks. The park hosts, however, a good variety of species typical of wet woodlands and dry fields, and should be generally productive in most seasons.

Apalachin Marsh (Map 4.2)

Rating: Spring***, Summer*, Autumn***, Winter*

Apalachin Marsh should probably be birded on the visit to the Vestal-Binghamton area. In spring, during periods of heavy migration, warblers are seen by the hundreds, including Tennessee, Magnolia, Black-throated Blue, Yellow-rumped, Black-throated Green, Blackburnian, Chestnut-sided, and Canada warblers, Yellow-breasted Chats, and American Redstarts. Of these, many nest in the marsh area. Area birders consider this one of their regional gems.

Apalachin Marsh is a great area little known to most New York birders, but definitely worth mentioning. It lies nearly five miles west of Vestal and covers some fifty acres. No permission is required to bird the site and depending upon one's aptitude for fence-climbing, it could be rated as easy to moderately difficult. If one's particular talents don't run to cyclone fence-climbing, the area can be less thoroughly birded from the outside. The Naturalist's Club of Broome County was instrumental in saving this lovely, wild, wet area from destruction, when the highway was under construction. It lies between the east- and west-bound lanes of NY 17. The area is now posted by the New York State Department of Transportation, but access is prohibited to hunters and snowmobilers *only,* and the posted warnings do not pertain to birders or hikers.

To reach the marsh, leave NY 17 at Exit 66 (Apalachin Exit). Gain NY 434 and proceed west for 2.5 miles. Pull off the road and park on the grassy shoulder of NY 434 opposite Nick's Marina. At this point, walk around the cyclone fence and up the creek bed through the tunnels under NY 17 — or if the water is too high, just walk across the highway. Proceed straight ahead through the clearing in the brush. Continue along the old dirt road that skirts the edge of Apalachin Marsh. At the end of the dirt road, there is a state-maintained cyclone fence. Climb over it and continue paralleling the fence to a dike-pathway that bisects the marsh. Climb the fence, cross the dike, and again climb the fence at the opposite end of the dike. All of this fence-climbing sounds a bit strenuous but, in fact, none of the fences

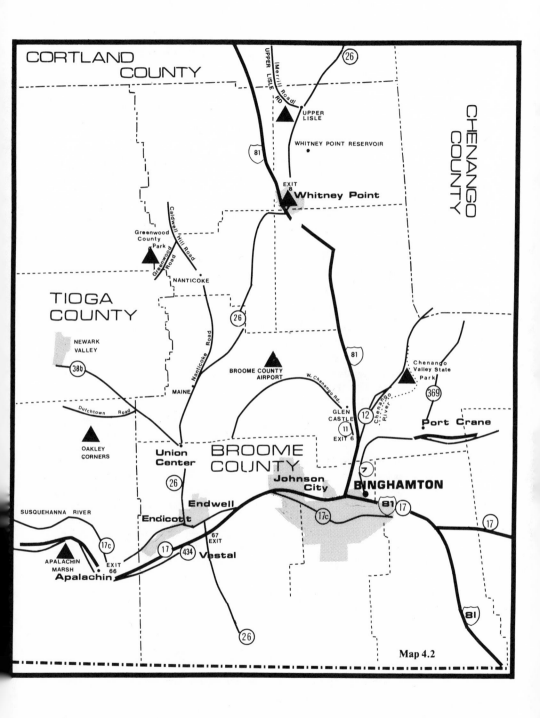

Map 4.2

are impossibly high. Follow along the edge of the marsh, which will return the birder to his starting point.

All along the edge of the marsh and the dike you should have been afforded excellent views of grebes; herons; waterfowl; rails; coots; flycatchers; Tree, Bank, and Barn swallows; Purple Martins, the previously mentioned warbler species, and various breeding sparrow species.

BROOME COUNTY
Chenango Valley State Park (Map 4.2)

Rating: Spring****, Summer**, Autumn****, Winter**

This pleasant state park, about ten miles north of Binghamton, has within its 900 acres, two lakes, a stretch of the Chenango River, and bird-rich stands of conifers and deciduous trees all accessible by well-maintained woodland trails. During the summer season, a small admission and parking fee is charged. During the off-season, admission is free. Birding the park is rated as easy. It must be done on foot for best results, but some productive areas can be covered by car or bicycle.

From downtown Binghamton and the junction of NY 17 and I-88 (NY 7), proceed north and east on I-88 (NY 7). Watch for exit signs directing to Port Crane and NY 369. Leave I-88, and turn left (north) on NY 369. Proceed north. Both I-88 and NY 369 follow quite close to the Chenango River for some miles. It will not be possible, however, to check out the river from I-88. After leaving the highway and gaining NY 369, stop wherever it is convenient and safe, and bird the river and its shoreline. Follow the signs to the entrance of Chenango Valley State Park.

Once inside the park go to the old fish hatchery. Here there is a road along sections of the former Chenango Canal and the Chenango River. Walk north on it through rewarding habitats. Then go to each of the lakes and hike the trails around them for lake and shore coverage. There is a state campsite within the park and a section with rental cabins. Behind Cabin number 12, a rough trail leads north through a mosaic of mixed woodlands. Bird this traditionally productive trail. Depending upon the season and habitat, the birder should be able to find herons, geese, Mallards, and Black Ducks, hawks soaring above the trees, owls, kingfishers, several picids, including Pileated Woodpeckers, White-breasted and Red-breasted nuthatches, Brown Creepers, and Winter Wrens. In addition, expect nesting Wood Thrushes and Veeries; breeding Blue-gray Gnatcatchers in the oaks; nesting vireos; resident Black-throated Green and Canada warblers and Northern Waterthrushes; Scarlet Tanagers; Cardinals; Rose-breasted and in winter, Evening, and Pine grosbeaks; Purple Finches; siskins; goldfinches; and Tree and White-throated sparrows. In summer Swamp and Song sparrows are resident.

Broome County Airport (Map 4.2)

Rating: Spring***, Summer***, Autumn**, Winter*

To reach the airport from downtown Binghamton, gain I-81, proceed across the Chenango River, and head north on the west side of the river. Leave I-81 at Exit 6, and turn left on NY 11. Proceed north on NY 11, turning left again at the junction of NY 11 and NY 12. At this point NY 11 is also named Castle Creek Road. Proceed to the village of Glen Castle. Turn left on West Chenango Road and proceed to the airport.

Examine all the fields on the airport perimeter for open country birds. Periodic stops should give views of Killdeers, American Golden Plovers, woodcocks, snipes, occasionally Upland Sandpipers, and possibly shorebirds in spring when the fields are wet. Raptors can be seen coursing over the fields, and in winter, be on the lookout for Snowy Owls. Other open country species regularly seen in these fields are Horned Larks, Water Pipits, Bobolinks, Eastern Meadowlarks, and several sparrow species, including Savannah and Field.

Whitney Point Flood Control Area and Upper Lisle County Park (Map 4.2)

Rating: Spring****, Summer***, Autumn**, Winter*

The Whitney Point Flood Control Area includes the Whitney Point Dam, the Whitney Point Reservoir, the flooded fields west of the village of Upper Lisle, and Upper Lisle County Park.

From the city of Binghamton, proceed north on I-81. Leave I-81 at Exit 8 (Whitney Point). Pass through the center of town and turn left (north) on NY 26. Proceed up the east side of the reservoir. Ample road shoulders allow parking, and in spite of occasionally heavy traffic, this area continues to be productive of water birds and shorebirds. More than 170 species have been recorded within the more than five miles of recreation area. Loons, Red-necked, Horned, and Pied-billed grebes, cormorants, herons, egrets, bitterns, Whistling Swans, about twenty-three species of waterfowl, several raptors, including Ospreys, Bald Eagles, Marsh and Rough-legged hawks, rails, and coots, over fourteen shorebird species, Bonaparte's Gull, Common and Black terns, Great Horned and Short-eared owls, several picids, flycatchers, Cliff Swallows, Long-billed Marsh Wrens, many thrush species in transit, Loggerhead Shrikes, numerous vireo and warbler species, both nesting and in migration, and a full complement of blackbird and sparrow species are all part of the area's extensive list.

Stop frequently to check out the reservoir and its shoreline. In this endeavor a telescope will be invaluable. The reservoir ranges from a quarter to half a mile wide, and along some of its protected grassy banks one sometimes turns up a surprise bird. Stop along the route at the Dorchester Picnic Area and search the site for migrating warblers and passerines in spring. At Upper Lisle, turn left on Upper Lisle Road (Merrill Road) and advance 0.3 mile. Here look over the extensive marsh between the road and river. Long-billed Marsh Wrens nest here. Turn left for a better view. Proceed with caution for this area sometimes floods in the spring, and the roads and bridge could be blocked. Spring, however, is the best time for this part of Whitney Point. Depending on water levels, shorebirds and waterfowl may be present. Bobolinks nest in the field on the right, and Virginia Rails, flycatchers and Swamp Sparrows nest close to this site. Proceed to Upper Lisle County Park and park the car. Walk along some of the short trails, which are especially productive during spring migration.

CORTLAND COUNTY
Lime Hollow, Chicago Bog and Tunison Laboratory of Fish Nutrition, and Wattenberg Sanctuary (Map 4.3)

Rating: Spring****, Summer**, Autumn***, Winter*

These four sites are very near each other, along the middle of the western edge of Cortland County. They are nearly the same distance from both Syracuse and Binghamton, and about three miles west of the city of Cortland. Birding all four areas can fill the better part of a day. Cortland County is not blessed with any major bodies of water and is at a high elevation. This area has some of the most exciting birding within the county, and is convenient to Ithaca and its "hot-spots," as well as the more consistently rewarding birding sites in *Kingbird* Reporting Region 5, to the north.

Approaching from either north or south, leave I-81 at Exit 12. Follow the feeder road straight ahead to NY 281. At the junction of NY 281, turn left and proceed south, about four miles, to where NY 13 intersects NY 281 from the left. You will have passed the Cortlandville Shopping Center on the left, just south of the Conrail railroad tracks. Just beyond the junction of the two roads, Lime Hollow Road comes in on the right. Turn right, and proceed west one-half mile. Park at the railroad crossing or at another convenient pull-off along the road.

This area is on the southern fringe of the Finger Lakes, and has that area's typical glacial terrain. Geologically, it is a kame-and-kettle area with numerous hillocks and depressions. Although marl is technically a mixture of clay and calcium carbonate, the term is often loosely applied to a wide variety of soils and rocks. The marl ponds here are quite excellent for shorebirds in autumn, and occasionally also in spring. They tend to dry down in summer and are then fringed with cardinal flowers. What appear to be sandy shores (or bottoms, when flooded) are, in fact, white calcium carbonate precipitate. This is rather ooze-like and difficult to walk through. It is advisable to stay on high ground.

Walk to the ponds located on both sides of Lime Hollow Road. On the north side is Stupke Pond, which takes its name from the road that heads northwest from Lime Hollow Road. Stupke Road is the first road coming in on the right, after the turn-off from NY 13. The pond should hold waterfowl in autumn and spring. The railroad tracks are no longer in use, and the right-of-way is being considered for use as a strip nature trail. You can safely walk along the railroad bed. As you do, look for Red-tailed Hawks and Barred Owls, which occasionally nest in this area. A variety of habitats can be reached by foot trails extending from the right-of-way. Bird the open fields, old pastures, and especially the woodlands close at hand. They host large numbers of many varieties of migrating passerines in spring and fewer breeders during summer.

By walking along the railroad bed on the south (left) side of Lime Hollow Road, birders can reach Chicago Bog, also known locally as Doughnut Pond. The ring-shaped acid bog and pond contains a continuous, circular mat of heath and sphagnum moss that is surrounded on all sides by water. The open water on the far

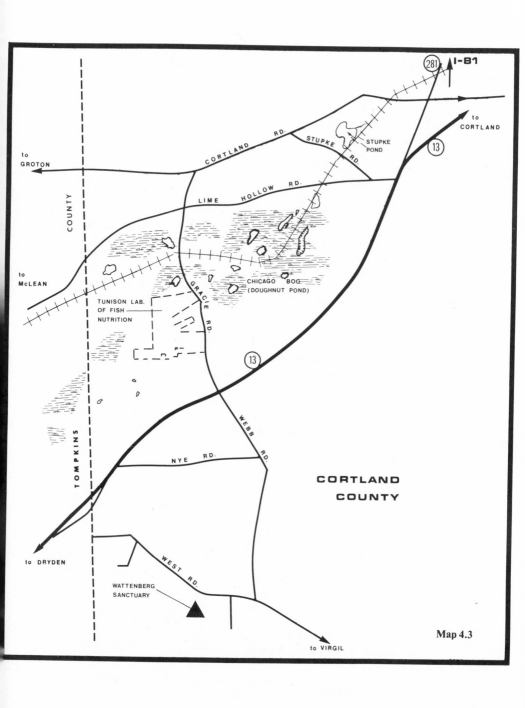

Map 4.3

side of the mat sometimes harbors nesting ducks and, during migrations, other waterfowl. The water girdling the mat on the close (road) side is acid, spongy, inadequately drained, and rich in plant residues. The entire bog and its surrounding environment is fragile, as well as treacherous, and should *not* be walked upon. Confine your birding to the lane and the wooded areas, from which you can thoroughly cover the area. Chicago Bog is surrounded by a hemlock woods on the south and mature oak-maple forest on the north, so it is not visible from the railroad tracks. The access to Chicago Bog is a path off Gracie Road.

From Lime Hollow Road walk south and west to the first big road that crosses the track bed. This is Gracie Road. Turn left here and walk a few yards to a clear path on the left, just south of the tracks. Turn left and walk east to the bog. Along the way listen for nesting Swamp Sparrows.

If you have decided to drive from Lime Hollow Road instead of walk, continue west on Lime Hollow Road to Gracie Road. Here turn left and proceed across the tracks to the first drivable road on the right. Turn right.

Continue to the parking lot of the Tunison Laboratory of Fish Nutrition. Here examine the large spring-fed experimental fish pools. The pools, the creek, and the wet woods that rim the hatchery provide an excellent setting for herons, Ospreys, American Woodcocks, and Common Snipes. Walk north on the path through the woods before crossing the road to reach Chicago Bog.

When you have thoroughly birded the area, drive south on Gracie Road to NY 13. Cross the highway and continue south on Webb Road for about two miles to West Road. Turn right and proceed one-half mile to Wattenberg Sanctuary.

The sanctuary is a ten-acre triangle of reforested pines and upland hardwoods, with a small stream crossing the south vertex of the triangle. The Cortland County Bird Club owns and maintains Wattenberg Sanctuary. Passerines in good numbers are easy to see here during migration. Bird the meadows of the farms adjacent to the sanctuary.

CHENANGO COUNTY
Rogers Environmental Center (Map 4.1)

Rating: Spring****, Summer***, Autumn***, Winter**

Rogers Environmental Educational Center encompasses more than 570 acres of deciduous and coniferous woodlands, willow-lined streams, ponds, swamps, marshes, and open fields, bordering the Chenango River in Chenango County near Sherburne. Throughout the various habitats, 4.5 miles of well-marked and well-maintained trails make the birding effortless. During May, at the height of migration, this center is at its best, especially for warblers. Mixed flocks occur regularly throughout late-April into May, but at the migration peak they are most numerous and most colorful.

To reach the Rogers Environmental Educational Center from I-90, take Exit 31 (Utica) and head south on NY 12. Proceed south some forty miles, passing through New Hartford, Waterville, and Hubbardsville, to the village of Sherburne. Sherburne is more than fifty miles north of Binghamton, so the route through Utica is unquestionably shorter. At the junction of NY 80 and NY 12, in Sherburne, turn right (west), and proceed one mile, crossing the railroad tracks and the Chenango River, to the main parking lot and interpretive buildings of the Rogers Environmental Education Center on the north (right) side of the road. If you come to Williams Road, you have gone too far. Turn back and locate the center by proceeding east along NY 80.

Stop in at the Center's interpretive building, which houses administration offices, educational displays, a library, and a window that overlooks a marsh. This observation point is equipped with outdoor microphones to bring in the sounds of the marsh, as well as spotting scopes.

To reach the Spruce Ridge Trail, go back to the parking lot, where there is an obvious access point on the lot's northeast side. This trail passes through about six acres of fairly dense spruce and pine plantations, that can contain relatively large mixed warbler flocks during migration. From the main trail one can branch off onto other trails passing through a variety of habitats. From these trails look for thrushes, vireos, and warblers—especially Blue-winged, Yellow, Cape May, Magnolia, Yellow-rumped, and Common Yellowthroat—as well as resident sparrow species.

A more productive area is located on what was once known as the Adams' Property. This plot belongs to the Rogers Environmental Education Center and comprises approximately 140 acres one mile west of the main parking lot. Either walk or drive to this site.

Proceed west from the main parking lot for one mile. Turn right (north) at Williams Road. Proceed 100 yards north and turn right again. Along this access road, near an old white storage building on the right, several cars can park.

Walk down the dirt access road and along the marked trails to the site's three ponds. Catwalks cover the most marshy spots, and the trails loop through open fields, mixed deciduous woods, along a streamside, close to ponds and along a lovely willow-lined canal. Sometimes the road is used by fishermen, who occasionally frequent the ponds. The resident avian population is large and varied, due to the quality and variety of the protected habitat. All along the trails look especially for herons and waterfowl in the wet habitats, raptors soaring overhead, rails and coots in their appropriate habitats, nesting Great Horned Owls in the willows, and Ruby-throated Hummingbirds feeding on the nectar of the abundant wildflowers lining the paths. Also expect Great Crested Flycatchers; Tree, Bank, Rough-winged and Barn swallows; wrens; Wood, Hermit and Swainson's (migrant, not resident) thrushes; Blue-gray Gnatcatchers; Yellow-throated, Solitary, Red-eyed, and Warbling vireos; Black-and-white, Blue-winged, Tennessee, Yellow, Cape May, Magnolia, and Blackburnian warblers. In addition, you should be able to find Northern Waterthrushes; Common Yellowthroat, Wilson's, and Canada warblers; and American Redstarts. Scarlet Tanagers, Rose-breasted Grosbeaks, Indigo Buntings and Purple Finches can all be found at this site.

OTSEGO COUNTY
Otsego Lake (Map 4.4)

Rating: Spring***, Summer*, Autumn**, Winter**

Otsego Lake is one of the many areas recognized by birders as a generally good, productive site in nearly every season; but it lacks specialties in any one. The 22-mile drive around the lake offers many vantage points along the way. The lake is surrounded by private land, except for a state park, Glimmerglass, at its north-eastern end. Access to the lake may be gained from the park and at the village of Cooperstown. Admission is charged at Glimmerglass during the summer season.

Otsego Lake lies just north of Cooperstown, and NY 80 runs parallel to the lake along its west side. Smaller roads complete the loop around the lake. Leave I-90 at Exit 30 (Mohawk-Herkimer). Proceed south on NY 38 for about thirteen miles, to its intersection with NY 80. Here turn right (south) and proceed south along the west side of Otsego Lake, stopping at convenient overlooks and vantage points to bird the lake and its surrounding greenery. At the south end of the lake, proceed east into Cooperstown, and then head north, up the east side of the lake, to Glimmerglass State Park. There is a campground within the park, and in the surrounding trees one can usually find warblers and other passerine species in May and early June.

Lordsland Conservancy (Map 4.4)

Rating: Spring****, Summer***, Autumn***, Winter*

The Nature Conservancy owns this eighty-acre tract, which is a combination of two approximately equal tracts, both of them farms in the nineteenth century. Here, the history of man and nature through the past 150 years is recorded in crumbling foundations, unpruned apple trees, a dug-out watering spring, widening fencerows, and a few surviving hopvines, which commemorate an entire economy. Stands of mixed woods, hemlocks, and hardwoods, successfully compete with old maples that shade out blackberry canes in favor of cinnamon fern. There are gullies lined with viburnums, hornbeam, and witchazel, through which streams flow in the spring then pass underground in the deep summer. A fragment of a cliff remains, containing fossils 375 million years old. On the property, a stream flows southward, widening in spring to water an acre of cowslips. The rare Jacob's ladder grows along the stream, and by mid-summer Joe-Pye weed, golden-rod, and asters fill the swamp, and willows, dogwoods, and alders grow.

Until about fifteen years ago, the western knoll and the nearly level south-ern twenty-five acres of the preserve were in hayfields or pasture. Now, natural succession of plants is beginning to become evident, varying, of course, with the soil and water conditions. In the southern corner of the preserve a parklike scene is created by white birches standing deep in bracken and ground pine. Elsewhere, su-mac, aspen, and maple edge out into weedy fields. Wildlife abounds on the pre-

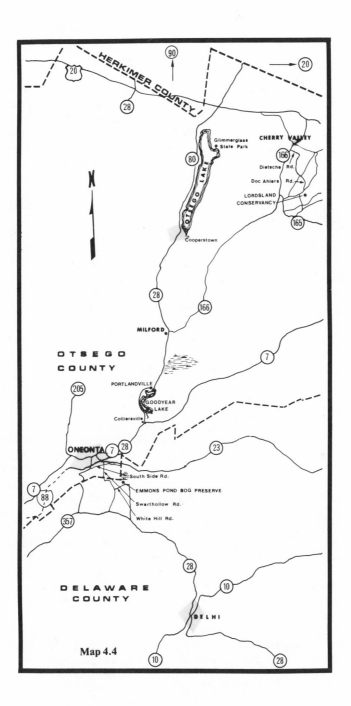

Map 4.4

serve. Signs of deer are everywhere, snakes sun themselves on the stone walls, hawks soar overhead, beavers maintain their dams across the poplar-bordered stream, which eventually floods the valley floor.

To reach Lordsland Conservancy, leave I-90 at Exit 29. Proceed south on NY 10 about eleven miles, to its junction with NY 20. Here, turn right and proceed west approximately 3.5 miles to the junction of NY 20 and NY 166. Turn left and proceed on NY 166 for two miles to the center of Cherry Valley. Continue south out of the village, being alert for Dietsche Road, which will come in on the left. At Dietsche Road, turn left and follow it east and southeast for about 2.5 miles. Dietsche Road becomes Doc Ahlers Road. Follow it to the Lordsland Conservancy entrance and park.

Once inside Lordsland, follow the trail through the upland habitat where you should be able to find Ruffed Grouse, woodpeckers, foraging swallows from neighboring farms, jays, crows, chickadees, nuthatches, thrushes, warblers, Scarlet Tanagers, Rose-breasted Grosbeaks, Indigo Buntings, finches, and sparrows. The water and brushy cover in this north-south valley attract migrating birds in great numbers. Many remain to nest in the swamp, including American Woodcocks; Alder Flycatchers; wrens; catbirds; Blue-winged, Yellow, and Chestnut-sided warblers; and Common Yellowthroat, to name but a few. Explore for as long as possible along the preserve trails.

Goodyear Lake and Portlandville (Map 4.4)

Rating: Spring****, Summer***, Autumn**, Winter*

Goodyear Lake is in the village of Portlandville, seven miles northeast of the city of Oneonta. Covering the lakeside is relatively easy. It can be reached by car or bicycle, and also may be covered from either. There are no admission fees, and for those indifferent to energy consumption, there are convenient points all along the shoreline where birders can park and safely walk to overlooks of the lake and down paths to the water. At several of the stops a telescope would be a real asset. The very best season to bird the lake and its environs is spring, when raptors, ducks, shorebirds, and warblers are found in peak numbers.

Oneonta is accessible by several routes, and an assortment of streamlined highways. From downtown Oneonta at the junction of NY 205 and NY 7 (NY 28), proceed east-northeast, on NY 7 (NY 28) through town, toward Colliersville. At Colliersville, turn left at the traffic light downtown and follow NY 28 north.

Proceed along NY 28, with Goodyear Lake on the right. Park along the roadway just above the dam and check for loons, grebes, herons, and bitterns in season and Pintails, Green-winged and Blue-winged teals, Wood and Ring-necked ducks in spring and autumn. A whole assortment of other waterfowl is easily visible in different seasons and might be called aesthetically pleasing but not particularly astonishing. In spring, the birder will be able to see many migrants just above the dam, but precious few of those species will remain to nest. However, one can easily find nesting Mallards and stunning Wood Ducks throughout the summer.

Additionally, the avid birder should be able to find Ospreys, Broad-winged and Red-tailed hawks, American Kestrels, Ruffed Grouse, and Ring-necked Pheasants in good numbers throughout spring and summer and well into the autumn season. Follow NY 28 northeast, stopping at the end of the lake just as you enter Portlandville. Check among the reedy areas and along the sandbar. Here a few nesting species of ducks and Pied-billed Grebes can be found. Look for active Osprey nests high in the trees bordering the lake. Proceed to the corner, where there is a quaint antique shop. Scan for Cliff Swallows that have plastered their gourd-shaped structures under the eaves of the shop. Barn Swallows commonly hawk insects in the same area. Proceed a short distance farther along NY 28, to the first right turn. Cross the bridge and park along the swampy area on the right. Both sides of the road are excellent here, as well as the bushy area on the street to the left. In the height of the spring migration birders from some distances around, congregate here to view northwardbound hawks and warblers. In the wet sloughs along the road one can frequently see rails and gallinules. Scan the sky for migrants passing over and listen for the familiar call of nesting Killdeers, Belted Kingfishers, picids and flycatchers.

DELAWARE COUNTY
Emmons Pond Bog Preserve (Map 4.4)

Rating: Spring****, Summer***, Autumn**, Winter*

Not far southeast of Oneonta is a 150-acre nature preserve containing a spring-fed pond encircled by a thick sphagnum mat, an upland hardwood forest, an abandoned pasture, a swamp forest, two or three small ponds, a pine plantation, rock outcroppings, and luxuriant patches of wildflowers, blueberries, herbaceous plants, and ferns. Emmons Pond Bog and the land surrounding it were purchased in 1970 by the Nature Conservancy, and it therefore enjoys environmental protection. The impressive serenity and lush beauty of this bog have been saved from the fate that has befallen so many bogs surrounded by rich agricultural and dairy land.

The open pond covers about seven acres. Those who follow the outlet on the north will discover that the stream flows over steep falls to the Susquehanna River. In the center, the pond is about twenty-five feet deep, and about twenty feet deep even close to the shore. The bog mat that surrounds the open water is approximately fifty feet wide and thirty feet across.

A close examination of the mat will reveal at least three species of sphagnum moss, a dense carpet of leatherleaf (whose name comes from its thick leathery leaves, which are perfectly suited to prevent evaporation), carnivorous plants like the beautiful pitcher plant (whose trumpet-shaped vase invites insects to crawl in, where they are dissolved by the plant) and sundew (another plant that consumes insects to provide itself with necessary nutrients). Also, bog cranberry, cottongrass, beautiful wildflowers like adder's tongue, bog laurel, bloodroot and meadow rue, delicate and willowy ferns, and finally, the magnificent orchid rose pogonia and trailing arbutus thrive here. All around the edge of the mat are high and low bush blueberry, pinxter azalea, and mountain holly.

A fortress of pines and spruces protects the bog on the north. Walking through this swamp forest will give the birder access to stemless lady slipper, fringed polygala, painted trillium, wild azalea, starflower, and mayapple.

From downtown Oneonta proceed along Main Street to the NY 23 crossing of the Susquehanna River. Follow it south over the river, then, instead of heading east on NY 23, proceed south a very short distance to South Side Road. Take South Side Road east a short way to its junction with Swarthollow Road, a dirt road that comes in on the right. Turn right and follow Swarthollow Road to White Hill Road, which comes in on the left. An old white schoolhouse at the junction of White Hill Road and Swarthollow Road is an excellent landmark. Turn left here and proceed to a sign for Emmons Pond Bog Preserve. The sign will be on the north (left) side of the White Hill Road. If you come to Coe Hill Road, you have gone too far; turn back to locate the preserve. Park along the shoulder of White Hill Road. Once inside the preserve it is imperative that all birders stay on the trails. This bog is — as are all bogs — fragile habitat whose success can only be ensured through visitors' care and consideration. Birds one might expect to see in and around the Emmons Pond Bog Preserve include: Pileated Woodpeckers, Wood Thrushes, Veeries, Yellow-throated, Solitary and Red-eyed vireos, Black-throated Blue, Black-throated Green, Canada, and Blackburnian warblers, Ovenbirds, and American Redstarts in the wooded hillsides; Blue-winged, Golden-winged, Nashville, and Chestnut-sided warblers on the brushier areas of open hillsides; Bobolinks, Grasshopper, Savannah, and Vesper sparrows in the open grasslands; and Great Blue and Green herons, bitterns, and Barred Owls around the bog. These species can all be found nesting in their appropriate habitats within the preserve.

West Branch Nature Preserve (Map 4.5)

Rating: Spring***, Summer*, Autumn***, Winter*

This Nature Conservancy-owned preserve of more than 445 acres is just outside the village of Hamden. Indeed, a glorious place to bird, it includes a twenty-acre stand of giant virgin white pine (the height of these trees is awesome. They rise high above the surrounding deciduous forest, and are visible from some distance); a fifteen-acre island on the West Branch of the Delaware River, one of the boundaries of the preserve; several large tilled fields bordering roads and the river; and heavily forested hills on the north end of the preserve. Obviously, this variety of habitats could provide a good mix of birds — and it does. The differences between the habitat types is quite pronounced and various field and edge habitats provide additional places to look for birds. In both spring and autumn, migrants pass through the preserve in large numbers. All of the passerines common to these habitats nest within the preserve, some in prodigious numbers. The birder should note that this is not an isolated pocket of natural vegetation but rather a protected area of considerable size, and as such, offers several extensive tracts of habitat suitable for supporting avian life. Old logging roads wind through the preserve, and they, as well as the trails branching off from them, can be explored to great advantage.

The village of Walton is accessible *via* NY 17, 30, and 206, or I-88 and NY 206. From Walton proceed east on NY 10, paralleling the West Branch of the Delaware River, about six miles. The preserve is six miles east of Walton and two miles west of Hamden. Watch for two stone houses set back from the road on the left (north) side of NY 10. Another clue is to look for a large red barn on the south side of NY 10, directly across from a large white house set close to the road on the north side of NY 10. Turn left into the drive-road of the preserve, and drive to the old logging truck turn-around at its terminus and park. Paths lead from the turn-around in several directions.

Some bird species one might expect to encounter at the preserve are: American Bittern, Wood Duck, Broad-winged and Red-tailed hawks, American Kestrel, Ruffed Grouse, Killdeer, American Woodcock, Common Snipe, Spotted Sandpiper, Whip-poor-will, Common Nighthawk, Ruby-throated Hummingbird, at least four picids, three or four flycatcher species, Tree, Bank, and Barn swallows, White-breasted Nuthatch, Brown Creeper, Brown Thrasher, Wood and Hermit thrushes, Veery, Yellow-throated, Solitary, Red-eyed and Warbling vireos, Nashville, Yellow, Magnolia, Blackburnian, Chestnut-sided, and Canada warblers, Bobolink, Eastern Meadowlark, Red-winged Blackbird, Northern Oriole, Scarlet Tanager, Rose-breasted Grosbeak, Indigo Bunting, and Purple Finch.

Bear Spring Mountain Wildlife Management Area (Map 4.5)

Rating: Spring****, Summer***, Autumn***, Winter*

Some 2.5 miles southeast of the village of Walton lies the sixth-largest wildlife management area in the state, with holdings totalling 7200 acres. Here the general public can pursue wildlife-related activities and birders can particularly profit from the array of habitats within the Bear Spring Mountain Wildlife Management Area. It combines most of the best features of Delaware County within its borders. The main habitat types are upland fields, brushlands, and deciduous woodlands. There are several pine plantations, as well as beaver ponds, several large and a few small ponds, roads and foot trails traversing the area with a generous mixture of picnic and camping areas. Members of the Delhi Bird Club have tallied more than 105 species within the area without any special effort to record or locate every species possible. It is however possible to see and hear 22 warbler species in a day, at the height of spring migration in May. Birding the area is rated from difficult to moderately easy, depending upon the amount of walking one does, and the steepness of the hills climbed. Good hiking shoes are recommended in every season, for only the main road is paved, and offshoots from it are gravelled or dirt. Insect repellent is advised in spring and summer, and poison ivy is present but scarce within the area.

The village of Walton is accessible *via* NY 17, 30, and 206, or I-88 and NY 206. Leave Walton on NY 206 and head southeast toward Downsville. Proceed to

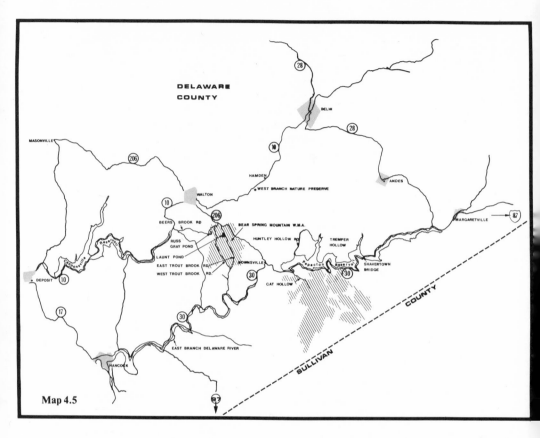

Map 4.5

the top of Bear Spring Mountain; there, opposite a large radio station tower, is the first entrance to the wildlife management area. If you turn right here, you will be on the northern, gravel extension of West Trout Brook Road and Beers Brook Road. The administration building is immediately on the left, and one might stop here to check road conditions and possibly to get a detailed map of the management area. There is another entrance to the area a quarter-mile farther down NY 206, on the right side of the road. It can be gained by turning right on East Trout Brook Road. The administration building can be reached from that entrance by taking the first right turn and proceeding west on the vehicle access road about one-half mile.

From the first entrance, proceed southwest on Beers Brook Road for 1.2 miles to Russ Gray Pond, on the left. There is plenty of room to park near the pond. From the second entrance proceed south on East Trout Brook Road approximately 0.7 mile to Launt Pond, on the right. These are the two largest ponds within the area, but there are also several smaller ones. Continue along the indicated roads (see accompanying map 4.5), stopping to bird frequently. Local birders often prefer to bird in the vicinity of picnic areas and campsites and have found Rustic Bridge Road and Spruce Grove Campsite especially productive.

Look especially for migrant Green-winged Teals, Pintails, Blue-winged Teals, Ring-necked Ducks, and Buffleheads in the ponds in both spring and autumn. Goshawks and Sharp-shinned and Cooper's hawks nest within Bear Spring Mountain, as do Red-shouldered, Broad-winged, and Red-tailed hawks. The nests of these raptors are, however, difficult to find. Other marvelous finds among nesting birds in the area are: Yellow-bellied Sapsuckers; Alder Flycatchers; Swainson's Thrushes; Golden-crowned Kinglets; Magnolia, Cape May, Yellow-rumped, Baybreasted, Prairie, and Mourning warblers. Look also for breeding Indigo Buntings; American Goldfinches; Rufous-sided Towhees; and Savannah, Grasshopper, Vesper, Chipping, Field, White-throated, Swamp, and Song sparrows.

Pepacton Reservoir (Map 4.5)

Rating: Spring***, Summer*, Autumn*, Winter**

Pepacton Reservoir is part of the extensive reservoir system constructed in the Catskill Mountains to provide New York City with fresh clean water. These impoundments have been constructed in picturesque settings with mountain scenery as a backdrop. They have been planned with a sizable buffer zone of natural and planted vegetation, usually trees, down to the water's edge. Some of the reservoirs lie in pre-glacial lakebeds and the surroundings are reminiscent of wilderness. Pepacton Reservoir lies between Downsville and Margaretville and covers some 7000 acres. It is nearly twenty-five miles long and about one-half mile wide on average. Its maximum depth is 200 feet. Brown trout, smallmouth bass, pickerel, yellow perch, rock bass, sunfish, and bullheads are the catch the shore fishermen will be pursuing.

Both Downsville, on the reservoir's western or dam site end, and Margaretville, at its eastern or inlet end, are easily gained *via* I-87 and NY 28, or NY 17 (Quickway) and NY 30.

NY 30 from Downsville to Margaretville spans both the southern and northern shores of the reservoir and some of the favorite spots to bird are along Cat Hollow, Huntley Hollow, Shavertown Bridge, and near Tremper Hollow (see map). There is a good deal of state-held land on the south side of the reservoir, and walking through some of it should be very productive.

Cannonsville Reservoir Watershed Area (Map 4.5)

Rating: Spring****, Summer****, Autumn***, Winter*

The Cannonsville Reservoir is the most recent of New York City's impoundments in the Catskill Mountains. It was constructed by the Army Corps of Engineers on the West Branch of the Delaware River, and encompasses approximately 450 square miles. The dam lies about three miles to the east of Deposit. The high hills of the western Catskills seem to fall away into the reservoir, which makes driving along its boundary roads an extremely pleasant experience. Birding the watershed is rated "easy," but local birders advise having insect repellent handy, especially in spring and early summer.

Leave NY 17 (Quickway), just south of Deposit. Gain NY 10 and proceed northeast for three miles, to the beginning of the reservoir.

One is not allowed access to the property immediately surrounding the water of a designated reservoir without a fishing license and a fishing permit. However, birders are free to walk all of the surrounding roads provided that cars are left in one of the many parking lots along the reservoir's periphery. One can also drive the side roads, where traffic is almost nonexistent. Stopping anywhere along these roads except at spacious, safe turn-outs is very ill advised. Deer, muskrat, opossum, raccoon, porcupine, fox, woodchuck, rabbit and snowshoe hare are some of the animals found in the area. Take care while driving the back roads at night, for deer and some of the more nocturnal mammals present a real hazard to the night motorist.

The wetlands and upland woods offer two diverse habitats in which to bird, and corresponding avian variety. Migrating loons stop on the reservoir in spring, along with Black Ducks, and Mallards, which remain to nest, as well as Canvasbacks, Redheads, scaups, Buffleheads, and Common Goldeneyes. Migrating Ospreys pass overhead near the dam. A search of the woods should produce migrating woodpeckers, flycatchers, thrushes, Eastern Bluebirds, vireos, and more than twenty warbler species. Scan all of the adjacent fields for upland birds both passing through and nesting. In summer and autumn, Great Blue Herons, Snow and Canada geese, waterfowl, mergansers, and Belted Kingfishers are common sights on the upper reservoir. Over the central fields, look for migrating eagles and nesting hawks. In winter, and right up through early spring, Bald Eagles, Red-tailed Hawks, and Great Horned and Snowy owls can be seen near the dam.

The prime time to investigate the back roads and dirt lanes is during the spring migration peak, when activity is high on the reservoir and in the deciduous woods and upland fields. Take along a picnic lunch and enjoy it at one of the tables by the reservoir.

ACKNOWLEDGMENTS

The author expresses thanks to the following people who supplied site information: Elva Hawken, Leslie E. Bemont, Cheryl Boise, Lynn Clark, John A. Gustafson, Shirley Hartman, Marion Karl, Jay Lehman, Jean Miller, Kathryn Wilson, and especially to Jay Lehman, whose reading of and comments on the Susquehanna chapter provided a valuable overview.

Oneida Lake Basin — Region 5

THE ONEIDA LAKE BASIN REGION contains all of Herkimer, Madison, Oneida, Onondaga, and Oswego counties, and the portion of Cayuga County north of the Seneca River. This totals more than 5100 square miles, most of which lies in the Lake Ontario Lowland and Oneida Lake Plain, along with lesser amounts in the Mohawk Valley and the western Adirondack Hills (Map 5.1).

The most important water features in this region are Lake Ontario, on which it borders; Oneida Lake and its tributaries; the Oneida Lake outlet, Oneida River; the Oswego River; two of the Finger Lakes, Lakes Skaneateles and Otisco; Onondaga Lake; Delta Lake (Reservoir); and the Fulton Chain Lakes and Stillwater Reservoir (both in north-central Herkimer County). The drainage of the region is divided between the Moose River, which empties the northeastern section, the Mohawk River which largely clears the eastern section, and the Seneca and Oneida Rivers (which together form the Oswego River), which drain most of the western half of the region. The northwest portion of Oswego County empties through the Salmon and Little Salmon Rivers and their tributaries, and the area around Hamilton, in south-central Madison County, drains to the south into the Chenango River.

The primary differences between a marsh and a swamp are that a marsh is treeless and only temporarily saturated with moisture, whereas a swamp is a low-lying tract of treed land that is permanently saturated. A swamp is usually overgrown with vegetation and is therefore a cross between an entirely aquatic and a marsh environment. The formation of swamps often takes place when lake basins fill up and the surface is so flat that run-off of rain water is exceedingly slow. Damp-soil vegetation usually grows up, helping to maintain the sodden swampy condition.

Within *Kingbird* Reporting Region 5, there are birdrich swamp areas that really ought to be explored regularly. The Ontario Ridge and Swampland lies east-southeastward from Oswego toward Oneida Lake and Syracuse. The area is poorly drained and characterized by numerous swamps and ridged ground moraine. All along the Lake Ontario shoreline from the western Cayuga County border north to the outlet of the Salmon River and farther to the Sandy Pond area, there are woods and cattail-fringed swamps that bear investigation. Sandy Pond

161

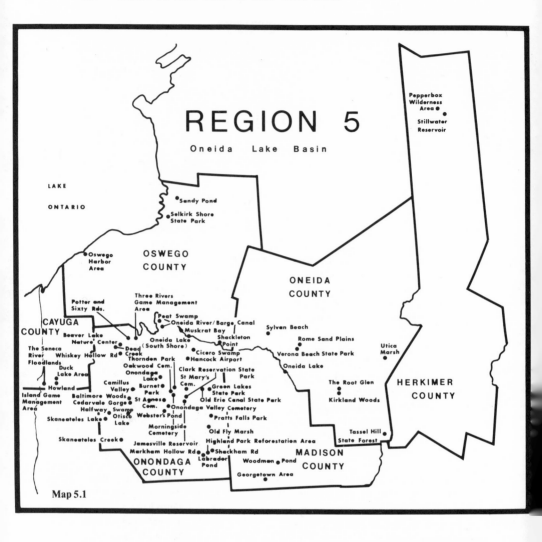

REGION 5

Oneida Lake Basin

Pepperbox
Wilderness
Area ●
Stillwater
Reservoir ●

LAKE

ONTARIO

● Sandy Pond

● Selkirk Shore
State Park

● Oswego
Harbor
Area

OSWEGO
COUNTY

ONEIDA
COUNTY

Three Rivers
Game Management
Area

Potter and
Sixty Rds. ●

● Peat Swamp
● Oneida River/Barge Canal

Sylvan Beach

CAYUGA
COUNTY
Beaver Lake
Nature Center ●
Muskrat Bay ●

Shackleton
Point ●

Rome Sand Plains

● Oneida Lake
(South Shore)

Cicero Swamp ●

Verona Beach State Park

Utica
Marsh ●

The Seneca
River
Floodlands
Whiskey Hollow Rd ● Dead
Creek
Thornden Park
Hancock Airport ●

● Oneida Lake

Duck
Lake Area

Oakwood Cem.
Onondaga
Lake ●

Clark Reservation State
Park

● Howland
Island Game
Management
Area

Camillus
Valley ●
Burnet
Park
St Mary's
Cem.

The Root Glen ●

HERKIMER
COUNTY

Baltimore Woods ●
Cedarvale Gorge ●
St Agnes
Cem. ●

Green Lakes
State Park ●
Old Erie Canal State Park

Kirkland Woods ●

Halfway
Swamp
Otisco
Lake ●
Websters Pond ●

Onondaga Valley Cemetery ●

Skaneateles Lake ●

● Pratts Falls Park

Morningside
Cemetery ●

Old Fly Marsh ●

Tassel Hill ●
State Forest

Skaneateles Creek ●

Jamesville Reservoir ●

Highland Park Reforestation Area

Markham Hollow Rd ●
● Shackham Rd

MADISON
COUNTY

ONONDAGA
COUNTY

Labrador
Pond

Woodman ● Pond

Georgetown Area ●

Map 5.1

itself has a host of large aquatic plant swamps and marshes in its immediate vicinity. Cicero Swamp, located between Syracuse and Oneida Lake, is a true swamp with adjoining sections of spruce-sphagnum bog. At various points around Oneida Lake there are heavily wooded swamps, and the nearly featureless plain south of Oneida Lake contains broad swamps and mucklands.

The eastern Lake Ontario Plain and the Oneida Lake Plain are dominated by low elevations, occasional sand deposits, areas of post-glacial drift, and good, rich agricultural soils. The natural vegetation most common in this region consists of forests characteristically containing relatively frequent stands of American elm and red maple, birch, beech, and hemlock, with oak and other northern hardwoods present but less abundant. In the highland areas of eastern and northern Herkimer County, spruce and fir are commonly found mixed with northern hardwoods; in fact, they outnumber the hardwood stands. Small remnant areas of oak-hickory forests can still be found in some areas abutting Lake Ontario and along some stretches of the southern shore of Oneida Lake.

HERKIMER COUNTY
Pepperbox Wilderness Area (Map 5.1)

Rating: Spring****, Summer***, Autumn***, Winter**

The jagged backbone of the Adirondack Park comprises a mountain complex that annually lures tens of thousands to its scenic mountains, fertile valleys, streams, lakes, meadows, watersheds, and wildlife. There are unexcelled opportunities for hiking, fishing, photographing, and birding within the park's six million acres. The impact of recreationalists on several of the most overused areas has prompted the inclusion of the Pepperbox Wilderness Area in this book. The dynamic and shifting condition of some of the more intensively used High Peaks, caused directly by cleated boots, improperly pitched tents, the uprooting and trampling of wildflowers and ferns, can certainly be lessened by changing the patterns of recreational use: that is, we must stop overusing the more fragile areas and encourage more use of some equally lovely, yet almost undiscovered, wildlands in the glorious heritage that is Adirondack Park.

The Pepperbox Wilderness Area lies in the west-central Adirondacks, in the high wooded slopes of tranquil northern Herkimer County. It is a remote area of true wilderness with quiet, bird-rich swamps and marshes tucked away in the deep woods. Beaver flows and pure, cold rivers and creeks are part of the wetland ecosystem that awaits the pack-and-binoculars-laden birder. The Pepperbox Wilderness Area lies north of Moshier Falls, the Beaver River, and Stillwater Reservoir. To the south rises Stillwater Mountain (elev. 2263 feet), and to the north stands Alder Bed Mountain (elev. 2039 feet).

To reach the wilderness area, approach from Watertown on its west; leave I-81 at Exit 45 and head south and east on NY 12. Proceed to Lowville. There in the center of the village cross NY 26 and proceed east on Number Four Road. Cross the railroad tracks and continue east through the hamlets of Watson, Bushes Landing, Petries Corners, and Crystal Dale. Continue to Number Four and the junction with Stillwater Road. Follow this east to the western end of Stillwater Reservoir. If approaching from the Utica area and points south, follow NY 12 (NY 28) northeast to Eagle Bay. There turn north (left) and follow Big Moose Road through Big Moose to the western end of Stillwater Reservoir.

You are now ready to hike through the area and bird its rare and invigorating, yet unoccupied, forests, streams, and basins. There are several narrow, ungullied, and relatively unmaintained trails which should pose a mild challenge for the birder. They can be followed; however, it would be especially prudent to carry along a copy of the topographic Number Four Quadrangle map, which shows the entire Pepperbox Wilderness Area. Insects swarm over the area in the breeding season, so appropriate repellent will be useful. Park your vehicle safely off the road near Stillwater Reservoir dam and outlet. Good waterproofed hiking boots are highly recommended.

Cross the outlet of the reservoir at the dam. Proceed due north from here for three miles to Sunshine Pond. For the first half-mile, the trail is closely lined with various low vegetation and hardwood tree species. There is an extreme diversity of wildlife all along the hike. In addition to raptors, cuckoos, owls, nighthawks, hummingbirds, and Red-headed Woodpeckers, and Yellow-bellied Sapsuckers, there are kingbirds, phoebes, Yellow-bellied and Least flycatchers, numerous Eastern Wood Pewees, and singing Olive-sided Flycatchers. Look carefully for fisher, mink, and otter tracks. Deer and deer mice tracks will be everywhere. After approximately one-half mile, the trail opens up into an area where various river drainage systems cross one another. These systems are open and very easy to follow as they conform to the rolling terrain. There are networks of beaver ponds and marshes, and lakes are rich in supporting the wildlife dependent on the productive swamp ecosystem. Continue due north exploring the fringes of the wettest areas. Look for nesting Common Loons, herons, bitterns, Wood Ducks, Green-winged Teals, Mallards, and Black Ducks, and the rarely reported nesting Red-breasted Merganser. Look overhead for Ospreys, Bald Eagles, Marsh Hawks and American Kestrels. Listen for Killdeers, American Woodcocks, and Common Snipes. As you reach Sunshine Pond, skirt its fringes, being watchful for the occasional rail secretively hunting along the shore.

Cross west to Deer Pond, skirting its southwest shore, and proceed on the trail south and west to the Moshier Ponds Chain. Here again you will find yourself traveling easily over open terrain dotted with unnamed ponds and remnants of beaver works. After thoroughly exploring this remote landscape, turn south following an inlet to Moshier Pond that drains the area from approximately Cropsy Pond.

Continue south through the open marshland, where you will remeet the original trail, approximately 1.3 miles north of Stillwater Reservoir. This can easily turn into an all-day hike, and a group of birders who recently did such a circuit spent six or seven hours covering it.

ONEIDA COUNTY
Utica Marsh (Map 5.2)

Rating: Spring****, Summer**, Autumn***, Winter*

Utica Marsh, located in the northern part of the city of Utica, encompasses 100 ± acres of freshwater wetlands. Its acquisition and subsequent status as a protected wildlife sanctuary was secured after nearly a decade of coordinated efforts of dedicated and persistent naturalists and state and local government agencies. Local officials had to be convinced that at least this section of the once enormous wetlands

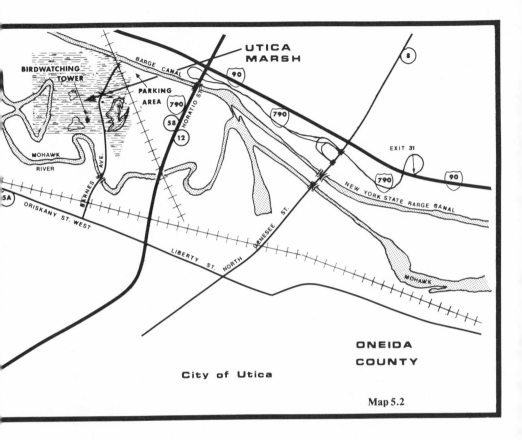

BIRDWATCHING
TOWER

PARKING
AREA

UTICA
MARSH

MOHAWK
RIVER

ORISKANY ST. WEST

LIBERTY ST. NORTH

EXIT 31

NEW YORK STATE BARGE CANAL

MOHAWK

ONEIDA
COUNTY

City of Utica

Map 5.2

of the Mohawk River floodplain should not be sacrificed to commercial interests; that the wetland should be managed according to physical and biological laws and not the laws of commerce and marketing. For nearly half a century the wetlands near Utica have been disappearing, as these unique and productive ecosystems have been filled and exploited. However, the Utica Marsh stands as a critical correction of past errors. It survives as a vulnerable reminder that the constructive channeling of human momentum can defend and renew and preserve our natural treasures.

To reach the Utica Marsh, leave the Governor Thomas E. Dewey (New York State) Thruway (I-90), at Exit 31. Turn left, thereby gaining Genesee Street. Proceed south on North Genesee Street to its intersection with Liberty Street. Follow Liberty Street to Oriskany Street. Turn right and follow Oriskany Street West to Barnes Avenue. Turn right (north), cross the Conrail tracks, then cross the Mohawk River, and proceed north to a convenient parking area on the right, not far from more railroad tracks.

You will be able to bird the marsh to best advantage if you've remembered to bring a spotting scope and, in late spring and summer, some mosquito repellent.

Because of its advantages over river boats and canal barges the railway soon surpassed those means of transport. By the mid-1800s, there were rail connections between New York City and lakes Ontario and Erie. At Utica Marsh alongside the tracks, there is a trail on old New York State Barge Canal holdings. Walk the nature trail, observing the wildlife in the wetlands below and to both sides. The rich alluvial soil characteristic of freshwater floodplains supports the vegetation upon which both vertebrate and invertebrate animal life depends. Water plantain, arrowhead, bulrush species, and blue and yellow flag thrive in the shallow-marsh. Orange day lily, swamp milkweed, elderberry, spotted and hollow Joe-Pye weed, late goldenrod, wild cucumber, and gray and red dogwood grow in the wet meadows. Walk through the uplands and note the black and red raspberry, wild strawberry, and thimbleberry patches among the ash, alder, and maple stands.

More than 140 bird species have been recorded in Utica Marsh as well as raccoons, muskrats, and deer. South and somewhat west of the railroad tracks are several deep clay pits, reminders that brick manufacturing was once profitable here. These excavations are filled with water, especially after spring flooding. Sampling of these pools by fisheries biologists found them amply stocked.

Early morning seems to yield the best results when birding the marsh. A trail leads from the parking lot to an observation tower. Two luxuriant marshes can be viewed from atop the tower. Follow the trail and tracks looking for Pied-billed Grebes and American and Least bitterns, all of which breed here. The six heron species recorded here have not been known to nest. Wood Ducks, Mallards, Black Ducks, and Blue-winged Teals nest in the marsh, and eight other duck species can be seen in either spring or autumn migrations. Ospreys are common spring migrants through the marsh, and they have often been seen overhead with bullheads or panfish in their talons. Look around the deep pools for Ospreys, by following the improved access route on the west side of Utica Marsh.

Marsh, Red-tailed, and Rough-legged hawks have often been seen in the area. American Kestrels nest within the marsh. Pheasants and Ruffed Grouse are common here, and Virginia Rails and Sora, as well as Common Gallinules, and American Coots nest here. Killdeers, woodcocks, snipes, and Spotted Sandpipers nest in Utica Marsh, while both yellowlegs species and Solitary, Pectoral, and Semipalmated sandpipers can be seen on migrations. Wilson's Phalarope has often been observed and some suspect it may some day be determined as a breeder in the marsh. This is one place to come to study the song differences of both Long-billed and Short-billed marsh wrens, as they hold forth long and loud, defending territories in the spring and early summer. Vireos and at least eighteen warbler species have been seen in spring migration. Nine sparrow species have been added to the Utica Marsh list over the years, but only Swamp and Song sparrows are confirmed breeders.

In summary, Utica Marsh is centrally located and easily accessible to those traveling east-west across the state. It deserves to be visited and almost always has surprising rewards, whether avian or botanical.

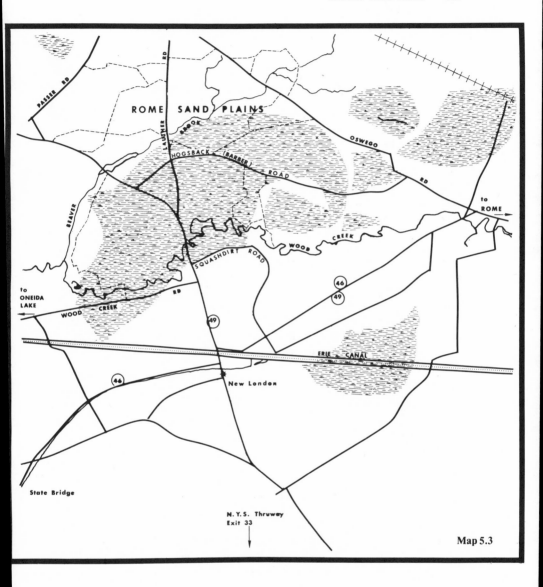

Rome Sand Plains (Map 5.3)

Rating: Spring****, Summer***, Autumn**, Winter*

This area slightly east of Oneida Lake and west of the city of Rome lies on the edge
of the Ontario Ridge and Swampland physiographic region; it is characterized by

infrequent sand deposits, areas of post-glacial drift and ridged ground moraine, and a few long narrow hillside gullies.

At the Rome Sand Plains, a wet, spongy bog is accessible over railroad tracks on its northeast side. A variety of habitat types occur within a small area here, and the more walking one does, the better the birding will be. Listen for nesting Whip-poor-wills. Other specialties to be watched for are small nesting passerines with northern, boreal breeding affinities. Look for nesting Red-breasted Nuthatches and Brown Creepers methodically taking insects and larvae from the tree bark. In the more open woodlands listen for the distinctive song of the Hermit Thrush. The Black-and-white Warbler, a ground or near-ground nester, can be found near the swampland edges and in mixed swamp-and-deciduous growth. Nashville, Yellow-rumped (Myrtle), Black-throated Green, Blackburnian, and Chestnut-sided warblers can be found near the northern-type coniferous forest-bordered bog, especially where there is a groundcover of sphagnum and labrador tea. Watch for Ovenbirds teetering along the open forest floor, especially in areas of abundant fallen logs, and for Northern Waterthrushes near the wooded streams and along banksides or ravine cuts. Along the marsh and swamp edges, as well as the edges of brushy second-growth thickets, watch and listen for nesting Common Yellowthroats. Rufous-sided Towhees and White-throated Sparrows can be heard singing near their nests as one approaches the bog, over the railroad tracks.

To reach the Rome Sand Plains, leave the New York State Thruway (I-90) at Exit 33. Follow NY 365 north a short distance to its intersection with NY 31. Turn left and proceed on NY 31 to the hamlet of State Bridge, where NY 46 intersects it. Turn right and proceed north on NY 46 to the hamlet of New London. Here, NY 46 and NY 49 meet. Follow NY 49 north for approximately 1.1 miles, crossing Wood Creek on the way to Lauther Road. Continue on Lauther Road for one-half mile to Hogsback Road and park the car safely off the road here.

Bird the woods on the left, reached by walking down the sandy road, which will be obvious. Then walk back on Hogsback Road, cross Lauther Road and walk east on Hogsback about one-half mile to a primitive road coming in on the left, which leads to a stand of spruce, cedar, balsam, and a few hemlocks. The road is skirted with second-growth deciduous birch and aspen. The walk is about one mile, and it includes a bird-rich wooded ravine and several water cuts.

Once back at the car, proceed up Lauther Road, cross Oswego Road (at this point Lauther Road changes its name to Humaston Road). Continue driving north on Humaston (Lauther) Road stopping once or twice to bird the possibly productive areas on the left side of the road. Park off the road where Humaston Road meets the railroad tracks and walk to the right (east), on the tracks, to reach the bog.

Tassel Hill State Forest (Map 5.4)

Rating: Spring***, Summer**, Autumn*, Winter*

Tassel Hill State Forest, on the southeastern corner of Oneida County, includes some 2500 acres of mixed hardwood-conifer stands; some shrubby, low, second-

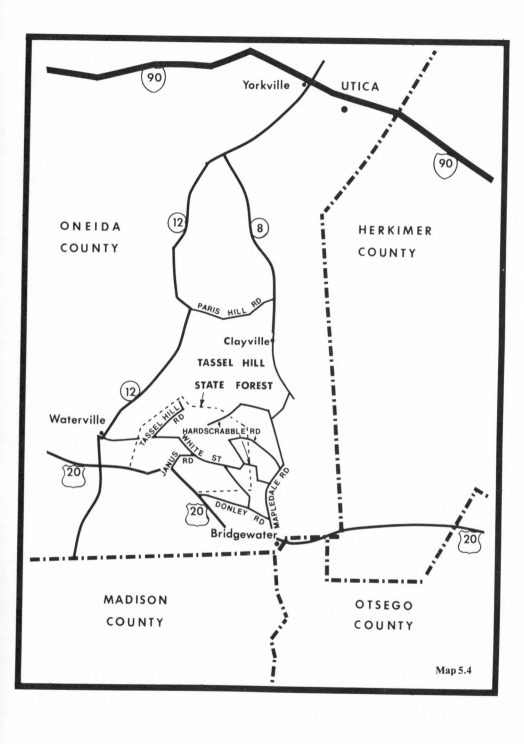

Map 5.4

growth deciduous trees; fields; numerous ponds and streams; some swampy areas; and a labyrinth of dirt trails giving access to every habitat type. These trails should not be driven, and they are too rough for smooth bicycling. Birding the area is rated "difficult to moderately easy," depending upon how taxing the birder wishes to make it. Hunting is allowed during the deer season, so birders should avoid the area from the middle of November through the first week of December. People who have birded the wildlife forest prefer it in spring and autumn migrations, in that order. For those who wish to do a little birding while cross-country skiing, much of the open area and many of the intersecting old roads lend themselves to it.

Leave the New York State Thruway (I-90) at Exit 31. Proceed to NY 790, a southward extension of NY 12. Turn left (south) on NY 790 and follow it into the city of Utica. At the junction of NY 790 and NY 8 turn left (southeast) and follow NY 8 south to the village of Bridgewater. From the Thruway exit to Bridgewater is about eighteen miles. In Bridgewater, NY 8 intersects NY 20. Here turn right (west) and proceed about five miles, to Janus Road, which comes in on the right. (If you come to the intersection of NY 20 and NY 12, you have gone too far and should turn around.) At Janus Road turn right and proceed about one-half mile to a convenient parking area off the road.

Walk through the wildlife area, taking any of the numerous roads that cross. It would be difficult to become lost in the forest so gear your birding to your available time and energy. The avian life is as varied as the habitats, and in spring this is an excellent place to see long-legged waders, ducks, Ospreys, accipiters, and buteos. Turkeys nest here as do picids, flycatchers, thrushes, vireos, and warblers. Although a cumulative list of species observed in Tassel Hill State Forest has not been rigorously kept, those who bird it praise it.

The Root Glen and Kirkland Woods (Map 5.5)

Rating: Spring***, Summer***, Autumn**, Winter*

These two areas are located generally about two miles west of the village of Clinton in Oneida County. The Root Glen is owned and maintained by Hamilton College, which administers the property as a place of beauty for public enjoyment and education, and to encourage conservation of rare or threatened plants. It is located between the campuses of Hamilton and Kirkland Colleges and is open daily from dawn to dark. No permission is required to bird the Root Glen. Kirkland Woods is a mixed hardwood–coniferous forest cut by a 2.5-mile trail. The woods is reached from the south end of the 7.5 acres of the Root Glen. All birding in these two areas must be done on foot (or in winter, on cross-country skis). There is no admission or parking fee at the Root Art Center, adjacent to the glen.

Leave the New York State Thruway (I-90) at Exit 32. Proceed south on NY 233 about 5.5 miles, just west of the village of Clinton. At the intersection of NY 233 and College Hill Road, on the right, and College Street, on the left, turn right (west) on College Hill Road. Proceed up College Hill Road for about one mile to the Root Art Center, which is on the left. Park in the art center parking lot.

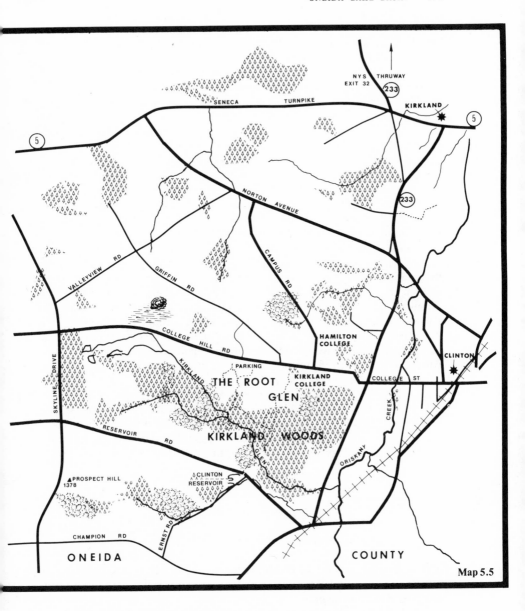

Map 5.5

The trail to the Root Glen begins at the parking lot. From there it is a very easy walk to the formal gardens, indicated by a white gazebo. Here one may sit on stone benches and observe birds in the garden below. This is recommended for the birder who does not wish to walk the one mile trail in the glen, and the 2.5 miles through Kirkland Woods.

The Root Glen represents the work of several generations of the Root family, who originally cultivated, preserved, and expanded this 7.5-acre wooded garden. This glen contains more than 55 species of trees, many of which are rarely encountered in central New York State. Hundreds of varieties of flowers, including several beautiful hybrid azalea, lily, laurel, peony, and heather species, can be found here. In the half-acre formal garden one can find most of the hybrid flower species.

Begin walking on the red shale path through the glen. Notice the enormous Norway Maple on the right. Its circumference is more than thirteen feet and it was planted in the mid-nineteenth century. Continue along the trail to the Hemlock Enclosure. These hemlocks were also planted in the mid-to-late nineteenth century. Be sure to go through the fern garden, where more than eighteen fern species appear. Throughout the glen notable white pine, black walnut, magnolia, white oak, and tulip trees are planted. Search those and the shrubs and hedges for some of the more than seventy-five bird species recorded in the glen.

Follow the red shale path to the foot of a slope, where a trail crosses and re-crosses a stream. Cross the stream in the southwest corner of the glen and continue up a wooded slope. The trail leads past a shrubby field. Continue to a trail intersection. Here follow the trail to the left and proceed along the two miles of path through Kirkland Woods. Emerging from the woods, follow the road north to the parking lot on the right. From here a trail leads back to the Root Glen.

The varied habitat of formal gardens, landscaped woods, fields of berried shrubs, young conifer woods, mixed-hardwood-coniferous forest, and streamside will provide the birder the opportunity to see a broad spectrum of nesting and migrating central New York State passerines. Look for Ruby-throated Hummingbirds, Pileated Woodpeckers, Yellow-bellied Sapsuckers, Eastern Phoebes, Eastern Wood Pewees, thrushes, vireos, Scarlet Tanagers, Rose-breasted Grosbeaks and Indigo Buntings, all of which nest here. In spring and early summer this is an excellent place to see many warbler species passing through. During the second and third weeks of May both the Root Glen and Kirkland Woods are at their peak.

MADISON COUNTY
Woodman Pond (Map 5.6)

Rating: Spring****, Summer**, Autumn***, Winter*

This small pond, its surrounding marshy and swampy area, and the nearby wooded hills combine to make the following detailed walk surprisingly productive in spring and during autumn migration.

Leave the New York State Thruway (I-90) at Exit 33. Follow NY 365 southwest to its junction with NY 5 in Oneida. Here turn right (west) and follow NY 5 a short distance to its junction with NY 46. Turn left (south) and follow NY 46 some thirteen miles to the hamlet of Pine Woods. Here NY 46 intersects NY 20. Cross NY 20 and continue south, crossing NY 26 and passing Leland Pond on the right, which is bisected by NY 26. Continue south a short distance to the small, easily

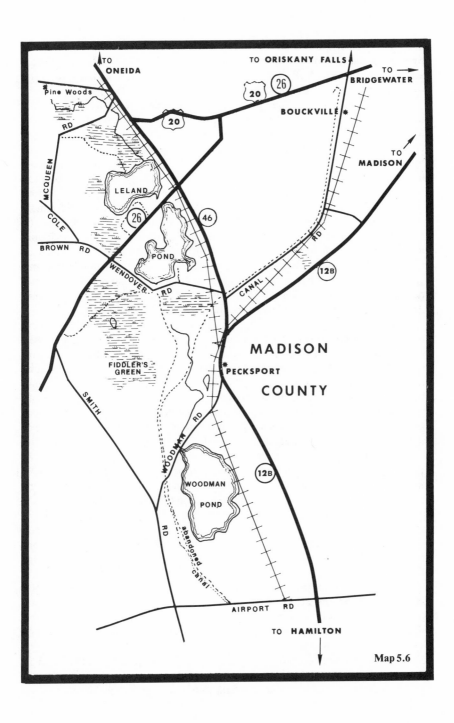

Map 5.6

overlooked settlement of Pecksport. Turn right at the first road coming in on the right, Woodman Road. Continue west and south on Woodman, crossing the now abandoned railroad bed. Look for a place on either the right or left side of the road to park, taking care not to block the road.

Walk south on Woodman Road to a trail that meets it on the right. Follow this trail until it connects with the abandoned canal towpath. Turn left (south) and follow the towpath south to Airport Road. Wood Ducks, Mallards, Ring-necked Ducks and Lesser Scaups can be found regularly on Woodman Pond, but more exotic sea ducks — namely, Oldsquaws, Black, Surf, and White-winged scoters — have been recorded here. Hooded Mergansers can regularly be flushed from the pond and its marshy rim. The Bald Eagle has been seen here in migration. Before walking in on the trail to the towpath, check the reedy edges of the pond for nesting rails. From either Woodman Road, at the northwest corner of the pond, or from the trail and towpath, the birder may be able to observe a family of Virginia Rails or Soras. While walking along the towpath toward Airport Road, watch for Horned Larks, migrant vireos, and a fine variety and impressive numbers of migrant warblers. All along the route, but especially along the short trail to the towpath, look for Rusty Blackbirds. These will probably be most easily discovered in April and September, west of the pond in the most swampy areas.

After reaching Airport Road, turn left and follow it east to the abandoned railroad bed intersecting it. This is the same railroad bed you have earlier crossed in your automobile on Woodman Road. At the railroad bed turn left (north) and walk up the east side of Woodman Pond to Woodman Road. Turn left and return to your car. All along the towpath, Airport Road, and the abandoned railroad bed, in addition to migrant vireos and warblers be watchful for Savannah, Vesper, Swamp and Song sparrows flushing from the brushy edges. Move fairly slowly to gain good views of these sparrows. A representative number of passerine species should pass through the area along this walk, as they work their way along the canal, feeding in the trees, shrubs, and reedy fringes.

Continue driving south on Woodman Road to its intersection with Smith Road, the first through-road. Here turn right and proceed on Smith Road until you find a convenient pull-off close to the wooded hills west of Smith Road. Walk up the hill and into those often productive woods. Great Horned Owls nest there, and Red-headed Woodpeckers and Yellow-bellied Sapsuckers have both been observed in the breeding season. Evidence that either or both picids are confirmed breeders here would be a valuable addition to the Breeding Bird Atlas Project.

After investigating the woods, drive up Smith Road to its junction with NY 26. Turn right and find a convenient and safe pull-off in which to park. Walk down (east) the road indicated on the map, which is usually impassable for cars. Turn right at the trailhead meeting the road. This trail runs south onto a narrow spit of land, higher than the Fiddler's Green swamp that surrounds it. In these woods Northern Waterthrushes and White-throated Sparrows breed. Hidden also in these woods is a bog and a bog-rimmed pond with typical bog-associated plants. This makes this part of the walk a delight to botanists as well as birders.

The Georgetown Area (Map 5.1)

Rating: Spring****, Summer***, Autumn*, Winter*

The southwestern section of Madison County is a sparsely populated area of open fields, reforested state land, a few active farms, and almost no traffic. Elevations run to more than 2000 feet, and although it has been relatively unexplored for its avian life, D. W. Crumb and others consider it one of the most productive late spring-early summer sites in the reporting region. The Georgetown Area has a high number of breeding species, including seven species of raptors and at least sixteen of warbler.

To reach the area from the village of Cazenovia, proceed south on NY 13, passing through the hamlets of New Woodstock and Sheds. At Sheds, NY 13 branches west and NY 80 heads southeast. Proceed southeast on NY 80 to Fire Tower Road, which comes in on the left.

Turn left (north) and bird up Fire Tower Road for approximately 1.2 miles to the top of a high hill overlooking a valley with both open fields and woods and to a large farm on the left. Dense evergreens and hardwoods border this farm and harbor nesting Sharp-shinned Hawks, Ruffed Grouse, Red-breasted Nuthatches, warblers, and sparrows.

Return to NY 80. Turn left (southeast). Proceed only a short distance to Chapin Road, which comes in on the right. Turn right (south) and drive up the steep hill through mostly farm habitat to the intersection of Chapman Road. The plateau at the top of the hill is an excellent area in which to find nesting Northern Harriers (Marsh Hawks) and American Kestrels, but especially nesting field birds — Bobolinks, Eastern Meadowlarks, Savannah, Vesper, and Field sparrows. Follow Chapman Road by turning left. Continue south and downhill watching for wild Turkeys, possibly with poults. Continue into an area of dense evergreen forest. Park along here and explore several of the fire trails that open into the woods. Cooper's Hawks, many warbler species, Dark-eyed Juncos, and Chipping Sparrows nest here.

At the intersection of Chapman Road and Muller Hill Road turn right (west). This is a good area to either park and walk or ride along and listen. Broad-winged Hawks nest in this vicinity. Approximately 1.4 miles along Muller Hill Road, there is an historical marker on the right. Drive (turn right) or walk into this lane lined with old maples. Shortly you will come to a clearing. If you have driven, you can park here in a "mini-park" with one picnic grill. Continue on foot straight ahead on an obvious trail. This trail curves around to the right and comes to a large shallow pond. Here look for Great Blue Herons, possibly ducks and flycatchers, vireos, and sparrows. Follow the trail as it continues right and returns to the "mini-park."

Regain Muller Hill Road heading west. At the next intersection, turn right and bird up to a microwave tower perched at 2100 feet. Watch and listen for Goshawks and Red-tailed Hawks, Barred Owls, Yellow-bellied Sapsuckers, Black-throated Blue, Chestnut-sided, and Mourning warblers — all of which nest in this area.

Continue north on this road, completing the circuit back at NY 80 just southeast of Sheds.

ONEIDA COUNTY
Oneida Lake, Sylvan Beach, and Verona Beach State Park (Map 5.7)

Rating: Spring**, Summer***, Autumn****, Winter*

Oneida Lake borders Oneida, Oswego, Madison and Onondaga counties. It is some 22 miles long and about 5.5 miles wide. It lies about eighteen miles south of the eastern part of Lake Ontario, and several small streams feed into the lake. The headwaters of the streams from Oneida County are close to the source of the Mohawk River, which empties into the Oneida River. The lakewaters enter Lake Ontario through the Oswego River. Oneida Lake terminates at its eastern end in a broad, sandy shore that runs in a straight line for about four miles, almost due north-south. The section of this shore lying north of the New York State Barge Canal, which enters the lake at about the midpoint of its blunt eastern end, is named Sylvan Beach.

The birding area described here encompasses 1.3 + miles of shoreline and over 725 acres of state park near the village of Sylvan Beach. The entire area is on public land and no permission is necessary, for public access roadways lead to the lakefront. An admission fee is charged at the Verona Beach State Park. Birding the area is rated easy but it cannot be birded from a car. Be mindful that this is a summer resort area, so many people will be walking the beach until after Labor Day weekend; but the shorebirding is excellent both before and after the summer vacation period.

Leave the New York State Thruway (I-90), at Exit 34 (Canastota), and head north on NY 13. This road runs along the eastern end of the lake, connecting NY 31 and NY 49. Proceed approximately 9.2 miles to the Barge Canal breakwater.

Although one may walk the stretch of beach beginning at the parking area near the Barge Canal breakwater, north to McLanathan Avenue, a distance of 1.3 miles, the very best shorebirding is usually found along the beach from Ron-Nell Avenue to McLanathan Avenue, a distance of 0.7 mile. To reach Ron-Nell Avenue, drive north on Main Street from the Barge Canal bridge and turn left on Ron-Nell Avenue toward the Kon-Tiki Restaurant and Cocktail Lounge. Here there is a parking area. Early morning light conditions, as well as fewer beach walkers, make that the best time to bird this beach. Walk to the beach and begin walking north.

The shorebird migration begins as early as the third week in July and continues until the last part of October, with a peak between late August and early September. Most species are gone by the end of September. On a typical late-July through August day, the most common shorebirds, usually present in considerable numbers, are: Semipalmated Plovers, Killdeers, Lesser Yellowlegs, Least and Semipalmated sandpipers, and Sanderlings. Fairly regular but less common are Black-bellied Plovers, Ruddy Turnstones, Greater Yellowlegs, Spotted and Solitary sandpipers, as well as Red Knot, Pectoral, White-rumped and Baird's sand-

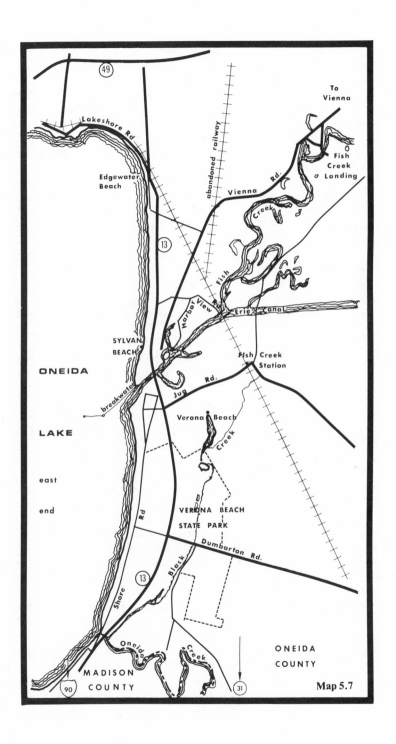

Map 5.7

pipers, Short-billed Dowitchers, Stilt and Western sandpipers. Very rarely, Buff-breasted Sandpipers occur here. Occasionally Hudsonian and Marbled godwits, and Red and Northern phalaropes can be found along this stretch of beach. The bird with the latest and most protracted migration, seems to be the Dunlin, which usually arrives in late September and is present through late October or early November. More than twenty-five species of shorebirds have been recorded here, making this a real "hot spot."

Although this area has gained fame for its shorebirds, Great Blue and Green herons are often seen on the beach in the early morning. Mallards, Black Ducks, and Blue-winged Teals are fairly common early in the season, with other migrating waterfowl appearing in late September and October. All through August and September, the sandbars and shallow pools near the beach hold large numbers of gulls and terns, including Herring, Ring-billed, and Bonaparte's gulls, and Common and Black terns.

When you are finished shorebirding, retrace your route to NY 13 and drive a short way south to the entrance of Verona Beach State Park. Bird the area east of NY 13, where early in the season, you may come across Chestnut-sided and Canada warblers that nest in the park, among others migrating through. This part of the park hosts a number of breeding and migrating sparrow species as well.

ONONDAGA COUNTY

This county encompasses only 795 square miles, which is considerably less area than many less populated New York counties. The largest city within the county is Syracuse, whose population is approximately 195,000. This is the fourth-largest city in the state, and it offers metropolitan facilities and pleasures. Within the last decade the restoration of downtown historic buildings and the creative use of midtown property has taken place here. The Canal Museum, located on Erie Boulevard East, is the only museum in the country devoted to canal history. It has changing exhibits, slide shows, movies, guided tours, and lectures, any of which most viewers will find enlightening. At the museum one comes to realize that the Erie Canal generated whole socioeconomic systems in the cities and towns that developed along its banks; Syracuse, of course, became a boom town because of the canal. To many of our great-grandfathers the canal must have represented the pinnacle of human achievement. The Canal Museum has attracted a colony of Purple Martins since Erie Canal times, and although the water is gone the martins return each year.

Syracuse also hosts six colleges and universities and is a cultural and intellectual center as well as a center of industry and transportation. Syracuse China, on Court Street, is the oldest and largest china manufacturer in the country, and tours through the facility are available. If the weather turns out to be unfavorable for birding, Syracuse can provide numerous diversions.

Onondaga County has been well covered by several varsity birders over the years. Much of the area can be covered on tours that combine several sites. The following tour descriptions have that aim. As a matter of convenience, I have arbi-

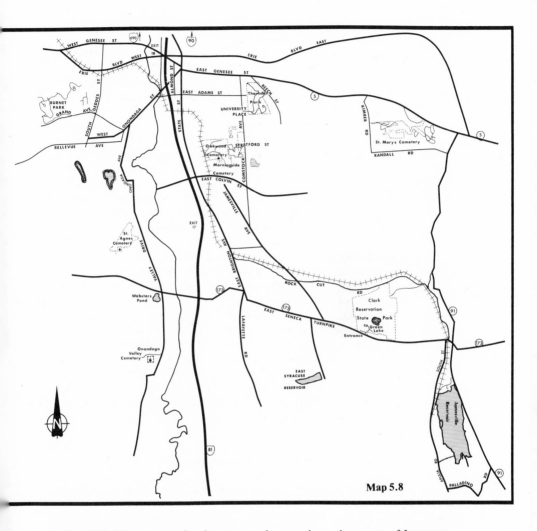

Map 5.8

trarily divided the county sites into two main groupings: those east of Interstate Highway 81 (both north and south of I-90), and those west of I-81 (again north and south of I-90). We'll begin in downtown Syracuse.

EAST OF I-81
Oakwood and Morningside Cemeteries and Thornden Park (Map 5.8)

Rating: Spring****, Summer*, Autumn***, Winter*

Leave the New York State Thruway (I-90) at exit 34A. Gain I-481 south and take it to Exit 4, which is the cloverleaf tie-up to I-690. Get on I-690 headed west. Stay to the left and gain I-81 headed south. Get off at the first exit on the left, which will be Exit 18. At the end of the ramp you will be on Almond Street, which parallels I-81 for some distance. Turn right on Almond Street. Proceed one or two blocks to Adams Street. Turn left (east) on Adams Street and follow it east eight or nine blocks to its intersection with Comstock Avenue. Here turn right (south) and follow Comstock Avenue about five blocks to Stratford Street. Here, turn right and follow this road into Oakwood Cemetery.

This cemetery continues to have one of the best concentrations of spring and fall migrating flycatchers, thrushes, and warblers, year after year. For those birders confined to the city with little birding time, this is the place to go. Wander the roads that wind through the cemetery and enjoy the spring migration for an hour or two before meeting business appointments. Morningside Cemetery is located at the southeast corner of Oakwood Cemetery and its stands of oak and briar bush fields also attract numerous migrants in spring and autumn. A ravine in the southern section of this cemetery has proven an especially good location for many sparrow species. Morningside Cemetery can be reached by regaining Comstock Avenue, turning right, and proceeding south to the cemetery entrance. It is advisable to bird these city cemeteries with a friend or two or to bird close to one's car. These areas suffer the same security problems that others do all over the state.

Leave the cemetery by the same route. Turn left (north) on Comstock Avenue and proceed several blocks to University Place. Here turn right (east) and continue east into Thornden Park. This city park lies between Comstock Avenue and Beech Street, is central to major business districts within Syracuse, as well as close to Syracuse University and the State University College of Environmental Science and Forestry, and offers the visitor to Syracuse an excellent opportunity to do a little morning birding before business hours. The walk from the rose garden to the water tower is especially rewarding. Along the way examine the tall, full evergreens that may harbor migrant flycatchers and warblers.

Clarence E. Hancock International Airport (Map 5.9)

Rating: Spring***, Summer*, Autumn***, Winter**

From the intersection of I-690 and I-81 in downtown Syracuse, proceed north on I-81. Near Exit 7, if traffic is not too heavy, glance over to the left where you may be able to see some of the Purple Martins hawking insects over the Canal Museum, where a colony exists. This is an extreme case of site tenacity, as the water from the Erie Canal, which originally attracted them, is no longer there. Continue north to Exit 27. Leave I-81 here and proceed east toward the airport on Airport Boulevard.

As one drives through the airport, look for Red-tailed and Rough-legged hawks coursing over this fine mousing area in season. American Golden, Black-bellied, and Upland plovers find the mowed fields on the airport grounds satisfactory to their needs. In winter scour the area thoroughly, scanning the horizon for Snowy Owls. Horned Larks, and Grasshopper Sparrows are commonly found on the airport grounds. Do *not* pull off along the main roads, for it creates traffic problems; park instead in the parking lot and walk back along the entrance road, investigating it more carefully. Alternatively, there is a service road on the right parallel to the main road as you drive in. This is one of the best scanning places, and it is safe to pull off here.

ONONDAGA AND MADISON COUNTIES
Oneida Lake – South Shore, Muskrat Bay, Shackleton Point, and Cicero Swamp (Map 5.9)

Rating: Spring****, Summer**, Autumn***, Winter**

From the intersection of I-90 and I-81, proceed north on I-81 for about six miles to Exit 30. Turn right off I-81 and gain NY 31 heading east. The second street on the right (and the first cross street) is South Bay Road. Here turn left (north) and proceed to its terminus. Park off the road.

Along the south shore of Oneida Lake many waterfowl rest in both spring and autumn. Loons, grebes, herons, an occasional Whistling Swan, geese, and ducks can be viewed from this and other shore sites before ice-out in spring and while migration is in full swing in September and October. Walk west along the shore and along Beach Road and scan Muskrat Bay, which is a good spot for water birds; in the spring also be alert along Beach Road for such swamp- and water-loving birds as Purple Martins, Warbling Vireos, Cerulean Warblers, and Northern Waterthrushes. Walk east along Lake Shore Road looking for waterfowl on the lake as well as for gulls.

Drive from South Bay east along Lake Shore Road to its junction with NY 31. In winter or late autumn, cross NY 31 and drive south on Eastwood Road to Bull Street. Here turn left (east) and follow Bull Street back to NY 31. All along Eastwood Road and Bull Street look carefully for wintering hawks, Northern Shrikes, and Common or Hoary redpolls. This is usually a highly productive stretch. Turn right on NY 31 and follow it across Chittenango Creek and the county line.

Once inside Madison County the first through-street encountered is North Road. NY 31 and North Road intersect in the middle of the village of Bridgeport. Continue east on NY 31 (Lake Road) to Petrie Road, which comes in on the left. Turn left here and check out the mixed conifer grove on the left side of the road. Continue driving up Petrie Road watching for possible wintering Marsh, Red-tailed, and Rough-legged hawks in the open country, on both sides of the road. Look all along the road, and especially in the conifer groves on both sides of Petrie Road at its intersection with Shackleton Point Road, for half-hardy passerines and owls. Mourning Doves will be common. Short-eared Owls might be seen flying at dusk. In the more open areas look for Horned Larks, American Robins, Northern Shrikes, and Eastern Meadowlarks. In the cattail swampy areas look for Swamp Sparrows and look even in the small open areas for Snow Buntings.

At the junction of Shackleton Point Road and Petrie Road, turn right and follow Shackleton Point Road, through a gate marking Cornell University property, to its terminus. Park the car.

This is Shackleton Point, which is a concentration point for migrant hawks in both spring and autumn. From the point, when Oneida Lake is not frozen, virtually any species of waterfowl that commonly occurs in central New York State

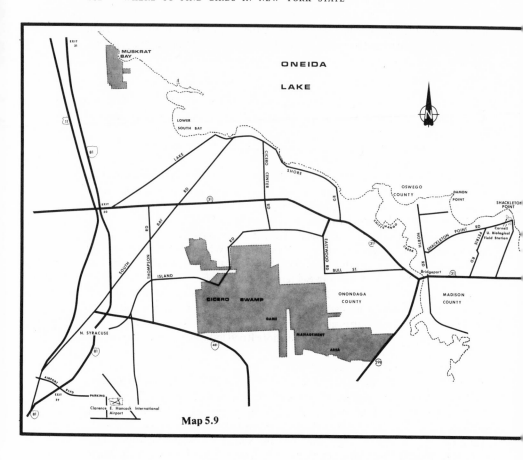

Map 5.9

can be seen. This is quite a fine place to look for Killdeers in the fields and Bonaparte's Gulls out over the lake. Scan the frozen parts of the lake carefully for Snowy Owls, which might be out on an ice ridge. On the walk to the point, one passes through mixed conifer stands and also a healthy fair-sized stand of deciduous trees. In either or both of those groves look for wintering Common Flickers, Red-bellied Woodpeckers, and Brown Creepers, in addition to those species already mentioned. The winter residents of Shackleton Point often maintain bird feeders, which attract the usual chickadees, nuthatches, and Cardinals, but which also host more exotic winter finches in flight years.

This is the site of the Cornell University Biological Field Station. The manager of the field station has indicated that it is not necessary to check in with him or others on the staff before entering the grounds. If a birder comes upon any of the staff while searching the grounds, it is sufficient to identify oneself and simply state that you are birdwatching. On the Cornell University property all of the groves and especially the large conifer plantations should be thoroughly checked for wintering or resident birds. The bird you flush may be a Ring-necked Pheasant but owls have been observed here. Spish for Black-capped Chickadees and thoroughly examine any "funny-looking" ones. Boreal Chickadees have wintered here. Listen and spish and you'll certainly attract nuthatches, Brown Creepers, Golden-crowned, and very rarely, Ruby-crowned kinglets. In spring migration, a simple Saw-whet Owl imitation may get response from the real thing.

Back to your car. Drive back down Shackleton Point Road (west) to North Road. Here turn left and follow North Road to NY 31. Turn right and take this west to its intersection with Bull Street. Turn left onto Bull Street. Cicero Swamp will be on your left (south). Again scan the open areas of this state game management area for hawks and shrikes. Turn right on Eastwood Road and follow it to Island Road, which comes in on the left, south of NY 31. Turn left onto Island Road and follow it slowly through the middle and western end of Cicero Swamp. In spring at dusk, there is an excellent chance you will see the American Woodcock going through its ritual courtship flights and of seeing, or at least hearing, Saw-whet Owls and Whip-poor-wills.

ONONDAGA COUNTY
Old Erie Canal State Park and Green Lakes State Park (Map 5.1)

Rating: Spring**, Summer*, Autumn**, Winter***

Leave the New York State Thruway (I-90), at Exit 34A, and proceed south on I-481 to Exit 4, where it is possible to merge right into I-690, headed west. Be alert for the first exit on I-690, which is Exit 11. Leave I-690 at Exit 11, gaining NY 290 (Manlius Center Road), headed northeast. Double back underneath I-481, and proceed eastward to Manlius Center and the junction of NY 290 and NY 257. Park the car in a convenient place near this junction.

By walking either east or west of Manlius Center, one can bird the Old Erie Canal State Park. A hiking and bicycle trail runs along the old canal, and by following it, quite a bit of birding can be done. Either drive east on NY 257 (Green Lake Road), or walk eastward along the Old Erie Canal State Park hikeway to the entrance to Green Lakes State Park. Both the old canalway and the state park are most productive in winter. The flats along the canal harbor half-hardies as well as hawks and Herring and Ring-billed gulls. Owls can be found along the treed sections of the canal and in Green Lakes State Park. Common Flickers and Pileated Woodpeckers can be found, the latter with more painstaking searching. Horned Larks, Common Redpolls, Song Sparrows, and Snow Buntings may be expected in their proper habitat at both locations.

Labrador Pond, Markham Hollow and Shackham Roads, Highland Park Reforestation Area, Old Fly Marsh, and Pratts Falls Park, a Tour (Map 5.10)

Rating: Spring****, Summer***, Autumn**, Winter****

The following tour should be started in early morning and might take as long as four or five hours, if the birder stops to give each place proper attention. It takes in what are easily some of the very best birding areas in *Kingbird* Reporting Region 5. If a visiting birder has only one day to give to birding in the Syracuse area, it would probably be most wisely spent covering this ground.

From the intersection of I-90 and I-81, proceed south on I-81 to Exit 14 (approximately 23 miles). Here leave I-81 and proceed east on NY 80. After a short distance (approximately 0.7 mile), Markham Hollow Road meets NY 80 on the right. Here, turn right.

Follow Markham Hollow Road south approximately 0.4 mile to a small unnamed road coming in on the left. Turn left (east) and pull off the road to park. From here one can bird Markham Hollow Road, on the west side of Labrador Pond, and from this unnamed road, the north end of the pond is visible. Don't, under any conditions, skip birding this area. Owing to its unique aspects, it has now been purchased by New York State and has been set aside as a wilderness area. It is completely open to hikers and birders and is superb for viewing migrants in spring. The area also supports a variety of nesting species.

In spring, walk along the roads spishing for foraging birds. Both Yellow-billed and Black-billed cuckoos have been recorded here. Turkey Vultures work the hilltops all summer and a pair of Red-tailed Hawks have a huge nest that has been in use for years that can be seen by scanning along the skyline. Listen for the call of the nesting Pileated Woodpecker. Picids and flycatchers of many species are regularly found here, with an Olive-sided Flycatcher appearing occasionally. Wood Thrushes and Veeries can be heard singing. Vireos and warblers pass through in waves.

Cerulean Warblers have nested here for several years as have several pairs of Northern Waterthrushes. The Hooded Warbler is another species that has been recorded as nesting in the immediate vicinity of Labrador Pond for many years. Look and listen for Scarlet Tanagers and many of the sparrow species that breed in moist habitats.

When finished birding from this parking location, continue driving east less than a quarter of a mile to NY 91. Turn right (south) and proceed south into Cortland County less than three-quarters of a mile to Shackham Road, which comes in on the left. Turn left and proceed north up Shackham Road, again crossing the Cortland-Onondaga county line. This road is another excellent aspect of this tour, and the birder should stop along it often to leave the car and spish for migrant passerines.

Shortly after recrossing the county line, one enters a New York State Reforestation Area. Turn right and park at the crossroads. Walk along the road looking for Ruffed Grouse, Turkeys, Hermit Thrushes, Golden-crowned Kinglets, many warbler species (including Magnolia, Yellow-rumped and Blackburnian), and Dark-eyed (Slate-colored) Juncos — all nesters in these woods.

Proceeding north, there is a stretch of fine habitat for productive field birding past the reforestation area. Stop along this length of Shackham Road and search for typical field nesting birds. Follow this road north to its junction with NY 80. Turn right (east) and proceed approximately 1.2 miles to Highland Park Road, which comes in on the right. Turn right (south) and follow the road into Highland Park Reforestation Area, a county park lying in the southeastern corner of Onondaga County. The elevations rise to about 1950 feet in the park, and this

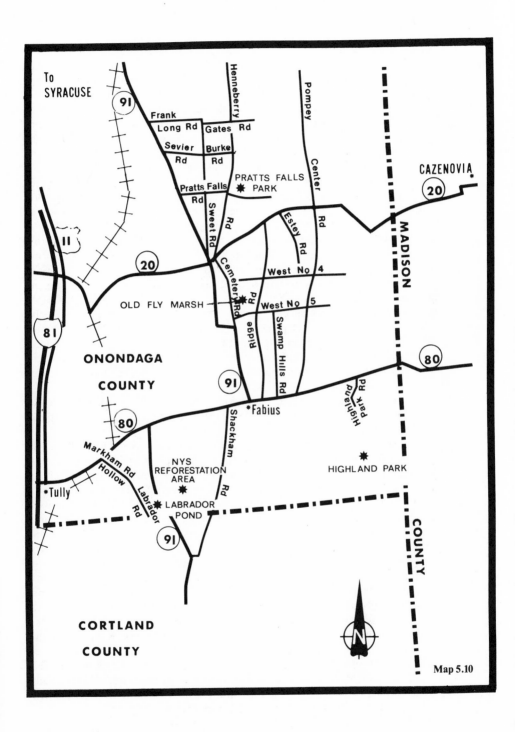

Map 5.10

relatively high country combines with heavy forests to attract many unusual nesting warblers in spring and summer and a whole variety of wintering finches in the colder months. Birding here is facilitated by well-marked county trails for hiking and it is well worth the time spent getting to and covering Highland Park.

Leave Highland Park by retracing the route in and turn left (west) on NY 80. Proceed approximately 0.75 mile to Swamp Hills Road, which comes in on the right, just east of the hamlet of Fabius. Turn right and proceed north up Swamp Hills Road, looking for Northern Waterthrushes, Canada Warblers, and White-throated and Swamp sparrows in the trees and bushes bordering the road. At approximately 0.75 mile, Swamp Hills Road meets a road called West No. 5. Here turn right and proceed to the next through-road, Pompey Center Road. Turn left and at the next road, West No. 4, turn left again. Within a very short distance, a road will meet West No. 4, coming in on the right. This is Estey Road. Note! This is a dirt road that can be muddy and impassable in spring. Here turn right and proceed north up almost the entire length of the road, birding all along it. Look for the Red-headed Woodpecker, which nests in this area. It is often seen sitting on fence posts or flying from fields to wooded edges. A remarkable concentration of Brown Thrashers nests along Estey Road, as do a few of the "winged" warblers. Louisiana Waterthrushes can be heard singing at various points along here in early spring. At Wise Road, which comes in on the left just below NY 20, turn left and then left again at its junction with Ridge Road. Proceed south on Ridge Road.

In the winter driving along all of these back roads, there are numerous open fields where farmers have spread manure. Search any or all of these fields for Horned Larks, Lapland Longspurs, and Snow Buntings.

Turn right at the intersection of Ridge Road and No. 5 Road. Proceed a short distance west and pull into the Old Fly parking lot on the right. *Do not* park along the road.

Old Fly Marsh, located just north of No. 5 Road, between Ridge Road (to the west) and Cemetery Road (to the east), has been purchased by an active conservation group in Onondaga County called Save The County. There are hiking trails on the south and east edges of the marsh. In this vest-pocket wilderness area Pied-billed Grebes nest and Great Blue and Green herons can be frequently viewed fishing. It is an excellent marsh for migrating waterfowl in both spring and fall and Canada Geese, Wood Ducks, and Hooded Mergansers nest on the pond. Look for Ospreys overhead, as well as migrating buteos in early spring. Common Gallinules nest here. Rarely, those who most often bird it, see a tern or two passing over at some height. Several owl species have been recorded within the immediate vicinity of Old Fly Marsh, as have several *Empidonax* flycatchers, including Willow and Alder flycatchers. Eastern Bluebirds have been seen in the area, and the marsh and its environs are quite excellent for numbers and variety of migrating fall warblers.

Once back in the car, continue west on No. 5 to Cemetery Road. Turn right here. Follow this road north and across NY 20, where its name changes to Henneberry Road. Continue north on Henneberry Road, looking for Eastern Bluebirds, which have nested along it. After approximately 0.5 mile, you will see Pratts Falls Road on the right. Turn right and the entrance to Pratts Falls Park is on the left. Enter the park and proceed to one of the park's two parking lots.

Pratts Falls Park is a good birding area in any season, but especially good in winter. Screech and Great Horned owls are permanent residents in the park. Black-capped Chickadees can be expected; however, in a good winter-finch year, look over every flock of chickadees for the rare but recorded Boreal Chickadee. Red-breasted Nuthatches, Evening Grosbeaks, Common Redpolls, Pine Siskins, Red and White-winged crossbills, Dark-eyed (Slate-colored) Juncos, and White-throated Sparrows can all be found wintering within the park.

This area of Onondaga County is really excellent for migrating fall warblers, for they fly across the relatively flat country of the Oneida Lake Plain and are suddenly confronted with the Pompey Hills, which reach elevations of 1700 feet. The birds feed up and through the hills, frequently stopping to feed and rest.

Leave Pratt Falls Park by regaining Henneberry Road from Pratts Falls Road. Turn right and proceed north on it for 0.25 mile. Turn left on Burke Road, which comes in on the left. Proceed west to the next intersection, which is Sweet Road. Cross Sweet Road. You will now be on Sevier Road. Park the car off the road, and walk to the southwest corner of Sweet and Sevier Roads, just below the WNYS television tower. This is a great place to view fall warblers in the early morning. Regain Sweet Road. Turn left on it, and proceed north a short distance, to the intersection of Gates Road. Turn left and immediately left again onto Frank Long Road to reach another good site for watching early morning migrants. Walk along Frank Long Road, west of Gates Road, and bird the willows lining the road.

Clark Reservation State Park and the Jamesville Reservoir (Map 5.8)

Rating: Spring****, Summer**, Autumn***, Winter**

Leave I-81 at Exit 17, south of I-690. Take East Brighton Avenue, east of I-81, south to NY 173 (East Seneca Turnpike). Proceed east on NY 173 for about two miles to the entrance to Clark Reservation State Park, on the left side of the road. Enter the park and proceed to a convenient parking lot. There is a small fee at both parks. Walk along several of the trails criss-crossing the park. It is an interesting area geologically and botanically, as well as for birding. Pileated Woodpeckers are often encountered and even more often heard. Some of the smaller woodpeckers can be seen in every season. The Winter Wren breeds in the park and its lovely song can be heard echoing down many of the park's limestone cliffs. Migrant warblers are found here in spring and autumn, and half-hardies can be found with a little active searching in the colder winter months. Be sure to check out Green Lake within the reservation, where unexpected birds sometimes appear.

After birding Clark Reservation State Park, regain NY 173 headed east. Proceed to the village of Jamesville and the junction of NY 173 and NY 91. Here turn right and proceed south along the east side of the Jamesville Reservoir.

This reservoir is an especially good place to bird in the early spring when ice-out is under way. Loons, grebes, geese, and an occasional Mute Swan can be seen near the northeastern end of the reservoir, as can migrant waterfowl of several species. To reach another fine viewing site, continue south on NY 91 to its intersection with Palladino Road, which comes in on the right. Take a right here and continue east by taking the next available right turn also. Within a short distance you will be at the junction of Palladino Road and Apulia Road. Here turn right and proceed north to the southwest side of the reservoir and the Jamesville Beach State Park. Enter the park and bird it and the reservoir from the most obvious places.

St. Mary's Cemetery (Map 5.8)

Rating: Spring****, Summer**, Autumn**, Winter*

St. Mary's Cemetery is the *best* spring migration area near Syracuse. Over the years birders have learned that it is here that the rarer warblers — Worm-eating and Kentucky, for instance — can be most frequently seen, as well as wave upon wave of regular warbler migrants. It is also possible to see mass waves of migrating thrushes here of five species. The best time of day is from dawn to about nine in the morning.

From Clark Reservation State Park or the Jamesville Reservoir: proceed north on North Street (NY 91) and Jamesville Road, from the intersection of NY 91 and NY 173. Continue north to the junction of Jamesville Road and East Genesee Street (NY 92). Turn left (west). The main entrance is just west of the intersection of Erie Boulevard East (NY 5) and East Genesee Street (NY 92).

ONONDAGA COUNTY, WEST OF I-81
Burnet Park, St. Agnes Cemetery, Webster's Pond and Onondaga Valley Cemetery (Map 5.8)

Rating: Spring****, Summer*, Autumn**, Winter*

These four places can be easily covered in an active morning of birding and are most profitable during the height of spring migration.

Leave I-690, at Exit 6, and turn right onto Erie Boulevard West. Proceed west on Erie Boulevard West approximately 0.6 mile to Geddes Street. Turn left. You will be on South Geddes Street. Proceed south about one-half mile to Grand Avenue. Turn right and follow Grand Avenue around to the entrance of Burnet Park. This is a good inner-city park for spring migrants in the early morning. However, the birder should not bird these areas alone. Groups of three or four people are advised by those who bird the city parks and cemeteries most often.

About four blocks south of the park is Bellevue Avenue, which runs east-west. Leave Burnet Park and gain Bellevue Avenue going east. Take Bellevue Avenue east for almost one mile and turn right on Onondaga Avenue. Follow Onondaga Avenue to South Avenue. Here turn right and follow South Avenue south, to Valley Drive. Turn left at Valley Drive and proceed to the eastern entrance to St. Agnes Cemetery. This small treed cemetery, like Burnet Park, is best at the height of spring migration.

Proceed south on Valley Drive, after birding St. Agnes Cemetery. On the west side of Valley Drive, just south of the Seneca Turnpike (NY 173), is Webster's Pond, which attracts migrating waterfowl. The pond is managed by the Angler's Association, a local organization. Just south of the pond is Onondaga Valley Cemetery, another good place to round out a morning's birding within the city of Syracuse.

Onondaga Lake (Map 5.1)

Rating: Spring*, Summer***, Autumn****, Winter*

The periphery of Onondaga Lake is very industrialized so although it is still an excellent place to view southwardbound migrant shorebirds from mid-August through late-September, really only one or two sites that afford viewing remain. Stopping along I-690 is quite unsafe, illegal, and certainly not recommended. The following is the most efficient way to bird what is left of the mudflats and to scan for waterfowl.

Headed north, leave I-81 at Exit 24, near the southeastern edge of Onondaga Lake. Gain NY 57 north. Proceed northwest of Syracuse and take the obvious pull-offs along this side of the lake. A telescope will be very helpful along this side of the lake. Almost every waterbird recorded in *Kingbird* Reporting Region 5 has been found on this lake, so it deserves investigation. After proceeding approximately 1.75 miles northwest on NY 57, turn around and return south on it. You will have passed the best and most useful pull-offs by that time.

Continue south on NY 57 to Hiawatha Boulevard. Turn right (west) and proceed west on West Hiawatha Boulevard to its junction with Erie Boulevard West. Here turn right and follow Erie Boulevard West to Willis Avenue. Turn right and follow Willis Avenue to State Fair Boulevard (NY 48). Turn left and follow State Fair Boulevard (NY 48) northwest to the community of Lakeside, and to Lakeside Road. Turn right and follow Lakeside Road to its terminus at I-690. Park the car.

Walk across the footbridge over I-690. Trails lead from the footbridge to the shore of Onondaga Lake. Proceed to the lakeshore and begin to walk southeast along it. The late summer and early autumn are times of peak shorebird abundance and variety along the mudflats on this, the west shore, of the lake. However, there can by quite a bit of upland shorebird activity here near the end of July and the flats should be checked then as well as later. Bird species one can expect to see, by walking along the shore include Killdeer, American Golden and Black-bellied plovers, Ruddy Turnstone, Pectoral, White-rumped, Baird's, Least, Stilt, and Semipalmated sandpipers. Sanderlings can be seen in late fall and right into winter. Walk approximately 1.5 miles southeast to the mouth of Nine Mile Creek. Here you will find the most productive mudflats. In winter scan the area on both sides of and up the creek for hunting Rough-legged Hawks and American Kestrels; large concentrations of gulls and some terns including occasional Lesser Black-backed, Laughing, and Franklin's gulls, and Caspian Terns, Snowy Owls, Horned Larks, Northern Shrikes, and of course, Lapland Longspurs and Snow Buntings.

Peat Swamp, Oneida River and Barge Canal,
Three Rivers State Wildlife Management Area, Potter and Sixty Roads,
Beaver Lake Nature Center, Dead Creek, East and West Dead Creek Roads,
and Whiskey Hollow Road, a Tour (Maps 5.11 and 5.12)

Rating: Spring****, Summer***, Autumn****, Winter**

The following tour could consume the better part of a good birding day. It does not involve a great deal of driving time — or much fuel. However, it does sample several fine spring and autumn migration bird viewing sites, as well as a wonderful steep-sided ravine in which a few uncommon warbler species nest. To enhance the birding one should start off on this tour early in the morning and carry along a lunch. There are few places to stop to get a quick bite to eat along the way. The more walking done on this tour, the better it becomes.

Leave I-90 at Exit 37. Proceed north on Seventh North Street, to its junction with I-481. Find a convenient place in this vicinity to pull off and park.

From this point you can scan Peat Swamp, and if you've begun the day at a fairly early hour, the traffic from I-481 should not hamper your birding. Search diligently for Virginia Rails, Soras, Screech and Great Horned owls, flycatchers, Long-billed Marsh Wrens, Common Yellowthroats, and Swamp Sparrows — all nest in or around Peat Swamp.

Continue north on Seventh North Street to its junction with NY 31. Here turn right (east). Proceed east to Caughdenoy Road, where you will turn left (north). Proceed north approximately 1.25 miles to Mud Mill Road. Turn right (east). Drive for 1.0 mile and park off the road.

Continue walking east along Mud Mill Road, toward US 11 and I-81. In the open field country just west of those two routes, a colony of Henslow's Sparrows has been thriving for some years. Check this out before continuing on the tour.

Once back on Caughdenoy Road, follow it north, over the New York State Barge Canal to its junction with Black Creek Road. Here, take a sharp left and follow Black Creek Road in a sort of semicircle back to its meeting with Caughdenoy Road.

All along this route, which closely parallels the Oneida River, watch for waterfowl, raptors, gulls, terns, and nesting Eastern Bluebirds. There will be many opportunities to pull off the roads and scan the river and open fields bordering it. Depending upon the season, migrant vireos, warblers, fringillids, and roadside-foraging sparrows can be found.

At Caughdenoy Road turn right and follow it around and west as it becomes Oak Orchard Road. At the intersection of Morgan Road and Oak Orchard Road, turn left (south). Proceed south on Morgan Road past Verplank Road to NY 31. Here, turn right (west) and proceed approximately 3.25 miles to River Road, just west of the Seneca River. Turn right and continue north 1.25 miles to the entrance of the Three Rivers State Wildlife Management Area.

As its name indicates, this area is near the meeting of the Oswego, Seneca, and Oneida rivers — actually, just west of them. It is an excellent area for land

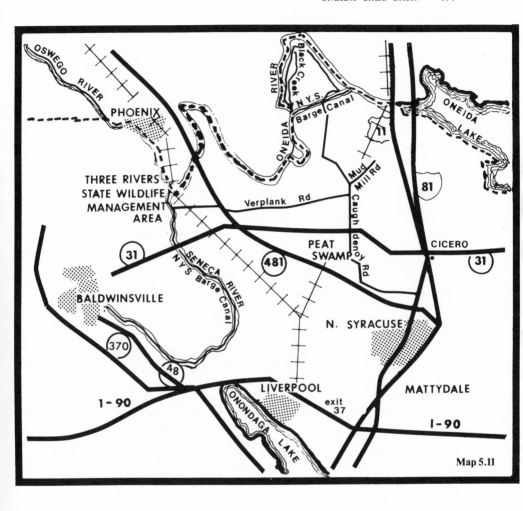

Map 5.11

birds, and one could spend a good deal of time flushing marsh and wet-grass dwelling species and spring and fall migrants, especially in edge habitats.

Drive north only a short distance (about one-half mile), on River Road to its junction with Potter Road. Turn left (west) and continue to its next crossroad, which will be Sixty Road. Turn right and find a convenient place to park the car.

Pileated Woodpeckers, Alder Flycatchers, and Golden-winged Warblers breed in the immediate vicinity. In autumn, the alder and willow swampy areas and the open fields come alive with migrating sparrow species, including Savannah, Grasshopper, Henslow's, Vesper, Chipping, Field, White-crowned, White-throated, Lincoln's Swamp, and Song sparrows. Walk north along this stretch of Sixty Road, testing your skills at identification of fall-plumaged sparrows. Once back in the car, turn around and head south on Sixty Road; advance a short distance to Kellogg Road. This road is slightly south of Potter Road and comes in on the right. Turn right (west) on Kellogg Road and follow it one mile to NY 48. Turn left (south) and follow NY 48 to its junction with Church Road. Turn right (west) and follow Church Road about two miles to East Mud Lake Road. Here turn left (south) and follow it to the entrance and parking lot of Beaver Lake Nature Center.

Common Loons, Red-necked, Horned and Pied-billed grebes, thousands of migrating geese, and more than twenty duck species use Beaver Lake as a feeding and resting stopover. The nature center has a system of cleared trails from which one can easily view migrant passerines. An excellent, tree-cleared overview of Beaver Lake can be reached by continuing south of East Mud Lake Road (instead of driving to the parking lot and entrance as described above) to its dead end at NY 370. Here, turn right; almost immediately, turn right again on Vann Road, which is the first road meeting NY 370 on the right. Proceed on Vann Road to the overlook on the right, which will be obvious.

When finished at this viewing area, continue on Vann Road to its end at Fenner Road. Turn right (north), and drive north to the intersection of Fenner Road and Church Road. Turn left (west) and proceed approximately 1.5 miles to a swampy forest located slightly east of Plainville Road. This area should be investigated, for it is often an excellent site to log migrants in both spring and autumn. Now, turn the car around and head east on Church Road. Drive east along it for about 4.5 miles to its junction with I-690 (NY 48). Turn right and head south on I-690. Continue south, passing by the first exit you will encounter (NY 370). Leave I-690 at the next exit, which will be NY 31. Turn right (west) on NY 31. After less than one mile, NY 31 makes a sharp left (south) turn. *Do not* make the sharp turn, rather continue west on what will now be Kingdom Road. As soon as possible safely pull off the road and park.

Walk back along the road to the north end of Dead Creek. Here and farther east several yards, at the north end of East Dead Creek Road, are marshy areas that harbor herons, ducks, and shorebirds in spring. After thoroughly birding these wet areas return to the car.

Turn left (south) on West Dead Creek Road, which is the first road west of Dead Creek. Continue driving south for approximately 2.5 miles to Whiskey Hollow Road. Turn right (west).

Whiskey Hollow Road is slightly more than one mile long. It is a lovely gulf road passing through a ravine which, in addition to providing nesting habitat for Red-headed Woodpeckers, Blue-winged and Golden-winged warblers, Cerulean, Blackburnian, and Mourning warblers, and several sparrow species, is a botanist's dream come true. The flora differs widely in composition and density on north-facing and south-facing slopes and is well worth the time to stop and investigate at any convenient pull-off.

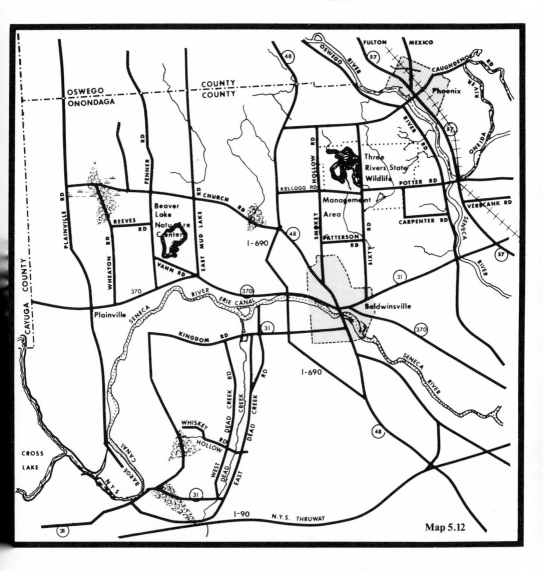

Map 5.12

Camillus Valley, Baltimore Woods, Cedarvale Gorge, Otisco Lake, Skaneateles Lake, Halfway Swamp, and Skaneateles Creek, an All-day Tour (Map 5.13)

Rating: Spring****, Summer****, Autumn**, Winter*

The following tour is quite long, but it passes through some of the most beautiful and productive birding habitat in central New York. This tour can also be extremely rewarding to those searching for summer resident breeding bird species.

Leave the New York State Thruway at Exit 39 (marked "Syracuse, NY 48 west"). Gain I-690 heading south. Proceed along the west side of Onondaga Lake on I-690 for about three miles to Exit 1. Here, leave I-690 and proceed south approximately 1.75 miles to where a well-marked fork indicates the westbound route toward Camillus. Keep right and follow this route. Follow it west and slightly south to the Hinsdale Road Exit. Leave the highway here and head north on Hinsdale Road. Drive north about a quarter-mile to the junction of Hinsdale Road and NY 173. Continue northwest on NY 173 for nearly one mile, watching for Thompson Road on the left. At Thompson Road, turn left and begin driving slowly south through the Camillus Valley, with Nine Mile Creek on your left.

Many birds, difficult to find elsewhere in central New York State, breed here in the Camillus Valley. As you travel the length of Thompson Road, look for both breeders and migrants. Red-tailed Hawks, American Kestrels, and an occasional Merlin may be seen hunting overhead. The marshy-edge habitat along Nine Mile Creek should harbor a few late-lingering shorebirds in the spring. Yellow-billed and Black-billed cuckoos have been recorded along this stretch. Stop all along this road to bird the various habitats. Common Nighthawks, Whip-poor-wills, and Belted Kingfishers are common in spring.

At the intersection of Thompson Road and DeVoe Road, turn left and proceed south just over one mile to Main Street in the village of Camillus. Turn left (east) and follow Main Street approximately 0.4 mile east to the edge of the village. Look for Munro Road on the right. At Munro Road, turn right (south) and continue to the first road coming in on the right, Lyons Road. Here, turn right.

Travel southwest along the entire length of Lyons Road, which is approximately 2.2 miles. Drive quite slowly, listening for Great Crested, Least, and Willow flycatchers. Leave the car at various stops to get good looks of the Blue-winged and Golden-winged warblers you will hear typically "buzzing." Check them out for "Brewster's" Warbler hybrid.

At the junction of Lyons Road and NY 174, turn right (north). Just north of this intersection, Orchard Oriole has been known to nest. Park and search on the right side of NY 174.

Continue north less than one-half mile and park on either the near or far side of the culvert, over which a one-line railroad track runs. Climb to the tracks and walk west (left) along them for tree-top views of breeding and migrant birds. Blue-gray Gnatcatchers and at least two or three vireo species breed along here. Expect to see Black-and-white, both "winged" warbler species, Black-throated Green, Cerulean, and Blackburnian warblers. Listen for the distinctive calls and songs of the Northern Oriole, Scarlet Tanager, and an occasional Rose-breasted Grosbeak.

Having birded from the tracks, continue driving north on NY 174 to its junction with NY 321 just beyond the railroad tracks. Here, turn left (west) and park on the right side of the road, just after making the turn.

From this parking area, one can walk north along NY 174, or west along NY 321. Either stretch is good for breeding and migrating birds. Also seek the following species on this, the northwest side of Camillus Valley: kingbird, phoebe, wood

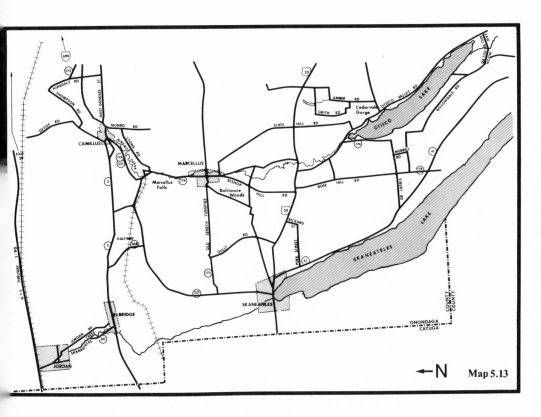

Map 5.13

pewee, Northern and Louisiana waterthrushes, Mourning Warbler, Yellow-breasted Chat, American Redstart, Indigo Bunting, and several breeding sparrow species.

Baltimore Woods, a preserve of 124 acres, is a recent purchase of the Nature Conservancy and the Onondago Nature Centers. It is headquartered in a small log cabin atop a scenic hill covered with mixed hardwoods and giant hemlock. Within the preserve there are open meadows, a small quarry and nearly three miles of cleared hiking trails from which the birder can view a whole array of nesting picids, flycatchers, vireos; Blue-winged, Golden-winged, Cerulean, Blackburnian, Mourning, and Hooded warblers; orioles, tanagers, and sparrows.

From the junction of NY 174 and NY 321, southwest of the village of Camillus (described above), proceed south on NY 174 toward the village of Marcellus. The center of the village is approximately 2.75 miles from the meeting of NY 174 and NY 321. NY 175 (West Seneca Turnpike) passes through the center of the village of Marcellus. From the intersection of NY 174 and NY 175, proceed south to the first possible right turn, which is called Bishop Hill Road. Turn right and drive approximately 0.7 mile to the top of the hill. Park the car and then bird the trails of Baltimore Woods, after checking in at the headquarters building, if it is open.

After leaving Baltimore Woods, regain NY 174 headed south. Continue birding along it, stopping to intensively bird the areas where Nine Mile Creek (now on your right), closely parallels the road. At the intersection of NY 174 and US 20, turn left (east) and proceed east approximately 3.25 miles to Amber Road. Turn right (south), and at the next available right, Smith Hollow Road, turn right again.

Proceed along Smith Hollow Road to its intersection with Smith Road, about one-half mile. Park and investigate the pond here; it is a good site for diving ducks, and its edges are excellent places to spot breeding flycatchers, warblers, and sparrows.

Turn left (south) on Smith Road and follow it as it turns into Amber Road and remain on it as it turns west, approaching the shore of Otisco Lake. This stretch of Amber Road completes its route through the Cedarvale Gorge. Stop at pull-offs and listen for nesting species as Wood Thrush, Veery, and Louisiana Waterthrush.

Amber Road ends at its junction with Otisco Valley Road. Here turn left and follow the eastern shore of Otisco Lake to its outlet and Spafford Creek. Stop at various points along the way to view migrating waterfowl. If the winds are right, there is a strong possibility that you may catch a hawk flight along this side of the lake in autumn. The Loggerhead Shrike has been seen in this general vicinity, so be alert for that rarely encountered species.

In the hamlet of Otisco Valley, Otisco Valley Road meets Saw Mill Road, which comes in on the right. Here, turn right and follow Saw Mill Road to a three-way fork. Do *not* make a sharp right turn, rather take Moon Hill Road, which veers northwest at this fork. Follow Moon Hill Road to Willowdale Road and here turn right. Proceed north up the west side of Otisco Lake, following Willowdale Road past Merrill Road to NY 174.

At the southwest end of the lake, the prescribed roads pass through habitat which offers good general spring birding with a high proportion of migrating and resident warblers and sparrows. Willowdale Road has stretches along it of county reforestation, which harbor woodpeckers, chickadees, nuthatches and some mimids and thrushes. The north end of the lake has varied habitat in which a variety of species can be found. Look in more open low fields for summering Grasshopper Sparrows.

At NY 174, turn left (west) and proceed west over it and Eibert Road to the eastern shore of Skaneateles Lake and NY 41. At NY 41, turn right (north) and advance as slowly as traffic will allow, scanning the lake and stopping at convenient and safe pull-offs.

In winter one can see geese and ducks from various points along NY 41. In spring, especially at the north end of the lake, it is possible to view waterfowl, gulls, and two colonies of Purple Martins. Turn right (east) on Pork Street in spring and summer to search out nesting Red-headed Woodpeckers. Where Pork Street meets Rickard Street, turn left and follow it a very short distance to US 20. Here turn left again and proceed about one mile to Gully Road (which may be called Fisher Road here). Turn right and continue up the length of Gully Road to its junction with NY 175 (Old Seneca Turnpike).

All along the length of this valley one can find nesting woodpeckers, tanagers, vireos, and warblers. Be sure to stop at a very productive pond located on the east side of Gully Road, about one-half mile from the turn onto it. Migrant water birds commonly use this pond as a stopover in spring. After 1.5 miles, there will be a swamp on the east side of Gully Road, which is usually a good place to find American Woodcocks and Common Snipes, as well as other swamp-associated species.

At the meeting of Gully Road and NY 175, turn left (west) and proceed to NY 321. Here, turn right again and follow this route northeast to NY 368. Here, turn left.

The west end of Halfway Swamp is located just east of the intersection of NY 368 and the railroad tracks, near the railroad branch, at the hamlet of Halfway. Proceed up NY 368 to this point, find a safe place to park the car, then get out to do some very enjoyable birding. Both Red-bellied and Red-headed woodpeckers nest here, and this is a fine place to learn not only their songs but also to learn to differentiate between the songs of both Willow and Alder flycatchers, since both species nest here. The Willow and Alder flycatchers were formerly considered different races of the same species (Traill's Flycatcher), but are now regarded as two distinct species. Because the two flycatchers look much alike, this location presents an ideal setting for separating them by voice and call notes. As an added treat, Northern Waterthrushes nest in the Halfway Swamp area, as do many more species.

After birding this site continue north on NY 368 to NY 5. Here, turn left (west). Proceed west to the village of Elbridge and the junction of NY 31C. Here turn right (north).

Skaneateles Creek is paralleled by NY 31C from the village of Elbridge north to the village of Jordan. The creek passes through a wooded gorge along part of this route and through bottomland thickets and deciduous woodlands. These various habitats make this an excellent stretch for easily finding migrant passerines in spring and nesting residents in summer. Follow NY 31C north, stopping to intensively bird shaded hillsides and suitable river-bottom habitats. The list of species recorded on this bit of creekside is long, and it includes most of the regularly occurring birds in central New York State.

CAYUGA COUNTY
The Seneca River Floodlands (Map 5.14)

Rating: Spring****, Summer**, Autumn**, Winter*

These floodlands are located on the western edge of *Kingbird* Region 5 in the town of Montezuma, near the Wayne/Cayuga/Seneca county lines. They lie north of the New York State Thruway (I-90) and east and south of the Seneca River. This is one of the finest areas in central New York for spring shorebirding, and the best time for numbers of species is usually from mid-March through the end of May.

To reach the area leave I-90 (the New York State Thruway) at Exit 40. Gain NY 34 headed south and take this to its intersection with NY 31. Turn right (west)

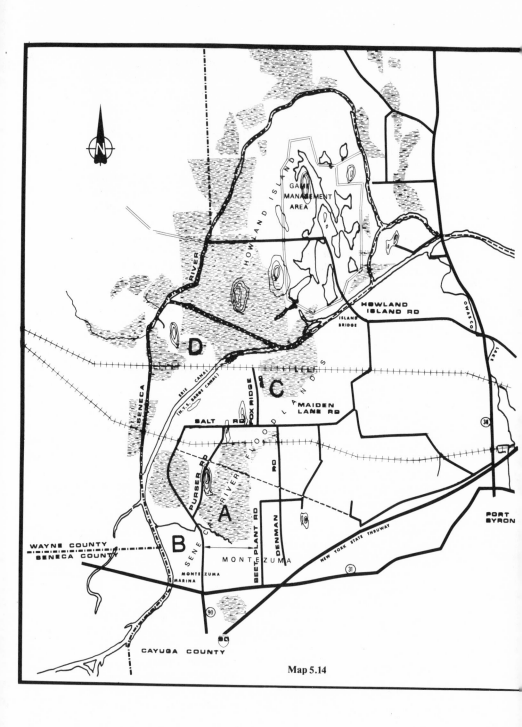

Map 5.14

and proceed on NY 31 about eight miles to Beet Plant Road, which is just east of the Seneca River. The floodlands have just a few streets and roads, and in order to follow them be sure to use the accompanying map.

Area B is best viewed from either Beet Plant Road or from the Montezuma Marina, where NY 31 crosses the Seneca River. If Area B is at a high flood stage, it will be packed with grebes and diving ducks. As the water becomes more shallow it will probably hold hundreds of wigeons including a possible European and swarms of American wigeons, and notable numbers of various diving ducks, as well as Pintails. If the area does not dry out too quickly it will have plenty of shorebirds in beautiful plumages among the waterfowl. As Area B dries out, hundreds and sometimes thousands of Snow Geese, of both white and blue morphs, and Canada Geese and Whistling Swans congregate to graze in this section and also in the section between Beet Plant and Purser Roads, indicated on the map as A. Very often the best place to view the masses of geese is from atop the Purser Road drumlin. Facing south here, birders can look down on the grazing throngs both east and west of Purser Road.

Watch for local and migrant raptors over the fields and marshlands. In the area indicated as C on the map are open marshes along Fox Ridge Road and Maiden Lane Road and old beet impoundments east of Fox Ridge Road. Walk along these roads searching for herons, an occasional Osprey or Bald Eagle, frequently seen Marsh Hawks and numerous Red-tailed Hawks. With luck you will probably turn up a rail or two along Maiden Lane Road.

The still-active railroad tracks are bordered in places with marshes that are alive with shorebirds in late April–early May. Drive to the northern tip of Fox Ridge Road and park off the road. Walk along the service road that roughly parallels the tracks, looking at shorebirds especially in the areas marked C and D. Listen for Blue-gray Gnatcatchers and a variety of warbler species in the scrubby trees bordering the marshes.

South of Maiden Lane Road there are two abandoned railroad beds that can be used as a triangular route on which to walk, away from the noise of traffic. Along this route it is quite easy to see and hear a variety of spring migrant passerines, the locally common Red-bellied Woodpecker, and in the brushy edges, a multitude of sparrow species. Park the car near the northwest bend of Beet Plant Road. Walk east along the abandoned railroad to Denman Road. Here turn south and walk along Denman Road, as far as the next abandoned railroad bed, which is now a power-line right-of-way. Turn right and walk northwest, coming out near Dwyer and Beet Plant Roads. This is quite an easy walk, but wear waterproofed boots or easily washed sneakers.

Howland Island Game Management Area (Map 5.14)

Rating: Spring****, Summer**, Autumn**, Winter**

Howland Island is surrounded by the Seneca River and the New York State Barge Canal. It lies directly north of the previously described Seneca River Floodlands and both areas can easily be covered on the same birding trip. The island contains approximately 300 acres of ponds, which are managed for waterfowl protection and are utilized by enormous concentrations of migrating geese and ducks. The whole management area contains 3000 acres of upland and floodplain forests, locust covered knolls, fields, and brushy habitats, as well as marshes, swamps, ponds, and mudflats. More than 230 species of birds have been recorded on the management area, and almost one-half of that number are local breeders.

In spring and throughout the summer, birders can expect to find herons, geese, ducks, Turkey Vultures, various hawk species, rails, gallinules, American Coots, shorebirds, resident owls, Common Flickers, Pileated and Red-bellied woodpeckers, and flycatchers, In late spring and early summer, the woods teem with wrens, thrushes, and vireos (including Yellow-throated) and Tennessee, Yellow, Cerulean, Palm, and Mourning warblers. Many sparrow species should be searched for in the fields, along edge habitats, and in the numerous brushy areas surrounding the ponds.

Hunting is permitted in autumn, so this is not the best time for birders to be wandering the management area's trails. Except for the main road, which leads to an overlook and picnic area, cars are prohibited. However, walking the extensive road system allows access to virtually all sections of the management area. The three most highly recommended walks are: (1) along the road on the east side of the base of the overlook, (2) along the road running east from the main road, just north of the entrance bridge, and (3) along the road running west from the main road, just north of the entrance bridge.

The manager of the game management area is housed just across the first dike. It is advisable to check with him as to whether any draining of the ponds has been done to provide feeding areas for shorebirds.

The following directions use I-90 as a reference point, in the event that one wishes to bird Howland Island Game Management Area before visiting the Seneca River Floodlands.

Leave I-90 (the New York State Thruway) at Exit 40. Gain NY 34 headed south and take this to its intersection with NY 31. Turn right (west), and proceed on NY 31 to the village of Port Byron and the junction of NY 38. Turn right (north) and proceed up NY 38 for approximately 1.9 miles to Howland Island Road. Here, turn left (west) and drive 5.5 miles to the bridge over the barge canal and the entrance to the island.

Duck Lake Area (Map 5.15)

Rating: Spring***, Summer**, Autumn***, Winter*

Duck Lake is almost directly north of Howland Island Game Management Area. Not only is the lake itself a productive birding site, but also a number of the roads surrounding it have swampy woods, hemlock and yew stands, wet creek crossings, brush-covered areas of high ground, muckland edges, spruce stands, a gravel pit, and open fields adjacent to them. These diverse habitats attract a fine mixture of breeding and migrant birds of both northern and southern affinities.

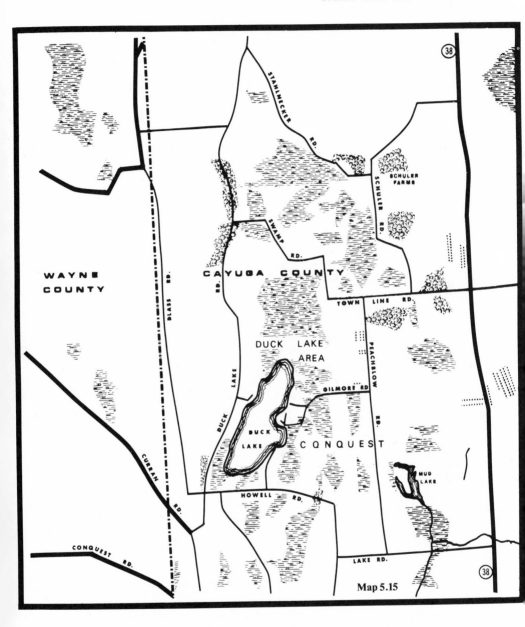

Map 5.15

On and around the lake, in the suitable habitats, look for these species: Common Loon, Great Blue and Green herons, Wood Duck, Turkey Vulture, Bonaparte's Gull, Barred Owl; Pileated and Red-bellied woodpeckers; Horned Lark, Bank Swallow, Brown Creeper, Water Pipit; Yellow-throated and Warbling vireos, "winged" warblers, Cerulean Warbler, Northern Waterthrush, Yellow-breasted Chat, many spring migrant warblers, Rusty Blackbird, Rose-breasted Grosbeak, Purple Finch, Savannah, Vesper, and White-throated sparrows.

To reach Duck Lake, use the same approach as described for Howland Island Game Management Area. However, do not turn west at Howland Island Road. At that point stay on NY 38 and continue driving north another 3.3 miles to Lake Road, which will come in on the left. Here, turn left (west) and follow Lake Road about one mile to Howell Road. Turn right and follow Howell Road around the southern end of the lake to Duck Lake Road. Here turn right (north) and begin to travel up the western shore of Duck Lake.

After approximately 0.8 mile, a road which allows the quickest access to the lake comes in on the right. At the end of it one has a wide view of the shoreline and the lake's surface. After checking that location, proceed north on Duck Lake Road to Swamp Road, which comes in on the right. Turn right and proceed slowly south and east on Swamp Road, passing through typical southern bottomlands habitats and also characteristic northern habitats. Traveling along its length should yield an interesting mix of bird species. At Town Line Road, turn left (east) and follow it to Schuler Road. Here turn left again and proceed up this road, being sure to check out thoroughly its southern end for migrant and breeding warblers. Farther north check the spruce stands on the Schuler Farms lawns. Stahlnecker Road will soon come in on the left. Turn left and proceed up (northwest) it to its intersection with Duck Lake Road. Stop all along this road to check for roadside sparrow species, and be sure to check the gravel pits near Duck Lake Road for Bank Swallow colonies. At Duck Lake Road, turn left and proceed to Swamp Road. Again turn left and continue to Town Line Road. Here, turn left and proceed only a short distance to Peachblow Road. Now turn right (south) and continue to Gilmore Road, the first road on the right. In early spring this road may be too wet and worn to drive over. In this case, simply park safely off the road and walk down it to the eastern shore of Duck Lake. Along the walk you will cross a creek, which should be carefully birded; Gilmore Road also passes through excellent warbler habitat, where notable concentrations of spring warblers have been recorded. The coves along the eastern shore of the lake often provide protection for large flocks of waterfowl.

OSWEGO COUNTY
Oswego Harbor Area (Map 5.16)

Rating: Spring**, Summer**, Autumn**, Winter****

Oswego Harbor, at the mouth of the Oswego River on the eastern side of Lake Ontario, serves as wintering waters for astonishing numbers of several duck and gull species. The harbor is protected from the force of breaking waves by both the Oswego River and the breakwater barriers built out into the lake. There is a power plant on the western harbor shore, whose warm discharge keeps the water open throughout the winter. On the eastern harbor shore, some distance from the mouth of the river, the Nine Mile Point Nuclear Generating Station's warm liquid outflow acts to keep the east side of the harbor ice-free. Sewage outflow from the Hammermill factory attracts gulls. There are excellent views from both east and west shores of the harbor, and the diligent birder should also check up and down the Oswego River for duck and gull species he may have missed elsewhere. Do not overlook searching the waves breaking into foam as they advance toward the shore or breakwaters. Oftentimes rare duck species can be detected in this dashing surf. But veteran birders will not only look lakeward; they will also check all rock ledges below the bluffs, buoys, piers, and vegetation growing along the bluff edges.

Watch for loons, Red-necked, Horned and Pied-billed grebes, Wood Ducks, wigeons, Gadwalls, Mallards, Black Ducks, Pintails, Canvasbacks, Redheads, Ring-necked Ducks, several thousand scaups, the uncommon King Eider, a few Oldsquaws, lots of Black, Surf, and White-winged scoters, Buffleheads, hundreds of Common and infrequent Barrow's goldeneyes, swarms of Common, Hooded, and Red-breasted mergansers, and of course, Ruddy Ducks. The anseriformes are obviously well-represented, but so also are the other orders.

Watch for raptors hunting in early morning and in mid-day. If an entire flock of gulls should suddenly rise and disperse, quickly scan the horizon for a raptor in flight. Mixed among the rafting ducks, there will probably be a few American Coots, relatively close to shore. Scan the shorelines for Killdeers, and the rocks and breakwalls for Purple Sandpipers. Do not be surprised to find a Red Phalarope in the dashing surf near shore or on the breakwaters. Glaucous, Iceland, Great Black-backed, Herring, and Ring-billed gulls can be found with very little effort. With luck you might find a Thayer's Gull. All one must really know is how to distinguish between them.

It will be well-worth the tireless searching it sometimes takes to turn up a Snowy Owl or two, either on the near breakwaters or along the bluffs. It is a frequent winterer at this harbor, so keep a sharp eye out for this marvelous yellow-eyed mass of feathers. Raptors are often seen hunting over the fields or perched in one of the conifers in or around the cemetery described below. Finally, be sure to look in the proper habitat for wintering Horned Larks, the occasional and exciting shrike around the dump, and Snow Buntings on the open fields, especially on the fort grounds.

To reach Oswego Harbor and the mouth of the Oswego River, leave I-81 at Exit 34. Proceed west on NY 104 about 28 miles to downtown Oswego. At the junction of East Second Street and NY 104 (here called West Bridge Street one block from the bridge over the Oswego River), turn right and follow East Second Street to Old Fort Ontario. This is an excellent site from which to view the harbor and breakwalls. Some birders who regularly monitor the harbor and its birdlife prefer the vantage from St. Paul's Cemetery. The first-time visitor, however, should first stop at Old Fort Ontario; then circle the fort to the fort's east side, on Mitchell Street, which is two blocks south of the lake. Proceed east on Mitchell Street.

If the birder is hell-bent on intensively birding the harbor, he may turn left at East Twelfth Street and proceed one block to the vicinity of the Hammermill

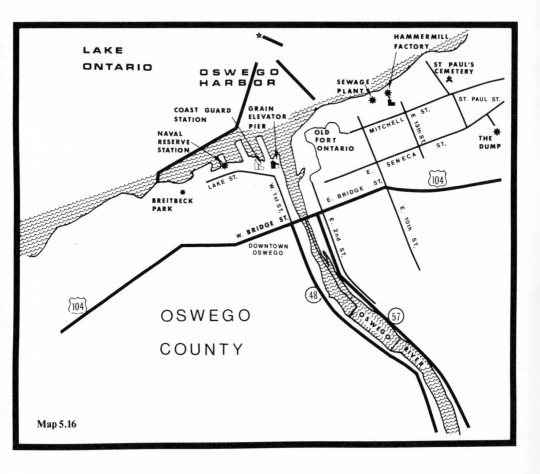

Map 5.16

factory and the sewage plant. From there, he may walk to the bluff edge and scan the outflow and the surf for the rare Harlequin Duck, Red Phalarope, or very rare gull.

Continue on Mitchell Street past East Thirteenth Street. The next street on the left will allow access to St. Paul's Cemetery. The dawn duck flight out of Oswego Harbor viewed from St. Paul's Cemetery can be spectacular at the very least, and from here one is afforded excellent views of the breakwalls, bluffs, rock ledges below the bluffs, as well as the fields to the east and the conifers within the cemetery.

After birding from either Old Fort Ontario or St. Paul's Cemetery, proceed on Mitchell Street to St. Paul Street, just beyond the access road to the cemetery. Here turn right and go to East Seneca Street. Turn left and advance only a short distance to the first street coming in on the right. Turn right on it and proceed to the town dump. Here, of course, one can get quite close to the feeding gulls and study the features that differentiate several gull species in a variety of age-dependent plumages. Before you leave this area, be sure to thoroughly search the trees that rim the dump.

Leave the dump and turn left on East Seneca Street. Proceed to East Tenth Street and there turn left. Go two blocks and turn right onto East Bridge Street (NY 104). Go across the bridge, and at West First Street turn right. Follow it about four blocks and make a left onto Lake Street. From here it is, in a manner of speaking, "duck soup."

Proceed along Lake Street stopping at the Grain Elevator Pier, the Coast Guard Station, Wright's Landing, and finally the United States Naval Reserve Station to get the best possible views of the harbor. Scan the stationary buoys to the left. Take as much time as necessary to scan thoroughly the long breakwater all the way out to the lighthouse—a telescope would be most helpful. If the winds are relatively calm, one might set up a telescope on the edge of the bluffs east of the Naval Reserve Station to scan more easily. If there are high winds, that would obviously be folly. Finally, proceed to Breitbeck Park, west of the station, to search the power plant's warm effluent. This area presents a marvelous opportunity for learning winter-plumage ducks and gulls of varying ages, and the serious birder should not pass it up.

Selkirk Shore State Park (Map 5.1)

Rating: Spring****, Summer*, Autumn****, Winter*

This state park of nearly 650 acres, is quite a fine lakeside stopover for the onrush of northbound birds in spring. In autumn too the brushy hedgerows and dense scrub harbor sparrows and songbirds. A visit here has all the potential of turning up a rarity or two, and it will pay to thoroughly search the pine groves and brush-covered dunes between Lake Ontario and the marshes of the Salmon River.

Leave I-81 at Exit 36. Gain NY 13 headed west. Follow this about three miles through Pulaski, to NY 3. Along the way stop to check those areas where the Salmon River parallels NY 13 closely on the right. Turn left (south) and continue on NY 3 for 1.5 miles to the state park headquarters.

A park trail map is available at the headquarters. Check the fields behind the headquarters building for sparrows and birds of open fields, then proceed to the scrubby marsh south of the parking lot. After birding here, proceed to the lake and bird the two main trails along it. These are actually the best paths from which to bird. Sometimes at the height of spring migration there is a wonderful intermingling of the songs of southern and northern-nesting warblers, some of which may not have been in full song as they passed through from farther south.

After birding the lakeside from these two main trails, drive out of the park and turn left on NY 3. Proceed approximately one mile to a road coming in on the left marked Pine Grove Road. Turn left. This road passes through a grove of white pines, where Pine Warblers nest, then it turns, paralleling the lake. Park.

Bird on foot throughout this woodland. In places there is a heavy understory and deep brushy tangles with a full complement of catbirds, thrashers, Cardinals, towhees, and sparrows. Climb to the top of the dunes where one has a clearer, almost eye-level view of the concentrations of passerines in the trees.

Sandy Pond (Map 5.1)

Rating: Spring****, Summer**, Autumn****, Winter*

Each of us has a birding site with which we've been irrevocably smitten and to which we are passionately loyal. Sandy Pond is such a place. It is a mecca for countless central New York State birders, and certainly not without good reason; it is a golden place of the eighteen-carat quality. Certainly, its variety of habitats — beach, sand dunes, lagoons, ponds, wet grass and cattail marshes, brushy thickets, shaded woods, and open fields — at least partially explains the diverse species recorded here; in no small part, it also explains the charm Sandy Pond has for its staunch supporters.

Some of the breeding birds of the Sandy Pond area include: American and Least bitterns, Wood Ducks, Green-winged Teals, Mallards, Black Ducks, Blue-winged Teals, several rail species, Common Gallinules, Killdeers, Upland Sandpiper, Black Terns, Long-billed and Short-billed marsh wrens, Bobolinks, Eastern Meadowlarks, and Savannah, Grasshopper, Henslow's, Vesper, Field, and Swamp sparrows.

During spring migration, many shorebird species, some in breeding plumage, can be viewed here in fair numbers; but, in the late summer and early autumn, the area is alive with milling flocks of skittering peeps and sometimes hundreds of shorebirds of many different species. Look for the following species from late summer through autumn: Semipalmated and Black-bellied plovers, Ruddy Turnstones, Whimbrels, Greater and Lesser yellowlegs, Red Knots, Pectoral and Baird's sandpipers, Dunlins, dowitchers, Semipalmated, Western, and Buff-breasted sandpipers, Sanderlings, and although never in great numbers, Red, Wilson's, and Northern phalaropes. In spring and autumn literally thousands of gulls and terns frequent the Sandy Pond area.

Although it has justly earned a reputation as a prime shorebirding site, Sandy Pond is equally fine for landbirds about mid-September, when winds are from the northwest and ranging between 13 and 24 mph (between 4 and 8 on the Beaufort Scale). Under these conditions, one can expect upward of fifty varieties of passerines in the area, including fourteen to eighteen warbler species. This is the ideal setting in which to either learn or sharpen your skills at really identifying those sometimes confusing fall warblers.

Leave I-81 at Exit 37, in the middle of the village of Sandy Creek. Proceed west on NY 15 out of the village and to its intersection with NY 3. Here, at Sandy Pond Corners, stay on NY 15. Cross NY 3 and continue to drive to the southwest corner of Sandy Pond. Park conveniently and safely off the road.

Walk a short distance to the lakeshore, then turn right (north) and bird along the sand spit, up to the Sandy Pond Inlet, which is approximately 2.5 miles.

On the return walk, go inland a short distance and walk south on a path that runs between the dunes and Sandy Pond. Be sure to stop and thoroughly bird a brush-edged, water-filled hollow, known to some as "warbler pool." A path that is sometimes indistinct skirts its north side, and from that path some remarkable rarities have been seen.

ACKNOWLEDGMENTS

The author would like to express gratitude to the following people for the information they supplied on the Oneida Lake Basin: Dorothy W. Crumb, Paul A. De-Benedictis, Bruno and the late Mrs. DeSimone (Dee), Marge Rusk, Chris Spies, and Dorothy Ackley. A very special thank-you to Dorothy Crumb for her extensive comments and reading of the Oneida Lake Basin chapter.

St. Lawrence — Region 6

THE ST. LAWRENCE REGION encompasses 5353 square miles and includes all of St. Lawrence, Jefferson, and Lewis counties. St. Lawrence County, largest of the state's sixty-two counties, contains 2768 square miles; its county seat is Canton and its largest city is Ogdensburg. Jefferson and Lewis counties are about equal in size and their county seats are also their largest urban areas: Watertown and Lowville, respectively. Watertown is the largest city in northern New York (Map 6.1).

The region's western boundary meets Lake Ontario and the scenic Thousand Islands, which begin near Cape Vincent, where Lake Ontario and the St. Lawrence River meet, and stretch northeast to the Ogdensburg area. Lake St. Lawrence and the St. Lawrence Seaway, with its intricate lock system, dam and power development complexes north of Massena, form the northern boundary. The eastern county lines of Lewis and St. Lawrence counties are also the region's eastern limits. The southern boundaries of all three counties comprise the southern regional line.

LEWIS COUNTY
Tug Hill Wildlife Management Area (Map 6.2)

Rating: Spring****, Summer***, Autumn****, Winter*

Tug Hill Wildlife Management Area comprises over 4985 acres in Lewis County. A trackless wilderness, desolate and rugged in both terrain and climate, this area's beauty and natural resources are intact. It is part of the Tug Hill Plateau, 775,000 acres of layered limestone, sandstone, and shale rising from Lake Ontario to a flat top over 2000 feet in elevation; on the east, its escarpment plummets to the Black River. The approximate boundaries of the plateau extend from Watertown on the north, to the Black River on the east, to Utica at the southeast corner and the western end of Oneida Lake at the southwest corner. This rise lies directly in the path of west-east storm fronts and experiences the heaviest snowfalls east of the Rocky Mountains. Snowfalls average 260 inches per year (the record is 355 inches), and

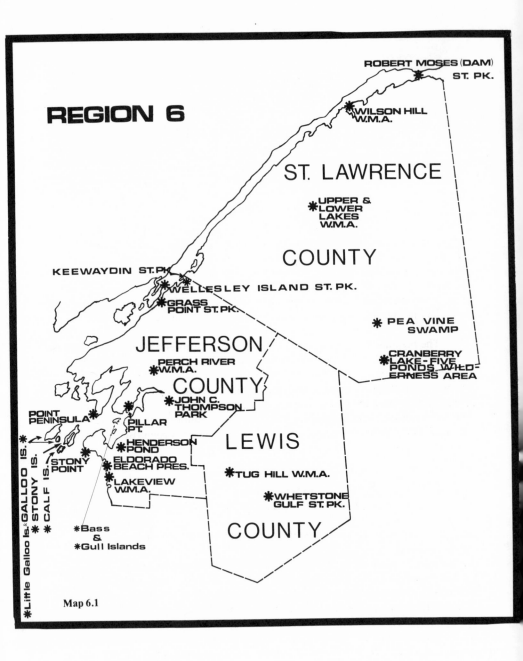

REGION 6

ROBERT MOSES (DAM)
ST. PK.

*WILSON HILL
W.M.A.

ST. LAWRENCE

*UPPER &
LOWER
LAKES
W.M.A.

COUNTY

KEEWAYDIN ST. PK.

*WELLESLEY ISLAND ST. PK.

*GRASS
POINT ST. PK.

* PEA VINE
SWAMP

*CRANBERRY
LAKE-FIVE
PONDS WILD-
ERNESS AREA

JEFFERSON

*PERCH RIVER
W.M.A.

COUNTY

*JOHN C.
THOMPSON
PARK

LEWIS

POINT
PENINSULA

*PILLAR
PT.

*HENDERSON
POND

*TUG HILL W.M.A.

*STONY
POINT

*ELDORADO
BEACH PRES.

*WHETSTONE
GULF ST. PK.

*LAKEVIEW
W.M.A.

COUNTY

*Bass
&
*Gull Islands

Little Galloo Is. &GALLOO IS.

*STONY IS.

*CALF IS.

Map 6.1

the average rainfall is 60 inches per year. Towns on the hill's perimeter rely on dairy farming, commercial lumbering, and outdoor recreational activities to support their economy.

Tug Hill Wildlife Management Area lies about six miles south of the settlement of Bellwood. This tract of public land can be reached and covered, in part, by car or bicycle. There is no admission charge or parking fee. Although the birding is best in spring and autumn, blackflies in late spring and early summer and deer flies and "punkies" from June to September make a hat, long-sleeved shirt, and considerable malodorous insect repellent a necessity. All roads through the Wildlife Management Area are graveled or dirt, so proceed with caution in April, for some roads may be in rugged shape after a typically severe winter. Deer hunters are active during the prescribed season. Cross-country hiking is allowed, and birding could produce a veritable spring of teal, a parliament of owls, or a fall of woodcocks; on the other hand, such a hike could also turn out to be a plague of blackflies and a defeat of directions. Only birders experienced at hiking with a compass and topographic map should attempt it. The woodlands are rich and vast, and the swamps, glacial ridges, and small ponds can yield great aesthetic pleasure; but Tug Hill's virtually roadless terrain, which has few distinguishing features for orientation, makes it a relatively easy area in which to get lost. However, the area can be easily birded from roads and trails without making a wilderness adventure of it. If you do decide to go off the trails *do not* go in alone; you could easily become lost.

Approaching from either north or south: Leave I-81 at Exit 42, proceed to Adams Center and pick up NY 177; proceed east for about 18 miles to the settlement of Bellwood; there take Sears Pond Road south (right turn) straight through to Flatrock Road (about 6 miles from the Bellwood turnoff on NY 177). *Note:* After about 4 miles, there is a four-way intersection, where Sears Pond Road angles off to the southwest, Rector Road comes in from the east and Parker Road continues south. Be sure to proceed on Parker Road south instead of veering off with Sears Pond Road at this point. At Flatrock Road there will be state signs designating the Tug Hill WMA.

The best and easiest way to bird this area is to drive to the first parking area (see accompanying map), leave the car, and walk along the dirt truck trails or old logging roads looking and listening for Yellow-bellied and Least flycatchers, kinglets, and Cedar Waxwings. In areas containing mixed stands of balsam fir, red spruce, and alders, look for both northern and southern breeders: for example, Hermit Thrushes; Golden-winged, Nashville, Blackburnian, Chestnut-sided, and Mourning warblers; White-throated and Lincoln's sparrows. After exploring the trails on foot, return to the car, advance to the next parking area, and repeat the same procedure.

In May and early June, wild woodland flowers are abundant everywhere; this is also the best time to spot kettles of northward migrating hawks overhead. Goshawks nest in the area so watch for them. Both birds and mammals abound here, so keep a sharp eye out for fisher, otter, mink, and white-tailed deer. Check open wet areas for better views of birds and check every beaver pond. Park and search at each opportunity, then return to NY 177 at Bellwood.

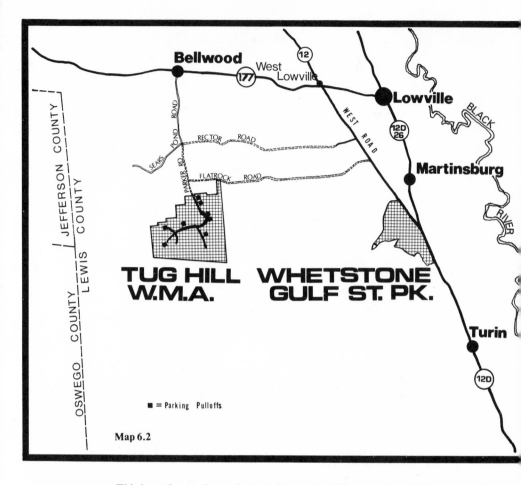

Map 6.2

This huge forested area dotted with nearly 1300 swamps and 900 very small ponds is essentially unbirded, and it holds all the allure of a valuable environmental resource, including beauty, clean air, and woodland peace. It is slowly gaining devotees in the birding community and should reveal its avian wealth through more extensive fieldwork. One more *caveat:* the area can be treacherous underfoot, and the birder should not enter it alone.

Whetstone Gulf State Park (Map 6.2)

Rating: Spring****, Summer***, Autumn****, Winter**

Whetstone Gulf State Park lies entirely within Lewis County and encompasses 1900 acres of publicly owned land. There was an admission charge of $2.00 in 1981, which included parking. At that time it was also possible to camp within the park for an additional fee of $4.50 per day. Although the park can be reached by car, it cannot be thoroughly birded by either bicycle or in a car. The birder must walk the trails, which are rated in some areas as difficult to moderately easy.

Whetstone provides extreme geological variety. It lies on the eastern edge of the Tug Hill Plateau, a few miles west of the Black River. Since the close of the Ice Age the chief geologic process in New York State has been erosion; here erosional resistance has been responsible for rugged benches and pinnacles that rise in columns reminiscent of the Dakota badlands. These buttresses are surrounded by gullies and ravines, which support vigorous vegetation.

The forested areas are primarily mixed, second-growth hardwoods. The elevation here rises to approximately 1900 feet. Within the park, Goshawks have been known to nest in the "bad lands" formations. The bird species nesting within the park's boundaries are those associated with beaver meadows and mixed balsam fir-red spruce-alder swamps. This is also a great place for finding ancient fossil rocks.

The two largest cities near Whetstone Gulf are Utica, through which the New York State Thruway (I-90) passes, and Watertown, through which I-81 passes. Approaching from Utica, take Exit 31 from the Thruway, and head north on NY 12. At the village of Boonville leave NY 12 and pick up NY 12D. Proceed north on NY 12D to the park entrance, which is about 8 miles north of Turin.

Approaching from the Watertown area on I-81, take Exit 45 and get on NY 12 going east and south and proceed to Lowville. At the junction of NY 12 and NY 26 in Lowville, take NY 26 south to the park entrance, about 2 miles south of Martinsburg. An alternate route would be to leave I-81 at Exit 42 (Adams Center). Proceed east across Tug Hill on NY 177 toward Lowville. Turn south (right) at West Lowville on West Road. Follow it south for 9 miles to the park entrance.

Once inside the park, walk the main trail around the Gulf to see and hear Great Crested, Alder and Least flycatchers; Tree, Bank, Rough-winged, Barn, and Cliff swallows, and Purple Martins; Winter Wrens; Yellow-throated, Solitary, Red-eyed and Warbling vireos; Tennessee, Nashville, Magnolia, Cape May, Black-throated Blue, Black-throated Green, Blackburnian, Chestnut-sided, Bay-breasted, and Blackpoll warblers. The profusion of warbler species nesting here is a delight.

The walk around the rim of the gorge is somewhat difficult and during the season, blackflies are relatively thick; but several who have taken the hike vouch that it is "really worth it."

JEFFERSON COUNTY
Lakeview Wildlife Management Area (Map 6.3)

Rating: Spring****, Summer****, Autumn****, Winter*

The Lakeview Wildlife Management Area (LWMA), is in Jefferson County, about four miles west of the village of Ellisburg. Its 3421 acres of publicly owned land include a barrier beach, Lake Ontario shoreline, extensive marshlands, dams, dikes, spillways, an observation tower, parking areas, and boat access points.

To bird this area properly, cover it by canoe or boat or on foot. Watch out for poison ivy inland and on the barrier beach. At the proper season mosquitoes on the barrier beach can be somewhat annoying. This area is generally moderately easy to cover, but it can be more difficult if one chooses to canoe through the marsh cattails.

Approaching from north or south on I-81, take Exit 36 (Pulaski) and get on NY 13 heading west. Take NY 13 to Port Ontario. From there, take NY 3 north for approximately 10.5 miles where signs will designate the LWMA. There are two parking areas off NY 3. The first is adjacent to the headquarters on the left side of the road. The second is approximately 1.5 miles north of the first, also on the left side of the road. From this parking site take a path due west to an observation tower overlooking an extensive marsh. The path continues north from the tower with a spur west to the edge of Sandy Creek and one east back to NY 3. Walking these trails should yield some interesting marsh birds. One cannot gain access to the beach from either of these parking areas.

There is a small boat launching site on the South Branch of Sandy Creek on NY 3 just inside the LWMA. From here one can launch a canoe and paddle north up the South Branch through the marshes to Lakeview Pond, passing through Goose and Floodwood Ponds. Another, more heavily used access site is to the north on NY 3. Go by the first road coming in from the left (South Skinner Road); proceed to the next road (North Skinner Road or Pierrepont Place) and turn left. This will take you to the edge of Lakeview Pond, where you can launch a canoe. Other boat access sites within LWMA are undeveloped — use them with caution. Launch at Lakeview Pond and explore the shoreline. Look for Ospreys and Black Terns, both of which can be seen fishing on the pond. Canoe south into Flood-wood Pond and investigate the cattails. American and Least bitterns nest there as do Black Terns. A large Black-crowned Night heronry is located nearby and that species can be commonly seen in the LWMA. Green-winged Teals, Mallards, Black Ducks, and Blue-winged Teals are all common in the area. Redheads and Ring-necked Ducks are uncommon here.

From the boat access site on Lakeview Pond a path leads to the natural beach along Lake Ontario. This beach is open to the public; however, current restrictions include no picnicking, no swimming, no camping, or firemaking, and no removing vegetation or other natural features.

LWMA includes a stretch of beach 4.5 miles in length, bounded on the north by the lands of Southwick's Beach State Park and on the south by the outflow of South Colwell Pond into Lake Ontario. A walk along this beach during spring migration can turn up a profusion of warbler species. During the fall southward migrating hawks, waterfowl, and an array of shorebirds can be seen in passage. This is the site upon which the carcass of a White-tailed Tropicbird was found October 26, 1954, following a hurricane. During the height of the spring migration Black-and-white, Tennessee, Orange-crowned, Nashville, Northern Parula, Yellow, Magnolia, Cape May, Cerulean, Bay-breasted, and Canada warblers have been recorded here, to name but a few. Another way to gain access to the beach is through the lands of Southwick's Beach State Park. The entrance to the park is located at the junction of NY routes 3 and 193. There is a parking fee at this state park from mid-May to late September.

Both LWMA and El Dorado Beach Preserve are only 45 minutes to an hour by car from Derby Hill hawk lookout in Region 5.

El Dorado Beach Preserve (Map 6.3)

Rating: Spring***, Summer***, Autumn****, Winter*

Through the combined efforts of member clubs of the Federation of New York State Bird Clubs, Inc., especially the North Country Bird Club of Watertown, and the Nature Conservancy, this unique shorebird migratory stopover area was purchased in 1969, and thus saved from residential development. The preserve encompasses 250 acres, most of which is a marsh and woods buffer zone for the more than 3500 feet of rocky peninsula shoreline so attractive to transient shorebirds.

There is only one practical entrance to El Dorado and that is from the north. In the past birders have tried—and regretted attempting to enter on the south end of the preserve. Access to the Bolton Road entrance from the south is by a rough private road and through a gate that is often locked. Directions here are, therefore, to the north entrance. From I-81, take Exit 41 (Adams). After exiting get on NY 178 headed west toward Henderson. Pass through Thomas Settlement and Roberts Corners. At Scotts Corners, approximately 7 miles, NY 178 turns north. Instead of following this route north, head due west toward Alexander Corners on Smith Road. Just after a stoplight Smith turns into Stony Creek Road at Alexander Corners and is very rough until the intersection of it and NY 3. At the intersection of NY 3 and Stony Creek Road, continue west on Stony Creek Road. You are now on a narrow black-top road that traverses grassy pastures and cedar groves. Follow it 1.3 miles to an unmarked dirt road (Grandjean Road) that forks to the left. Bear left on it and drive 0.9 mile, always bearing left, to a gateway. Pass through and continue to bear left on this grassy two-track road for 0.2 mile and then park. Do not turn right where a sign says "Private Road," and do not inquire at the cottages for directions. In the vicinity of the parking area is a blue-and-orange sign announcing El Dorado Shores. The birder who exercises forethought will have remembered boots and raincoat or windbreaker. The short, low peninsula is often subjected to strong winds and precipitation.

There are two schools of thought on just how to bird El Dorado. Some prefer to walk the path from the parking lot to the beach, turn south (left), bird down the beach, and return to the parking lot by the old dirt road. However, I take another approach, as do the following directions.

Walk south on the dirt road ahead of you and turn west at its end toward Blind Point, which is identifiable by a thicket of red cedar trees surrounded by weedy vines. This is nearly the southernmost boundary of the preserve. Turn north and bird up the stretch of shoreline facing you. Work all of the wave-cut benches and notches along the beach face. These low-lying formations contain various detritus, much of which is thickly interlaced with a dense, slimy layer of algal vegetation and shallow pools, where surface and subsurface insects and larvae abound and into which shorebirds probe. This alliance of freshwater plants, without roots, stems, or even leaves, thrives in this zone. The rich plant and insect life in these pools is the source of the area's attraction to migrating shorebirds. Bird up to and past a collection of glacially deposited boulders and north beyond the rocky point that marks the end of the major shore-birding section of the preserve.

If one considers a bit of the biology basic to shorebirds, it becomes immediately apparent when the birding is prime at El Dorado. The majority of species in the Charadriidae and the Scolopacidae families breed in the Arctic, with the relatively short breeding cycle rarely exceeding six or seven weeks (from incubation through fledging). Egg laying occurs immediately after the spring thaw and the shortness of the season appears to preclude double clutching or clutch replacement in almost all species of these families. The distance covered on migration by these species is quite commonly from the northern Arctic to southern South America. In most species the adults leave the breeding grounds before the young, thereby insuring that the food source remaining will be utilized by the young to build up pre-migratory deposits of fat. Additionally, richer food, found south of the Arctic, is required by the adult birds for their annual molt.

Consequently, by the middle of July, adult Semipalmated Plovers, flocking Killdeers, Ruddy Turnstones, some Common Snipes, Spotted Sandpipers, Greater and Lesser yellowlegs, White-rumped, Least and Semipalmated sandpipers, Short-billed Dowitchers, and Stilt Sandpipers may be seen at El Dorado. Numbers of these adults build up, and by mid-August a few Pectoral and Baird's sandpipers may appear. In the last half of August almost all adult-plumaged birds have disappeared. Black-bellied Plovers in every conceivable plumage variation, usually in small groups or singles, will be visible from late August until the first freeze. Occasional Red Knots and Western Sandpipers may be observed at this time. American Golden Plovers are often found from the beginning of September through mid-October, and are almost without exception immature birds. Whimbrels and Solitary Sandpipers are rare migrants here and may occur singly July 20 through September 10. Dunlins and Sanderlings arrive in late August and early September. The Upland Sandpiper is an occasional visitor, usually found away from the shoreline. All three phalarope species occur here but never in large numbers. The sequence in which they can be expected is: Wilson's, July 15 through August; Northerns, mid-August through early October; and Reds, rarely but regularly in October and November. Extremely rare occurrences of Buff-breasted Sandpiper, Marbled Godwit, and Hudsonian Godwit have been recorded in this shorebirders' heaven.

Once the really cold weather sets in and the ground and mud pools are frozen, all shorebirds depart. The time to observe the greatest numbers and variety is in the last week of August or the first week of September, in northwest winds, just after a front has passed through. Hot, windless, summer-like days are generally not as rewarding as brisk days with energetic wave action.

Although this preserve has earned its fame as a special shorebird area, it contains 250 acres and its shoreline is less than one mile in length. Encircling the beach face is a dense border of shrubs and vines edged with clumps of jewelweed, yellow-gold slender-stalked plumes of hybrid goldenrods, and various other grasses and weed species. The heads of several species are filled with seeds in late summer and fall, and white-footed and deer mice fiercely compete with several sparrow and finch species for the profusion of seeds released in autumn.

Behind the hedgerow a narrow ledge parallels the shore and rises about four feet above the water. This ledge, varying in width from eight to thirty feet, was formed by repeated dissipation of waves and backwash along the backshore. Sediment, sand, and vegetation are constantly shifting along this raised platform, which can sometimes be a wonderful place to find probing birds.

Henderson Pond (Map 6.3)

Rating: Spring****, Summer*, Autumn****, Winter*

Stony Point and High Cliffs (Map 6.3)

Rating: Spring****, Summer***, Autumn****, Winter***

All three of these areas can be covered on a single day and they are best birded in spring and autumn. Henderson Pond, one mile east of Henderson, includes about a thousand acres, most privately held. From I-81, take Exit 41 (Adams) and get on NY 178 headed west. Proceed approximately 8 miles, being careful to remain on NY 178 heading northwest at the traffic light at Scotts Corners where Smith Road (NY 152) meets NY 178. Proceed to Ayles Road, which comes in on the right. Turn on Ayles and travel down the road approximately 1 mile; Henderson Pond will be visible on the left. Park the car just above the first farm on the right side of Ayles Road. A telescope is recommended for excellent views—all birding must be done from the road only. Trees may obstruct the view of the pond when they are completely foliated.

This pond is a major spring waterfowl resting and feeding site. In the peak of the migration over 10,000 Canada Geese, along with Wood Ducks, hundreds of Mallards, Black Ducks, Pintails, numerous Blue-winged Teals, and occasional Snow Geese can be seen. Overhead in spring, migrating or hunting resident Turkey Vultures, Sharp-shinned, Broad-winged and Red-tailed hawks may be seen. Very rarely observed is a migrating Red-shouldered Hawk, but there are records from this general vicinity. Both Green Herons and Black-crowned Night Herons nest in the area. Look for nesting Chimney Swifts, Barn Swallows, and Purple Martins. From early September through October migrating Evening Grosbeaks, Purple Finches, Common Redpolls, Pine Siskins, crossbills, juncos, White-crowned and White-throated sparrows, and large flocks of Snow Buntings are easily seen.

From Ayles Road return to NY 178. Turn right, and proceed northwest to Henderson. Stay on NY 178. Proceed west across NY 3. Go past the Henderson boat access site. Turn left on the next road that comes in from the left, Lighthouse Road. Follow this dirt road to its end, where you will find a lighthouse. Park and walk down the short dirt road to the lakeshore. You are now at Stony Point.

This is an excellent place to sit down on a large rock and take a look through the scope. At first the area may seem birdless, but with a little patience large numbers of diving ducks may be seen in both spring and autumn. Double-crested

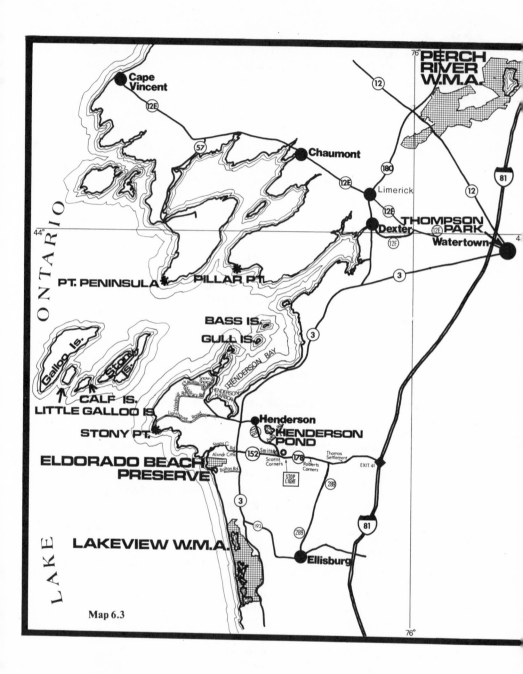

Map 6.3

Cormorants and gulls may be seen overhead. Here one is able to see good numbers and a rich diversity of shorebirds. During migration watch for birds cutting across the point to short-cut the main spit of land. Sitting at the point the surf and wind noise may preclude hearing migrating warblers overhead. If so, step back into the cedars where they can be seen. Having thoroughly birded the point return by Lighthouse Road to the first road coming in on the left, Windmill Road. Turn left on it and follow it to its junction with Military Road (concrete); turn left and go to the end of Military Road. Turn right on North Schoolhouse Road (dirt), which takes you to Snow Shoe Road (blacktop). Turn left and proceed to Snow Shoe Point and park. The High Cliffs begin here and extend south along the Lake Ontario shore. The scenery is quite spectacular. Looking south you can see Stony Island. Sit on the "Cut" (bridge and boat channel access) and watch for waterfowl and migrating raptors.

If you are birding this area in the late fall or early winter, be on the lookout for Goshawks, and the occasional Cooper's Hawk as you work your way back, wandering along the backroads, to NY 3 and NY 178.

Little Galloo, Calf, Gull, and Bass Islands (Map 6.3)

Rating: Spring***, Summer***, Autumn**, Winter*

The only access to these four islands, located at the eastern end of Lake Ontario, is by a boat capable of enduring open lake conditions. To go out in the small car-top type is dangerous and foolish. The most centrally located boat launching site is on Henderson Harbor. Little Galloo and Calf islands lie about 11 miles to the west and Gull and Bass islands about 7 miles north of Henderson Harbor. Little Galloo and Calf islands are on the western side of Stony Island, Bass and Gull islands are northeast of Association Island. The total acreage of the Bass and Gull islands is around 60 acres.

From I-81, take Exit 41 (Adams) and get on NY 178 headed west. Follow NY 178 through Henderson and across NY 3 to Military Road. About 0.5 mile down Military Road is the boat launching site. Navigation will be much easier if you first obtain U.S. Hydrographic Chart No. 21 or Cruise Nautical Chart No. 3." Cruise Nautical Chart No. 3 was available from the Department of Parks and Recreation at a cost of $7.50 in 1981.

Little Galloo is the site of the largest Ring-billed gullery in the northeast, numbering in excess of 75,000 breeding pairs. On all four of these islands Double-crested Cormorants, Black-crowned Night Herons, Pintails, Red-breasted Mergansers, Herring Gulls, and Common Terns nest. Covering the four islands in one day is entirely feasible and would be an extremely pleasant freshwater "pelagic" trip.

Little Galoo and Calf islands are owned by the Sealrite Division of the Phillips Petroleum Corporation. It also owns or leases most of Stony Island. Gull and Bass Islands are privately owned. *Under No Circumstances Should Birders Land on Any Of These Islands.* They support species especially sensitive to human disturbance. These islands are, in effect, natural laboratories, studied by biologists interested in colonial nesters. Any disturbance might prove devastating to nesting success. Birding should, obviously, be done from a boat.

Point Peninsula, Pillar Point, and Sherwin Bay (Map 6.3)

Rating: Spring**, Summer*, Autumn***, Winter***

All of these areas could have been included in the previously described freshwater "pelagic" trip and can be covered by boat; however, because there are many cottages along the lakeshore, the human population is dense in the summer and the area is subject to heavy traffic and quite a bit of noise. Birding is best done in the late fall and winter. All can be easily covered by car; walking is not absolutely necessary. But, of course, coverage will be enhanced if some of the area is done on foot.

In the late fall, these areas are excellent for loons, grebes, diving ducks, gulls, and occasionally shorebirds until the waterside freezes. Toward the land, from autumn well into winter, look especially for Goshawks and Rough-legged Hawks, Snowy and Short-eared owls, Gray Partridges, pipits, longspurs, and Snow Buntings. The terrain is level and suitable for farming, although much of it is abandoned. In irruption years, hawks and owls are highly visible in these areas.

Sherwin Bay and Pillar Point lie about ten miles west of Watertown; the area includes nearly eight square miles. From Watertown head west on NY 12F (I-81 Exit 46 for Watertown International Airport) for 8 miles, then turn right onto NY 180 toward Dexter. In Dexter turn left onto Lakeshore Drive and follow "Pillar Point" signs to South Shore Drive. Continue along the lakeshore, noting migrant or wintering waterfowl in Black River Bay. Sherwin's Bay, just east of Pillar Point, has some marsh and mud flats which, unless there is no open water, should be investigated for late or lingering shorebirds. Continue around the southernmost part of the peninsula, which is Pillar Point, and up the west side of the peninsula. Check out the cross roads on Pillar Point for wintering raptors. Middle Road will provide a good view of the uplands.

Point Peninsula is a seven-square-mile tongue of land about eight miles south of Cape Vincent. From Watertown head west on NY 12E. Note that NY 12E is not directly accessible from I-81, but it is easily found once you are in Watertown. Pass through the village of Three Mile Bay on NY 12E. The next road coming in on the left will be NY 57. It should be marked with a sign reading "Long Point State Park" or "Point Peninsula." Turn left onto that road and follow it south to South Shore Road and then right to the point. If time permits, it might be profitable to search the more protected waters of Chaumont Bay from Long Point State Park at the north end of the peninsula. The entrance to the park is on State Park Road, and it is well marked. Driving in this region may be quite difficult in winter and every sensible precaution should be taken. Do not undertake going to Point Peninsula or Pillar Point following heavy snowfall without first checking conditions with local birders in Watertown or at the town offices: Town of Lyme,

Chaumont, telephone (315) 649-2789; and Town of Brownville, telephone (315) 639-6604.

John C. Thompson Park (Map 6.3)

Rating: Spring***, Summer***, Autumn***, Winter**

The John C. Thompson Park (Thompson Park) is a small, 600-acre public park within the city of Watertown. It is birded by local people, especially during spring and autumn migrations. The following species are some of the more outstanding recorded there: Snowy Owl, Alder and Least flycatchers, Bohemian Waxwing, Mourning Warbler, Pine Grosbeak, and Red Crossbill.

A checklist of all of the birds recorded there can be obtained from the Watertown Chamber of Commerce, 46 Public Square, Watertown, New York 13601.

From I-81, take Exit 45, Arsenal Street. Follow Arsenal east around the town square and turn onto State Street. Follow State to the entrance of the park, which is well marked. There is also an entrance off Gotham Street, if one is coming from the south on NY 12.

Perch River Wildlife Management Area (Map 6.3)

Rating: Spring****, Summer***, Autumn****, Winter*

About eight miles northwest of Watertown, still in Jefferson County, lies the 7000-acre Perch River Wildlife Management Area. Although it is entirely on public land, access is restricted to open areas (owing to waterfowl production and refuge zoning) and special areas opened to waterfowl hunters in the autumn. At the discretion of the management personnel areas are made off limits, and their restrictions should be respected. This should not, however, prevent any birder from using it, for the area is never completely closed. It can be reached by automobile and can be covered by either car or bicycle. It is not necessary to seek permission to enter the area, and there is neither a parking nor admission fee. Special regulations *are* in effect during the autumn hunting season, however, so check with the Region 6 DEC office at that time. The address is: Department of Environmental Conservation, Bureau of Wildlife, State Office Building, 317 Washington Street, Watertown, New York 13601. Coverage of the area by those who use it is rated as easy. Because a great deal of the area consists of fresh marshes and pond and lake habitat, a telescope will prove helpful.

To reach the area from the north, leave I-81 at Exit 48, NY 342. Proceed west on NY 342 to its termination at the intersection of NY 12. Turn right on NY 12 and head northwest toward Clayton. Approaching the area from Watertown or south of it: leave I-81 at Exit 47, get immediately on NY 12 and head northwest on it. From the intersection of NY 342 and NY 12 drive 4 miles and watch for signs reading "Allen Road" and "Vaadi Road." The best access points are off Vaadi

Road and, secondarily, Allen Road from parking areas off each of these roads. The Allen Road parking area is reached by turning left off NY 12 and the Vaadi Road facility (where there is also a boat launching ramp) is reached by turning right off NY 12. *Nota Bene:* If you arrive at Gunns Corners, where NY 180 crosses NY 12, you have gone too far, and you should turn back and watch more carefully for Allen and Vaadi roads.

A great variety of wetland species can be found on the PRWMA. The following species, somewhat unusual for this region, have been found there: Great Egret, breeding Least Bitterns, Glossy Ibis, Mute and Whistling swans, Brant, Green-winged Teal, Mallard X Black Duck hybrid, Redhead, Ruddy Duck, Osprey, Bald Eagle, various raptors, Hungarian Partridge, Black Terns, and Red-headed Woodpecker, as well as many species of passerines in migration.

Because the area is replete with mammalian wildlife as well, be on the lookout for otters, muskrats, raccoons, and white-tailed deer.

Grass Point, Wellesley Island, and Keewaydin State Parks (Map 6.1)

Rating: Summer****, Autumn****, Winter*

Grass Point, a modest municipal park of 27 acres in the Thousand Islands on the St. Lawrence River, is the nesting site of a colony of Henslow's Sparrows. It is also a site to check out if you are traveling through the area during the height of the autumn warbler migration; predicting just *when* it will be good is impossible, however, and a special trip there simply to see migrants is not advisable. It is included here mainly as a place to look for Henslow's Sparrows.

The park is about three miles southwest of the seven-mile island-hopping Thousand Island Bridge, one mile east of Fisher's Landing. Its entrance is at the very northern end of NY 180. Proceed north on NY 180 past its intersection with NY 12, and go directly to the park's entrance. In the fallow fields about 70 to 100 yards from the entrance and on both sides the sparrows have been found.

When birding the area, check any or all of the parks nearby, especially Wellesley Island State Park, reached by taking Exit 51 from I-81, two miles north of the Thousand Island bridge. Keewaydin State Park, which has its entrance off NY 12 approximately five miles north of Grass Point State Park and one mile west of Alexandria Bay, is also of particular interest. Both parks have hosted notable rarities in the past.

ST. LAWRENCE COUNTY
Bonaparte Lake, Peavine Swamp, Cranberry Lake, and the Five Ponds Wilderness Area (Map 6.4)

Rating: Spring****, Summer***, Autumn***, Winter*

Cranberry Lake is actually within the six-million-acre tract known as the Adirondack Park. Located in the southeastern section of St. Lawrence County, it is rated

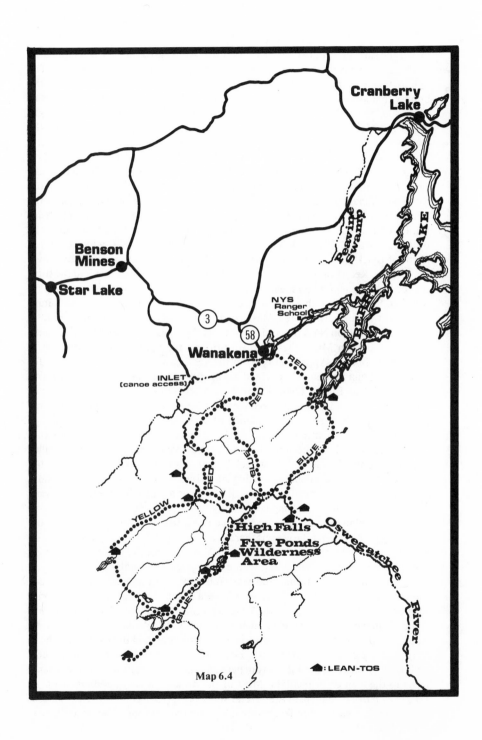

Map 6.4

:LEAN-TOS

one of the state's most pristine wilderness areas — due, no doubt, to its remoteness. Cranberry Lake is the northern gateway to the Five Ponds Wilderness Area. The rich avifauna of the Cranberry Lake area is simply a tantalizing foretaste of the bounty that awaits the visitor. On the east side of the lake Bear Mountain rises to over 2400 feet; and to the south Long Tom Mountain towers, over 2600 feet high.

The most successful birding around Cranberry Lake is found on foot on the marked trails from Wanakena. To reach the area from the north (Massena-Potsdam-Canton), take NY 56 to NY 3; 8 miles southwest of the intersection on NY 3 is the village of Cranberry Lake. If one stays on NY 3, still headed southwest, after another 6.6 miles a small road comes in on the left. This is County Road 58, Wanakena Road. Take it to the hamlet of Wanakena. Wanakena is the home of the New York State Ranger School, and its direction signs will prove helpful. If you find yourself at Star Lake, you have gone too far. Turn back. From the southwest, leave I-81 at Exit 48. Head east on NY 342 to NY 3, which comes in north of Black River. On this route, you will bypass the confusion of downtown Watertown. Keep heading east on NY 3 for about thirty miles. Here find the turn-off for Lake Bonaparte on the left. A small detour to bird this lake can be very rewarding. Take the South Shore turn-off and not the North Shore or Hermitage Road turn-offs. South Shore is the first turn-off on NY 3 heading northeast.

Once back on NY 3 proceed east another 30 miles to Benson Mines. About 4 miles past Benson Mines there is a turn-off on the right (County Road 58) to Wanakena. Take it. Approaching from the southeast, leave I-87 at the exits for Lake Placid and Saranac Lake. Head north and west on NY 73 to Saranac Lake. Follow NY 3 to Tupper Lake. From there head west on NY 3 to the village of Cranberry Lake and then proceed another 6.6 miles to a left hand turn-off (County Road 58). Take it to the hamlet of Wanakena.

There is excellent birding on NY 3, 3.1 miles south of the village of Cranberry Lake, and about 3.5 miles before the Wanakena turn-off at a place called Peavine Swamp. This is an Adirondack bog of considerable size and interest, but have boots, map, and a compass if you plan to bird the swamp far from the road. I suspect Peavine Swamp will become well known as a profitable area as it becomes more frequented by the birding community.

If you plan to canoe the Oswegatchie River, note that it is not navigable immediately at Wanakena. However, there is a two-mile hiking trail, indicated with round yellow trail markers, that leads from Wanakena south to Inlet, a boat access site. This trail, Moore's Trail, follows the Oswegatchie River along one of its most interesting stretches of rapids and waterfalls. Inlet can also be reached by car, thereby eliminating canoe portage over Moore's Trail. To reach Inlet by car, drive west on NY 3 for 2.5 miles past the Wanakena turn-off to Sunnylake Road, which comes in on the left. It is the first road meeting NY 3 after Wanakena Road (County Road 58). Turn left on it. Follow it to a tee intersection and turn left again there. Driving to the end of this road, named Inlet Road, will bring you to a parking area and the boat access site at Inlet. From here one can canoe upstream as far as twenty miles to the area of Beaverdam. Most of the distance is easy paddling against a very slow current, but there are several short portages around rapids. Canoeists will encounter beaver dams frequently. There are five lean-tos adjacent to the river for canoeists' use. A map with details on camping at these lean-tos is available from a Forest Ranger at either Wanakena or in the village of Cranberry Lake.

Back at Wanakena, take the hike to the Cranberry Lake, Five Ponds Wilderness Area. Leave the village and hike south on a truck trail marked with round red discs. This truck trail is open to the public for hiking only, and vehicular use is prohibited. After walking for 1.5 miles along the red-marked trail, you will notice a blue-marked foot trail branching off to the left, but stay on the red-marked truck trail. Although that trail is a shorter route to High Falls, the birder doesn't really want to take it. The red-marked trail leads through one of the finest stands of old-growth mixed timber in the wilderness area and along an extremely interesting and productive stretch of the Oswegatchie River. Walking 7.6 miles from Wanakena on the red-marked trail, you will come to High Falls which is easily one of the most spectacular landmarks in the wilderness area. At High Falls two open Adirondack lean-tos are provided for the hiker; one on either side of the river. This is an excellent place to camp overnight, and hikers can bird the Five Ponds area and return to Wanakena the following day. The entire distance from Wanakena to High Falls by this route is 7.6 miles. If the birder had taken the blue-marked trail short-cut, the distance would have been 5.9 miles — but the red-marked truck trail is every bit worth the 1.7 mile greater distance.

The relatively long walk to High Falls (and Five Ponds later) will be rewarding. Northern Three-toed Woodpeckers are rare in the state; but, among the white pines, balsams, and spruces there are indications on occasional large dead trees that this woodpecker dwells in the vicinity. Goshawks nest near the lake and should be carefully watched for. The Spruce Grouse is a permanent resident in the area as is the Gray Jay. Common Ravens have been seen from Cranberry Lake. The Boreal Chickadee breeds near the lake and has been seen from the truck trail. In addition to the Northern Three-toed Woodpecker, Yellow-bellied Sapsuckers and Hairy and Downy woodpeckers breed in the area. The list of warbler species breeding in the general area should be inducement enough to take the long hike and to brave the bugs. Black-and-white, Nashville, Northern Parula, Black-throated Blue, Yellow-rumped (Myrtle), Black-throated Green, Blackburnian, Chestnut-sided, Bay-breasted, Mourning, and Canada warblers have not only been either sighted or heard singing on the truck trail, but have also been recorded on the latest two Sabattis Breeding Surveys run, not ten miles from the southern inlet of Cranberry Lake. In addition, Wood Thrushes, Golden and Ruby-crowned kinglets, Cedar Waxwings, and Rusty Blackbirds nest in the area.

The incomparable and melodious songs of the breeding Winter Wrens are a particular delight along the trail. If the hiking birder chooses to camp at one of the lean-tos and makes an owl foray after dark he can easily hear Great Horned and Barred owls, and if he is especially lucky he may discover a Saw-whet. In 1973, a first St. Lawrence County breeding record for Philadelphia Vireo was provided at the State University Biological Station at Cranberry Lake. The Lincoln's Sparrow also breeds in suitable habitat there. These are but several of the species to be found in the Cranberry Lake, Five Ponds Wilderness Area. It is not, to date, a well birded area and the *Kingbird* Region 6 editor enthusiastically welcomes records and sightings from the area and encourages photographic substantiation.

To reach Five Ponds from High Falls, retrace the truck route for 1.75 miles, passing by the Plains Trail (coming in on the right), to the intersection of the Leary Trail. This trail is marked with blue discs. Turn left on it and head south. Proceed for 3 miles to Five Ponds. The five ponds are Big Five, Little Five, Big Shallow, Washbowl, and Little Shallow. There are lean-tos at Big Shallow and Little Shallow ponds. There are several remarkable stands of large white pine in this area and you will have passed through an old-growth stand of large white pines and spruce along the way. This is an excellent birding area.

After birding, retrace the route you took in back to the intersection with the red-marked truck trail. At this point you could decide to continue on the blue-marked Leary Trail for 2.7 miles and meet the truck trail 1.5 miles outside of Wanakena, or you could turn left back onto the truck trail and retrace the route taken the day before. Another alternative is to turn right and proceed one mile to where the blue-marked Plains Trail comes in on the left. Take that trail through entirely different plains-like habitat for 2.2 miles. A trail marked with red discs comes in on the left. Take it 0.8 mile to the southeastern end of Dead Creek Flow. Follow it around the tip of the Flow, which will be another mile. After a short distance you will come to a red-marked truck trail; the distance from there to Wanakena is 2.2 miles. The total miles from the intersection of the Five Ponds Trail with the truck trail returning completely on the truck trail west and north to Wanakena is 5.9 miles; proceeding on the Leary Trail and then truck trail to Wanakena is 4.2 miles; and proceeding east to the Plains Trail and then north and west to Wanakena is 7.2 miles. Each route takes the birder through different habitat and the choice is his.

This walk should probably be done in either May or early June or in early autumn for the biting and bothersome insect density is truly prodigious in June, July, and early- to mid-August.

Upper and Lower Lakes Wildlife Management Area (Map 6.5)

Rating: Spring****, Summer***, Autumn****, Winter*

The Upper and Lower Lakes Wildlife Management Area two or three miles west of Canton is one of the largest wetlands in the northeast, encompassing 8781 acres. This is on state public lands and access can be gained to a great deal of the area without permission; however, to those areas for which a permit is necessary, apply to the Department of Environmental Conservation office in Watertown (address same as Perch River WMA). During the late spring and summer, the areas with nesting waterfowl are closed. There is neither an admission charge nor a parking fee. Again mosquitoes can be somewhat troublesome during summer months as the management area provides attractive breeding habitat.

To reach the Upper and Lower Lake Wildlife Management Area from Ogdensburg, take NY 68 heading east, for approximately 10.5 miles, where you will see signs designating the area and the headquarters building. Coming from Canton, take NY 68 west toward Ogdensburg for about 3 miles to signs designating the area and headquarters. There are numerous parking fields along NY 68 and NY

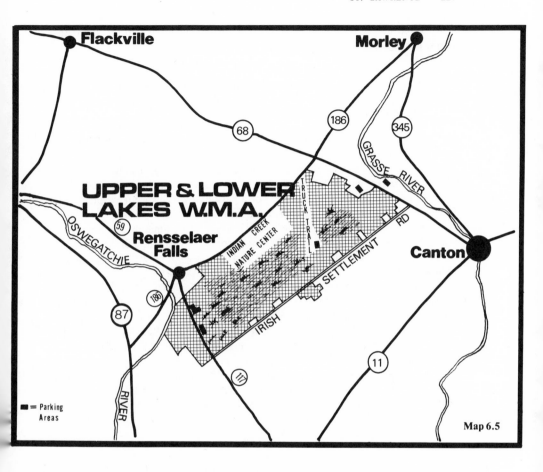

Flackville

Morley

68

186

345

GRASSE RIVER

UPPER & LOWER LAKES W.M.A.

TRUCK TRAIL

OSWEGATCHIE

59

Rensselaer Falls

INDIAN CREEK NATURE CENTER

RD

186

SETTLEMENT

Canton

87

IRISH

RIVER

117

11

■ = Parking Areas

Map 6.5

186 and NY 117. There is a truck trail just north of the Indian Creek Nature Center Area, off NY 186, which leads to a parking lot. Walking along the truck trail takes the birder out into the marsh to a point from which birding should be excellent. This particular parking lot is open to permit holders only in the fall.

Birding will be more profitable with a telescope. The best method of covering the area is to drive the perimeter, stopping to bird at parking areas, as well as using the trails and observation blind on the Indian Creek Nature Center Area off NY 186. Water and water-associated species can be found in considerable numbers as well as species normally found in open fallow fields. Owing to its size and proximity to the St. Lawrence River, the area consistently produces interesting observations. Black Terns and Henslow's Sparrows nest in the area. In 1974, the second breeding record for Ring-necked Duck, for the St. Lawrence valley in New York State was established on this refuge. Most of the puddle-ducks and a few of the divers utilize the area on both north- and southbound migrations.

Interested and observant birders can also view fishers, otters, snowshoe hares, beavers, and white-tailed deer.

Wilson Hill Wildlife Management Area (Map 6.1)

Rating: Spring****, Summer***, Autumn****, Winter*

Of the 3415 acres contained in the Wilson Hill Management Area, the portions open to the birding public can be easily covered in car or on bicycle or foot. The area is about seven miles west of Massena and is north of the small town of Waddington. Access is gained from NY 37, which runs along the St. Lawrence River. There is neither an admission charge nor a parking fee. This is an excellent area from which to photograph waterfowl and marsh birds.

From Massena, proceed south on NY 37 for approximately 10 miles. From Waddington proceed north on NY 37 for approximately 10 miles. Signs along NY 37 indicate NY 131 and the Wilson Hill Island Causeway. Follow those signs to the WMA.

This is primarily a shorebirding and waterfowling area with such species as Canada Goose, Wood Duck, Gadwall, Pintail, Blue-winged Teal, Northern Shoveler, and Redhead breeding. Shorebirds viewed here are transients and include most of the species that occur along the St. Lawrence River at other points. There is a parking area at both the east and west entrances of the refuge. From these a marked footpath leads along the bay side of the island. There is sufficient cover here for the birder to photograph. The observation towers on NY 131 and the Causeway (Willard Road) are recommended as particularly good vantage points. There is a newly built nature trail here especially for the use of birders.

St. Lawrence Seaway/Robert Moses Power Dam (Map 6.6)

Rating: Spring**, Summer*, Autumn***, Winter****

Please refer to the map while reading this description. A more detailed map of the Robert Moses State Park, the largest in the Thousand Islands Park Commission, can be obtained at the Robert Moses Power Dam and may be of considerable use to the birder. The area detailed on the accompanying map lies about seven miles northeast of Massena and encompasses nearly five square miles (3067 acres) much of which is water and the rest of which is publicly owned land. Permission is not required to gain access; there is neither an admission nor a parking fee. In summer a large colony of Cliff Swallows nest on the dam. Although this is primarily a winter birding area, it certainly should be avoided when temperatures fall below 10°F, because then there is much fog and thus poor visibility. Judging from recent *Kingbird* winter season reports, the Massena area is *the* place to bird in winter in Region 6.

To get there: from downtown Massena take NY 37 northeast toward the Massena Airport. Just past Highland Road and just past the airport turn left on

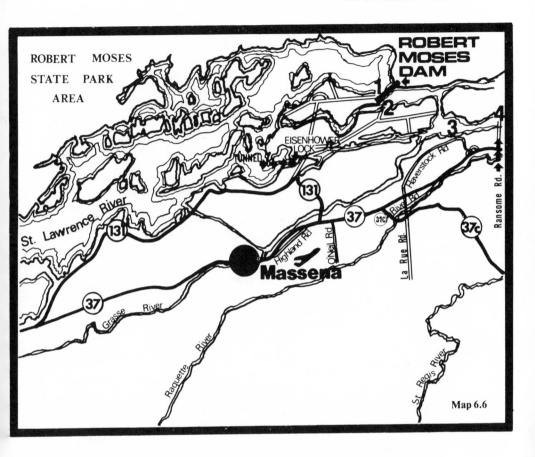

Map 6.6

NY 131. Follow signs for the Robert Moses Power Dam. Pass under Eisenhower
Lock (by the tunnel) and continue past the information booth to the overlook
number 1 (see map). To reach Hawkins Point overlook (number 2), proceed as indicated for the Robert Moses Power Dam *but* turn right at the information booth
and then proceed to Hawkins Point. In order to reach the other two overlook
points indicated on the map, go back to square one: downtown Massena. From
there again head northeast on NY 37 past the airport and past O'Neil Road. Exit at
the junction of NY 37C and then take the first left on River Road. At the first intersecting road, which is LaRue Road, turn left and follow that road north and
east (it will become Haverstock Road) to the overlook number 3. This is the Rey-
nolds Aluminum Plant. By following Haverstock Road around to the traffic circle, it is possible there to regain NY 37. Head a bit farther east on NY 37 to the first
road that comes in on the left, which is Ransome Road. Take it to its terminus to
the Ransome Road overlook number 4.

Admittedly, the above directions may prove confusing for some. However, as consolation, it would appear that the rewards outweigh the labors of the labyrinth. Once through the maze, quickly scan the horizon over plowed fields for some of the numerous raptors that winter here. No fewer than five Rough-legged Hawks, a Screech Owl, a Great Horned Owl and seven Northern Shrikes frequented the Massena area December 1977 thru January 1978. A group of 1026 Common Goldeneyes was seen at the Robert Moses Power Dam on February 26, 1978. Two adults and one immature male Barrow's Goldeneyes were seen from Ransome Road throughout the latter third of January 1978. Gadwall, Redhead, Bufflehead, and Hooded Merganser were all noted during that winter from these overlooks. The very considerable 4000 Common Mergansers noted from the dam December 1, 1977, dwindled to 102 by February 25, 1978. During the same winter, gulling in the area was gilt-edged: 14 Glaucous (two mature), 18 Iceland, 700 Herring, and 1000 Great Black-backed gulls were viewed from Hawkins Point. In December 1977 at Hawkins Point, J. Van Riet discovered an immature Black-legged Kittiwake, which many later observed. To top off his winter of astounding observations, Van Riet discovered an immature Ivory Gull in early December 1977 and then again in January 1978 from Hawkins Point. Which all goes to prove that *everything* comes to him who waits.

If you get bored, however, contact one of the enthusiastic local birders and find out where the 11 Bohemian Waxwings were found north of Massena in January 1978 and hunt up a few yourself.

ACKNOWLEDGMENTS

To the following people go the author's thanks for their contributions to the St. Lawrence section: Lee Chamberlaine, John Belknap, Frank A. Clench, J. Van Riet, and Robert C. Walker. Special gratitude is extended to Lee Chamberlaine not only for his critical reading of the St. Lawrence chapter but also for the indispensable information he supplied at several points along the way.

Adirondack-Champlain — Region 7

T HE ADIRONDACK-CHAMPLAIN REGION encompasses all of Clinton, Franklin, Essex, and Hamilton Counties. Within this area birders can find the unique natural qualities and unspoiled primitive conditions enjoyed by generations of colonial period nature lovers as far back as 1650 and by the American Indians for uncounted generations before that. Admittedly, the astounding mobility available to the recreation-minded public for the past few decades has caused the depredation of large tracts of pristine habitat. In many areas a profusion of commercial attractions has sprouted with more accompanying land alteration and abuse in the name of progress. However, throughout the 6290 square miles of this region there are still rocky slopes, rugged peaks, lakes, streams, waterfalls, ice caves, extensive zones of flat and swampy land, giant white pine forests, superb red spruce and yellow birch stands, black spruce bogs, and even meadows and farmland — and mile after mile of unspoiled hiking trails (Maps 7.1 and 7.2).

The opportunities for the birder to enjoy the avifauna in the unexploited areas are limitless. Additional studies of this area should add considerably to the ornithology of New York State. The High Peaks, Lake Champlain, and the Champlain Valley provide some of the most exciting birding in the northeast. For the serious observers of birdlife, it is as near perfection as is possible in this age.

HAMILTON COUNTY
Brown's Tract Ponds and Shallow Lake Trail (Map 7.3)

Rating: Spring***, Summer***, Autumn**, Winter
(Roads are usually not passable in winter.)

Brown's Tract Ponds are located in the town of Inlet, in the northwest section of Hamilton County, about 2 miles west of the village of Raquette Lake. They lie at elevations of 1780 to 1820 feet, surrounded by the eastern end of Cascade Mountain (*ca.* 2000 feet) on the northwest, West Mountain (*ca.* 2085 feet) on the northeast, Estelle Mountain (*ca.* 2530 feet) on the southeast, and Black Bear Mountain (*ca.* 2450 feet) on the southwest. These two ponds lie just within the Raquette River

231

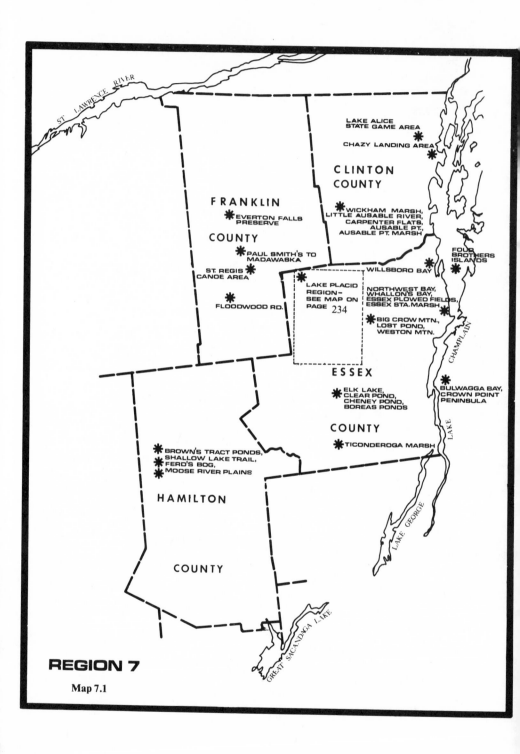

ST. LAWRENCE RIVER

LAKE ALICE
STATE GAME AREA

CHAZY LANDING AREA

CLINTON
COUNTY

FRANKLIN
※ EVERTON FALLS
   PRESERVE

COUNTY

※ WICKHAM MARSH,
LITTLE AUSABLE RIVER,
CARPENTER FLATS,
AUSABLE PT.,
AUSABLE PT. MARSH

※ PAUL SMITH'S TO
   MADAWASKA

FOUR
BROTHERS
ISLANDS

ST. REGIS
CANOE AREA

WILLSBORO BAY

LAKE PLACID
REGION—
SEE MAP ON
PAGE 234

NORTHWEST BAY,
WHALLON'S BAY,
ESSEX PLOWED FIELDS,
ESSEX STA. MARSH

※ FLOODWOOD RD.

※ BIG CROW MTN.,
LOST POND,
WESTON MTN.

ESSEX

CHAMPLAIN

※ ELK LAKE,
CLEAR POND,
CHENEY POND,
BOREAS PONDS

※ BULWAGGA BAY,
CROWN POINT
PENINSULA

COUNTY

LAKE

※ BROWN'S TRACT PONDS,
SHALLOW LAKE TRAIL,
※ FERD'S BOG,
※ MOOSE RIVER PLAINS

※ TICONDEROGA MARSH

HAMILTON

LAKE GEORGE

COUNTY

GREAT SACANDGA LAKE

REGION 7

Map 7.1

watershed, which feeds into Raquette Lake, then joins waters from other north-western Adirondacks lakes and rivers and eventually empties into the St. Lawrence River. This area contains giant white pine, superb red spruce, and yellow birch trees; both rugged and gently rolling terrain; outstanding trout-filled crystal lakes and streams—and prime birding habitat. The area is rated as "moderate to easy" to bird. Remember that mosquitoes may be bothersome from late spring to mid-summer, and blackflies especially in June. Take along bug repellent.

The village of Raquette Lake can be reached by way of NY 28 from the southwest, I-87 and NY 28 from the northeast or southeast, or NY 12 from the northwest. From Raquette Lake village proceed 2 miles west on NY 28. Look for signs designating the Brown's Tract Ponds Campsite. Follow these to the camping headquarters building. Park here and obtain permission to leave your car; also inquire about directions to the hiking trail to Shallow Lake. The ponds have two fine beaches, which, although usually duckless, do provide a place for a cool swim.

Walk the roads within the campgrounds and take the trail through the boggy region leading to Shallow Lake. This is an excellent area for both leisurely and more rugged birding, and it should yield such boreal breeding specialties as the Spruce Grouse, Black-backed and Northern three-toed woodpeckers, Yellow-bellied and Olive-sided flycatchers, Gray and Blue jays, the Boreal Chickadee, abundant Winter Wrens, and Wood, Hermit, and Swainson's thrushes. The songs of the breeding warblers and vireos will be a special and indescribable pleasure. The rare vagrant, the Blue-gray Gnatcatcher, has been sighted here. Solitary, Red-eyed and Warbling vireos; Black-and-white, Tennessee, Nashville, Northern Parula, Magnolia, Black-throated Blue, Yellow-rumped, Black-throated Green (abundant), Blackburnian, and Bay-breasted warblers all breed in the area; watch for them and listen attentively for their songs.

Rusty Blackbirds nest in the low, wet areas. Rose-breasted and Evening grosbeaks will also be found. Flocks of migrating White-winged Crossbills can probably be found here from mid- to late October. The number of sparrow species breeding in the area is marvelous; Dark-eyed Juncos, Chipping, White-throated, abundant Lincoln's, Swamp, and Song sparrows can all be encountered here.

At night, walk along the Uncas Road listening for both Barred and Saw-whet owls. The Great Horned Owl is an uncommon, permanent resident within the Adirondack Park but it has not recently been reported from this particular area. Its detailed distribution in *Kingbird* Reporting Region 7 is not well known. Therefore, the editor of this region would welcome any location records of this species.

Ferd's Bog (Maps 7.3 and 7.4 and 7.5)

Rating: Spring****, Summer***, Autumn***, Winter
(Roads are usually not passable during winter.)

This typical spruce-tamarack bog is named after the birder who in 1969 discovered its unique avifaunal treasures: Ferdinand LaFrance. This is one of the only locations in New York where both Black-backed (*Picoides articus*) and Northern

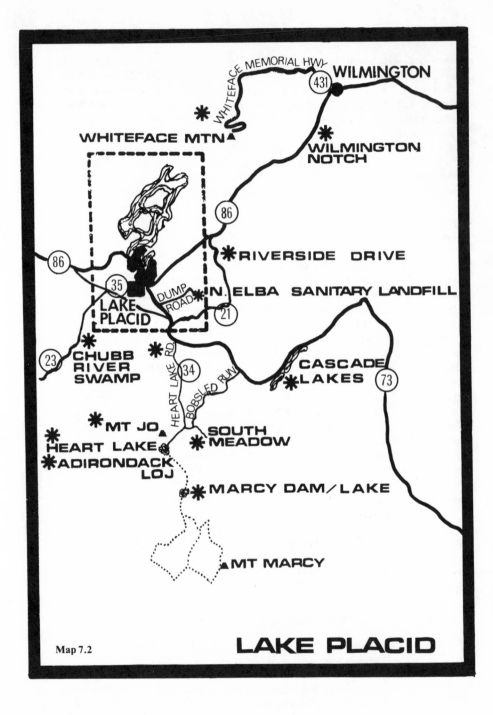

WHITEFACE MEMORIAL HWY

WILMINGTON

431

WHITEFACE MTN▲

WILMINGTON
NOTCH

86

86

35

DUMP ROAD

RIVERSIDE DRIVE

N. ELBA SANITARY LANDFILL

LAKE
PLACID

21

23

CHUBB
RIVER
SWAMP

HEART LAKE RD.

134

BOBSLED RUN

CASCADE
LAKES

73

MT JO▲

HEART LAKE

ADIRONDACK
LOJ

SOUTH
MEADOW

MARCY DAM / LAKE

▲MT MARCY

Map 7.2

LAKE PLACID

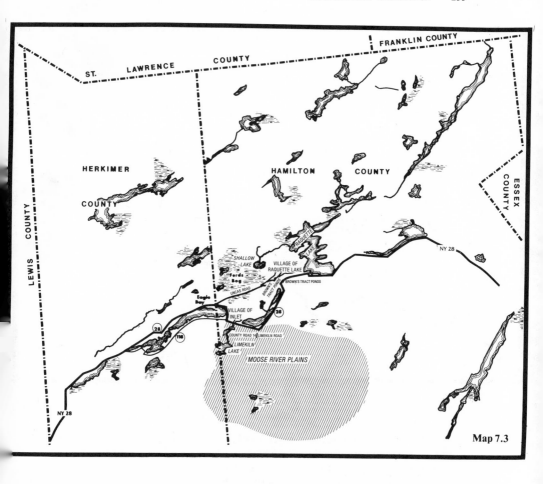

Map 7.3

(*Picoides tridactylus*) three-toed woodpeckers breed. It is essentially an area of soft, wet, spongy ground consisting chiefly of decayed and decaying vegetable matter. It all started as a relatively shallow stagnant lake or pond, which filled in slowly with sphagnum moss. The growth spread out from the shores toward the center of the pond as a floating mat on the surface. Over time, this sphagnum moss mat, which reaches all the way to the bottom, covered the entire pond. This mat makes what is sometimes called a 'quaking bog', for it quakes under foot. In it a few new conifers and live black spruce grow; they are most numerous at the edges of the pond. Peat now fills the pond. In addition to sphagnum moss, pitcher plants by the thousands, sundew, bog laurel, leatherleaf, and cranberries make up the vegetative layer of the bog. Ferd's Bog is surrounded by a ring of dense pine, spruce, American Larch, and Balsam Fir. Beyond that one can see the steep, hardwood-covered slopes of the Cascade Mountains.

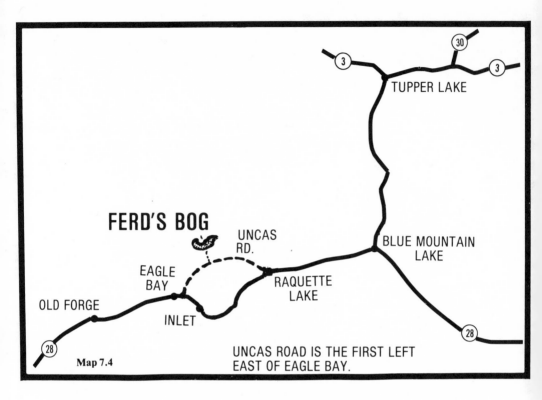

FERD'S BOG

UNCAS RD.

EAGLE BAY

OLD FORGE

INLET

RAQUETTE LAKE

BLUE MOUNTAIN LAKE

TUPPER LAKE

UNCAS ROAD IS THE FIRST LEFT EAST OF EAGLE BAY.

Map 7.4

The main roads one must travel to reach Ferd's Bog always suffer throughout the winter and are usually not passable until the second week of May. This is the best time to see three-toed woodpeckers for the blackflies and mosquitoes do not reach their peak numbers until late May through mid-July.

The village of Eagle Bay can be reached by way of NY 28 from the southwest, I-87 and NY 28 from the northeast or southeast, and NY 12 from the northwest. From Eagle Bay, travel northeast on NY 28 to the first road coming in on the left. This is Uncas Road, a sand-and-dirt road that is sometimes impassable until late May. Uncas Road begins about 2 miles northwest of the village of Inlet. Proceed 3.6 miles on Uncas Road, crossing an abandoned railroad bed on which is now located a series of private hunting camps. Watch for a camp on the left named Camp Buck Horn. Park once you have found it, taking care not to block access in or out of the private campsites along Uncas. The trail to Ferd's Bog begins behind Camp Buck Horn. The trail and the bog can only be reached on foot.

Walk down to the bog on the trail located behind Camp Buck Horn, about a ten- to fifteen-minute walk. Listen along the way for Goshawk, Red-shouldered, Broad-winged, and Red-tailed hawks. If you don't locate them on the way to the bog, be sure to search for them over the peaks of the Cascade Mountains once you

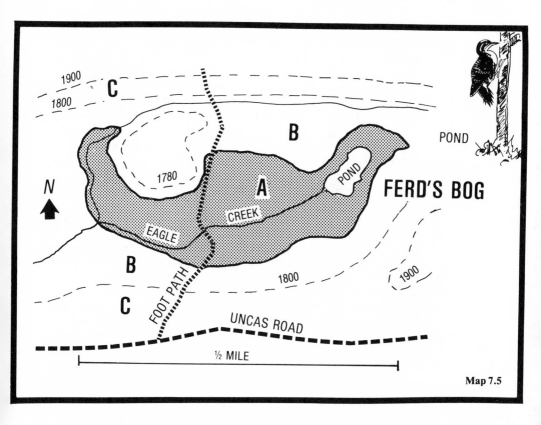

Map 7.5

reach the bog's edge. Walk quietly, listening for the drumming of Ruffed Grouse. Search all of the deciduous trees and even the dead tree hollows in the low wet woodlands adjacent to the bog for Barred Owls. In addition to the three-toed woodpeckers you should be able to spot Common Flickers, Pileated Woodpeckers, Yellow-bellied Sapsuckers and Hairy and Downy woodpeckers. These are common breeders in the 150 acres of the Ferd's Bog area. As you approach quite close to the bog you should be able to rouse Yellow-bellied Flycatchers, Eastern Wood Pewees and Olive-sided Flycatchers. Gray Jay and Boreal Chickadee, White-breasted and Red-breasted nuthatches, Brown Creeper, Winter Wren, Eastern Bluebird, Golden-crowned Kinglet, Cedar Waxwing, Solitary and Red-eyed vireos, and at least twelve warbler species (including Tennessee, Nashville, Magnolia, and Bay-breasted warblers) can easily be found on the way to the bog or around its perimeter. Rusty Blackbirds, Dark-eyed Juncos and Chipping, White-throated, Lincoln's, Swamp, and Song sparrows all nest within the area.

Once you have reached the bog opening follow this procedure: find a fallen tree on which to sit and just sit down and wait for a while. If you have come to the bog in early-to-mid-May you will not be bothered by either mosquitoes or black-flies. Listen for the sounds of the three-toed woodpeckers digging nests. The birds can be located by following the loud tapping sounds. The bog opening is quite wide and visibility is excellent from its edges. Both three-toed woodpecker species are permanent residents of the bog area but, if you really want to see them you should go to the bog before July 20. John Bull gives the earliest egg date of Northern Three-toed Woodpecker as May 14 (May 18 for Black-backed) and the latest fledgling date as July 24 for Northern (July 23 for Black-backed). Therefore, go to the bog to see this specialty between early May and mid-July. The birds tend to wander more following the fledging of the young and could be missed after mid-July. If rising from your fallen tree seat seems advisable in order to see the birds, follow along the edges of the fragile and treacherous bog in order to protect both yourself and it.

Most bogs, like Ferd's Bog, are very fragile. All that supports a person walking on it is a mass of roots and vegetation. Once this mat is broken through, the surface will not support much weight. Ferd's Bog is well over twenty feet deep in some places, so traversing it can be dangerous. The paths crossing the bog will break through with continued traffic. Along some stretches they have already done so, but, have been strengthened with poles placed side by side to form a makeshift bridge. These bridges have settled into the sphagnum moss and are not visible, so most people using the paths are not aware of the damage that has been done. *It is not a good idea to cross the bog at all; moreover, it is a very bad idea to attempt crossing it without a guide who is entirely familiar with its network of paths and deep, silty, weak areas.*

Moose River Plains (Map 7.3)

Rating: Spring***, Summer***, Autumn***, Winter
(Roads are usually closed during the winter.)

The Moose River Plains encompass thousands of acres in the west-central section of Hamilton County. Included are the lands drained by the Moose River watershed, whose waters flow into the Black River and empty into Lake Ontario. The Black River drains 1918 square miles of the southwestern Adirondacks, and the Moose River is one of its two chief tributaries. The flood plain of the Moose River has been formed from deposits of sediment carried down by the river. Each time the river rises and overflows its banks, the water spreads over the flood plain, depositing a layer of silt. Over time the flood plain has risen, being highest near the river at its natural bank, or levee. The silt deposited in the plains is extremely rich and supports a wide variety of plant and animal communities. Within the plains lie the high peaks of Fawn Mountain (el. 2500 feet), Mount Tom (el. 2800 feet), and Little Moose Mountain (el. 3630 feet); the pristine waters of Limekiln Lake, Fawn Lake, Lost Ponds, and Mitchell Ponds to name but a few; dense stands of mixed growth woodlands; agricultural flat lands, open woodlands, and Black Spruce bogs.

This area has not yet been heavily birded; however, because of its variety of habitats and remoteness, it will certainly not disappoint birders eager to discover new, extremely productive sites.

To reach the Moose River Plains go to the village of Inlet, which can be reached by way of NY 28 from the southwest, I-87 and NY 28 from the northeast or southeast, and NY 12 from the northwest. Inlet is about 2.5 miles southeast of the village of Eagle Bay. From Inlet travel 3 miles southeast on NY 28 to County Road 14, Limekiln Road, which comes in on the right. Take Limekiln Road for about 3 to 3.5 miles to the gate of Limekiln Lake Campsite. Here it is possible to get a detailed map of the area. From the campsite one can drive throughout the plains on relatively primitive roads, or the remote ponds can be reached on foot by those who wish more exercise while birding. A four-wheel drive vehicle will be necessary to cover the area well in very early spring or in late fall.

Migrating Hooded Mergansers have been recorded in large numbers here from mid- to late October. Common Mergansers probably nest within the area. The Moose River Plains offer the vast, secluded habitat essential to many nesting raptors. Ospreys, Goshawks, Sharp-shinned, and Broad-winged hawks nest on the plains. Migrating Golden Eagles have been seen in the area during early to mid-October. American Kestrels nest on the plains. Historically, Spruce Grouse have been sighted in the area, but recent reports of this species are very scarce. This may, however, be simply a function of few birders frequenting the plains and not indicative of low grouse populations. Several Picids have been seen on the plains. Possibly, the Northern Three-toed Woodpecker and historically the Gray Jay have been recorded here, as have several species of vireos, Magnolia, Black-throated Green, Bay-breasted, and Blackpoll warblers. Rusty Blackbirds are commonly seen here, as are Evening Grosbeaks, Red Crossbills, and several sparrow species.

More field investigations throughout the plains will undoubtedly reveal a long list of breeding and migrant species.

FRANKLIN COUNTY

In addition to the wonderful birding sites in Essex County, there is marvelous birding to be experienced farther inland throughout Franklin County.

Floodwood Road (Maps 7.6 and 7.7)

Rating: Spring****, Summer**, Autumn**, Winter*

The Floodwood Road area is about twenty-five miles west of Lake Placid and fourteen miles north of Tupper Lake in a heavily glaciated area containing numerous small ponds and water-filled depressions. In this wilderness, otter, fisher, mink, and black bear can still be seen from Floodwood Road.

To reach the area, proceed west from the village of Lake Placid on NY 86.

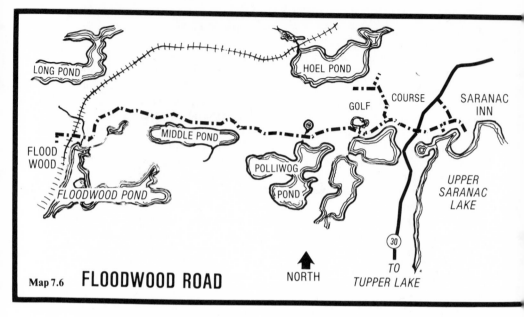

Map 7.6   **FLOODWOOD ROAD**   NORTH   TO TUPPER LAKE

Pass through the hamlet of Ray Brook and continue to the village of Saranac Lake. Inside the village, NY 86 takes on the name of Lake Flower Avenue. Follow NY 86 as it turns left (west) and becomes River Street. Continue west as NY 86 becomes NY 3 (George LaPan Highway) and follow it to Edgewood Road. Here turn right and follow Edgewood Road to Ampersand Avenue. Turn right on Ampersand Avenue and then take the first left onto NY 18. Continue on NY 18 to the junction of it and NY 86 (NY 30). Here turn left and follow the road past the golf course on the right. Take a sharp right on Floodwood road, which will be the next road coming in on the right.

There are a number of private hunting camps on the left-hand side of the road. Pass those and begin birding immediately at Green Pond, which is on the left. Bird all along the five miles of road from here to the railroad grade at the hamlet of Floodwood. Stop wherever there is open water. Walk long stretches of the road searching for nesting picids, flycatchers, vireos and warblers. Listen and watch for resident Common Loons on Floodwood Pond. Between Middle and Floodwood ponds, look down the bank on the south side of the road to the spectacle-shaped ponds, on which you may very well see a female Hooded Merganser with young. Red-shouldered Hawks have been known to nest nearby. After dark, all along the road the birder should be able to hear, and perhaps see, Barred and Saw-whet owls. All along the road flickers, Pileated Woodpeckers and Yellow-bellied Sapsuckers may be seen and heard carrying on their nesting activity. In the height of the nesting season it is no trick to tally breeding Winter Wrens, Hermit, Swainson's, and Gray-cheeked thrushes, Veeries, vireos, Black-throated Blue, Black-throated

Green, Backburnian and Chestnut-sided warblers. Call up or "spish" for Canada Warblers where the two small alder-covered brooks cross the road at Polliwog and Floodwood ponds. Check all the suitable habitat for nesting Rusty Blackbirds.

At Floodwood Pond the road effectively ends except for four-wheel drive vehicles. However, birders would do well to hike the trail to Long Pond that begins on the north side of the road in the vicinity of the railroad tracks. It is a rewarding walk that will expose the birder to many songs of the Adirondack breeding species; although it is sometimes a fairly damp, "buggy" walk, appropriate footwear and insect repellent help, and it is a rich hike indeed.

After birding along the length of Floodwood Road, turn around and head east on it and out onto NY 86/NY 30. Proceed a short distance, following signs for the Saranac Inn Public Boat Launch. At the first tennis court on the right, stop and check the white birches close to the highway on the left. Here there has been an active Broad-winged Hawk's nest. Stay a fair distance from the nest tree to avoid harrassing the adult birds.

Paul Smith's to Madawaska (Map 7.7)

Rating: Spring****, Summer***, Autumn**, Winter*

Along this route, in the height of the breeding season, the birder can hear and observe species in full breeding plumage and in full song that are, in other seasons and in other parts of the state, only briefly glimpsed passing to or from the breeding grounds. Be especially alert for the uncommon-to-rare local breeder, the "fool hen," or Spruce Grouse, whose statewide range is confined to the Adirondack boreal forest, bogs and plains and High Peaks. Populations of this grouse, also known as the "spruce" or "swamp partridge" have declined so much that the species is a remnant of the Adirondack avifauna. In New York State, this is a protected species and may not be hunted. Quite unlike its close relative, the Ruffed Grouse, it is ridiculously tame. Birders have often approached within a few feet of this species before it spooked and flushed. Some of the best habitats in which to seek the Spruce Grouse, are along dirt roadsides, along abandoned railroad beds and in the immediate vicinity of gravel pits. In these sites the bird seeks dust bowls and grit, as it, like other Galliformes, is an inveterate dust bather. This elaborate activity of deliberately kicking fine soil and dust up onto the back and letting it sift down through the plumage, probably functions to remove lice and other parasites and to maintain bedraggled and soiled plumage. If you are fortunate enough to happen upon a Spruce Grouse dust bathing, approach the bird quietly. At close range you will easily observe the unbarred, lighter-tipped tail, the conspicuous black and white underparts of the male, the feathering down to the toes, and the lack of neck ruff in both sexes of the Spruce Grouse that distinguish it from the fan-shaped, distinctively dark-banded tail, the unfeathered lower leg, and the conspicuous ruffed neck of the Ruffed Grouse.

From Lake Placid proceed to the village of Saranac Lake (see directions for Floodwood Road, above). Proceed north on NY 96 through Saranac Lake to NY

192A, then to NY 192 heading northwest. Follow this route to the village of Paul Smith's. Here, at the junction of NY 30 and NY 192, proceed straight across NY 30 onto the Keese's Mills Road. Bird all along this road for the next 6.5 miles. Keese's Mills Road parallels the Saint Regis River and eventually turns into Blue Mountain Road. At 6.5 miles, turn right at the Rockefeller estate gate house. Birding is excellent all along this road. After 6.0 miles park and follow the woods road on the right approximately 1.5 miles to a large shallow pond: Madawaska Pond. Spruce Grouse are often seen along this road or along its edges. Gray Jay is frequently seen here. For those looking for Boreal Chickadees, this place is a must.

Everton Falls Preserve (Map 7.7)

Rating: Spring****, Summer***, Autumn***, Winter*

This glorious preserve of 530 acres, in the town of Santa Clara, Franklin County, belongs to The Nature Conservancy, and is preserved as a natural area for the enjoyment and education of everyone. The preserve features about a mile and a half of frontage on the East Branch of the St. Regis River, in the northwest side of the county, at nearly the northern boundary of the Adirondack Park. The preserve is closest to the hamlet of Saint Regis Falls, approximately eighteen to twenty miles from the city of Malone.

This preserve has a history. About 1830 a tavern, an inn, and a sawmill were built on this site. Several years later a regular community named Everton grew up around these three buildings, and logging of the conifer forests supported a sizable population. The forest of the river valley continued to be extensively and intensively lumbered until, in the 1890s, the marketable timber simply gave out. Once thriving, now depauperate, Everton collapsed and was hastily abandoned. Today barely a trace of it remains and the forest has reclaimed the land.

Everywhere in the area the dominant conifer is balsam fir, with lesser amounts of white cedar, white pine, hemlock, tamarack, and red and black spruce. Upland hardwoods are evident on the slopes of Conger Mountain to the south and Mutton Ridge to the northeast. Infrequently, there are significant stands of yellow birch and sugar maple. Scattered here and there one sees black cherry, white ash, red maple, beech, and an understory of striped maple.

The long spiry balsams line the outer edges of the marshy floodplain above the falls along this part of the stillwater, which stretches for nearly ten miles upstream. The basin of the river was, at one time, filled by a postglacial lake. The elevation ranges approximately 350 feet from a base elevation of 1400 feet. The diverse flora and fauna are impressive, and include deer, beaver, and an astonishing array of bird species. The preserve appears to be remote enough and large enough to easily sustain the population levels of animals and plants now inhabiting it, and there is every reason to believe that these populations will naturally increase.

At Everton Falls water plunges twenty feet over a ledge of Precambrian rock. Here the densely wooded banks close in and a series of rapids begins, as the valley broadens and grows steeper. This is the approach to the confluence of the

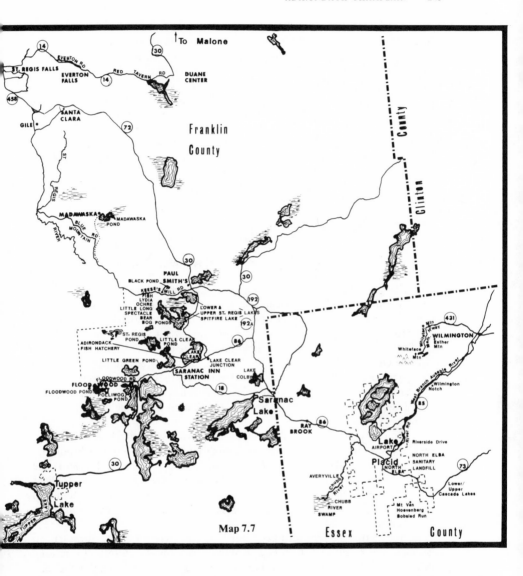

Map 7.7

Middle Branch of the Saint Regis River. Time spent here in the marshes and mixed conifer-upland hardwood forest and in the swamps and beaver meadows will prove a bonus after the long trip to the preserve.

To reach the Everton Falls Preserve from the Madawaska area: follow the basic directions for Paul Smith's to Madawaska just above. Follow Keese's Mills Road past the turn-off at the Rockefeller estate gate house. Proceed north on Keese's Mills Road as it turns into Blue Mountain Road, and continue north to its junction with NY 458. You will be about 2 miles west of the hamlet of Santa Clara. Turn left onto NY 458 and drive about 5 miles to the village of Saint Regis Falls. Look for the junction of NY 14 (Red Tavern Road). Turn right (east) on Red Tavern Road and proceed east for 6 miles to the entrance of the Everton Falls Preserve. Park safely off the road and walk into the preserve.

To reach the area from northeast or northwest of the village of Saint Regis Falls, leave the Adirondack Northway (I-87) at Interchange 42. Take NY 11 west through Mooers, Ellenburg, and Chateaugay, and then southwest to Malone; or from the northwest, follow NY 37 to the village of Malone. Pass through Malone on NY 11 to its intersection with Rockland Street (Finney Boulevard). Turn left and follow this route until it becomes NY 30. Proceed south on NY 30 for 15 miles to the intersection of it and NY 14 (Red Tavern Road). Turn right. Proceed 8 miles west to the entrance of Everton Falls Preserve.

Saint Regis Canoe Area (Map 7.7)

Rating: Spring***, Summer****, Autumn**, Winter**

This trip, primarily by canoe, requires at least a day, and probably two or three days with night camping in the canoe area. Birders must pack their own food and stoves. Firewood may be scarce at some of the campsites. Blackflies, deer flies, and "no-see-ums" could make the trip fairly unpleasant without plenty of insect repellent. Be sure to carry life preservers and a spare paddle. Sign the trail register at Little Green Pond before launching on this marvelous adventure.

The several miles of lake and pond waterways described here traverse some of the most pristine natural wildlife areas the state has to offer. Canoeists can put-in just east of the tiny hamlet of Saranac Inn Station and come out almost any-where.

To reach the area: follow directions to the village of Lake Placid under Essex County. From it, proceed west on NY 86 through the hamlet of Ray Brook and continue to the village of Saranac Lake. Inside the village take NY 86 (Lake Flower Avenue) as it turns west (left) and becomes River Street. At the stoplight near the Town Hall, turn right on Main Street. After one block bear left and turn onto Broadway. Follow Broadway (NY 86) north straight out of town. Pass Lake Colby on the left.

Bear left at the fork for Lake Clear Junction. Here NY 86 also turns left, so it should not be difficult to follow. Continue west on NY 86 (NY 30), passing through Lake Clear Junction and Lake Clear on the right. Turn right on the road indicated to the Department of Environmental Conservation's Adirondack Fish Hatchery. This loop road passes through the grounds of the fish hatchery. Turn right off the loop road onto a dirt road. Cross the railroad tracks. Continue straight a short distance and then turn right into the parking lot. Here there is a public launching site on Little Green Pond.

Pack every bit of gear into the canoe and get set for some real fun. This venture should be the highlight of any spring or summer. Prepare yourself for some lovely sights, smells, and sounds; you will enjoy every minute of it.

Cross Little Green Pond east and north into Little Clear Pond. Along the west shore stop to investigate a marshy inlet with flooded dead trees in the northwest corner of the pond. This will be just before the portage. Here you should expect to find nesting Hooded Mergansers, Black-backed Three-toed Woodpeckers, and Olive-sided Flycatchers. Of course, long before you reach this point you will have spotted and heard several warbler species. Keep glancing overhead for herons, waterfowl, buteos, and accipiters. Believe me, you will not be disappointed. Now portage 0.6 mile to a plank bridge and dock on a marshy inlet of the St. Regis Pond. Canoe into the languid pond. Explore as you so choose. This area can stand a whole lot of exploration. Make camp for the night at the lean-to at the point on the left, or at one of the other campsites on the islands, or even around the shore.

From this base camp birders can leisurely explore St. Regis Pond at night and in the early morning. No matter how apparently weary, do not be talked into going to bed early. Stay up and take a dusk (or later) walk around the area, especially to the tiny kettle at the southwest end of the pond. Here you can amuse yourself watching the antics of indescribably busy beavers. Listen for Great Horned, Barred, and Saw-whet owls; they will answer your imitating calls. See how many species of frogs, toads and tree frogs you can identify by their very distinctive calls, croaks, and trills, before you retire. If you should catch one of the singing performers, note the peculiar vocal sac distended in song.

In the early morning explore the flooded west end of the pond near the fish dam. Go to the narrow inlet leading to a tiny pondlet at the southeast end. All the while be on the alert for nesting Common Loons. Do not approach their nests too closely, for this is a species experiencing special nesting difficulties; do not add to their burden. Hooded and Common mergansers should be encountered as well as several raptor species. Be alert for fishing Ospreys. Spotted Sandpipers nest in the general area and a keen observer should see one or two behaving territorially. Black-backed Three-toed Woodpeckers, *Empidonax* species, especially Olive-sided Flycatchers, a myriad of Adirondack breeding warblers, and Rusty Blackbirds all occur here.

In the late-nineteenth century, the route of the Seven Carrys was a popular trip for guests and guides from the fashionable hotels at Saranac Inn and Paul Smith's — which are located at opposite ends of the ten lakes and ponds and seven short portages that make up the entire Seven Carrys. The Adirondack Park Land Use and Development Plan prohibits the use of motor boats and snowmobiles in the canoe area and so, at least for this stretch, one is spared jarring noises from the civilized, horsepowered world.

From Saint Regis Pond birders would do well to make several day canoe/hikes either west to Ochre Pond, Fish Pond, and Lydia Pond or north and east to Little Long Pond or Spectacle Pond. It is an easy day trip through Green, Little Long, Spectacle, Bear and Bog ponds to Upper Saint Regis Lake.

An alternative, of course, would be to use two cars and complete the route of the Seven Carrys. A reverse trip (again using two cars) could easily be started from Black Pond, west of Paul Smith's, crossing the Keese's Mill Road and then canoeing up the outlet to Lower Saint Regis Lake, through the marshy slough, to

Spitfire Lake, and then into Upper Saint Regis Lake, and from there into the Saint Regis Canoe area, at the southern end of the tour.

ESSEX COUNTY

Specific directions to the next seven birding sites begin from Lake Placid village. The basin of Lake Placid was formed by prehistoric glacial gouging in what is a fault (fracture) zone. Here fractured rock has been displaced from adjacent rock masses. This basin of bedrock was then modified by erosion and is kept full by to-day's rains and snows. Lake Placid village is a fashionable, modern summer and winter resort nestled among the heavily forested Adirondack peaks. It had, in 1981, a permanent population of more than 2700, although at the peak of both summer and winter tourist seasons it is at least double that and probably triple. Lake Placid is easily reached by way of I-87, NY 86, and NY 73 from the northeast and southeast, NY 3 and NY 73 from the west, and NY 28, NY 30, NY 3, and NY 73 from the southwest.

Riverside Drive and North Elba Sanitary Landfill (Maps 7.2 and 7.7)

Rating: Spring***, Summer**, Autumn***, Winter*

Along and just off this paved mountain road there is good birding for those who are willing to do a little exploring.

The junction of NY 86 and NY 73 in downtown Lake Placid is marked by a stoplight. From there, proceed northeast on NY 86 for 3.4 miles to a bridge cross-ing the West Branch of the Ausable River. Cross the bridge and immediately turn right on Riverside Drive (also known as River Road). The river roughly parallels Riverside Drive and at 0.4 mile from the turnoff of NY 86, the river comes right to the road. Stop here. At this point on the road, an active Osprey nest can be viewed in a distant towering dead tree, up the river on the far bank over the tree tops. Drive slowly and stop frequently to check the brushy streamside edges, which are highly attractive to a variety of warblers. Park and walk along the riverbank or hike into the woods to explore any of several paths leading into the forest. Check out openings leading to the river. As you approach NY 73, there are open fields, where you can easily find Bobolinks and Savannah Sparrows. Follow Riverside Drive to NY 73; the junction is opposite the Olympic Ski Jump. A sandpit on the right, across from the jump, affords views of nesting Bank Swallows in summer. Here turn right and follow the road toward Lake Placid. After a short distance you will see the Lake Placid Municipal Airport on the right. The open fields near this airport may have shorebirds in them, following late summer or early autumn storms. Check them out. After 0.9 mile from your turn onto NY 73, you will see the dump road coming in on the right. Turn right (east) on the dump road. Follow it to the North Elba Sanitary Landfill. An Iceland Gull was sighted here in late May, 1975. Herring and Ring-billed gulls will be found here and perhaps even a black bear. Take note of the nesting Bank Swallows in summer. Ravens may be found here along with crows. In winter check out the spruces around the dump area for winter finches, owls, and for the occasional surprise. After leaving the dump, turn right on NY 73 and follow it one-half mile into Lake Placid.

Whiteface Mountain Memorial Highway and Wilmington Notch (Maps 7.2 and 7.7)

Rating: Spring**, Summer***, Autumn**, Winter
(The road to the summit is usually closed from mid-November to mid-April.)

Russell M. L. Carson stated in *Peaks and People of the Adirondacks* (1927): "It is to be hoped that the road on to Whiteface will never be built, and that the hosts of unknowns of the future, who shall attain the summit by strength of leg and sweat of brow, may always find there their inspiration for a prayer of thanks for an untarnished mountain top." Preservationists fought the building of this road, which stretches from Wilmington to within 276 feet of the summit of Whiteface Mountain, but advocates argued that there should be at least one Adirondack peak accessible by car. Advocates of the highway won. Recreationists have also constructed ski slopes on Whiteface Mountain, which boasts the state's fifth-highest peak — 4867 feet. Summer climbers and hikers are welcome to try the wealth of well-marked trails leading to the summit.

To reach the Whiteface Mountain Memorial Highway from the village of Lake Placid, proceed northeast on NY 86 from its junction with NY 73 at a stoplight in the village. After 3.4 miles you will come to a bridge crossing the West Branch of the Ausable River. Cross the bridge and proceed northeast approximately 3.2 miles on NY 86, along the West Branch of the Ausable River, to Wilmington Notch. Get out and look at this spectacular gorge. Migrating Golden Eagles have been recorded here in mid-September. Be sure to check out the nesting Red-tailed Hawks and Common Ravens here. Before the decline of the Peregrine Falcon in the eastern United States, there was a pair that nested regularly on the precipitous rock face on the Whiteface Mountain side of the gorge. Proceed northeast for approximately 5.6 miles to Wilmington. At the junction of NY 86 and NY 431 take a left. Follow NY 431 west for 3 uphill miles to the highway toll booth. Pay toll and continue up the eight-mile mountain road to the summit with its lookout tower. The total rise from the beginning of the toll road to the summit is 2500 feet, so of course, the ascent involves passing through various altitudinal "layers" of Adirondack birdlife. Birders should stop at each of a series of scenic pull-offs and take note of the vegetation changes from dense large spruce to cripplebrush to thick stunted growth at the highest elevations. Gray-cheeked Thrush, reported to be the small Bicknell's race, occurs above 3000 feet, although its song is usually heard only briefly at dawn and at dusk. At the summit, if visibility is clear, the birder will be afforded excellent views of the Saranac River, Lake Placid, Esther Mountain, Lake Champlain, Mount Marcy, and the High Peaks.

Good flights of migrating hawks may occur here. Establishment of manned hawk watches was attempted in fall and spring of 1974 and 1975. They were discontinued owing to thick autumn fogs and snow depths of over five feet in April. Local observers have moved their lookouts to lower elevations near Lake Champlain's western shore.

Chubb River Swamp (Maps 7.2 and 7.7)

Rating: Spring***, Summer**, Autumn**, Winter****

This is a low-lying tract of land permanently saturated with moisture, overgrown with vegetation of the spruce-tamarack-cedar association and heavy-laden with wildflowers, ferns, and lichens (*Usnea* spp.). The Chubb River is chock-full of trout, and with the exception of the blackflies, which are especially thick in June, it is altogether a birder's paradise.

In preparing to bird the Chubb River Swamp, one cannot over-stress the indispensability of waterproofed footgear. However, there is another school of thought, which insists that "Old sneakers are best for hiking into Chubb River Swamp in spring and summer. One birder who wore high fishing waders had to be rescued after stepping into a deep sphagnum-covered hole in the trail and having them fill with water. Others have left boots behind, stuck in the mud." My *caveat* remains: wear waterproofed footgear and bring along dry socks and a change of shoes or boots. Those who have birded Chubb River Swamp by canoe insist that is the very best way to cover it.

In this swamp you may find resident breeding Spruce Grouse, Saw-whet Owls, Black-backed and Northern three-toed woodpeckers, Gray Jays, Boreal Chickadees, Yellow-bellied Flycatchers, Ruby-crowned Kinglets, Rusty Blackbirds, and Pine Siskins. Mike Peterson's discovery of a female White-winged Crossbill building a nest here February 22, 1975, provided only the second documented nesting in the state; the first breeding record also occurred in Essex County in 1958. In addition, Adirondack-type breeding warblers have been sighted in spring and summer in the swamp. Recently, a Cape May Warbler was recorded here in June.

To reach the swamp from Lake Placid, go south on Main Street past the Olympic Arena to the stoplight at the junction of NY 86 and NY 73. Turn right on NY 73. Follow it to the Old Military Road. At Old Military Road (County Road 35) turn right. At 2 miles, turn left onto Averyville Road. Continue on this road for 2 miles to a bridge crossing the Chubb River. Park the car on the right shoulder or in the small parking area at the start of the trail to the left on the near side of the bridge. The trail leaves the parking area and follows the river into the swamp.

Follow the trail a quarter-mile to a bend in the river, near a prominent large rock. Immediately past this bend is an opening to the spruce and tamarack-lined Chubb River. Canoeists and winter hikers can travel from here upriver, avoiding the difficulties of the trail. Summer and late-spring birders will note that near this spot Pine Siskins and White-winged Crossbills have attempted nesting the past few years. In recent years both species of three-toed woodpeckers have been sighted here in summer as well as winter (mid- to late December).

The trail continues for several miles through thick stands of spruce past the first of three major "blowdowns." The first blowdown is another easy-viewing site for the Northern Three-toed Woodpecker. The trail curves right, past a second blowdown, and becomes impassable near the third blowdown at the river's edge. This location is a favored site of the Boreal Chickadee. Hooded Mergansers and Ruby-crowned Kinglets are nesters, and Northern Parula is a commonly found breeder here in late spring and early summer. All along the trail watch for Spruce Grouse and Gray Jays.

The Chubb River Swamp is a fragile habitat and preservation of the bog mat should be uppermost in the minds of birders. Stay on the trail and avoid making new openings in the sphagnum layer.

Lake Placid to Cascade Lakes (Maps 7.2 and 7.7 and 7.8)

Rating: Spring***, Summer**, Autumn**, Winter***

This is a ten-mile stretch of road that can provide very profitable birding, especially in spring and winter. Traffic is sometimes heavy and slow; and during winter, ice, blowing snow, and high winds make driving through the Cascades hazardous. But this only makes it more exciting. Birders should certainly enjoy all that this route has to offer in the way of spectacular scenery and marvelous birdlife; but, because of the weather hazards and the narrow shoulders on this road, drive with great care.

The junction of NY 86 and NY 73 in downtown Lake Placid is marked by a stoplight. From here take NY 73 southeast. At 0.5 mile outside of the village take a left onto dump road (see Riverside Drive and North Elba Sanitary Landfill, above). After investigating the North Elba Sanitary Landfill proceed southeast on NY 73. Investigate the sandpit on the left opposite the Olympic Ski Jump for nesting Bank Swallows in summer. In winter look for visiting Fringillids (December through February).

Cross the bridge over the West Branch of the Ausable River and continue south on NY 73. At 2.5 miles Heart Lake Road comes in on the right. Stop just before this road. Get out and search the large open fields for Savannah and Vesper sparrows. From this vantage point there are glorious views of Mt. Marcy, Mt. Colden, and Avalanche, Algonquin, and Indian passes. Continue for about 3 miles to Bobsled Run Road, which comes in on the right. This road leads to the internationally famous Mt. Van Hoevenberg Bobsled Run, built in the early 1930s. If you've never seen it, it is worth a short detour down Bobsled Run Road. Continue on NY 73 for approximately 2.3 miles to Upper then Lower Cascade Lakes on the right. Take a sharp right, down a steep hill, to the small picnic area between the lakes. Bird the area, which should be most productive in late May and June and from December to February. A pair of Common Ravens nest on the cliffs across from Lower Cascade Lake and can be easily seen in May and June.

Over long periods of time the relentless forces of weathering cause rock to disintegrate. This broken rock builds up and gradually begins to move downslope;

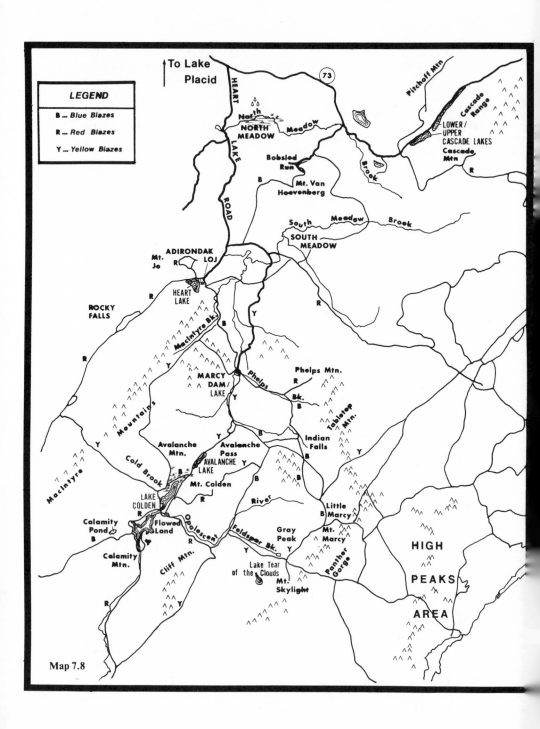

LEGEND

B ... Blue Blazes
R ... Red Blazes
Y ... Yellow Blazes

To Lake Placid

73

Pitchoff Mtn

Cascade Range

LOWER/
UPPER
CASCADE LAKES

Cascade
Mtn

HEART LAKE ROAD

North Meadow
NORTH MEADOW
Meadow

Bobsled Run

Mt. Van
Hoevenberg

South Meadow Brook
SOUTH MEADOW

Mt. Jo
ADIRONDAK LOJ

HEART LAKE

ROCKY FALLS

MacIntyre Bk.

Phelps Mtn.

MARCY DAM/ LAKE

Phelps

Phelps Bk.

Tabletop Mtn.

Mountains

Avalanche Mtn.

Avalanche Pass
AVALANCHE LAKE

Indian Falls

Cold Brook

Mt. Colden

River

MacIntyre

Little Marcy

LAKE COLDEN

Calamity Pond

Flowed Land

Feldspar Bk.

Gray Peak

Mt. Marcy

Panther Gorge

Calamity Mtn.

Cliff Mtn.

Opalescent

Lake Tear of the Clouds

Mt. Skylight

HIGH

PEAKS

AREA

Map 7.8

rolling, sliding, creeping. Gravity exerts a constant drag on this disintegrated rock, moving it and assorted litter, to lower and lower levels. The entire process is called "mass-wasting." Rockfalls and landslides are the most extreme examples of the phenomenon.

In the early 1860s, a landslide blocked the bed of what was then known as Long Pond, dividing it in two. The resulting ponds are classic examples of landslide lakes. Thereafter, they became known as Edmund's Ponds, after a local resident. In 1878 a summer resort hotel was built between the lakes (where the picnic area is now) and they were renamed Cascade Lakes. Long Pond Mountain became Cascade Mountain, whose peak rises to 4092 feet, and which can be seen rising behind the site of the old hotel. The area was severely burned over in the infamous forest fire of 1908 and is now mostly forested with second-growth birch. It is impressively beautiful in all seasons. Across NY 73 (north) opposite Cascade Mountain is Pitchoff Mountain. The deep valley between them was caused by strong glacial erosion, stream erosion, and landslides.

Common Ravens are probably more frequently seen along this stretch of road than anywhere else in the Adirondacks, and they are regularly sighted in all months of the year. Pine Grosbeaks and other winter finches are common most years.

Heart Lake Road and South Meadow (Map 7.8)

Rating: Spring****, Summer***, Autumn**, Winter*

Heart Lake and Mt. Jo (Map 7.8)

Rating: Spring***, Summer**, Autumn***, Winter*

Nearly fifty years ago the naturalist Aretas Saunders wrote: "The traveler will do well to *walk* the Heart Lake Road, at least once, particularly choosing for the occasion a morning in early summer." This is still excellent advice although the road can easily be driven. The Heart Lake Road was burned over in the forest fire of 1903, but by 1926, although charred stumps were still visible, the area was covered with aspen and fire cherry, young maples, spruces, and balsams. It has been returning to its formerly attractive state and today the slopes are mainly covered with deciduous second-growth.

Walking the Heart Lake Road around Heart Lake and the South Meadow is rated as generally quite easy. Waterproofed hiking boots are recommended as the trails may be slightly damp — wet in late spring and early summer. Blackflies can be bothersome in June and carrying an effective insect repellent is advisable. The hike to Mt. Jo (cannot be driven) is relatively steep and is conservatively ranked as difficult. Although it is more taxing than ordinary birding, the hike is highly recommended; birders will be rewarded with the variety and abundance of birdlife, not to mention the gee-whiz mountain scenery.

From the stoplight at the junction of NY 86 and NY 73 in downtown Lake Placid, drive southeast on NY 73 out of the village, past the Olympic Ski Jump. At 2.5 miles, Heart Lake Road comes in on the right. The open fields at the beginning of Heart Lake Road are usually productive. A Snowy Owl was seen here in January 1981. The Red-headed Woodpecker has also been seen in the area. Eastern Bluebirds may be seen around the farmhouse on the right and in the open fields Savannah and Vesper sparrows are commonly found. Proceed down the road. Swallows, including Cliff, can be found around the barns on the left. At 1.3 miles you will come to North Meadow Brook on the left and a spruce-covered hill (formerly Sheep Pasture Hill) behind it. A Ruddy Turnstone was discovered along the brook in September 1976, and Boreal Chickadee and Black-and-white, Tennessee, Nashville, and Black-throated Blue warblers have been spotted on the hillside. Proceed along the road and enter the woods — coniferous at first and then deciduous. You will note numerous passerine species during the breeding season. At 3.5 miles, a sign on the left marks the road to the South Meadow. Turn left onto South Meadows Road, which ends after a mile. At its terminus, park on the left and follow the trail through the meadow and over the bridge that spans South Meadow Brook. Look especially for streamside-loving species; if you are lucky, you will encounter Alder (fee-bee-o) and Least flycatchers.

The South Meadow is the site of a former glacial moraine lake. A lake formed upstream behind ridges of rock fragments and boulder litter that were deposited by a retreating glacier. Such glacial lakes were originally shallow. In the last 15,000 to 20,000 years they have filled with sediment and vegetation, and their drainage systems have been rearranged. Over time, they have converted to meadows of which South Meadow is typical.

Return to the Heart Lake Road and turn left. Continue to Heart Lake and Mt. Jo, stopping along the way to listen and look for birds.

From the junction of Heart Lake Road and South Meadows Road proceed 1.2 miles south on Heart Lake Road. It ends at Adirondak Loj. Park in the large lot on the left adjacent to the stone Campers' and Hikers' Building, which is maintained by the Adirondack Mountain Club. Pay the minimal day's parking fee and walk back across Heart Lake Road. Locate the red Department of Environmental Conservation (DEC) trail markers. Follow them, heading west past several private lean-tos around the north shore of Heart Lake for approximately 0.5 mile. It continues up a slight rise and onto state land to Rocky Falls (a hike of 2 miles) and to Indian Pass (a hike of 5.5 miles), both of which are in the MacIntyre Mountains. You will be on the Indian Pass Trail which is well marked and easy to follow. Indian Pass Trail is, however, a relatively rough one up a stream bed, and requires some rugged climbing. Walk for as long or short a period as you choose. In late May the songs of flycatchers, thrushes, vireos, and warblers are glorious. A night spent at Heart Lake will probably reward you with the calls of Great Horned and Barred owls, as well as Whip-poor-wills.

Heart Lake is technically known as a kettle pond. That is, it is a rounded pond formed by the melting of a gigantic glacial ice block that was either totally or partially covered by rock fragments — ranging in size from clay particles to giant boulders. The rock waste has washed downslope since glacial times.

If you have decided to climb Mt. Jo, use the following route. At 0.10 mile from the Heart Lake Road you will see a sign on the right pointing to the trail to Mt. Jo. After proceeding 0.25 mile one can choose between a Short Trail (turns right here) and a Long Trail (turns left here). The Short Trail climb is approximately 1.43 miles, the Long Trail being 0.28 mile longer (but less strenuous than the Short Trail). Proceed to the summit, which is at an elevation of 2876 feet. The view from the summit is wonderful, and this is one of the lowest elevations where Blackpoll Warblers can regularly be found in the breeding season. Notice on the east side of the summit the pitted and grooved depressions known as weather pits. Also take note of the prominent quartz veins; a result of slow weathering on a highly resistant, stable material. On the descent and return circuit from Mt. Jo, most of the breeding birds that occupy separate altitudinal "layers" of the Adirondacks can be heard and, foliage permitting, seen. This is a short, strenuous, but enthusiastically recommended hike.

Adirondak Loj to Marcy Dam/Marcy Lake (Map 7.8)

Rating: Spring***, Summer**, Autumn**, Winter*

The following account details a moderate climb of 2.2 miles involving a gradual ascent of 187 feet. Much of the trail (Van Hoevenburg Trail) follows level ground. It passes through conifers, along streamsides, through mature deciduous forest and second-growth vegetation. Marcy Dam is on the site of an old logging dam at an elevation of 2366 feet. A wide variety of woodland birds can be seen and heard especially in May and June. Look especially for Yellow-bellied Sapsuckers, Olive-sided and Yellow-bellied flycatchers, and Solitary and Philadelphia vireos. Philadelphia Vireos have recently been found nesting just below Marcy Dam. In late spring and early summer look also for Blackpoll, Mourning, and Canada warblers. Singing Blackpoll Warblers have been discovered in recent years at Marcy Dam, which is a relatively low elevation for that species. In addition, an attentive birder should be able to find breeding Scarlet Tanagers and Rose-breasted Grosbeaks.

If the early spring has been especially wet, the trails may be muddy with several stretches of slippery corduroy. Blackflies may be bothersome. Wear waterproofed boots and carry insect repellent. Be sure to bring a trail lunch (including something to drink) and be sure to sign the trail registers at both Adirondak Loj and Marcy Dam. This is for your own protection.

To reach the beginning of the trail from downtown Lake Placid and the junction of NY 86 and NY 73, proceed southeast on NY 73 out of the village. At 2.5 miles, Heart Lake Road comes in on the right. Turn right here. Follow this road to its terminus (about 5 miles south). You are now at Adirondak Loj. Park in the large lot on the left adjacent to the Campers' and Hikers' Building, which is maintained by the Adirondack Mountain Club. Pay the minimal day's parking fee.

The Van Hoevenberg Trail is one of the oldest routes to the top of the highest mountain in New York State—Mount Marcy, whose elevation is 5344 feet. It is the shortest route. The entire trail (a part of which the birder will follow to Marcy Dam/Lake Marcy) from Adirondak Loj to Mount Marcy is 7.46 miles and it ascends 3166 feet.

The Van Hoevenberg Trail begins at the Department of Environmental Conservation (DEC) trail register. It is clearly marked with blue markers and is additionally labeled every half-mile with blue numbers. *Before beginning, be sure to sign the register.* Follow the blue markers 2.2 miles to Marcy Dam. Cross the dam and sign the DEC trail register there. There are lean-tos and campsites around the shores of Marcy Lake; however, if you plan to spend the night there bring your own shelter and do not expect to find firewood at this highly regarded location. Along the hike you will have crossed eight small bridges and the boulder-strewn Marcy Brook. These should provide productive birding sites.

Marcy Dam/Marcy Lake to Colden (by way of Avalanche) (Map 7.8)

Rating: Spring****, Summer***, Autumn**, Winter*

The following trip should be taken by varsity birders only who prefer their birding mixed with large doses of the outdoors and who don't in any way mind sleeping within ear-shot of a cascading cataract or struggling with an early morning campfire in order to cook breakfast. In short only those who want an adventure should attempt it.

This is at the very minimum an overnight trip and, depending upon your mood, it could be extended to a three to five day venture. Bring along your own food, stove, sleeping bag, tent, etc. Take care to prevent hypothermia. Avoid inappropriate clothing, exhaustion, and becoming chilled. Under no circumstances should you bring your pet. The most accomplished of dogs would have real problems scaling the ladders at Avalanche Lake and Lake Colden.

From Marcy Dam (see Adirondak Loj to Marcy Dam/Marcy Lake above), follow the yellow trail markers from the Marcy Dam Hikers Register up Marcy Brook, which progresses gradually uphill. After 1.0 mile cross a log bridge; continue upward along a tributary for 0.25 mile, where it bears left and then enters a steep switchback into the Avalanche Pass. The small waterfall on the left in Avalanche Pass (at the highest point of land) flows northeast to the St. Lawrence and southwest to the Hudson. This steep pitch, at the 3000 foot level, may produce Yellow-bellied Flycatchers. At this elevation Blackpoll Warblers become common. Continue through the pass and down to Avalanche Lake. This is a spectacular section of the hike.

At one time lakes Avalanche and Colden were joined, but, were separated by land slides. The slides, whose results are now visible on Mt. Colden took place in 1869 and 1942. The second bridge skirting Avalanche Lake will interest you. The cleft of the great Colden Dike is best seen from there; it is a wall-like mass of igneous rock cutting through a different kind of rock; in this instance, metagabbro rock. Dikes of metagabbro are dark gray, weathering to a rusty color. This is the most famous of all of the Adirondack dikes, although there are many less dramatic dikes scattered throughout the region.

Be on the lookout for beavers in Avalanche Lake. Before reaching the lake you will pass through a wooded section which hosts Yellow-bellied Flycatchers, Solitary and Philadelphia vireos, and Chestnut-sided, Blackpoll, and Canada warblers. Warblers, in fact, are abundant along the rugged trail. Along the right shore of the lake, Boreal Chickadees are easily seen and heard, although they are ordinarily quite secretive during nesting season. Breeding White-throated Sparrows can be heard all around the lakeshore. Follow the trail, which crosses an outlet at the foot of the lake. It then turns right and at about 3 miles reaches a fork. Follow the blue markers to the right. Where the trail skirts the swampy upper end of Lake Colden one can find Black-backed Three-toed Woodpeckers, Northern Waterthrushes, and Rusty Blackbirds. Bear left past the ranger cabin and Beaver Point to the ladder above Colden Dam. Here at Lake Colden and adjacent Flowed Land (elevation 2750 feet), one can find the Gray-cheeked (Bicknell's) Thrush, singing at dawn and at dusk in early June. At dusk enjoy the noisy beaver activity in Lake Colden or listen to coyotes calling as the moon rises over Mt. Colden.

Before 1840, Avalanche Lake, Lake Colden, and Flowed Land were entirely fishless. In the 1840s and 1850s they were stocked with brook trout. By 1930 these waters were well known among the aficionados of fly fishing as the best in the eastern United States. The waters were very little fished; but when they were, the catch was marvelous for two or three decades.

Today the unthinkable has happened. Over these past few decades, the pH values in these formerly pristine lakes have gone from marginal to lethal for trout. The marvelous brook trout of the 1930s have been literally exterminated. With the hope that the water quality would improve, helicopter stocking of the lakes continued until a few years ago. The result was an abundance of dead fish, so the program has been discontinued. After much technological searching, the cause has been determined: air pollutants. These pollutants have been introduced by precipitation. They originate many miles away and are carried over the Adirondacks by the clouds. The resulting acid rain or snow substantially increases acidity and thus reduces pH values in the lakes—which were once thought inviolate.

From Lake Colden-Flowed Lands, which is a popular camping area in the heart of the High Peaks Wilderness Area, a variety of return trips can be planned from Colden. Birders could take the red-marked trail up the Opalescent River. This is an unspeakably beautiful trail and is marvelous Boreal Chickadee country. Follow the red markers to a point at which yellow markers indicate for the hiker to bear left. Follow the yellow-marked trail up Feldspar Brook to the tiny source of the mighty Hudson River: Lake Tear of the Clouds. Here, above 4000 feet, the Gray-cheeked Thrush is regularly encountered in early June. This is a wonderful place to camp overnight. The following day can be spent continuing to the summit of Mount Marcy on the yellow trail, and then down *via* the blue-marked Van Hoevenberg Trail (or its variants) to Marcy Dam (Marcy Lake) and from there to Adirondak Loj. Certainly this circuit can be completed in three days or less. Those who wish to bird along the trails should plan for a three- to five-day trip, advising friends beforehand and signing registers accordingly.

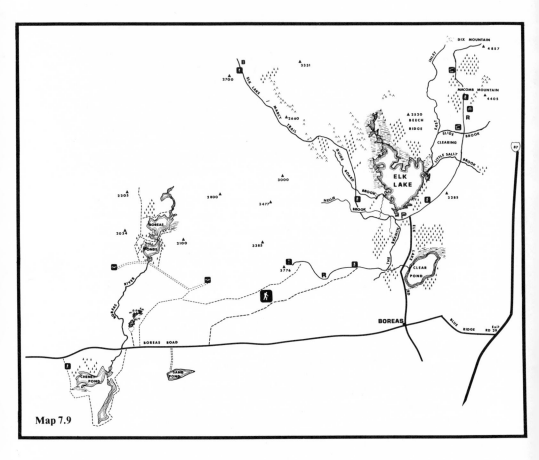

Map 7.9

Elk Lake, Clear Pond, Cheney and Boreas Ponds (Map 7.9)

Rating: Spring****, Summer****, Autumn**, Winter*

The Elk Lake area is, without reservation, one of my best-loved birding sites in the Adirondacks. Elk Lake and Clear Pond are at the northernmost reaches of the Hudson River watershed. They are in Essex County and lie about nine miles north-west of the village of North Hudson. The area includes approximately five miles of town and dirt road and more than twenty miles of hiking trails. Elk Lake, with its beautiful little islands, lies in the center of a private forest preserve, where deer, black bears, and coyotes reside. Clear Pond lies at the base of Clear Pond Moun-tain, which has an elevation of 2700 feet. From Elk Lake (el. 1993 feet) one looks east and north and sees the summits of Mt. Colvin (el. 4057 feet), Nippletop Mountain (el. 4620 feet), Dix Mountain (el. 4857 feet) and McComb Mountain (el.

4405 feet), to name but a few. The conifers and hardwoods throughout the entire area add to its scenic beauty. Elk Lake is indeed one of the most beautiful sites in the entire Adirondack Park, and a visit here to bird and hike can only be enthusiastically recommended.

Leave the Adirondack Northway (I-87) at Interchange 29, which is the exit for the village of North Hudson. Proceed west on the Blue Ridge Road (Boreas Road) for approximately 4.1 miles. Turn right (north) at the sign for Elk Lake Lodge. You will now be on a rough dirt road which terminates after 5.2 miles at Elk Lake. The intersection of Blue Ridge Road and Lake Road is a good spot to begin birding. Continue to make stops all along the road to the lake in the varied forest habitats and along the side trails; they will especially heighten the experience.

Travel north on the town part of Lake Road for approximately 2.3 miles, where a trail comes in on the right leading to the south shore of Clear Pond. Investigate this trail, looking for breeding Common Loons, and Hooded and Common mergansers on the pond. Search the conifers rimming the pond for nesting Sharp-shinned, Red-shouldered, and Broad-winged hawks. Don't be surprised to see Ruffed Grouse dusting in the road. Nesting American Woodcocks have been recorded here. Even an accidental Northern Phalarope was found here by Geoffrey Carleton in 1951.

Once you have birded the pond, proceed north again up the road. A short distance past Clear Pond the road becomes dirt and remains so to the lake. Park in the *upper* lot on the right hand side of the road next to the Department of Environmental Conservation (DEC) trail register, at the end of the road. Be sure to sign the trail register, giving your destination. Also be sure to sign out on your return. Most of Elk Lake property is privately owned, but birders are welcome to use the trails, providing they do not build fires, camp, fish, or hike off the trails. All of the trails are closed to birders/hikers during the hunting season in the autumn. Dogs are not allowed in certain areas described below.

There are a number of different trails one can take at Elk Lake. Generally the habitat has some of the most pristine boreal-montane forest remaining in the Adirondacks. Ornithologically, it is interesting because the nesting bird species include Common Loon, historically Pied-billed Grebe, currently American Bittern, Black Duck, Hooded and Common mergansers, Goshawk, Sharp-shinned, Red-shouldered, Broad-winged, and Red-tailed hawks, and numerous family groups of Ruffed Grouse. There is some conjecture that Black Terns nest at Elk Lake and any breeding evidence should be submitted to New York State Avian Records Committee (NYSARC). Breeding Great Horned, Barred and Saw-whet owls, Chimney Swifts, Ruby-throated Hummingbirds, Common Flickers, Pileated Woodpeckers, and Yellow-bellied Sapsuckers, and both three-toed woodpeckers have been sighted in the area in recent years. Eastern Kingbird, Great Crested, Olive-sided, Yellow-bellied, Least and Alder flycatchers and Eastern Wood Pewees have been recorded here in nesting season, as well as Tree, Bank, Rough-winged, and Barn swallows. Boreal and Black-capped chickadees nest in the area, as do Red-breasted Nuthatches, Brown Creepers, and Winter Wrens. On a good morning in June a diligent birder should be able to hear or see Hermit and Swainson's thrushes, both kinglets, Cedar Waxwings, Solitary and Red-eyed vireos, probably at least fifteen warbler species, including Northern Parula, Magnolia, Black-throated Blue, Yellow-rumped, Blackburnian, Bay-breasted, and Blackpoll warblers. Also Rusty Blackbirds, Rose-breasted Grosbeaks, Pine Siskins, Red and White-winged crossbills, and Chipping and Swamp sparrows can be seen. An experienced field observer can note about sixty to seventy species of birds per day from early June through early July, and about fifty to sixty species per day into early August. Historically, Bufflehead, Solitary Sandpiper, Northern Phalarope, Gray Jay, Philadelphia Vireo, and Evening Grosbeak have been recorded in the area in either June or early July and should always be looked for. The *Kingbird* Region 7 editor and NYSARC, as well as the High Peaks Audubon Society would gladly welcome fully documented sightings of species uncommon to the region for inclusion in state and regional reports.

Having reached Elk Lake the birder would be well advised to follow one of the following trails in order to fully bird the area.

Across from the parking lot there is a Department of Environmental Conservation trail register and the beginning of the Elk Lake/Marcy Trail, which is designated with blue trail markers. Walk over a suspension bridge, which crosses the outlet (Branch) to the lake. Follow the blue discs west and north along the west side of the lake through a second-growth conifer stand. Continue on the trail. When you have walked 0.5 mile, the trail meets an old gravel lumber road and follows it, turns left, and crosses Nellie Brook. After only a few hundred yards the trail leaves the road by turning right, climbs relatively steeply for a very short distance and then levels off. Almost immediately, it makes a sharp left onto an old lumber road and soon thereafter turns right onto a gravel lumber road, which it now follows for 1.0 mile, crossing Nellie Brook again, up and over an easy-to-climb ridge, and down and across Guideboard Brook. You will now have walked 1.75 miles and after birding the vicinity thoroughly you can reach your point of departure by retracing your route over the blue disked trail.

An alternative route for birding the Elk Lake region starts at the parking lot and the DEC register and follows the trail leading to the summit of Dix Mountain. You will not travel the entire 7.3 miles of trail, but only the first mile. The Dix trail is designated with red discs. Follow the red-marked trail from the parking lot east and north along the east side of Elk Lake. After walking 0.5 mile note that a tote (hauling or lumber) road comes in on the right. The trail follows the tote road for the remainder of the hike. Proceed northeast, crossing Little Sally Brook and reaching a mini-summit after traveling over a nearly level grade for 0.49 mile. Be sure to bird actively all along the trail.

After reaching this height of land, the birder may wish to continue northeastward. At this point, however, you must make a decision. If you are not equipped with comfortable waterproofed boots that can withstand the rigors of a wet, soggy corduroy stretch of trail, turn back and retrace the incoming route to the point of origin. From this point through the next 1.3 miles, the habitat becomes considerably more diverse, and you would probably hear additional bird species as you pass through the habitat changes. But, although the grade is moderate from here to Slide Brook lean-to, about one mile distant, be forewarned that there are several soggy stretches and probably one or two short mudhole sections through which you must hike to reach Slide Brook lean-to. By way of temptation, let me inject an entirely subjective point of view. Slide Brook lean-to is an absolutely idyllic place to lunch. It is on the north side of Slide Brook and on the left edge of Slide Brook Clearing. In 1947, a gigantic slide came down from Macomb Mountain, creating several branches up the mountain and of course, several various good-sized brooks indicating ready access to the path up the slides.

Whichever of the trails and sub-trails you decide to follow, I am confident that you will not be disappointed in any way. To bird in such productive and pristine habitat at least once every spring season is a breath-taking and memorable experience no serious birder should deny himself.

If, after leaving the Elk Lake/Clear Pond area you plan to travel west, be sure to stop at both Sand Pond and Cheney Pond to check out the birdlife there.

Proceed south on Elk Lake Road to its intersection with Boreas (Blue Ridge) Road. Note that wintering Cooper's Hawks have been recorded in this area in early January. Turn right (west) on Boreas Road and proceed for about 4.8 miles to a turn-off onto a dirt road on the left. Park and walk down the dirt road to Sand Pond, which lies at the base of Sand Pond Mountain.

Continue west on Boreas Road another 2.7 miles, crossing the bridge over the Boreas River. Just beyond the bridge there are paths on both the right (heading north) and the left (heading south) sides of the road. Both trails have their own rewards. The birder should follow both if he has sufficient time. The one to the left leads to Cheny Pond, which can be rewarding.

In the area the birder should be able to find easily three or four picid species, Eastern Kingbird, Great Crested Flycatcher, Rough-winged Swallows, and Purple Martins, thrushes, and Veeries, vireos, and several warbler species. Watch along the road and in the open fields for Bobolinks, meadowlarks, Indigo Buntings, and three or four sparrow species, including Savannah, Chipping, and Field. Two Buffleheads were recorded recently on the Boreas River in this area during June. Be on the lookout for this species. Historically a small colony of Great Blue Herons was known to breed in the marshes adjacent to Boreas Pond, which is reached by following the trail on the north side of Boreas Road. American Woodcocks, although rarely recorded at Boreas Pond, should also be watched for.

Big Crow Mountain, Lost Pond, and Weston Mountain (Map 7.10)

Rating: Spring***, Summer***, Autumn**, Winter*

Because so many of the peaks in the high country of Adirondack Park receive a disproportionate amount of use, I propose the following birding area as an alternative interior Adirondack site, with relatively easy-to-hike trails, wooded summits, and yet numerous outlooks and overlooks, and a diversified avifauna. It lies in the beautiful northeastern quadrant of the Adirondacks and affords an almost completely forested walk in diversified habitats. There is much in the birdlife resident here that is yet unknown and that only needs a group of skilled birders to further explore and reveal its riches.

Big Crow Mountain is located just east of the village of Keene. From the

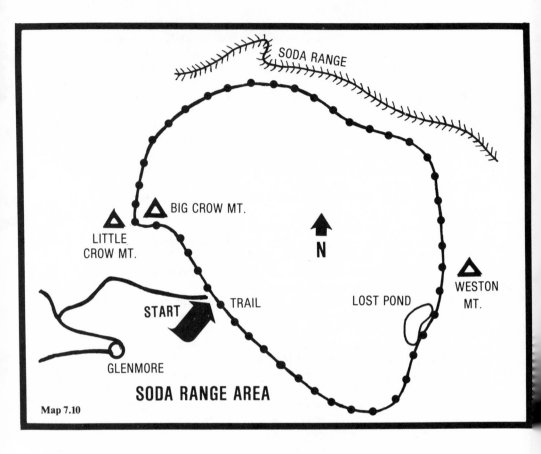

SODA RANGE

BIG CROW MT.

LITTLE CROW MT.

N

WESTON MT.

START

TRAIL

LOST POND

GLENMORE

SODA RANGE AREA

Map 7.10

Adirondack Northway (I-87) and Interchange 30, take NY 73 north and west through Saint Huberts and Keene Valley and proceed 5 miles to Keene. From the Adirondack Northway (I-87) and Interchange 31, take NY 9N west through Elizabethtown, drive 12 miles west (turning right at the junction of NY 73), and proceed into Keene. At the junction of NY 9N (NY 73) and NY 13 (East Hill Road), turn right (east) onto East Hill Road. Drive about 3 miles to a sharp lefthand turn, which is O'Toole Road. Drive up this little dirt road for about 1.0 mile to its end. Here you will find the Big Crow Mountain Trailhead. Park well off the road.

Proceed to the beginning of the trail to a sign that indicates the way to Lost Pond. Instead of following that trail, however, turn left and follow the Adirondack Mountain Club trail due north. Progress for 0.5 mile through an area of balsam fir and white and yellow birch. You will be walking on what seems a veritable carpet of red bunchberries. You will walk through some red pine and aspen stands. On nearly the entire length of this 0.5 mile section of the trail you will find numerous vantage points that offer breathtaking views of the mountains. Look and listen for bird species typical of the boreal-montane forest: Yellow-bellied Sap-sucker, Hairy Woodpecker, Yellow-bellied and Olive-sided flycatchers, Ruby and Golden-crowned kinglets. As you approach the peak of Big Crow Mountain listen for Blackpoll and Yellow-rumped warblers. At the summit (el. 2800 feet) you will find bare rock outcroppings interspersed with luxurious growths of featherbed-like mosses. Look for Common Ravens soaring over the cliffs and listen for their hoarse croaks.

This is a superb place to sit down and look out across Keene Valley at the Great Range of the Adirondacks. Twenty-eight major peaks are visible from here. In almost every direction panoramas unfold. To the southeast is over three miles of unbroken forest expanse culminating at the 3687 foot peak of Hurricane Moun-tain, below which the Peregrine once maintained an eyrie; to the north are Oak Mountain and Potash Mountain; to the west lies the Cascade Range, with many peaks over 4000 feet elevation; farther north one sees the Soda Range (Nun-da-ga-o Ridge). Practically due east is the summit of Weston Mountain, which is locally called Rocky Spur Ridge. After spending some time at the summit return *via* the trail on which you ascended.

Walk past the beginning of the trail and now follow the signs designating the way to Lost Pond. This is a very well-defined 1.5 miles trail through the valley viewed from the summit of Big Crow Mountain. It is marked with red Adirondack Mountain Club markers. Cross Gulf Brook at 0.35 mile and stay left. Be on the lookout for Sharp-shinned Hawks and Barred Owls. Eastern Kingbirds and Great Crested Flycatchers nest in the more open and damp areas along the trail. Investi-gate the bank of the brook in the ravine below on your right. The birder should certainly encounter the Eastern Wood Pewee and the Olive-sided Flycatcher. After 0.56 mile the trail takes a sharp left turn and leads to an old lumber clearing. After a short steep climb the trail levels off.

Watch for Veeries in the moist deciduous undergrowth as you approach Lost Pond. Search the white birches you pass through just before coming to the fireplace at the foot of the pond. In and along small openings in the forest look for Bay-breasted Warbler. Along woodland edges look and listen for nesting Northern Orioles. Look for Evening Grosbeaks and Purple Finches. Bird from the south to the north end of Lost Pond and continue on the trail toward the peak of Weston Mountain. The trail now rapidly gains elevation for approximately 0.5 mile, lead-ing to the 3195 foot peak of Weston Mountain. Along the ascent the birder will probably see Black-capped and Boreal chickadees, Red-breasted Nuthatches, Brown Creepers, Winter Wrens, Solitary and Red-eyed vireos, and a variety of warbler and sparrow species. Once again, at the summit, be on the lookout for ravens.

Return by the same route to your car at the foot of Big Crow Mountain.

ESSEX COUNTY

Along the easternmost edge of Essex County, east of the Adirondack Northway (I-87) and along Lake Champlain, there are several notable birding sites that have been, especially within the last decade, continuously monitored, year-round, by the vigorous members of the High Peaks Audubon Society and other birders. These sites have yielded high returns and the knowledge gained there has added appreciably to the ornithology of New York State.

Ticonderoga Marsh (Map 7.11)

Rating: Spring****, Summer**, Autumn**, Winter*

Lake Champlain is a 130-mile waterway along the state's eastern border, with restored forts and crumbling breastworks dating back to the French and Indian Wars. The narrow southern end of the lake was a hotly contested territory during the Revolutionary War and forts sprang up at almost every strategic inlet and narrows. The names have lived on and in some cases, the forts have, too. Fort Ticonderoga has been reconstructed to pre-Revolutionary War specifications. It houses a military museum and a treasure of artifacts from the 1700s. It perhaps is the most famous of New York's forts and holds a record for changing hands during the war.

Below Fort Ticonderoga is Ticonderoga Marsh, called "Ti Marsh" by those who most often bird it. The area encompasses approximately fifty acres of private land, but permission is not necessary to bird it, unless the iron gate at the entrance to the fort road is locked (which it is when the fort is not in operation). The fort is usually open to visitors from early May through late October. To gain permission to walk past the locked gate, birders should call the main office of Fort Ticonderoga at (518) 585-2821. It is perfectly acceptable for birders to walk past the locked iron gate. Cows are pastured in the orchard and grove below the fort much of the time and the railroad leading out over the marsh *via* a bridge is an active line over which several trains pass daily. Water levels of Lake Champlain may inundate the marsh some years, so birders should be prepared with waterproofed footwear. Covering this area on foot is rated moderately easy, so it is accessible to nearly everyone.

From the Adirondack Northway (I-87) and Interchange 28, head east on NY 74. At Ticonderoga village follow signs for Fort Ticonderoga to the east on NY 74. You will reach a stone gatehouse and iron gate marking the entrance to the fort road. Bear right on it. This road into the fort has an especially fine stand of hardwoods near the site of the former French lines, and provides excellent woodland birding for flycatchers, thrushes, warblers, Scarlet Tanagers, and Rose-breasted Grosbeaks. Drive up the road and at the fort park on the right shoulder next to the fence or park in the main parking lot.

Walk down through the pasture and orchard area below the fort, and then west through a deciduous-hawthorn grove and over a fence to the railroad embankment. During migration, huge waves of small land birds fill the trees and shrubs, especially warblers. The marsh itself, best viewed from the shoreline or railroad embankment, holds Green Herons, Least and American bitterns, Common Gallinules and Black Terns. Look especially for flycatchers, swallows, Long-billed Marsh Wrens, Mockingbirds, Blue-gray Gnatcatchers, Warbling Vireos, and

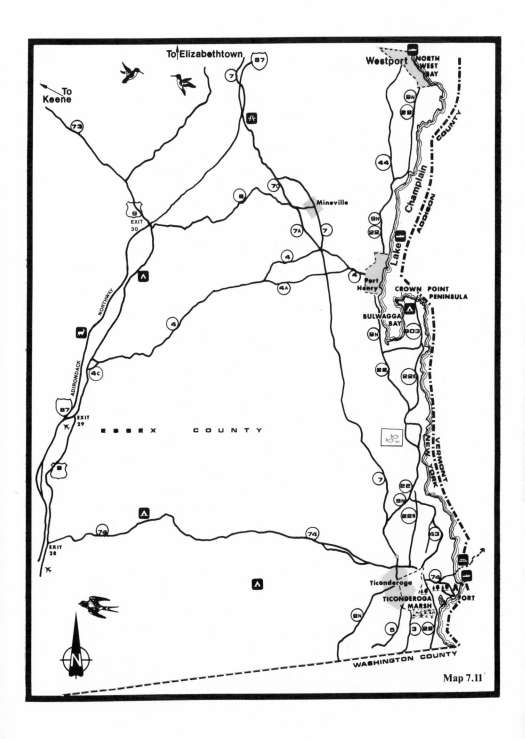

To Elizabethtown

To Keene

Westport

NORTH WEST BAY

Mineville

Port Henry

CROWN POINT PENINSULA

BULWAGGA BAY

Lake Champlain

ADDISON

VERMONT

NEW YORK

ESSEX COUNTY

NORTHWAY

ADIRONDACK

EXIT 30

EXIT 29

EXIT 28

Ticonderoga

TICONDEROGA MARSH

FORT

WASHINGTON COUNTY

N

Map 7.11

Tennessee, Cape May, Chestnut-sided, Blackpoll, and Wilson's warblers, along with Northern Waterthrushes walking the shoreline and the frequently occurring Orchard Oriole. Waterfowl may be numerous in migration and hawks migrate along the ridge of Mount Defiance (el. 853 feet) to the southwest. A Semipalmated Plover occurred here in May 1980, and a Ruddy Turnstone was seen on the shoreline below the great stone fortress in August 1975. Least Sandpipers have occurred as recently as May 1980, and other completely unexpected migrants may occur.

Ti Marsh may be explored by canoe, by launching near the railroad station on NY 22 (southwest of the fort). After launching, paddle east and slightly north across the narrow bay up into the marsh, and bird from the boat.

Bulwagga Bay (Maps 7.11 and 7.12)

Rating: Spring***, Summer*, Autumn***, Winter*

Crown Point Peninsula (Maps 7.11 and 7.12)

Rating: Spring****, Summer**, Autumn***, Winter*

Bulwagga Bay lies about one mile south of the village of Port Henry and Crown Point Peninsula lies about two miles southeast of Port Henry. The two areas can easily be birded within a few hours, and one should not be sacrificed for the other. Both are excellent birding sites, but, especially in spring migration (April and May) and in the autumn migration (September through early November). Together these two areas comprise approximately one hundred acres of both privately held and public land, including several miles of shoreline and the expanse of Bulwagga Bay with flooded wetlands and deciduous trees. Permission is not required to bird throughout the area.

These sites are near Coot Hill Hawk Watch, and the early-rising birder could cover all three areas in an active day's birding (see chapter on Hawk Watches for directions to Coot Hill).

Birding Bulwagga Bay is rated as easy and Crown Point is rated moderately easy. Wear rough clothing and waterproofed boots as barberry thorns are rampant on Crown Point and may snag clothing; poison ivy is also rather thick along the western side of the peninsula, especially near the old overgrown railway embankment. Here also, the ground tends to be somewhat damp seasonally. Do not bring dogs along, for there are several active red fox dens in the vicinity. Waterfowl may be well offshore or in areas not easily reached from the road, so a spotting scope is recommended.

If you are coming from the south, leave the Adirondack Northway (I-87) at Interchange 29. Head east a short distance, following signs to US 9. Turn left (north) at US 9 and follow it to NY 4C and NY 4. Turn right (east) here and follow NY 4 and NY 4A into Port Henry. If coming from the north, leave the Adirondack

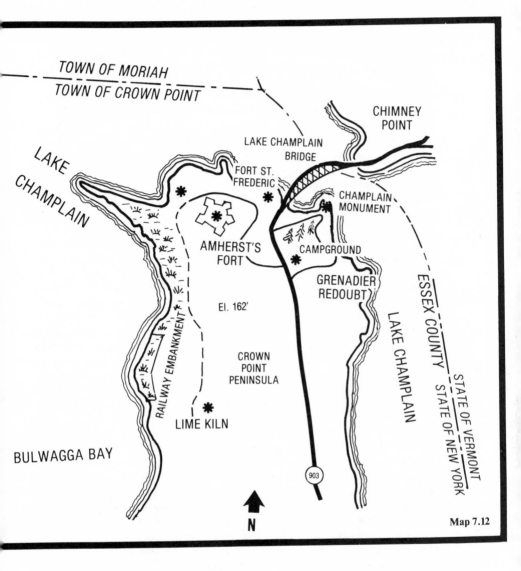

Map 7.12

Northway at Interchange 30. Head east on NY 6 (Mineville Road). Follow it into the village of Mineville. Turn right (south) on NY 7 and follow it to its intersection with NY 4A (NY 4). At that junction, turn left (east) and follow NY 4A and NY 4 into Port Henry.

From Port Henry, drive south out of the village on NY 9N (NY 22). Look for a sign indicating the Bulwagga Bay Campsite. Turn left (east) at the sign. Drive across the railroad tracks and past the entrance booth. Park in the picnic area on the right. Walk down to the water's edge and check the bay and marsh shoreline and sandy spits to the right for waterfowl and resting gulls. Look diligently for the occasional rare shorebird. Return to NY 9N (NY 22). Turn left, heading south again. Within a short distance you will come to a restaurant named Captain Blye's Pines. Stop at the overlook across from it to scan the open bay with a telescope. Continue south on NY 9N (NY 22) through the rock cuts. Turn left (east) at the Crown Point Bridge turnoff and park.

Check the flooded areas and shrubby edges for grebes, herons, ducks, geese, the occasional rail, and any shorebirds. Raptors are frequently seen in the adjacent fields from here all the way to the bridge, as well as in passage over Bulwagga and Big Hollow mountains, to the south, and over Coot Hill, immediately west. Double-crested Cormorants often gather in the bay or are seen off the tip of the peninsula in late summer. Remarkably, Caspian Tern has been recorded here in May.

Continue driving east and north on NY 903, toward the Crown Point Bridge (which spans a narrow section of Lake Champlain into Vermont). Bird all along this road searching for raptors as well as passerines. Just before the toll bridge turn left opposite an old grenadier redoubt, onto an access road leading to the French and British fort ruins. Park near the small museum. If this access road is closed, park on the shoulder in front of the toll booth and walk into the fort ruins area. These ruins are remnants of the French Fort St. Frederic, built in 1731, nearly twenty-five years before the construction of Fort Ticonderoga; and of the largest fort ever built (1759) in the American colonies by the British. Across the road, the Champlain monument contains a small bust of Rodin's "La Belle France" set into its north side; you really shouldn't miss it. Birding in the fields near the toll booth can be very rewarding, often providing a variety of raptors, perhaps even a Bald Eagle.

In the conifer groves near the monument and around the campsite across the road from the fort ruins, look for Great Horned and Barred owls. From the south wall of General Amherst's fort, large clumps of barberry bushes extend south and west. Walk west past the gate of Amherst's fort through an abandoned farmyard. Continue over a grassy field and some limestone ledges to the barberries and continue south to the ruins of an old lime kiln. Blue-gray Gnatcatchers nested below this kiln in 1977 and along a path to the north in 1980. From there, bird west to an abandoned railway embankment that runs beside a sometimes flooded stand of mature deciduous trees. This stand borders Bulwagga Bay. Herons, waterfowl, and raptors can easily be seen here. Whip-poor-will may be heard at dusk. Flycatchers, swallows, thrushes, vireos, and especially warblers abound in the height of migration. Yellow, Cape May, Yellow-rumped, Prairie (singing males present in 1976 and 1977), and Palm warblers are often abundant in waves from mid- to late May. Species of blackbirds, orioles, and sparrows are abundant when migratory waves funnel northward up the peninsula to this narrow northern tip.

Before leaving the peninsula, hike over to the bridge and search the water, especially in the narrows underneath the bridge, for waterfowl specialties present in both spring and autumn.

North West Bay (Maps 7.11 and 7.13)

Rating: Spring**, Summer***, Autumn***, Winter****

Whallon's Bay and "Essex Plowed Fields" (Map 7.13)

Rating: Spring***, Summer*, Autumn***, Winter**

Essex Station Marsh (Map 7.13)

Rating: Spring***, Summer**, Autumn**, Winter*

The following tour, if it is birded thoroughly, should take anywhere from four to six hours, with a short stop for lunch. The best birding is either from ice-out until June or in September and October. Each of these places is rated as easy to bird. Although North West Bay deserves a special additional visit in December, the tour should be thoroughly productive during migration seasons.

Leave the Adirondack Northway (I-87) at Interchange 31. Proceed east on NY 9N for 4 miles to the village of Westport. Follow the village's main street, NY 22 and NY 9N left (at North Main Street) to the New York State boat launching site at the north end of the village. Here there is a parking lot overlooking the northern expanse of North West Bay. Birding the bay varies considerably from year to year. Waterfowl numbers tend to fluctuate widely and the water levels in Lake Champlain noticeably influence the availability and quality of the shorebird habitat. Loons, three grebe species, five puddle duck species, ten diving duck species, and all three merganser species frequent the bay. Red-throated Loon, Eared Grebe, and Black Guillemot have been rare and welcome bonuses here. Use of a spotting scope will bring its own rewards.

After birding carefully and thoroughly at this vantage point, return to the car and drive south on NY 22 and 9N. Turn left just across the bridge and proceed to the village beach, where there is an excellent lookout over the south end of the bay. This is a marvelous place to see shorebirds, as more than twenty species have been recorded on the mudflats here in summer. Search the flats near the sewer outlet just north of the beach. Ruddy Turnstones, Willets, Red Knot, White-rumped, and Baird's sandpipers, as well as Wilson's and Northern phalaropes have been recorded here. From the road overlooking the bay, above the marina, look for one or two of the seven gull species recorded here (including Glaucous, Lesser Black-backed, and Little gulls). The more rare gull species will be seen primarily in December and early January. When you have birded the beach, the flats, and the marina, return to your car.

Regain NY 22 heading north. Just past the north end of Westport proceed northeast on NY 22J (Lake Shore Road). In the trees lining Lake Shore Road between Westport and Whallon's Bay, there were flocks of Bohemian Waxwings for three of the four winters before 1981. Drive approximately 7.3 miles on NY 22J to an intersection with Whallon's Bay Road on the left and Albee Road on the right. Albee Road and its rugged dirt extension curve along the south side of Whallon's Bay, all the way out to Split Rock Point and Split Rock Lighthouse. Historically, Split Rock Point was the boundary between the Algonquin and Iroquois Indian holdings and, following the 1713 Treaty of Utrecht, marked the boundary between the French and British interests in North America. Lake Champlain is deepest at this point (400 feet).

Except along the shoreline, Whallon's Bay is too deep for waterfowl to feed. Most species, especially the geese, raft here and feed in the extensive cornfields to the west. Hundreds of geese remain in the bay usually until late January, when it freezes. Waterfowl, including flocks of up to 100,000 geese per day, may pass directly up or down the lake—often stopping at Whallon's Bay—during the height of migration. By late April and early November, Whallon's Bay is usually covered with up to 3000 Canada Geese, joined sometimes by Snow Geese, including the blue form. A recent mid-April saw 5000 to 10,000 ducks (mostly Blacks), five species of diving duck, and mergansers join the geese. A famous Bald Eagle eyrie is located just south of Split Rock and although it may be no longer productive, adult Bald Eagles are reported at the bay almost every summer.

Now return up Albee Road to its intersection with NY 22J. Cross NY 22J so that you are now on Whallon's Bay Road headed west. Continue west for 1.5 miles to the top of a ridge (nearly to NY 22M) where you will find plowed fields (usually in wheat or corn). This area is locally known as "The Essex Plowed Fields." Scan the fields, from mid-August through early September, for American Golden Plovers, Black-bellied Plovers, and Buff-breasted Sandpipers. On September 15, 1979, hurricane Frederic brought a storm-blown, immature, dark-phase Parasitic Jaeger to the fields, to the enjoyment of many local birders.

In winter this is an exceptional area for raptors, both hawks and owls. Scan the fields and overhead for wintering Rough-legged Hawks, Screech, Great Horned, Snowy, Barred, Long-eared, Short-eared, and Saw-whet owls, all of which have occurred here in recent years. Great Gray Owl has been recorded three miles to the north. Search the bare winter fields for Horned Larks, Lapland Longspurs, and Snow Buntings. Continue birding along NY 22K to Whallonsburg. Cross the railroad tracks and the Bouquet River and you will see the junction of NY 22K and NY 22.

Turn right onto NY 22. Proceed north 2.7 miles where NY 22 turns nearly ninety degrees to the east. Turn with it. Continue another quarter-mile, but do not cross over the railroad tracks. Park on the north (left-hand) side of the road near a small bridge. This is Essex Station Marsh.

Check the marsh and adjacent pasturelands from the bridge. This small, very productive wetland, like others smaller but similar to it, are quite limited in this part of the Champlain flyway. Walk down the dirt access roads on both sides of NY 22, bordering the wetland. Species found here in recent years include Great Blue and Green herons, Cattle Egret, American Bittern, Canada Goose, Green-winged Teal, Mallard, Black Duck, Blue-winged Teal, Virginia Rail, Sora, Common Gallinule, Semipalmated Plover, Killdeer, Common Snipe, Greater and Lesser yellowlegs, Upland, Spotted, Solitary, Least and Semipalmated sand-

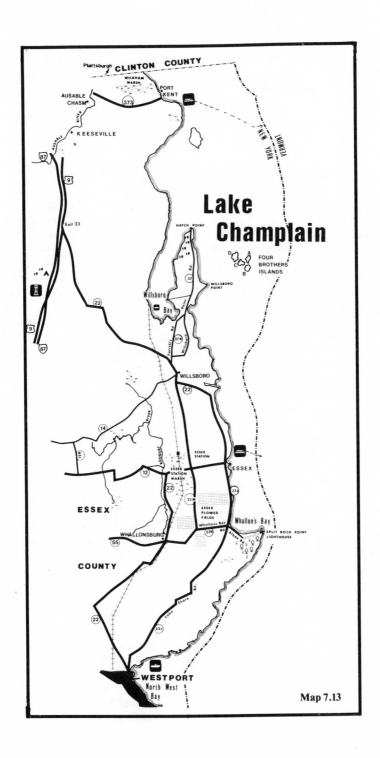

Map 7.13

pipers, Black-billed Cuckoo, owls, picids, flycatchers, Tree, Barn and Cliff swallows, as well as a variety of warblers and fringillids, including Savannah and Swamp sparrows. Carefully examine the north side of the marsh with a telescope, especially when birds may be partially hidden by high grass at the far end.

Willsboro Bay (Map 7.13)

Rating: Spring***, Summer*, Autumn***, Winter***

Birding Willsboro Bay involves covering about four miles of Willsboro Point Road, access roads, and walking trails, through private farms, estates, and summer camps from the New York State Boat launching site outside the village of Willsboro north to Hatch Point at the northernmost tip of the peninsula.

Leave the Adirondack Northway (I-87) at Interchange 32. Proceed north and east on NY 12, 12B, and 14 to the village of Willsboro. If coming from the north, an alternative is to leave the Northway at Interchange 33. Proceed south and east on NY 22 to the village of Willsboro. Drive out of the village heading north and take an immediate right on Farrell's Road. Bear left at the fork of Farrell's Road (NY 27B) and NY 27. Follow Farrell's Road to the stop sign at its junction with Willsboro Point Road. Turn left and proceed to the next road coming in on the left, where you will see signs for the New York State Boat Launching Site. Turn left and follow the access road to the parking lot.

Walk down to the bay off the launching site and check it and the broad expanse of Willsboro Bay to the west for loons, grebes, and waterfowl. Walk back to the entrance at Willsboro Point Road. Cross it and proceed to the Lake Champlain side of the peninsula. Here you will be crossing through privately owned property; keep in mind that every birder respecting the rights of these property owners helps ensure access for birders who follow. Having birded both sides of NY 27 (Willsboro Point Road) here, continue driving north on it. The road becomes increasingly narrow. Park at the loop at the end of the road. Here there is a trailhead. Follow the short trail through the woods to the rock ledges of the Hatch Point promontory. Bird the open lake and mouth of Willsboro Bay from here. Assemblages of 100 Common Loons raft here in late autumn. Mergansers collect here in early spring and winter. This is an excellent place to study various plumages of loons and mergansers on both migrations. The vast bay and open lake waters attract a wide variety of Anseriformes; in fact, this vantage point is one of the best in *Kingbird* Region 7.

Four Brothers Islands (Map 7.13)

Rating: Spring****, Summer****, Autumn**, Winter*

The Four Brothers Islands lie in Lake Champlain about 2 miles east of Willsboro Point in Essex County. The total area of the four islands is approximately six acres. In 1977, the University of Vermont acquired the islands and it is holding them as a management resource within the boundaries of the Adirondack Park. These low rocky islands support a few conifer and hardwood stands, frequent patches of stinging nettle, wild mustard and wild raspberry, as well as several species of grasses.

Historically, the islands have supported a Ring-billed gullery comprising in 1980 about 7500 nesting pairs, approximately 250 nesting Herring Gull pairs, and a few pairs of Great Black-backed Gulls. The islands have substantial heronries of Cattle and Snowy egrets as well as Black-crowned Night Herons. Among species recorded on the islands in recent years, in addition to the above-named gulls, egrets, and herons are Canada Goose, Gadwall, Mallard, Black Duck, White-winged Scoter, Common and Red-breasted mergansers, Ruddy Turnstone, Spotted Sandpiper, Short-billed Dowitcher, Long-tailed Jaeger, Iceland (Kumlien's) Gull (*Larus glaucoides kumlieni*), and a great variety of passerine species. Tree Swallows still nest in the crevices of the cliffs as they did as far back as 1887.

Travel by boat to the islands, 2 miles from Willsboro Point, after launching at the New York State Boat Launching Site on Willsboro Bay (see the immediately preceding article for directions to the launching site). Boats may be rented at the Willsboro Bay Campsite Marina, located off Willsboro Point Road. Lake Champlain can be dangerous, so the birder should check all weather forecasts before attempting this trip. Avoid the reefs and shoals near the islands' shores.

Nota Bene: Birders may not land on the islands without express permission, which must be obtained from the University of Vermont, Environmental Program, Burlington, Vermont 05401. The University is very reticent to give access permission during nesting season, owing to the dangers of disturbing nesting adults or stampeding chicks over the cliffs. Undue mortality to colonial nesters and waterfowl must be avoided. However, this should not dissuade birders from going out to the islands. All species can be well seen from a boat.

For ease of reference, the Vermont Fish and Game Department has unofficially designated the islands in the following manner: "A" (the northeast island), "B" (the southeast island), "C" (the central island), "D" (the northwest island). The accompanying map has followed this scheme.

ESSEX COUNTY AND CLINTON COUNTY
**Wickham Marsh, Little Ausable River, Carpenter Flats, Ausable Point,
and Ausable Point Marsh** (Maps 7.13 and 7.14)

Rating: Spring****, Summer**, Autumn**, Winter***

These birding sites are located about one to three miles northwest of the village of Port Kent and seven to ten miles southwest of Plattsburgh. They can both be covered in several hours of active birding, and completely covering both sites is considered easy. Wickham Marsh is in Essex County and Ausable Point and Marsh are in Clinton County. Each of these sites is on public land, so permission to bird them is not necessary. They border Lake Champlain and are some of the most productive and pristine birding areas in the region. The roads into and fields bordering the marshes may be wet in spring and waterproofed footwear is advised. In any season winds off Lake Champlain can be bitterly cold and the birder should be prepared.

To reach Wickham Marsh, leave the Adirondack Northway (I-87) at Interchange 34. Proceed northwest on NY 9N, which is named Pleasant Street in Keeseville. From the village of Keeseville, take US 9 north (Front Street), crossing the Ausable River at Ausable Chasm, turn right on NY 373, following signs for the Port Kent Ferry. Follow NY 373 along the southern boundary of the village and turn left on Lake Street (NY 17A), just as Lake Champlain comes into view. Lake Street meets NY 373 on the left only and is the street immediately preceding the railroad tracks. Do not cross the tracks, or if you already have, turn back and take a right, heading north on Lake Street. Lake Street parallels the lakeshore as it passes through the hamlet. Proceed north on it out of town until you see the sign for Wickham Marsh on the left.

By crossing the railroad tracks and continuing to the terminus of NY 373 (after a sharp right on Quay Street), one reaches the Port Kent Ferry dock. This is an excellent place to view waterfowl in early January, and should be checked out. In winter look for Surf Scoters as well as the occasional white-winged gull feeding on the pack ice.

From the Wickham Marsh sign and entrance, walk in the chained-off dirt road. Note the interesting variety of ferns lining the damp woods road, along the southern border of the marsh. Follow the road through a low deciduous forest. Singing Yellow-throated Vireos are present in good numbers in these woods in late May. Migrant waves of Yellow-rumped Warblers may feed here from late April through mid-May. Proceed along the trail until you meet an open field on the right (north). This field meets the marsh shore. Bird the marsh and return to the trail and road. Walk north to another excellent vantage point, from which the marsh can be seen. On a good spring morning an experienced birder should be able to find nesting Least Bitterns and Blue-winged Teals, migrant Marsh Hawks and Ospreys, and perhaps even a Peregrine Falcon, nesting Virginia Rails, Soras, and Common Gallinules. In winter, when the marsh may be frozen, climb the railroad embankment across the road from the marsh to view waterfowl. In early January, Canvasbacks, Buffleheads, goldeneyes, and mergansers congregate here on Lake Champlain by the hundreds and sometimes thousands.

After birding Wickham Marsh, proceed north less than a mile, heading toward Ausable Beach. Take the first right and park almost immediately on the left near the railroad crossing. From here walk north on the tracks into the Ausable Marsh. More common wetland species can be seen here, as well as breeding waterthrushes, Mourning, and Canada warblers.

To exit, head west on this same road to its junction with US 9. There turn right heading north. Proceed a very short distance until you see a large bridge crossing the Ausable River. Park before crossing the bridge and bird the area on the right (east), which is known as Carpenter Flats. It is worthwhile to visit this site

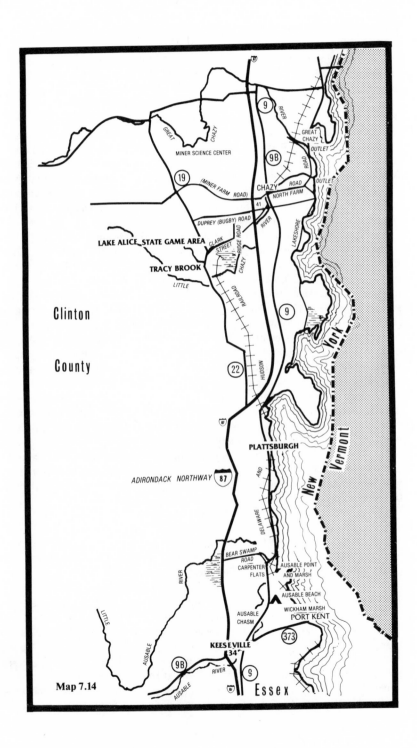

Map 7.14

before the spring foliage is at its luxuriant peak. A keen birder should be able to see resident swallows, thrushes, vireos, and at least fifteen or sixteen warbler species, including Tennessee, Nashville, Magnolia, Blackburnian, and Chestnut-sided. Carpenter Flats is in Clinton County, and it is one of the top sites in the county for viewing warblers in breeding plumage.

Once back on US 9, again head north for a very short distance. Just before reaching Bear Swamp Road (which comes in on the left), turn right (east) on the dirt road on the right immediately south of it. All along this road, which, in fact, parallels the Little Ausable River, birding is excellent in spring, most notably for marsh birds and warblers. Walk along the road and dike, birding underneath the railroad trestle and in the willows along the railroad tracks. This is a marvelous place for warblers and sparrows, so give it sufficient time.

Again return to US 9 and proceed north past Bear Swamp Road for 0.2 mile to the road with a prominent sign for Ausable Point Campsite, which comes in on the right. Turn right on it, and bird all along this road to Ausable Point. As you pass along the road, note the grebes, herons, bitterns, and waterfowl in the marshes on either side. Be on the lookout for some or any of the ten species of raptors recorded here. During any season check Allen's Bay to the north for waterfowl species. While crossing over the bridge over the Little Ausable River, take note of the Rough-winged and Barn swallows hawking insects above it. A good many of these nest underneath the bridge. During the spring migration periods check the hedgerows and trees lining the road. Several vireo and warbler species can be viewed on a good day here.

Once past the park toll booth, proceed to the picnic area. If you have come in early morning in spring you should see and hear breeding Pine Warblers in the pines. In addition, these pines usually hold the more common, more northerly breeding winter finches, as the weather becomes colder.

Do not leave the area without exploring the marshes south of the campsite entrance road. Herons, bitterns, waterfowl, rails, flycatchers, marsh wrens, and sparrows abound throughout the floodplains and river bottoms, making this one of the most productive birding sites in the county.

Chazy Landing Area (Map 7.14)

Rating: Spring****, Summer**, Autumn**, Winter***

The Chazy Landing Area includes a three-mile stretch of road along the lakeshore, just southwest of the outlet of the Chazy River and south of the outlet of the Great Chazy River. The land here is primarily agricultural, and in the fields birders can view geese, dabbling ducks, and the occasional shorebird on migration; in the winter they will find raptors, owls, longspurs, and buntings.

Leave the Adirondack Northway (I-87) at Interchange 41. Proceed east on Miner Farm Road to US 9. Here turn right (south) and go south a very short distance to North Farm Road, the first road coming in on the left. Turn left, and begin to bird the lowlands through which North Farm Road passes. Continue east to Lakeshore Road. Turn left and slowly progress northward, stopping often and regularly to check out the fields bordering Lakeshore Road. Cross the Little Chazy River outlet. Proceed about one mile where you should see a small red farmhouse on the left. In the extensive fields behind this house, geese in the thousands, shorebirds, and ducks gather to feed in migration. In winter Red-tailed, Rough-legged, and Marsh hawks, Gray Partridges, gulls, Snowy and Short-eared owls, Horned Larks, Lapland Longspurs and Snow Buntings are easily seen. Across the road from the red farmhouse note the small state game area. This is one of the places that is quickly becoming known as a hotspot for rare herons, ducks, and shorebirds.

Proceed north all along the Lakeshore Road and follow it as it veers northwestward, paralleling the Great Chazy River. Cross over the river to NY 9B and follow it west to US 9. Here turn left (south) and proceed south to Miner Farm Road, where you will be all set to follow directions to the next site.

Lake Alice State Game Area (Map 7.14)

Rating: Spring****, Summer***, Autumn**, Winter*

Lake Alice State Game Area is in the northernmost section of Clinton County, approximately seventeen miles north of Plattsburgh and two miles southwest of the village of Chazy. The Delaware and Hudson Railroad passes through the southwest part of the game area, and the birding is good along the tracks.

It is state-owned land, so permission to bird the area is not required. The best time to bird the area is in spring, but keep in mind that the mosquitoes and blackflies are thick from late-May through June. Birding the area is rated "moderately easy" and to thoroughly cover it, the birder should walk the roads and trails, and especially along the railroad tracks.

To reach the area: leave the Adirondack Northway (I-87) at Interchange 41. There pick up NY 19 (Miner Farm Road) and proceed west on it. At the Miner Science Center, which you will see on the right side of the road, turn left and proceed south on Ridge Road. From the intersection of it and Duprey (Bugby) Road the Lake Alice State Game Area is visible to the south.

Bird all over the management area, especially along Ridge and Duprey Roads. The prime and most productive area is along the railroad tracks southeast of the game area, west of where they cross Ridge Road, and also where they cut across Clark Street. Virginia Rails, Soras, Black Terns, Alder Flycatchers, and Long-billed Marsh Wrens nest here. Drive southwest on Clark Street to its intersection with NY 22, birding carefully the north side of the street, where Tracy Brook runs close to it. Bird the mixed deciduous trees along Duprey Road north of the lake. Expect to see here nesting Broad-winged Hawks, pheasants, Barred Owls, Pileated Woodpeckers, Yellow-bellied Sapsuckers, vireos, warblers, and tanagers. This is an especially productive section of the game area.

ACKNOWLEDGMENTS

The author wishes to express her special gratitude to the following people who contributed information on the Adirondack-Champlain region: John M.C. Peterson, Geoffrey Carleton, Ferdinand La France, Philip H. Warren, Theodore D. Mack, Bruce McP. Beehler, Patricia Taber, Esther Ann McCready, and Elsbeth Johnson. Sincere thanks to Mike Peterson and Geoffrey Carleton for their reading of and comments on the Adirondack-Champlain chapter.

Hudson-Mohawk — Region 8

THE HUDSON-MOHAWK *KINGBIRD* REPORTING REGION includes eleven counties. From north to south they are Warren, Washington, Saratoga, Fulton, Montgomery, Schenectady, Rennsselaer, Albany, Schoharie, Columbia, and Greene, an area of approximately 6770 square miles. At its northern reaches, the region takes in the southern part of the Adirondack Park; the southern portion of the region takes in the northern part of the Catskill Park. In addition to these ranges, the Grafton Mountains, an extension of the Berkshires, enter northeastern Rensselaer County. The Taconics extend along the eastern border of Rensselaer and Columbia counties. The Helderberg Escarpment lies diagonally across Albany County, in some places reaching elevations of 1600 feet. The highest point in the region is the summit of Hunter Mountain (4025 feet). This peak lies in Greene County, within the Catskill Forest Preserve. The Hudson River cuts the entire region from north to south and the Mohawk River crosses most of it in a generally east to west direction. Literally hundreds of lakes dot the region; among the largest are Lake George, Sacandaga Reservoir, Saratoga Lake, and Round Lake (Map 8.1).

In the more mountainous areas there are extensive stretches of forest — although none of it is virgin timber, for lumbering has been an important and thriving industry throughout most of this area at some period or another in the state's history. Most of the forests are of the oak-hickory and hemlock-northern hardwoods associations, with northern coniferous forests of spruce and balsam at the highest elevations.

One of the most interesting but, unfortunately, exploited ecological treasures in this region is the Albany Pine Barrens, which covers some 4000 acres from just within the city limits of Albany to the adjacent towns of Guiderland and Colonie.

In addition to the dominant mountain and water formations, there are numerous bogs, remnant forests, cattail marshes, dramatic ravines, and carefully protected preserves, which every birder ought to visit.

Many of the outlying areas remain relatively unexplored by the modern army of bird watchers, so strike out on your own to find good productive birding sites to share with all of us.

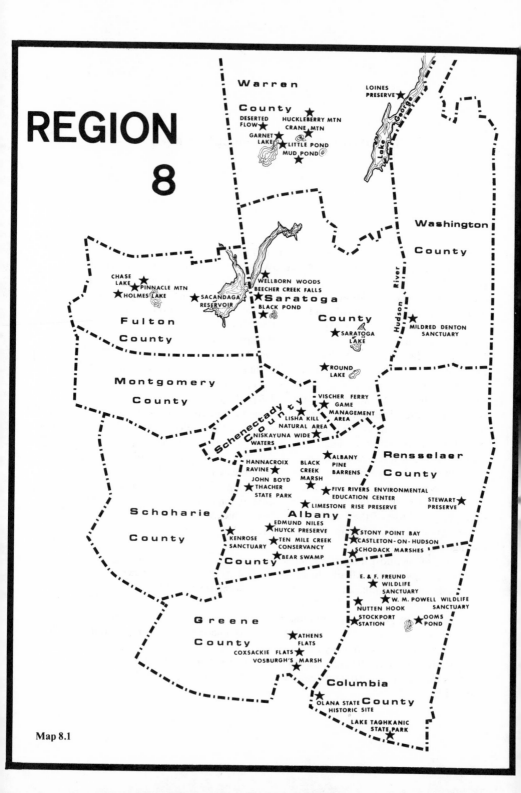

Map 8.1

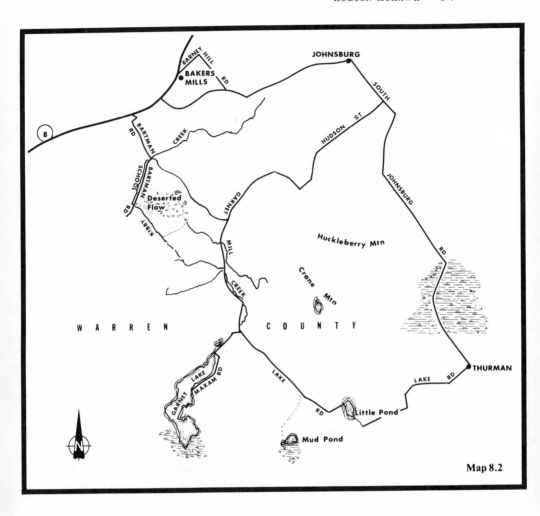

Map 8.2

WARREN COUNTY
Deserted Flow (Map 8.2)

Rating: Spring****, Summer***, Autumn**, Winter*

In the northern part of Warren County there are some unsung birding sites that are badly in need of investigation. They are relatively easy to reach and yet have the advantage of being off the beaten path. They hold tremendous potential for the adventuresome birder or tireless Breeding Bird Atlas worker. One such place is Deserted Flow. It was, at one time, the site of an extensive beaver colony, whose activity and energy can only be judged by the maze and diameter of dead trees left behind, because the beavers were trapped out, probably as recently as ten to fifteen years ago.

The pond itself is only a quarter-mile long and quite narrow, but, there is a huge area to the east of the pond (much larger than the pond itself), that is flooded and covered with large standing dead trees. The road to the pond passes through varied habitats: open old farmland with stone walls bordering a now deserted apple orchard, a plantation of large pines, and quite a dense stand of spruce.

With mixed woodlands and marshy flow habitats one might look for three-toed woodpeckers nesting as well as some of the more northern breeding flycatchers, vireos, and warblers. The hike to Deserted Flow is not physically taxing, but during spring and summer, you should be sure to bring insect repellent. The remaining beaver dam plugs a branch of Kibby Creek, which some consider pure potable water; however, I would advise that you carry in some drinking water.

To reach Deserted Flow, leave I-87 (Adirondack Northway) at Exit 25 (NY Route 8, Chestertown, Hague, Schroon Lake, Brant Lake). Follow US 9 south to its junction with NY 8. Turn right (west) on NY 8 and follow it west to the hamlet of Bakers Mills. Pass through Bakers Mills and begin looking for Bartman Road, which comes in on the left, south of the hamlet and about 1.4 miles west of the center of Bakers Mills. Turn left and follow the twists and turns of Bartman Road for approximately 2.2 miles. After about a mile, the road will make a sharp right-hand turn, but stay on Bartman Road and *do not* turn left. Approximately 0.2 mile north of the intersection of Bartman Road and Bartman School Road there is an old logging road that comes in on the left. Look for it immediately after crossing Kibby Creek, which, up to this time, has been on the right side of the road. Park the car off the road and walk in on the old logging road.

The trail is less than a mile long, and it leads east into forest preserve land. This road traverses the various habitats mentioned above. Just as the trail appears to end, near the edge of the beaver flow, look for yellow markers on the trees. These indicate the path around the flow to the pond, and to its northwestern shore, where there is a small and inviting spruce-covered ledge. .

Crane and Huckleberry Mountains, Garnet Lake, Mud Pond, and Little Pond, a Tour
(Map 8.2)

Rating: Spring****, Summer****, Autumn*, Winter*

The following tour samples several different habitats: streamside, lake, brushy marshes, high elevation birch stands, open fields, wet meadows, steep slopes onto which only lichens and mosses seem able to hold, and climax stands of oak, maple, pine, and spruce. Crane Mountain rises to over 3200 feet and possesses many dramatic stretches of partially exposed ridge and sheer cliffs. Combined with the other habitats represented, this tour is designed to be a birder's delight.

To reach the beginning point of the tour, leave I-87 at Exit 25, as described in the previous write-up (Deserted Flow). Follow US 9 to NY 8. Turn right (west) and follow NY 8 to the hamlet of Johnsburg. Here turn left at South Johnsburg Road and follow it south 1.1 miles to Hudson Street. Turn right (west) and proceed for about 2.75 miles to its intersection with Garnet Lake Road.

Along Hudson Street several places are good birding stops: at 1.5 and 1.75 miles, where spurs of Kibby and Mill creeks cross the road; and at 2.2 miles, where a small bridge crosses Paintbed Brook. There is room to park off the road east of the bridge. Stopping along Hudson Street also allows one excellent views of the massive bulk of Huckleberry Mountain to the south.

Turn left at Garnet Lake Road and after 0.75 mile stop for a really breathtaking view of the rock hulk—metasedimentary, interspersed with gneisses—of Huckleberry and Crane mountains to the left. Proceed south 2.75 miles stopping at several points along the way to bird, especially where Mill Creek comes close to the road. Just before you come to the turn-off to Garnet Lake there is a brushy, marshy area on the left side of the road that is an excellent place to investigate. At 3.5 miles make a right turn on Maxim Road toward Garnet Lake.

Proceed down the east side of Garnet Lake. The drive itself is quite beautiful. Approximately 0.5 mile in on Maxim Road there are idyllic uncrowded picnic sites, and a little bit farther on, a sandy beach. This is a marvelous place for a quick swim. In the 1930s, Garnet Lake was dammed, which more than doubled its previous size and flooded a huge swampy area south of it. That explains why the southern end of Garnet Lake is dotted with stumps both above and below the water's surface. Much of the lake is covered with water lilies and pond weeds. The beauty of the lake may be directly attributed to the dramatic mountains surrounding it. On the west, Ross Mountain rises to 2625 feet; on the southwest, Mt. Blue is imposing at about 2900 feet. Examine the gently sloping wooded forests along the eastern shore of Garnet Lake. Approximately 0.8 mile south of the dam, there is a path that follows an old logging road into the woods. Follow this road a short way, looking and listening for nesting warblers.

After birding Garnet Lake, return to Garnet Lake Road. Here turn right and follow the road for 2.1 miles. Park off the road and walk into the woods on an old logging road on the right. It is only a short walk to Mud Pond, which is circled by a quaking bog. Sundew, *Myrica Gale,* rose mallow and pitcher plants all grow on the sphagnum mat and the shallow pond itself is thickly blanketed with water lilies. After walking in 0.5 mile, the path turns south for the remaining 0.2 mile to Mud Pond. This is a fine place to search out nesting woodpeckers, as well as kingbirds, flycatchers, possibly Boreal Chickadees, wrens, and breeding vireos and warblers.

After walking the 0.7 mile back to the road, proceed 1.1 miles to Little Pond, which is on the left. Stop here and investigate the birches rimming the pond. Continue driving east for 1.2 miles. Turn left and after 0.4 mile follow the road to the right. From there continue to the tee at the hamlet of Thurman. Here, at South Johnsburg Road, turn left (north) and follow the road northward. At 1.0 mile on South Johnsburg Road pull off and walk east to a marshy expanse to find breeding or late migrants. Then, continue north for about 5 miles, completing the tour. Again, there are several places one might wish to stop to do a little roadside birding along the way.

Loines Preserve (Map 8.1)

Rating: Spring****, Summer***, Autumn**, Winter*

The Loines Preserve comprises 25 acres of mature beech, oak, white pine, and ash, with birch covering the gentle slopes and hemlock predominating on the steeper slopes. There is one impressive white pine in the preserve with a trunk circumference of 11.0 feet. The preserve is a narrow, almost rectangular strip of land lying between the Northwest Bay marsh of Lake George and spanning both sides of NY 9N. The marshy vegetation within the preserve includes cattail, burreed, various pond-and-water weeds, wild celery, spikerush, and pickerelweed. The northern end of the preserve contains alder, sensitive fern, highbush blueberry, shadbush, black birch, and sheep laurel. In early spring one can enjoy the new trailing arbutus, partridgeberry, wintergreen, shinleaf, Christmas fern and marginal shield fern. Within the preserve there are shady rocks lush with liverworts and mosses and sunny rocks with lichens and reindeer moss.

To reach Loines Preserve: leave I-87 at Exit 24 and follow Bolton Landing Road east to Federal Hill Road. Turn right (south) and take Federal Hill Road to the hamlet of Bolton Landing. From the stoplight in the center of Bolton Landing and the junction of NY 9N, turn left and proceed north 5.1 miles. Park here along the shoulder of the road and enter the Loines Preserve.

The Nature Conservancy owns and maintains Loines Preserve and although there are no formal trails or paths through the property, it is easily traversed. The presence of the marshy habitat and the rich diversity of plant life makes it an attractive migration stopover for more than 65 passerine species, which can be viewed at leisure as you walk through the evergreen groundcover.

FULTON COUNTY
Holmes Lake, Pinnacle Mountain, and Chase Lake, a Tour (Map 8.3)

Rating: Spring***, Summer****, Autumn***, Winter*

This is a tour in the southern fringe of the Adirondack Park, northeast of the hamlet of Caroga Lake. Largely unknown and unvisited, these peaks and lakes offer an alternative to the overused higher peaks farther north. As a bonus, these peaks are easier to hike than the northern Adirondacks. The area described here is crisscrossed with old logging roads, cut through before the state acquired the land; they are now grassy lanes which make ideal footpaths. The vegetation in general is open and not congested with the "blowdowns" that one finds in the more northern sections of the park. The area contains upward of twenty wild lakes and ponds and is totally surrounded by state lands. Although the trails and paths are not as well-

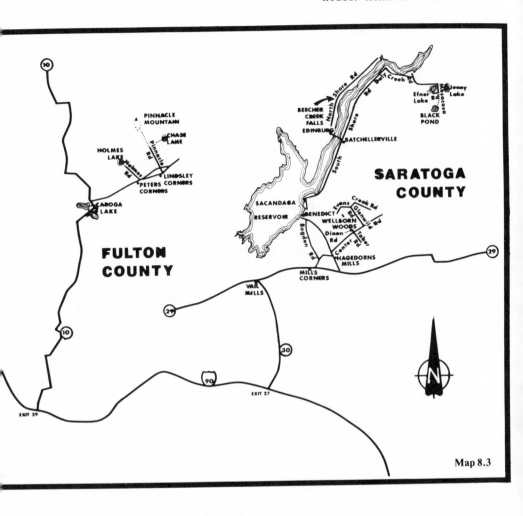

Map 8.3

marked as those in the High Peaks, the going is considerably easier and blazes are not usually necessary to simply get from one place to another. Nonetheless, it is advisable to take a compass and topographic map (the Caroga Lake and the Jackson Summit USGS quadrangles) if you plan to take a full day's hike or longer. The birdlife one will encounter is typical of that found at 2520 feet and lower elevations in birch, pine, and spruce forests: streamside nesting species; species of open meadows at high elevations; species frequenting secluded crystal lakes; species fond of second-growth around long-forgotten human settlements; as well as those species preferring the ubiquitous marshy areas of upland plateaus.

To reach the beginning of the tour route, leave I-90 (the New York State Thruway) at Exit 29 and head north on NY 10. Follow NY 10 for 18 miles to the hamlet of Caroga Lake. From here there are two excellent access routes which are described below. However, there are also several other fine access trails in the vicinity, which you may prefer to try. Be sure to submit avian records from any or all of these routes to the *Kingbird* Region 8 editor.

Just north of Caroga Lake hamlet and West Caroga Lake, a county road comes in from the right, with signs indicating the route to Benson. Turn right here and proceed eastward about 4 miles to the hamlet of Peters Corners. Turn left (north) on Holmes Road and drive toward Holmes Lake to the end of this road, a distance of approximately 2.8 miles.

At the end of Holmes Road, park and look for the trailhead leading northward to Holmes Lake. Walk to the lake, birding all along the way. You will be paralleling the Holmes Lake outlet on the drive up Holmes Road and also while walking toward the lake. Expect an appropriate variety of birds along both routes.

After you have thoroughly birded this area, return to Holmes Road. Drive south to Peters Corners, then turn right (east) and head toward Benson. After 2 miles you will reach the hamlet of Lindsley Corners. In the center of the hamlet, Pinnacle Road comes in on the left. Turn left (north) on Pinnacle Road and drive about 3 miles to its end. Park here.

A jeep trail heads north from here to the summit of Pinnacle Mountain. This is an easy walk, rich in birdlife. There is another trailhead south of the jeep trail, however, leading to the east — a 2-mile easy hike to Chase Lake. Although both trails are well worth hiking, I prefer the Chase Lake route.

Chase Lake covers 66 acres. It has, on its eastern side, an irregular shoreline with interesting little coves and peninsulas to investigate and bird to your heart's content. Its forested surroundings, and the varied habitat on the walk in, make Chase Lake a birder's dream-wish realized.

SARATOGA COUNTY
Wellborn Woods Conservancy and Beecher Creek Falls (Map 8.3)

Rating: Spring****, Summer***, Autumn***, Winter*

The Great Sacandaga Lake is a man-made flood control and power reservoir created by damming in 1930. Most of the land along the lakeshore is leased from New York State and privately developed: lodges and private campsites dot its shores, several very small communities have developed on its southwestern and southern shores, and there is a crossroads hamlet at its northernmost tip. There is a fine small birding site, both picturesque and productive, on the western shore of the reservoir, quite close to the hamlet of Edinburg. The following directions are to Beecher Creek Falls, by way of Wellborn Woods Conservancy.

Leave I-90 (the New York State Thruway) at Exit 27 and head north on NY 30. Turn right (east) on NY 29 at Vail Mills. Drive about 6 miles, then take the first left turn, 1 mile past Mills Corners. After about 0.7 mile, turn left at a paved county road. This road will pass through Hagedorns Mills to Center (or Center Line)

Road (approximately 1.2 miles from the turn). Turn right and drive approximately 2.9 miles to Dixon Road and an unpaved trail on the left (north). Taber Road will be on the right (south). Park well off Center Line Road.

Walk into Wellborn Woods Conservancy on Dixon Road (north). This tract of about 100 acres is owned and maintained by the Nature Conservancy. The land is gently rolling, heavily-wooded, and it is crossed by Evan's Creek, which stays open year-round. Until the early twentieth century, the area was farmed, but it has since reverted to forest. The mixed forest contains a rich assortment of northern and temperate zone tree species, including conifers, white pine, spruce, balsam fir, Canada yew, and hemlock. As one walks through the woods there is plenty of evidence of porcupine-killed hemlocks. Walk to the dammed, triangular pond in the center of the preserve by following Evans Creek east. The pond is approximately 110 feet wide and extends upstream approximately 160 feet. It is nearly fourteen feet deep near the dam, so nonswimmers, take care. Note the huge wooden slab 500 feet north of the pond. This serves as a home for many of the preserve's porcupines. Look for a small clearing on the side of the pond, where a seventy-foot-deep well with a manual pump provides pure drinking water. An old-fashioned wooden outhouse completes the preserve's facilities.

Eleven fern species, six species of clubmosses, and the beautiful wood horsetail grow in the forest as well as such a large assortment of wildflowers that only a few examples can be given: at least eight members of the lily family, six species of orchids, dwarf ginseng, pipsissewa, wild bergamot, turtle-head, and the carnivorous sundew. There are many mosses, liverworts, and lichens.

Walk through Wellborn Woods looking for breeding and migrating woodpeckers, flycatchers, vireos, warblers, and sparrows.

When you are finished birding Wellborn Woods, return to the paved county road north of Hagedorn Mills and turn right. Proceed to Northampton and turn right to Batchellerville along the road that parallels the shoreline.

Between Northampton and Batchellerville there are numerous places to pull off the road and scan the lake for migrant waterfowl in spring and autumn. At Batchellerville, turn left (west), crossing the eastern arm of the reservoir, and on the opposite (western) shore, turn right, in the hamlet of Edinburg, onto North Shore Drive.

Proceed 0.5 mile out of Edinburg, where North Shore Drive makes a sharp bend as it crosses Beecher Creek. Drive about 250 or 300 yards to a spot where there is sufficient room to pull off the road.

At this site you will find Beecher Creek Falls, a lovely covered bridge downstream from it, and a picnic table a short distance up the slope, between the road and the falls. A steep hemlock-covered slope shades the falls. These trees, as well as the grove near the picnic table, should be searched for transient and breeding warblers. Walk down to the covered bridge and cross it to an unused road on the opposite side. Walk along the roadway, birding as you go. Follow the small footpath on the shoreline east, past the ruins of an old mill. In the early to mid-1800s, a series of dams along this stretch of the creek provided power for several mills. Bird the stream and forest below the falls on the far side. In addition to several gratifying avian species, look also at the unusual growth of mosses, liverworts, and ferns in the deep shade on the rocks there. Beecher Creek Falls is one of those sites hungering for some careful attention from both local and visiting birders.

Black Pond (Map 8.3)

Rating: Spring****, Summer****, Autumn***, Winter*

Black Pond is included here, not because it has traditionally proven to be an excellent birding site, but because it represents a typical northwoods pond in the southern Adirondacks. It *should* provide wonderful birding, if it is thoroughly and consistently investigated by local and visiting birders.

This area is easily reached from Beecher Creek Falls (see above). Cross the eastern arm of the Great Sacandaga Lake (Sacandaga Reservoir), on the bridge extending from Edinburg to Batchellerville. At Batchellerville, turn left and follow the South Shore Road north, which parallels the east side of the reservoir. Proceed to Daly Creek Road and signs that indicate the way to Eggleston Falls and Efner, Jenny, and Hunt Lakes. Here, turn right and follow Daly Creek Road to Eggleston Falls and the junction of West Mountain Road. Turn left on West Mountain Road and proceed east to Mesacosa Road, which comes in on the right, east of Efner Lake and west of Jenny Lake. Turn right on Mesacosa Road and follow it to its dead end, 0.75 mile south. Leave your car here and begin walking. The hike to Black Pond is 4 miles, so the round-trip is a walk of 8 miles over easy terrain.

Walk east nearly 100 yards and follow the jeep trail as it turns south. From here the way to Black Pond is so well-marked that it would be impossible to not find your way. The trail moves up a hill and then somewhat westward; shortly, it passes an area of small saplings, obviously burned over; it then descends to a clearing; from there the trail branches eastward (right) to the more open section of Black Pond and westward (left) to the outlet of Black Pond.

Saratoga Lake and Round Lake (Map 8.4)

Rating: Spring****, Summer*, Autumn****, Winter*

Saratoga Lake, about three miles to the southeast of Saratoga Springs, covers about six square miles. Its shores include both public and privately held land. The area can be birded by car, by bicycle, or on foot, but use of a bicycle is not advisable, owing to the narrowness of the road around the lake's perimeter. Birding it has been rated as "easy" by those who most closely monitor it. Many access points to the lake involve passing over private lands; some have been used casually by birders. Here discretion is urged; if you have any doubt at all, take the time and trouble to seek the private owner's permission.

The major reason for the inclusion of this lake in the guide is its attraction for sea ducks as well as locally uncommon waterfowl, gulls, and terns. Red-

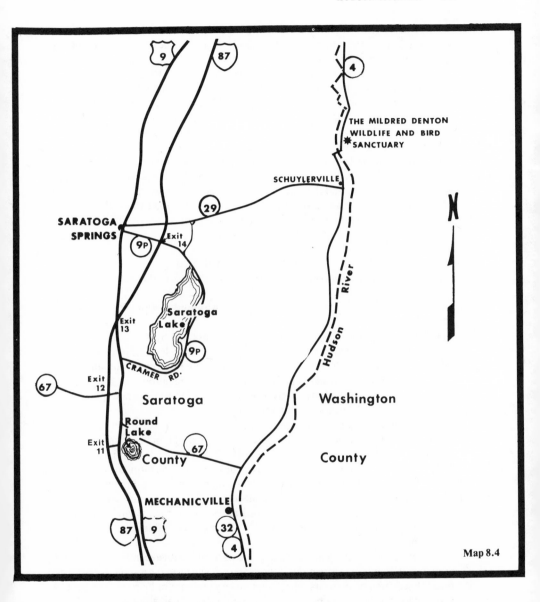

Map 8.4

throated Loons and Red-necked Grebes are regular visitors to the lake. The more locally rare species—Whistling Swan, White-fronted Goose, Gadwall, Pintail, and Northern Shoveler as well as Oldsquaw, Black, Surf, and White-winged scoters, and Ruddy Duck frequent Saratoga Lake. Bonaparte's Gulls and Black Terns frequent the lake, which are also otherwise locally uncommon.

A visit to Round Lake, approximately 5 miles to the south of Saratoga Lake and quite visible from the surrounding roads, is often productive and frequently yields some avian surprises.

Leave I-87 (the Adirondack Northway) at Exit 11. Proceed east to US 9. Cross US 9, headed north to NY 67. Turn right and go east on NY 67, stopping where Round Lake is visible from the road. A scope would be most useful in birding both Round and Saratoga Lakes. Keep making right turns at every crossroad until you have circled Round Lake and are back at US 9. Get back on I-87 at Exit 11 and proceed north to Exit 13.

Leave I-87 at Exit 13. Take US 9 south to NY 9P, which may here also be labeled Cramer Road. Turn left. Proceed east and north on NY 9P, working your way around the south and the east sides of Saratoga Lake, taking advantage of all of the access points, making sure to avoid pulling off the road where the shoulder is narrow. Again use discretion when gaining access over private lands.

WASHINGTON COUNTY
The Mildred Denton Wildlife and Bird Sanctuary (Map 8.4)

Rating: Spring****, Summer***, Autumn**, Winter**

This 370-acre wildlife preserve is owned and maintained by The Nature Conservancy, and it lies in an area replete with scenic beauty and historic interest. It is bisected both by US 4, which is otherwise known as the Burgoyne Trail, and by the Old Champlain Canal, which parallels US 4. About a mile south of the preserve, on the west side of the Hudson River, is Stark's Knob; the volcanic plug below which the Battle of Saratoga took place in 1777.

The underlying bedrock throughout the preserve and surrounding countryside is Normanskill shale; you can see a good exposure of it in the windy Shale Hill section of the preserve, near the entrance off NY 4. The shale was mined until the early 1970s, and since then the bared hill has developed a vegetation of well-spaced aspen, gray birch, and sumac. Poverty grass, haircap moss, and many lichens as well as many earth star (*Geaster*) puffballs lie prominantly about on the exposed shale. Surrounding the recently disturbed areas are groves of white oak, white pine, scotch pine, hawthorn, and shad. White pine and red maple dominate the taller woods.

There is a good deal of open land, pastured until recently, and second-growth woodlands in this section of the preserve. The preserve is so large that a detailed description of all of its representative habitats cannot be presented here. A very long list of birds has been recorded within the 370-acre preserve.

There are no formal trails throughout the preserve, however, hikes in any direction from the entrance should yield a good abundance and variety of birds.

Leave I-87 (the Adirondack Northway), at Exit 14. Proceed east on NY 9P for 0.5 mile. Turn left (north) on Gilbert Road and follow it a short distance to its intersection with NY 29. Turn right (east) and follow NY 29 for approximately 7.0 miles to its junction with US 4 in Schuylerville. At US 4 (NY 32 here), turn left and proceed north for approximately 2.0 miles. Here US 4 turns right and crosses the

Hudson River. Bypass the River Road turnoff on the left after crossing the Hudson, and continue east and north on US 4. As previously mentioned, US 4 roughly bisects the sanctuary. Continue driving and be watchful for a dirt road, coming in on the right, which leads to a shale pit from which shale was mined until 1972. Turn right on this dirt road and park without blocking the access to US 4. Once parked, walk either east or west of US 4, throughout the preserve until time or inclination sends you back to your car.

SARATOGA COUNTY
Vischer Ferry Game Management Area (Map 8.5)

Rating: Spring****, Summer***, Autumn***, Winter*

The Mohawk River has its headwaters in Lewis County. From there it flows south, east, and southeast, curving and twisting until it enters the Hudson River at Cohoes, about nine miles north of Albany. Cohoes Falls, in the northeast part of the city of Cohoes, is where one can see the Mohawk River drop seventy feet. Further, it is the site of a glacial pothole where the complete skeleton of an American Mastodon (*Mammut americanum*) was discovered in the early 1880s. This browsing mammal is a relative of the Wooly Mammoth (*Mammuthus primigenius*), which is sometimes pictured as an elephant wearing a wool overcoat. The Hudson and Mohawk River valleys were primary migration routes for moose-elk, mastodon, caribou herds, etc. It is easy to imagine a mammal as heavy as a mastodon sinking in an isolated bog or thick glacial deposit of clay or even dying above the falls and being washed into a sinkhole destined to be worn through at the foot of a waterfall.

The Mohawk River is the longest and largest of the tributaries of the Hudson River, totaling 175 miles in length. The Mohawk Valley is renowned for its scenic beauty and for the fertility of its soil. The valley was home to the easternmost tribe of the Iroquois League (or Six Nations); the Mohawk Indians. Reputedly they were the fiercest in combat and the most skilled warriors of all the tribes in the Iroquois League. In the Mohawk Valley, Tories as well as patriots, allies and enemies of the British, lived during the American Revolutionary War.

Vischer Ferry Game Management Area lies along the historic and scenic north shore of the Mohawk River, north and a little bit west of Albany. Along its opposite shore are extensive marshes, traditionally known as the Niskayuna Wide Waters (see the first entry under Schenectady County).

The Vischer Ferry area and the Niskayuna marshes should be visited on the same day. Among the nesting species and migrants recorded are Great Blue Herons and Great Egrets in late summer; Least and American bitterns nesting regularly; breeding Wood Ducks; and in spring and autumn, migrating American Wigeon. Mallards and Black Ducks nest in both places, while Pintails and Blue-winged Teals are recorded only as migrants. Hooded Mergansers can be found nesting on the Vischer Ferry side. Several species of raptors are common in the area in any and every season. Regular breeding species also include Virginia Rail, Common Gallinule, and American Coot. Gulls, terns, and Belted Kingfishers can be found hunting widely over the swamps and shallow ponds. The assemblages of

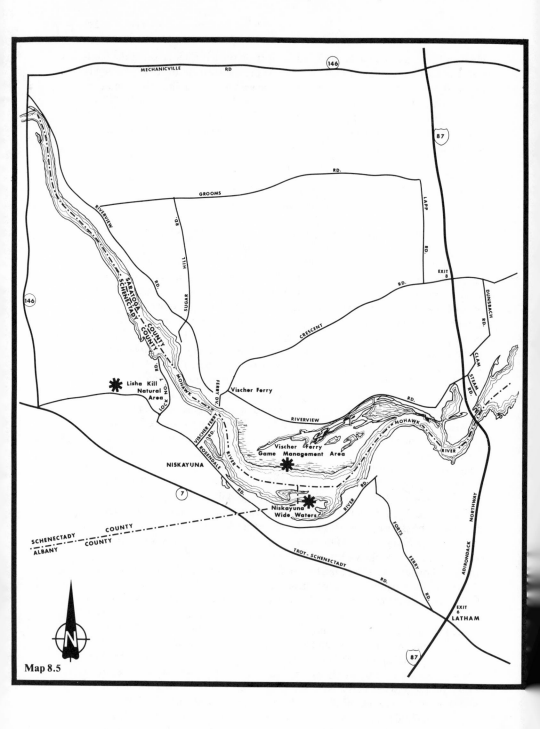

MECHANICVILLE RD.
146
87
RD.
GROOMS
RIVERVIEW
RD.
RD.
LAPP
146
SUGAR
11TH
RD.
SARATOGA COUNTY
SCHENECTADY COUNTY
EXIT
8
RD.
CRESCENT
DUNSBACH
RD.
CLAM STEAM RD.
Lisha Kill
Natural
Area
MOHAWK
LOCK NO. 7
RD.
FERRY DR.
Vischer Ferry
RIVERVIEW
RD.
MOHAWK
VISCHER FERRY
ROSENDALE
RD.
RIVER
Vischer Ferry
Game Management Area
RIVER
NISKAYUNA
7
Niskayuna
Wide Waters
RD.
FORTS
FERRY
ADIRONDACK
NORTHWAY
SCHENECTADY COUNTY
ALBANY COUNTY
TROY - SCHENECTADY
RD.
RD.
EXIT
6
LATHAM
87

N

Map 8.5

Tree and Bank swallows in late-summer and early autumn are absolutely spectacular. Long-billed and Short-billed marsh wrens nest on both sides of the Mohawk in the described areas, as do legions of Red-winged Blackbirds and a miscellany of sparrow species.

To reach the Vischer Ferry Game Management Area, leave I-87 (the Adirondack Northway), at Exit 8. Immediately gain Crescent Road headed east and take it less than 1 mile to Dunsbach Road. Turn right (south) here and follow Dunsbach Road to its end at Clam Steam Road. Turn right (west) and follow Clam Steam Road south and west to its junction with Riverview Road. A sharp right turn here will take you over I-87, well south of Exit 8. Stay on Riverview Road, proceeding west and roughly parallel to the Mohawk River.

Continue very slowly west approximately 1.6 miles to the entrance of the game management area, on the left. Here one may park. Stop to investigate all or any of the ponds or marshy stretches that appear interesting along the road to the entrance. This will take time. Some days it has taken me what seemed *hours* to get as far as the entrance.

It is a good idea to walk all along the dirt roads and wet grassy paths leading toward the river. Work your way along the old Erie Canal, which can be so productive in places that you may be standing quite still just looking for twenty minutes or more at a time. But try to move on. There is still plenty to see.

Back in the car (of if you would rather hike), proceed west on Riverview Road another 2 miles to Ferry Drive, which is just west of the firehouse station. Turn left and proceed to the river, where the drive dead-ends. This affords a nice view of the river. Again, driving west on Riverview Road, go as far as Sugar Hill Road—about another 1.8 miles. Turn left and follow it to its end. This affords a fine birding vantage over the dam, Lock Number 7, on the opposite shore, and Goat Island, upstream.

At this point you can choose one of several options: (a) either continue northwest on Riverview Road (paralleling the river), to its junction with NY 146; there turn right (north) and follow NY 146 north and east back to I-87, meeting it at Exit 9. Or (b) you may simply reverse directions, and follow Riverview Road south and east, passing through the game management area marshes and ponds again to Clam Steam Road, and north on Dunsbach Road, west on Crescent Road to I-87 at Exit 8. Or (c) you may decide to continue straight north on Sugar Hill Road to its junction with Grooms Road. Here turn right (east), and follow Grooms Road through entirely different habitat to Lapp Road. Then take Lapp Road south (right) to Crescent Road. A left turn here will bring you back to I-87 at Exit 8.

SCHENECTADY COUNTY

Niskayuna Wide Waters and Lisha Kill Natural Area (Map 8.5)

Rating: Spring****, Summer***, Autumn***, Winter*

The Niskayuna Wide Waters is just the other side of the coin; these marshes on the south side of the Mohawk River essentially complement those of the Vischer Ferry Game Management Area.

For what birds can be expected along and in these marshes southeast of the city of Schenectady, see those mentioned in the previous description.

To reach Niskayuna Wide Waters, leave I-87 at Exit 6 and head west on NY 7. Proceed a very short distance to Forts Ferry Road, which is the first road on the right, west of I-87. Turn right and follow Forts Ferry Road north for 2 miles to its junction with River Road. Turn left and follow River Road for the next 4 miles.

Employ the same technique of birding as described for the north side of the Mohawk River. There are all sorts of interesting and inviting marshes, ponds, roads, and trails to explore. Obviously, a scope would be extremely helpful, but it is not absolutely necessary on this tour. Continue birding along River Road as it becomes Rosendale Road. At Vischer Ferry Road, be sure to turn right and proceed to a fine overlook of the Mohawk. Lock Number 7 Road, the next road coming in on the right also has a fine overlook. At its end, one has splendid views of the river both above and below the dam. Do not miss it. Back on Rosendale Road, be sure to stay left where it forks. From Lock Number 7 Road to the Niskayuna Firehouse No. 2 and the Grange Hall, both of which are on the left side of Rosendale Road, is approximately 0.9 mile. If you come to the Iroquois Junior Hill School, on the right side of Rosendale Road, you have gone too far and must turn back.

Turn left between the firehouse and the Grange Hall and park behind the firehouse. No vehicles are allowed in the Lisha Kill Natural Area.

The trail to the woods can be found between the firehouse and the Grange Hall. The walking distance around the main trail, including the return to the entrance, is approximately 1.5 miles. Other trails have been cleared for hiking throughout the $110 \pm$ acre sanctuary, and birders are requested not to leave the trails. There are some steep sections, and streams are usually bridged by stepping stones and logs. Of course, be especially careful along the edge of the Fly Kill Cliff within the preserve.

This is unquestionably one of the most satisfying of the Nature Conservancy's preserves, with its wooded highlands, lesser ravines, forty-acre tract of old-growth forest (dominated by mature white pine, hemlock, and oak), various tracts showing progressive stages of forest successional recovery from farmland and pasture, streams, cliffs, wildflowers, ferns, and varied terrain. There has been no major human disturbance to the area for well over 50 years and the old-growth tract retains, in a splendid state of preservation, the natural conditions and flora characteristic of forests found in the Mohawk Valley in colonial and precolonial times.

More than sixty species of birds have been identified in the preserve, as well as several reptiles, Pisciforms, and an ungulate or two, and tree toads and frogs. Listen especially for picids (at least 5 species have been recorded), flycatchers, Brown Creeper, Winter Wren, and the songs of several thrush species. Vireos and warblers (approximately 10 species) nest in the quiet of these forest types and be sure to look lively for the odd finch and also for the unexpected sparrow species.

ALBANY COUNTY
The Five Rivers Environmental Education Center (Map 8.6)

Rating: Spring****, Summer***, Autumn***, Winter*

This environmental education center is located just outside of the village of Delmar, which is only a short distance southwest of Albany. The types of habitat on the 260 acres are diverse, and birders will find several miles of trails meandering through the property. With the varied habitats there is correspondingly diverse birdlife, as one might expect.

The center is operated by the New York State Department of Environmental Conservation. The grounds are open daily all year, and an Interpretive Building and exhibit room are open weekdays from 9:00 a.m. to 4:30 p.m. A current list of birds sighted on the property is kept in the Interpretive Building. There is neither an admission charge nor parking fee at Five Rivers.

The property can be easily birded from a number of trails that are cleared and maintained year-round. There is some poison ivy, but if one stays on the trails, there is virtually no danger. Wild mammals abound on the grounds and birders should keep a sharp eye out for the bonus white-tailed deer, red fox, mink, raccoon, flying squirrel, or muskrat they may sight.

To reach the center, leave I-87 at Exit 23 and head south on US 9W. Proceed south on US 9W to its intersection with NY 32. Turn right and follow NY 32 to Elm Avenue. Here turn right and proceed a short distance to its intersection with Delaware Avenue (NY 443). Bethlehem High School will be on your left. Turn left and proceed 1 mile to Orchard Street and turn right. Proceed to Game Farm Road, the first available left turn. The entrance to the center will be visible within 0.25 mile.

After checking the bird list inside the main building, choose among a number of trails to bird. The Beaver Tree Trail and the Woodlot Trail are both relatively short, and the trailhead for each lies just outside the center entrance, on opposite sides of Game Farm Road. The Old Field Trail passes through abandoned fields of grasses and goldenrods, and one can see young gray-stemmed dogwood, willow, conifers, and wild rose thriving. Additionally, there is an abandoned apple orchard, which illustrates various stages of plant succession along the Old Field Trail.

The Vlomankill Trail follows the stream as it courses through a hemlock-shaded ravine. Although this may not be the most productive trail in terms of numbers, the Louisiana Waterthrush nests along the banks of the ravine. If one did not encounter a single other species, but did hear the high, wild ringing of the Louisiana Waterthrush's musical song, the walk would have been well worth it.

The North Loop Trail proceeds due north, swings west and south and then veers southeast, passing for 2 miles through fields and forests around the rim of the center's property. This is a very productive trail and samples a good deal of the varied and abundant plantlife within the center.

Upwards of fifty species breed at Five Rivers. Around any of the five ponds, especially those three having marshy cattails and rushes, look for Great Blue Heron, which is a spring and autumn migrant. Be watchful for nesting Green Herons and Canada Geese. In the ponds surrounded by stands of willow, speckled alder, and mixed deciduous forests, look for Wood Ducks, Mallards, Black Ducks, migrant Blue-winged Teals, and an occasional Osprey passing through. In winter scan the old fields and swampy edges of ponds for Red-tailed and Rough-legged hawks, and of course, American Kestrels. Ruffed Grouse, Killdeers, American Woodcocks, and Spotted Sandpipers all nest on the grounds. Search the lawns, which have scattered groupings of Norway spruce, for the cooing Mourning Dove brooding its young. Screech and Great Horned owls are permanent residents in the pine-hemlock forests. In late summer look overhead for kettling Chimney Swifts. There are plenty of good vantage points around various ponds from which you can watch Belted Kingfishers hunting all spring and summer, providing ample food for their young.

House Wren, Mockingbird, American Robin, Wood Thrush and kinglets can be heard vociferously tuning up for the breeding season as spring days grow longer. In autumn, listen for the high-pitched buzzing of the Cedar Waxwing migrating through. In winter, be alert for an occasional Northern Shrike sitting quite silently on an almost bare tree limb.

Many vireo species pass through on spring migration; but, the only one to stay and nest is the Red-eyed. A profusion of more than 25 warbler species can be found in spring in the spruces, flitting along the brushy edges, and in the mixed forests of beech, birch, maple, oak, and hickory. Watch especially for Yellow and Chestnut-sided warblers, Common Yellowthroat, and American Redstart, which all stay to nest.

Search the abandoned open fields for breeding Bobolink and Eastern Meadowlark and the marshy edges of pools for Red-winged Blackbird. Toward mid-May listen for the unmistakable song of the nesting Northern Oriole.

One of the most notable bird species on the center's long list is Henslow's Sparrow. Furthermore it breeds in the area of the old fields. Dark-eyed Junco and Tree Sparrow are strictly winterers; however, Chipping and Field sparrows nest. White-throated Sparrow can usually be found in late autumn and winter, while Song Sparrow is one of the hardy permanent residents of The Five Rivers Environmental Education Center.

The Albany Pine Barrens (Pine Bush) (Map 8.6)

Rating: Spring****, Summer***, Autumn***, Winter*

Originally, the Albany Pine Barrens was a broad expanse of sandy dune formations, thickly covered with low scrub oak and pitch pine forests, stretching from Saratoga on the north to Kinderhook, in Columbia County, on the south and west along the Mohawk River, to (and perhaps past) Schenectady.

Today, owing to the ravages of "progress," the pine barrens, so similar to the same physiographic and biologic provinces found on Long Island, at Cape

Cod in Massachusetts, and in the southern counties of New Jersey, are limited to a mere 4000 acres, lying between Colonie and Guiderland and stretching from the northwest section of the city of Albany westward.

In truth, "Barrens" is a complete misnomer. The name hardly reflects the botanically rich composition of the area. Contrasting with the dune areas, streams course through wet ravines surrounded by ridges of dwarf oaks, Indian grass, wild orchids, jersey tea, sweetfern, sandberry, huckleberry, juneberry, blueberry, and ferns of several species. Obviously, the settlers of the mid-1800s considered the area a wasteland, and exploited it mercilessly. But the wooded swamps and forest understory support a growth of well over 100 species of herbaceous plants.

The strange beauty characterized by the unusually stunted and scrubby growth is, in every sense, beyond duplication. Although, for much of its history it has been ignored, repeatedly parcelled and badly manipulated; some 2000 of the remaining barrens are in what, conservationists consider, salvageable tracts. There are several regional plans abroad which emphasize preservation of relatively large blocks of the Pine Bush. Yet, tangible results seem largely dependent upon materialization of state, local, private, and federal funding.

Relatively little concentrated ornithological field work has been conducted in the Albany Pine Barrens, although it is much needed in so unique an environment. Known breeders include: Northern Harrier, Blue Jay, Mockingbird, Gray Catbird, Brown Thrasher, Wood Thrush, Yellow, Pine and Prairie warblers, Ovenbird, Common Yellowthroat, Rufous-sided Towhee, and Chipping and Field sparrows. More than thirty species of birds use the pine barrens in which to feed, roost, hunt, etc. A quiet walk through the area in mid-spring would certainly reveal itself as a treasure to the naturalist and student of plant succession, as well as to the birder.

A further reason for defending the pine barrens from more indiscriminate development is the preservation of the tiny blue butterfly, the Karner Blue (*Lycaeides melissa samuelis*), sometimes referred to as the Melissa Blue. When the Pine Barrens were much more extensive, lepidopterists noted that numbers of this species seemed "uncountable" around Albany. Today, with luck and diligent searching, perhaps one or two per day can be found in the sandy Pine Bush. The Karner Blue can survive only in specialized environments. Its life cycle is intimately tied to profusions of wild Blue Lupin (*Lupinus perennis*), for it is the only plant its caterpillar normally eats.

Interestingly, the Karner Blue is double brooded in the Albany Pine Barrens; adults appear in late May-early June and then again in late July-early August. This is not the only butterfly species that breeds in the barrens; elfins and skippers also share the breeding grounds of the Karner Blue. The Xerxes Society, an organization dedicated to the preservation of butterflies in North America, estimates that there may be as few as twelve to fifteen colonies of the Karner Blue in the entire Northeast. It would be a rare treat to come across one or more of these creatures while birding in the Pine Barrens. The Karner Blue has a bluish hue margined with orange on the upper surface of its wings and the underside of the hindwing has prominent sparkling orange spots.

To reach the Albany Pine Barrens, leave I-87 (the Adirondack Northway) at

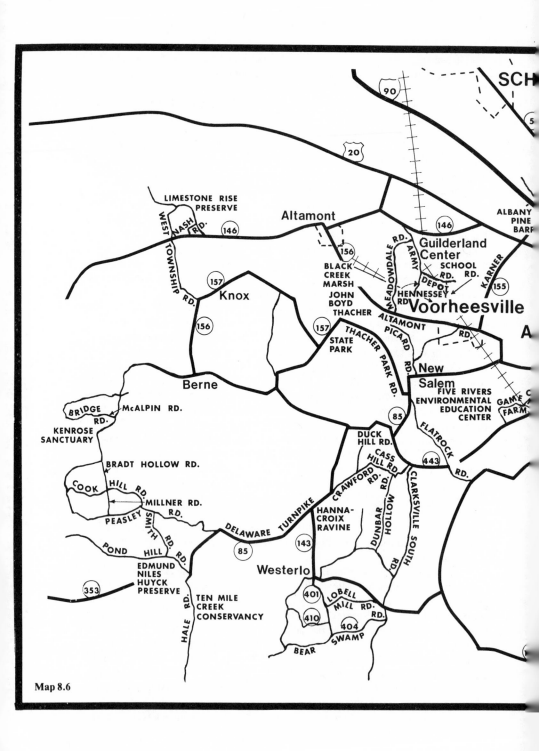

Map 8.6

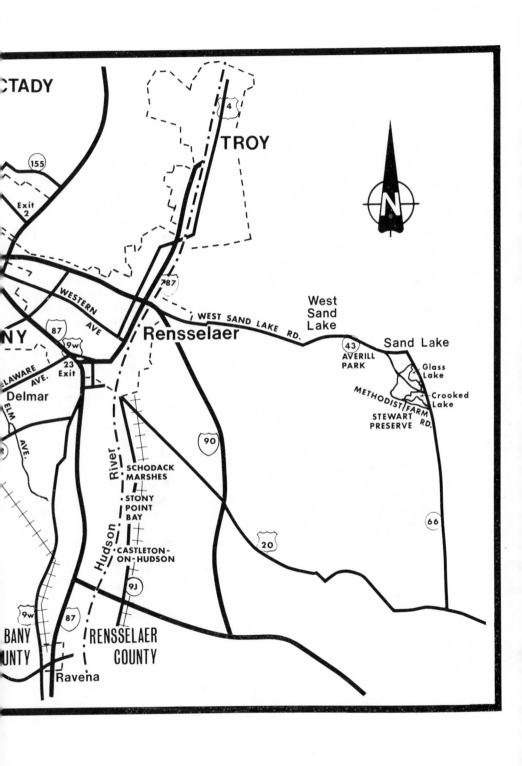

Exit 2 and head west on NY 5. Proceed to the junction of NY 155 (Karner Road). Turn left and proceed south to US 20, turning off Karner Road at any of the roads coming into it from the right or left. From these winding side roads, the barrens can be birded.

Black Creek Marsh, John Boyd Thacher State Park, Limestone Rise Preserve, a Tour
(Map 8.6)

Rating: Spring****, Summer***, Autumn**, Winter*

Black Creek Marsh lies about 2 miles west of the village of Voorheesville and 6 miles southwest of the city of Albany. It encompasses approximately 700 acres of both private and publicly held land, although gaining permission to bird the private areas is not required.

Birding here is rated as "moderately easy," although the area cannot be covered by car. One must go on foot to find the avian riches of Black Creek Marsh.

The Hudson-Mohawk Bird Club conscientiously monitors the marsh and reports that it is the most productive site in *Kingbird* Reporting Region 8 for Virginia Rail and Sora.

A visit to the area in spring should yield Green Heron, Least and American bitterns, several duck species, Osprey, Red-shouldered, Red-tailed, and Marsh hawks, Ruffed Grouse, Ring-necked Pheasant, American Woodcock, the *Rallidae* mentioned above, Common Gallinule, Common Snipe, Pileated Woodpecker, Willow Flycatcher, swallows, Long-billed and Short-billed marsh wrens, Eastern Bluebird, Blue-gray Gnatcatcher and Swamp Sparrow — every one of these species nests in and around the marsh.

If the water level is low, there may be additional migrant shorebirds of many species and in high numbers probing and feeding in the muddy water.

Leave I-90 at Exit 24. Follow signs for US 20 (Western Avenue). Turn right (west) on US 20 and proceed about 5 miles to the junction of NY 146. Here turn left and follow NY 146 west to the hamlet of Guilderland Center. At the junction of NY 202 (immediately past the traffic light) turn left. At the next available left turn, which will be Army Depot Road (NY 201), turn left. Follow Army Depot Road (NY 201) for approximately 1.0 mile to Hennessey Road, which comes in on the right. Turn right and proceed to the top of the first hill. This has been a fine place to view migrating hawks. After scanning for hawks, proceed south on Hennessey Road to the railroad tracks. Park the car here.

Walk east (left) on the railroad tracks above the marsh for about a mile, past thickets, open ponds and pastureland. This should be an extremely rewarding hike. At School Road (NY 208), retrace your steps and work your way back to Hennessey Road.

If you have sufficient time before leaving the area you might continue walking west on the tracks a short distance to Meadowdale Road. This is not as bird-rich as the walk to the east; however, the area tends to be more boggy and could (and has) produced species not encountered east of Hennessey Road.

John Boyd Thacher State Park and Limestone Rise Preserve have been in-cluded in this tour not only because they are close to the Black Creek Marsh, but also because to bird the three areas in one day in late spring would be a memorable experience. J. B. Thacher State Park comprises over 1130 acres of some of the most spectacular scenery and heavily forested cliffs along the rim of the Helder-berg Escarpment. Along the exposed rim, the elevation is approximately 700 feet and there are panoramic views in every direction.

The variety of breeding birds within the park is essentially what makes it a birder's delight. At the park's lowest elevations near a small stream known as In-dian Ladder Creek, Least and American bitterns nest along with Common Snipes and marsh wrens, in the nearly impenetrable, overgrown marsh. In the more open woods, also near the creek, Alder Flycatchers and Louisiana Waterthrushes breed and can be heard singing their distinctive songs as one hikes to the higher elevations.

Red-shouldered and Red-tailed hawks nest throughout the park in their pre-ferred habitats. The Red-shouldered in low, watered woods, where it builds its nest thirty to sixty feet up, and the Red-tailed in higher, more dense stands of trees, where it builds its nest forty to seventy feet up. The Ruffed Grouse nests through-out the park. Its nest is a leaf-lined depression at the base of a stump or tree, some-times near a water source. The nest oftentimes holds up to 14 pale-buff eggs. Listen for the measured thumping as the male grouse drums its amorous message to the female while warning other males off its territory. It is certainly one of the most pleasurable sounds of the deep forest. The Pileated Woodpecker, the log-cock, nests throughout the park, unrestricted by elevation.

In the dense stands of hardwoods and the shady areas containing hemlocks, the woods reverberate in spring with the sounds of nesting singing warblers. At lower elevations breeding birds include Black-and-white, Blue-winged, Golden-winged and Worm-eating warblers. Farther upslope, listen for breeding Brown Creepers, Winter Wrens, Hermit and Gray-cheeked thrushes, kinglets, and Black-throated Green, Blackburnian, and possibly Chestnut-sided warblers, as well as Dark-eyed Juncos and White-throated Sparrows.

The list of breeding birds within the park is certainly not limited to those few named; however, the numbers and species one hears and sees is entirely depen-dent upon the extent of roaming one is willing to do and the amount of climb-ing one decides to undertake. It is never a strenuous climb, even on the steepest ascents.

From the last place you have parked the car (Hennessey Road just north of the railroad tracks, prior to the walk over the Black Creek Marsh); drive south on Hennessey Road, across the railroad tracks, and a short distance to NY 156 (here also Altamont Road). Turn right and proceed past Meadowdale Road, which comes in on the right. The next road on the left is Picard Road. Here turn left and follow it south. Continue south on it as it becomes NY 85A, and proceed to the junction of NY 85 and the hamlet of New Salem. Turn right and stay on NY 85 as it goes slightly west and then turns south. The next road coming in on the right will be Thacher Park Road (NY 157). Here make a sharp right. Follow NY 157 as it makes another sharp right, and proceed along it to John Boyd Thacher State Park. Old Indian Ladder Trail, inside the park, is an excellent trail to hike and sample the birdlife at various elevations at this point along the Helderberg Escarpment.

Birding Limestone Rise Preserve should provide the perfect ending for one of the most perfect birding days to be spent in the Helderberg Escarpment region. The name refers to the limestone ledge outcroppings found in the southern section of the preserve. This Manlius Limestone belongs to the same formation as the rocky cliffs at John Boyd Thacher State Park. Approximately 100 feet east of the trail, south of NY 146, is a ledge which is an excellent place for the interested visitor to locate rocks to study, and in which can be found characteristic fossils of the local Manlius Limestone exposure. Do study them but, *do not* remove them, as fossils are decidedly a non-renewable resource. If everyone visiting the preserve took only one fossil away, the area would be entirely depleted within a remarkably short time. There is another excellent exposure for studying fossils approximately 375 feet west of this site. Three types are especially abundant here: *Tentaculites* (descriptively called "carrots"), an extinct type of marine animal, represented by a small cone-shaped shell with encircling ridged rings 0.25–0.50 inch long; *Hermannina* (formerly *Leperditia*) shells, a small crustacean, appearing like a grain of puffed wheat; and *Howellella* (formerly *Spirifer*), a small brachiopod.

The parallel arrangement of mud cracks, mud pebbles, shells, and shell impressions (including those of the *Tentaculites*) and the faint ripple marks in the rock, all indicate that there were tidal flat conditions during the formation of Limestone Rise in the Devonian period, about 350 million years ago. Exercise caution while walking through the preserve. Many years of erosion have widened the joints and cracks in the rock. Cracks up to eighteen inches wide and more than six feet deep have been observed. In some places the underlying rock has been dissolved more than the surface rock and as a result the surface rock has collapsed. This forms small "sinks" or large "sink holes." All of the phenomena associated with limestone and its complicated breakdown may be present, including underground caves and streams.

Plant life at Limestone Rise is rich with rocks covered with mosses and an almost solid groundcover of herbaceous plants. Barren strawberry and acute-lobed hepatica form a blanket covering wide areas, interspersed with small islands of other plants. In the southwest part of the preserve there is a huge covering of white trillium. There are at least seven fern species found here, including the limestone-loving walking fern.

The forest canopy consists mostly of sugar maples and hop hornbeam, with an understory of witch hazel and striped maple. It is a real challenge both to keep your eye on where you are walking and to search the canopy for the abundant birdlife here. Hawks nest at Limestone Rise, so watch for them. Be sure to check out the swamp across NY 146 and the open fields and woods beyond the swamp, for their varied avifauna. These two additional ecosystems constitute the northern section of the preserve and lie north of NY 146. Birdlife is most abundant in this part of the preserve; breeders include ducks, Ruffed Grouse, American Woodcock, vireo species, and several warbler species.

From Thacher State Park and NY 157, proceed northwest on NY 157 to its junction with NY 156. Turn left and follow NY 156 west, through the hamlet of Knox, and to the point where it makes a sharp left (south) turn. *Do not* turn here toward Berne; rather, stay on Knox Road heading west. A short distance farther, West Township Road comes in from the right. Here turn right (north) and follow West Township Road to its junction with NY 146. Turn right (east) and watch on the right (south) side of the road for signs indicating the property of the Nature Conservancy. Soon you will come to the entrance, which can be gained by turning left (north) off NY 146 onto an old woods road, less than 1 mile from the turn onto NY 146. It is marked, you can't miss it. The name of the road is Nash Road, although it may not be marked with its name. Proceed north on Nash Road approximately 680 feet, to the preserve trailhead. Park here and enter at this point.

Hannacroix Ravine, Bear Swamp, Ten Mile Creek Conservancy, The Edmund Niles Huyck Preserve, Kenrose Sanctuary, an All-day Birding Tour
(Maps 8.6 and 8.7 and 8.8 and 8.9)

Rating: Spring****, Summer***, Autumn**, Winter*

To complete this tour, set aside a full day, get an early start, pack a gigantic lunch, take along a couple of cold beers, and don't plan to get home until late. The ideal time to do this tour is immediately after the height of the spring migration when there is still a lot of full song and defining and defending of territories is still going strong. This tour should be extremely rewarding right through the close of the first week in June. From then on, the tour becomes less productive. If the weather prediction includes rain or high winds, postpone the trip. Ideally, this trip should be done on a relatively still day, and if it is slightly damp and somewhat overcast that's perfectly okay. Be sure to start out near sunrise so that you will be hitting the second spot, Bear Swamp, early enough in the day to bird bitterns, rails, and marsh-wrens.

To reach Hannacroix Ravine, leave I-87 at Exit 23, and head south on US 9W. Proceed on US 9W to its intersection with NY 32. Turn right and follow NY 32 through the hamlets of Bethlehem Center and farther southwest, Feura Bush. Continue right on to Flatrock Road, which comes in on the right. Here, turn right and proceed a short distance to NY 443. Turn left and into the hamlet of Clarksville. You will come to the junction of NY 443 (which turns right and proceeds north), NY 312 (also called Clarksville South Road, which turns left and proceeds south), and Cass Hill Road (which bears straight ahead, more or less, and continues west). Gain Cass Hill Road and proceed 2.2 miles past Dunbar Hollow Road on the left, and past Duck Hill Road on the right. Cass Hill Road passes over the Hannacroix Creek, and there is an unmarked trail along the east edge of the ravine (for the hardy only) just east of the creek. From there, one simply walks south following a rough and unworn footpath to a marvelous waterfall. However, an alternative is to continue approximately 200 feet west of the creek, to a parking pull-off on the left (south) side of the road, near the entrance gate. There one can leave the car and proceed south on an abandoned secondary road (running south from Cass Hill Road), on which the walking is considerably easier.

The Hannacroix Ravine is a holding of the Nature Conservancy, and it in-

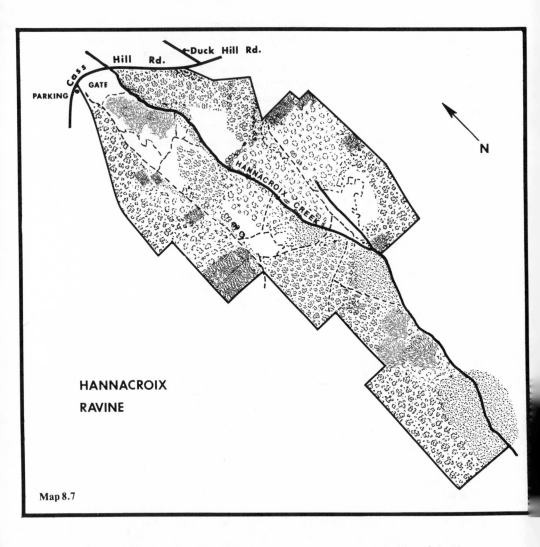

Map 8.7

cludes more than 320 acres stretching downstream along both sides of the Hanna-croix Creek. Cass Hill Road used to be the route of the Albany-Schoharie Turn-pike in colonial times, when stagecoach was the primary means of getting from one place to another. Stagecoaches stopped at Reidsville (just southwest of the preserve), which was then a thriving center for "slate" (actually sandstone) quar-ries. This is the origin of the "slate" used to pave the old sidewalks of Albany.

   Hannacroix Creek flows southeast to Alcove Reservoir and ultimately to the Hudson River. It drains an area well-covered with glacial drift. A deep ravine has, however, been cut through the shale and sandstone here, which varies in depth to a maximum of about eighty feet. The gorge extends for nearly three miles. All along, there are many first-rate waterfalls. It is a real tonic to watch the stream tumble over the resistant sandstone. Along the creek old hemlocks tower above the beech-maple-birch-hickory forest. Along the cliffs a fern-phreek can find polypody, shield fern, Christmas fern, and sensitive fern, along with mosses and liverworts. It is a haven for those birders who also favor wildflowers.

The uplands support a mixed conifer-hardwood community of varying ages. That is, there are white and red pine plantations approximately 35 years old, and some of the upland that was once tilled and pastured more than a century ago has mostly reverted to forest.

The "official" trail is the one described running south from the entrance gate. The other paths are not really regularly cut and cleared trails. What could be more marvelous for the adventuresome? Follow the old lumber road paralleling the abandoned secondary road, on its east side. Walk along the grass and juniper covered ridges. Look for birds near the old foundations, in the fields, throughout the evergreen plantations, and even near the shale pits. There is the possibility of simply following the stream bed during the dry season, or walking the length of the stream, on the east side of the preserve, in spring.

The birdlife is abundant within the Hannacroix Ravine. It is of a rather specialized nature; but, those experiencing the musical songs of the Winter Wren, boreal breeding warblers, and finches with more northern affinities, will certainly not be disappointed.

Now move on to Bear Swamp.

Bear Swamp is a veritable living museum extending over more than 280 acres, and featuring two wooded swamps, a quaking bog, one of the most northern relict outposts for Giant Rhododendron, ferns in considerable variety and abundance, giant pine stands, an enormous breeding bird population — including nesting Goshawks, Red-shouldered Hawks, Pileated Woodpecker, several vireo species, and nineteen warbler species, as well as numerous members of the Icteridae and Fringillidae.

There is so much to see and hear at Bear Swamp that you should make sure that you arrive there early and apportion enough time to bird it well. Do overcome the psychological barrier of rising early and be rewarded by seeing and hearing many brightly colored wood-warbler species.

Owing to the possible threat posed by the United States Air Force acquiring a large part of the Great Bear Swamp for an electronic installation, about 1957, the Nature Conservancy frantically accelerated its negotiations to purchase the area. The swamp is now a Registered Natural Landmark of the Department of the Interior.

From the entrance to the Hannacroix Ravine preserve on Cass Hill Road, continue driving southwest as Cass Hill Road becomes Crawford Road and then becomes the Delaware Turnpike (NY 85). Watch for the junction of NY 143, which comes in on the left. Turn left at NY 143 and proceed south to the hamlet of Westerlo. In Westerlo proceed left around a traffic circle and then immediately turn right (south) on NY 401. Continue south past Lobell Mill Road, which comes in on the left, and past NY 410, which comes in from the right. The next road crossing NY 401 will be Bear Swamp Road (NY 404). Here, turn right.

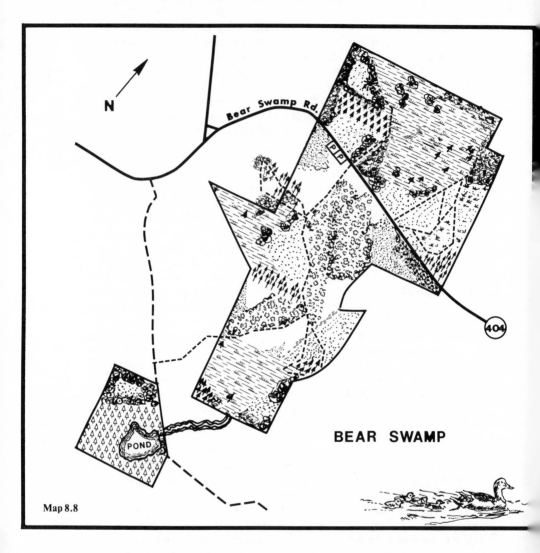

N

Bear Swamp Rd.

PIP

404

BEAR SWAMP

POND

Map 8.8

Bear Swamp Road was once a lumbering road, and it now divides the swamp into Little Bear Swamp, on the north side of the road, and Great Bear Swamp, south of the road. Watch for the parking facility on the south (left) side of the road. Pull in here and look for the old woods path heading north to a clearing under a stand of large pines. The end of the path is near the bank overlooking Little Bear Swamp. Here there is a memorial plaque fastened to a boulder, commemorating ten local conservationists instrumental in preserving the swamp. The beginning of this path marks the entrance to the preserve.

South of Great Bear Swamp there is a separate pond (Source Pond) approximately 0.75 mile from it, which feeds into the swamp. Surrounding both sections of the swamp and pond there are woods of red maple, yellow birch, black ash, and elm; there are white pines, hemlocks, and Balm of Gilead on the less moist sites. Source Pond is reached *via* an old dirt road indicated on the map.

In and around the bog there are pitcher plant, sundew, labrador tea, winterberry, withe rod, high bush blueberry, and other plants especially adapted to the acidic, nutrient-poor and cool conditions of bogs.

The plants of the bog are growing on approximately 20 + feet of peat which is certainly indicative of the depth of the pond left by the receding glacier some 13,000 years ago. There were many such stagnant ponds left at the end of the period, but most of them ultimately filled with plant matter that did not decay.

The pride of this swamp is its remarkable rhododendrons. Hundreds of Giant Rhododendron (*Rhododendron maximum*) plants bloom here from around June 25 to July 15. You may know this primarily southern bog plant, by one of its two other common names: Great Laurel or Rosebay. Bear Swamp is one of the northernmost relict locations for this shrub, and is therefore all the more valued as a constituent of the flora. It is supposed that this amazing stand is maintaining itself owing to sufficient shade and a genuinely stable groundwater level.

Walk the remnants of a corduroy road, originally built to keep horses from sinking into the bog. This is part of the trail system of the preserve. Some of the most prized fern species can be found flourishing along it – *e.g.,* the three members of the flowering fern family: royal fern, interrupted fern, and cinnamon fern.

Red and scotch pines have been planted here, to contrast with the native white pines. Look for nesting warblers in and around all of the pine stands.

Bears? Is that the origin of the name? Well, there was actually a bear sighted here twenty-five years ago, but rest assured, snow-shoe hare and white-tailed deer are decidedly more common in the swamp.

Now on to the Ten Mile Creek Conservancy.

After birding Bear Swamp, retrace the same route in, gaining NY 401 going north and follow it to NY 143, also heading north. At the junction of NY 143 and NY 85 (Delaware Turnpike), turn left (west) and proceed west on NY 85 to the village of Rensselaerville. Turn right on Main Street. Note the mileage on the odometer at this point. Ten Mile Creek Conservancy is 1.2 miles from this point. Continue driving west. NY 85 ends soon, at the bridge crossing Ten Mile Creek. Continue on Main Street, which becomes NY 353 just outside the village. Look for Hale Road, which soon comes in on the left. Turn left and proceed south on Hale Road until the odometer has clocked 1.2 miles. Park safely on the road shoulder.

Cross to the east side of Hale Road. The 110 + acres of the Ten Mile Creek Conservancy include overgrown former farm fields near the road, and wooded areas on the west slope of Ten Mile Creek. From the road, the view embraces the valley of the Ten Mile Creek, with alternating forest and cultivated fields, and, to the south, three rather impressive peaks of the Catskill Mountains.

The field on the east side of the road seems almost impenetrable, filled with blackberries, grapes, sumac, apple, fire cherry, juniper, young maples, ashes, and birches. It is also a paradise for wildlife. Be sure to do a little birding here from the outside. Calling or "spishing" should bring some birds into view. Now walk south along the sturdy stone wall until you sight an opening. There are various entrance-ways and paths that meet the wall.

Once inside, walk east through the brush to the mature forest on its other (eastern) side. Here, on the slope above and near the stream, the forest floor is car-peted with blue cohosh and leeks. Here and there are Jack-in-the-pulpit, wild ginger, red trillium, waterleaf, toothwort, baneberry, and hepatica. There is an un-derstory of witch hazel and striped maple. The taller trees are beech, hemlock, and sugar maple.

At the bottom of the slope is the stream responsible for carving the ravine. It is a year-round trout stream. Birding along the stream is allowed at any time. Downstream, approximately 1.75 miles, the stream is open to public fishing. How-ever, here upstream, birders hiking streamside should not run into any fisherman. The birds one might expect to encounter are those species typical of thick brush-land, overgrown farm fields, and mature forest and streamside.

After birding the Ten Mile Creek Conservancy, reverse your direction on Hale Road, so that you are facing north. Drive to the junction of Hale Road and NY 353. Turn right and proceed to the bridge crossing Ten Mile Creek. Park in the vicinity of the bridge being careful to not block access to the road.

The Edmund Niles Huyck Preserve and Biological Research Station encom-passes 1400 acres, including most of the land on either side of the creek above the bridge. From the bridge, follow either of the trails located on opposite sides of the creek and walk upstream (west) for approximately 100 yards to Rensselaerville Falls. Here, Ten Mile Creek cascades over sedimentary rock formations of Devo-nian age, resulting in a beautiful waterfall that has cut a deep ravine.

From here, take the trail that continues from the foundation of an old mill on the south side of the creek up through the ravine. It then passes through stands of mixed hardwoods, most of which are of second-growth, then through spruce plantations about fifty years old. The trail then goes through old fields to the southern end of Myosotis Lake.

The trails throughout the preserve range from easy to moderately difficult to hike. Bridges interspersed throughout the grounds make stream-crossings sim-ple. The trail from the falls to the southern end of Myosotis Lake has a half-mile section over some fairly steep inclines, so take this into consideration before at-tempting it.

Lake Myosotis was impounded in approximately 1800 to provide water for the townspeople, and to power the mills of the village of Rensselaerville, which was then a prosperous industrial center, with shops, mills, and flourishing farms on the surrounding countryside. The 100-acre lake remains the source of water for the village dwellers.

Trails go around Myosotis Lake and continue north to Lincoln Pond. There is a trail system around both sides of Lincoln Pond, meeting at its north end and from there continuing north for some distance.

Lincoln Pond covering about ten acres, was impounded to provide power for a small sawmill and farm at the southern end of the pond; the farm's cottage and barn have since become the center for the Biological Research Station. Biological research at the preserve has been supported continuously since 1938, although the preserve is independent of other educational or research institutions. Graduate and postgraduate students of the natural sciences here pursue their research projects, most of which are pertinent to this region, either as year-round or as summer residents. Visitors are requested not to enter any of the laboratories or other buildings around Lincoln Pond and are further asked not to disturb in any way signs, identifying markings, instruments, or equipment set out by students or research scientists.

Myosotis Lake and Lincoln Pond may also be reached by car. About 100 yards east of the bridge over Ten Mile Creek (on Main Street, as described above), Pond Hill Road meets Main Street, coming in on the left (north). Turn left (north) on Pond Hill Road and proceed along it for approximately 0.75 mile. There, by turning left, one can park near the lake and begin hiking the trails from that point. An alternative is to proceed up Pond Hill Road for 1.5 miles to Lincoln Pond, where parking is allowed on the left.

The trail around Lincoln Pond passes through stands of tall hemlocks and skirts dense thickets and an old marsh at the north end of the pond.

Smith Road (Grevatt Road) originates northeast of the boundary of the preserve and meets Pond Hill Road just east of Lincoln Pond. Walking north about 150 yards on Smith (Grevatt) Road, birders will pass part of a fifty-year-old pine plantation and come to the trailhead, on the left side of the road, of another route, which provides access to more of the varied habitats found on the preserve. This trail, in a labyrinthine manner, meanders through pine and spruce stands, natural and reforested mixed deciduous woodlands, and old fields and orchards.

Bobcat, mink, beaver, deer, and (extremely rarely) bear, can be found on the grounds.

In summary, this is a unique and active research station open throughout the year to the public, and a visit to it should substantially enrich the experiences for the birder. The list of migrant and breeding species recorded on the 1400 acres of the preserve is too long to enumerate, but a sampling includes nesting Goshawk, Pileated Woodpecker, Winter Wren, Hermit Thrush, Magnolia, Black-throated Green, and Blackburnian warblers, and Louisiana Waterthrush.

It is a simple matter to reach the Kenrose Sanctuary from the Huyck Preserve and Biological Research Station. When you have finished birding at Huyck, drive north on Smith (Grevatt) Road, which is just east of Lincoln Pond. Follow Smith Road to its junction with Peasley Road. Turn left (west) and continue west on Peasely Road, keeping right where it forks. You will now be on Cook Hill Road. Follow this road west a short distance, to its intersection with Millner Road. Turn right and stay on Millner Road going north as it becomes Bradt Hollow Road. Stay on Bradt Hollow Road as it turns right. As you proceed northeast on Bradt Hollow Road watch for Bridge Road coming in from the left. At Bridge Road turn left and advance about 800 feet, where one can pull off the road and safely park a car on the left.

This large, beautiful preserve consists of 360 acres in various stages of succession. Large sections once devoted to agriculture were last cultivated more than 25 years ago. Uplands once cleared of hemlocks and oaks for tannin, elms for

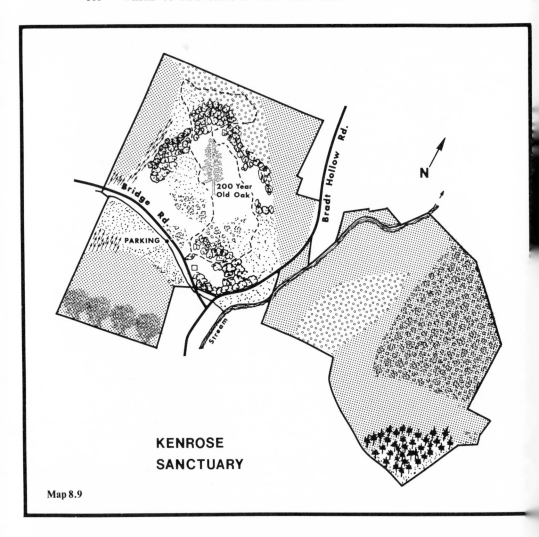

**KENROSE
SANCTUARY**

Map 8.9

charcoal, black cherry and walnut for furniture, white pines for construction of homes, etc., have been replanted with locusts (as early as 1937), white pine, elm, and ash. There is an old beech forest on the property, and, among the stands of oaks, there is one white oak about 200 years old and a red oak about 150 years old. Several ravines enrich the property. One was the site of an old sawmill, whose foundations are still in evidence. The shales and sandstones exposed in the ravines are of Devonian age and contain fossils of marine organisms that lived nearly 350 million years ago.

The entrance to the cleared and maintained trail is easily found. It is on the north side of Bridge Road, just at the eastern edge of a big stand of hardwoods, and just west of the McAlpin house. From the parking site, walk a few hundred feet east to locate it. This preserve was a gift to the Nature Conservancy of Mrs. McAlpin in memory of her husband, Dr. Kenneth McAlpin, who served on the medical faculty of Columbia University for many years before his death. A quantitative and mapping study of the vegetation of the property has been continuing since about 1970, as a baseline against which to measure environmental change. This explains the permanently marked quadrants one may come across while birding. Take care to not disturb any of these staked-off areas.

The birdlife at Kenrose Sanctuary includes a long list of migrants and breeders. The breeding species (about sixty-five) include Red-tailed and Broad-winged hawks and American Kestrel, Ruffed Grouse, Killdeer in the most open areas, American Woodcock, Screech and Barred owls, Whip-poor-will, Ruby-throated Hummingbird, Common Flicker, Pileated Woodpecker, Yellow-bellied Sapsucker, Hairy and Downy woodpeckers, Eastern Kingbird, Eastern Phoebe, Tree and Barn swallows, Black-capped Chickadee, Gray Catbird, Brown Thrasher, several thrush species including Eastern Bluebird, at least two vireo species, and approximately twelve to fifteen warbler species, including Black-throated Blue, Black-throated Green, Blackburnian and Chestnut-sided. Of course, Bobolink, meadowlark, blackbirds and Northern Oriole breed on the property as well as several members of the Fringillidae family, including Indigo Bunting, very possibly Evening Grosbeak, and Vesper, Chipping and Field sparrows.

After parking the car on Bridge Road, one can walk across Bradt Hollow Road along McAlpin Road (east) and north of McAlpin Road into the roughly 100 acres of the preserve located there. Presently this is a trailless section of the sanctuary, but, it is not difficult to walk through the open meadows and hardwood stands.

RENSSELAER COUNTY
Stewart Preserve (Map 8.6)

Rating: Spring****, Summer****, Autumn**, Winter**

Stewart Preserve, located east of the city of Rensselaer, has several unique features that make it well worth a special trip to bird it.

From I-87 (the New York State Thruway), just south of Albany, at Exit 23, get on I-787, which skirts Albany on its east side, and runs north along the Hudson River. Proceed north on I-787 to the Rensselaer ramp of the Parker Dunn Memorial Bridge. Cross the Hudson River here and, from the ramp, gain Third Avenue (NY 43). Proceed east on the Third Avenue extension of NY 43 (in some places Defreestville Road), out of Rensselaer and its immediate suburbs. Continue on NY 43 as it becomes West Sand Lake Road. Continue through the villages of West Sand Lake and McLarenville. Proceed to the village of Averill Park and the junction of NY 66 and NY 43. Turn left and follow NY 43 and NY 66 to the hamlet of Sand Lake. Here, the two routes turn right (south). Proceed south on them, stopping to check out both Glass Lake and then Crooked Lake, on the right (west) side of the road.

Continue to the first road coming in on the right, which will be Methodist Farm Road. *Nota bene:* From Averill Park to Methodist Farm Road is a distance of 3.5 miles. Proceed west on Methodist Farm Road for 0.8 mile, to a dirt road on the left (south). This is Stewart Lane (Coon Road). In spring the ground is often spongy and if that is the case, Stewart Lane should not be driven. Park off Methodist Farm Road and walk the 0.3 mile length of Stewart Lane.

If the road is not too muddy, however, it is fine to drive in and park on Stewart Lane. At the end of Stewart Lane, follow a primitive trail indicated with red tape-markers fastened to trees. At this point, some birders are reluctant to follow the markers, as it appears as if they are now trespassing. Do not hesitate. Walk through the yard, on the other side of which is the trail leading to the preserve. Go to the old stone wall and the entrance to the more than 120-acre sanctuary.

On the right you should be able to find remains of an old stone cellar hole, left from another era, when the area was all part of the original, extensive Van Rensselaer land holdings. Kiliaen Van Rensselaer, a wealthy Dutch merchant, was granted 700,000 acres of land, spanning both sides of the Hudson River, after the river was discovered by Henry Hudson. Although Hudson was an Englishman, he was sailing for The Dutch East India Company at the time of the discovery, and consequently, Holland claimed the entire Hudson River Valley.

The eastern fifty acres or so of the preserve has not been lumbered for the past forty years and has not been cultivated for almost sixty years. The western seventy+ acres had seen some lumbering and grazing activity until thirty years ago. No bedrock outcroppings have been found on the preserve, although the area contains quite a bit of glacial debris. The entire Sand Lake district is well-known for its elliptical hills, composed of unsorted glacial material, all left behind by the last receding ice cap. The long axes of these groups of drumlins lie parallel to the direction of the flow of the ice, which, here, is northeast-southwest.

Much of the preserve is low-lying land, so birding the preserve during the early spring should be done with waterproofed footgear. Poison ivy thrives in some parts of the preserve, and one is urged to learn to recognize it and then avoid it.

Walk into Stewart Preserve on the entry path and notice the profusion of day lilies near the entrance. Walk on beneath white ash and sugar maple stands, on a carpet of gill-over-the-ground, jewelweed, and moneywort. Bird for a short distance more and you will find yourself in a birch-pine woods, with a groundcover of haircap moss and partridge berry. You are now in for a marvelous surprise. Continue through the woods looking for trees of enormous girth and widely spreading branches. There are several giant oaks (with one measuring about thirteen feet in circumference) as well as beeches and maples. The precise age of these trees is not known; however, it is very likely that they antedate the white settlers in this area. When European settlers first described the local scenery, they detailed dense, massive stands of forest. They wrote of "oak-openings," which were grassy expanses interspersed with the open-grown oaks, whose crowns were often wider than the trees were high. The Indians resident in the area at the time of first settlement by whites, were observed to selectively burn tracts of land. This served to improve hunting for the Indians, while also maintaining the "oak-opening" sites. The thick, corky bark of the oak enabled it to survive these fires.

Descend the hemlock-covered slope to the moist lowland below. Under the beeches and pines notice the parasitic beech-drops and Indianpipes. Here, in addition to supporting a broad spectrum of birdlife, there are many species of wildflowers. Look for several species of violets, hepatica, miterwort, foam-flower, baneberry, dwarf ginseng, starflower, Indian cucumber-root, Solomon's seal, false Solomon's seal, and twisted-stalk. For the fern fanatic there are oak fern, spinulose shield fern, sensitive fern, Christmas fern, lady fern, and maidenhair fern. This is a rich, quiet, cool place with so many varied treasures that birders will wish to spend considerable time exploring it.

The Schodack Marshes, Stony Point Bay, Castleton-on-the-Hudson, a Tour (Map 8.6)

Rating: Spring****, Summer***, Autumn***, Winter*

These areas comprise the best birding along the Hudson River in Rensselaer County; although finally deciding upon their respective names was rather unsettling. The Schodack Marshes are also known as the Castleton Marshes and Stony Point Bay is also known as Schodack Creek. Lest that information further confuse the reader, suffice it to describe the area as a tour along the east side of the Hudson River, from below the city of Rensselaer on the north to the village of Castleton-on-the-Hudson on the south.

From I-87 (the New York State Thruway), at Exit 23 just south of Albany, get on I-787, which skirts Albany on its east side, and runs north along the Hudson River. Proceed north on I-787 to the Rensselaer ramp of the Parker Dunn Memorial Bridge. Cross the Hudson River here, and from the ramp, gain Columbia Avenue (US 9 and US 20), headed southeast. Just before reaching the Rensselaer city limits, turn right on South Street (NY 9J) and proceed south.

There are several things to keep in mind while doing this tour: (a) birding is generally better at low tide than at high tide; (b) traffic along NY 9J interferes with singing birds in the marshes, thus birding in the early morning would make the venture more productive; (c) many birds will be "out-of sight," owing to the rank growth of vegetation in the "bottomless" marshes; (d) some side roads will be wet and muddy, difficult at best and impassable at worst, in early spring; (e) although traffic is rarely heavy and wide shoulders do border the marshes, NY 9J is always slightly hazardous, owing to fast traffic; (f) at some points one's car ought to be parked on the shoulder of NY 9J, while one crosses or walks the railroad tracks through the marshes. At these points watch and listen for oncoming trains.

With those *caveats,* let me again emphasize that this is not a physically taxing birding tour and further, that it yields many more avian returns than my above-listed cautions might indicate. The intrepid birder will always prosper! Remembering that is of paramount importance.

From the turn-off onto NY 9J (River Road) proceed approximately 0.9 mile south to a small pond on the right. The marsh is relatively narrow here, just south of the Rensselaer city limits. Although much of the pond has been destroyed by unsightly fill, it is often worth a stop, in both spring and summer. Least Bittern adults and young, Green-winged Teals with young and Common Gallinules are often highly visible here.

Notice that River Road follows the base of a ridge forming the east side of the Hudson River Valley. Continue along NY 9J for about 1 mile south of the pond. Here the marsh is wider and more open and there is a narrow farm road on the right leading to cultivated fields. These low-lying fields on Papscanee Island abut the marshes and creeks and have a long history of being cultivated. Much of this land has been planted in corn. This is one of the best spots to park the car and enter the marsh on foot. All along this stretch of the Hudson River there are areas where farmers have posted their property. There are birders who monitor this area on a regular basis and who have never been so much as questioned regarding their activities.

Should you decide to walk into the marsh on any of the small roads, in addition to being aware of the avifauna present, look for star-nosed mole, masked (common) shrew, big short-tailed shrew (mole-shrew), meadow mouse (vole), muskrat, northern meadow jumping mouse, fox, raccoon, weasel, mink, and white-tailed deer frequenting the marsh and its edges.

Walking down this particular road has usually been enormously productive in the pre-dawn light, as a chorus of bitterns, rails, and gallinules can be heard. American Woodcocks are active here at this time of day and the whistle of Common Snipe can be heard overhead.

Continue south to Hays (Hayes) Road. From here there is an excellent view extending north and south across NY 9J. The marsh is covered chiefly with cattail, purple loosestrife, and occasionally woody saplings. The undergrowth is amazingly dense and almost impenetrable, but, this is just the habitat that harbors secretive marsh birds. Observe the marsh from the western road shoulder of NY 9J. Undoubtedly, if you have come at daybreak, you will see or hear breeding Green Herons and Least Bitterns. Watch for Snow and Canada geese migrating north and passing overhead in large flocks. Wood Ducks, Green-winged and Blue-winged teals breed in this part of the marsh, as do Mallards and Black Ducks. You will be astonished by the noises being made by nesting Virginia Rails, Soras, Common Gallinules, American Woodcocks, and Common Snipe. Breeding Willow Flycatchers and various swallow species are present in good numbers. Long-billed Marsh Wrens, Red-winged Blackbirds, and Swamp and Song sparrows will all be vocally defending their territories.

Continue south approximately 0.4 mile past Hays Road. This is another good stop, near the south end of the broad open marsh. After this, it gives way to thin woods. In addition to watching the marsh, keep an eye on the wooded ridge on the east side of the road. The forest here is mostly mixed deciduous, with hemlock and pine in some of the ravines. This is a fine area for migrating and breeding raptors. Ospreys may be moving north out over the river. The Red-tailed Hawk is a permanent resident and hunts the marsh year-round. Occasionally there are small concentrations of Red-tailed and Rough-legged hawks hunting the abundant marsh-dwelling shrews and voles. Look along the wires and fences for American

Kestrels and Belted Kingfishers at almost any time of the year, but, especially in the spring and summer. Possibly Blue-winged and definitely Golden-winged warblers sing from some of the hillside openings on the east side of NY 9J. The density of nesting Red-winged Blackbirds is quite high along the entire length of the marsh. In the autumn, scattered groups of up to fifty Rusty Blackbirds descend on the marsh. Glance over to the wooded ridge before you leave this spot; there is every chance that you will see a colorful male Rose-breasted Grosbeak fly far out over the marsh and then reverse itself and return to its usual woodland haunts. Swamp and Song sparrow numbers in the marshes are prodigious. Some unknown percentage of each species regularly remains to winter among the cattails.

Continue driving south for approximately 0.6 mile, where Statts Island Road comes in on the right. This is an unsurfaced road but is usually negotiable, at least for a short distance. Turn right and proceed to where the road crosses Papscanee Creek. There is a small bridge crossing the creek. Often enough, this area is lacking in avian interest; however, it is primarily a tidal creek and it sometimes abounds with migratory waterfowl in March and April, shortly after ice-out. In some years, depending upon variable local conditions, shorebirds are fairly numerous in the marshes visible from Statts Island Road. Birders can walk west on this road, across the creek, through the marsh, across the railroad tracks, and on a few more yards, to gain the Hudson River, if they choose.

Regain NY 9J (River Road) headed south. Proceed approximately 0.6 mile where the Papscanee Creek runs right next to the road. During the peak of the waterfowl spring migration this area contains high numbers and many species of ducks. These are easily observable from the car. In addition to the already named ducks there will be numbers of American Wigeons, Gadwalls, rarely the Eurasian form of the Green-winged Teal, Pintails, Canvasbacks, Redheads, Ring-necked Ducks, scaups, Buffleheads, Common Goldeneyes, Hooded, Red-breasted, and Common mergansers, and Ruddy Ducks.

Follow the creek south for about 0.9 mile to the bay at Stony Point Road (NY 108), where the creek passes under a small railroad bridge, enters the woods and is lost to view. Park the car along NY 9J. Walk across the bridge and along the tracks to any access point along the creek. One could also walk downstream along the tracks paralleling the creek. Along this part of Papscanee Creek 1 mile north of Stony Point, is prime habitat for Horned and Pied-billed grebes, Great Blue and Green herons, and Black-crowned Night Heron, as well as diving ducks and mergansers.

From Stony Point, drive an additional 2 miles into the village of Castleton-on-the-Hudson, on NY 9J. Park on the rise between Seaman and Scott Avenues, looking west along the Hudson River. This is an excellent vantage point. The river is not only a flyway for migrating ducks and geese, but also for migrating raptors. Look for Ospreys and the rare Bald Eagle. A fine variety of gulls is regularly seen along the river. You will doubtless be able to identify at least three or four species from this point.

At the southern edge of Castleton-on-Hudson, where a small creek, Vlockie Kill, flows west to the Hudson, a dirt road to the right leads out to the south along Schodack Island. This road may be impassable in wet seasons and after heavy rains but its northern stretch parallels the main channel of the Hudson River, giving fine views of any river ducks present. The one remaining farm road takes one through good birding areas for a total of about 4 miles. Return to NY 9J along the same route rather than risking any of the other roads which are sometimes impassable.

COLUMBIA COUNTY
Nutten Hook and Stockport Station (Map 8.11)

Rating: Spring****, Summer***, Autumn***, Winter*

Nutten Hook and Stockport Station are both mentioned as fine Hudson River birding areas in Columbia County. Nutten Hook is an old brickyard site, south of the village of Stuyvesant. Although it is unfortunately quite littered, it offers a good vantage of the river. Stockport Station is six miles north of the village of Hudson and encompasses about 100 acres of river and railway frontage. Both sites are best birded on foot. At the Stockport Station site, one must be alert for Conrail passenger trains passing between Albany and New York City every two or three hours.

From the village of Castleton-on-the-Hudson, drive south on NY 9J for 8 miles, passing into Columbia County. Proceed through the village of Stuyvesant. After approximately 2.6 miles, there will be blue historical markers indicating the way to Nutten Hook. Look for these on the right side of NY 9J, just after passing Sharptown Road, which meets NY 9J on the left. Turn right at the markers and follow the dirt road across the railroad track and to the river. This is Nutten Hook and the ruins of the old brickyard. Here, one has an unobstructed view of the river. Expect to see the previously mentioned migrating waterbirds as well as small flocks of Double-crested Cormorants.

Proceed south on NY 9J as it veers eastward, away from the river. Pass Brickyard Road on the right and Rossman School Road on the left. At approximately 2.6 miles from the Nutten Hook turnoff, US 9 comes in from the left and merges with NY 9J. Just before this merging point, a road marked "Dead End" meets NY 9J on the right. This road may not be labeled, but it is a county road: NY 59A. Turn right here and follow NY 59A to its end at Judson Point. Park the car. The river is straight ahead, and to the west one can see the Athens Flats and the Catskill Mountains.

Try to time arrival at this point to coincide with low tide. If you succeed, there will be flats exposed along with an island in mid-river. Grebes, (including Red-necked Grebe), herons, Anseriformes, possibly raptors, some shorebird species and several larids, including Bonaparte's Gull, should be present. Walk across the railroad bridge trestle and wander south along the tracks for about a mile. Note the abundance and variety of birds in the marshes adjacent to the tracks. After walking that mile one reaches the most productive wetland at this site. The birders who most regularly cover this section of the river call this "the marsh." At this point one ought to see species not encountered farther north along the tracks.

This is Stockport Station. Just south of here Stockport Creek empties into the Hudson River.

After thoroughly covering "the marsh" area, walk north back on the tracks to Judson Point, where the car has been parked.

Olana State Historic Site (Map 8.10)

Rating: Spring****, Summer*, Autumn****, Winter (not open)

Olana is the name of the estate on which stands an elaborate 37-room castlelike mansion, designed and built by the nineteenth century landscape artist, Frederic E. Church. Today the 250-acre estate is owned by New York State and enjoys its status as a National Historic Landmark. It is open May through October. There is neither admission nor parking fee. Although insufficient information is available regarding nesting species on the grounds, one suspects that with the diversity and age variation within the vegetative profile, nesting avian species composition must reflect that heterogeneity. The estate has been birded in both spring and autumn migration periods and it is reportedly a marvelous place at which to catch waves of passerines as well as migrating raptors.

Leave I-87 (the New York State Thruway), at Exit 21. Follow feeder roads and signs to NY 23 and the Rip Van Winkle Bridge. Proceed east on NY 23 and cross the Hudson River on the Rip Van Winkle Bridge. Continue east on NY 23 to its junction with NY 9G. Here turn right (south) and continue 0.8 mile to the entrance road to Olana.

From NY 9G, turn left and drive up the winding approach to the mansion. This drive is an excellent place to begin birding. The roadway is lined with flowering dogwood, maples, Carolina hemlock, eastern hemlock, pine, basswood, and birch. Stop along this approach and check the trees and shrubs bordering it for migrants.

There are four distinct trails leading through essentially different types of habitat marked and maintained on the estate. Walking any and all of the trails is considered easy and each trail has its unique virtues.

The 0.5-mile Victorian Garden Trail is distinguished by green trail markers. It begins along the brick path in front of the mansion. Check out the pink dogwood and the eastern hemlock on the front lawn as well as the large yew hedge and flowering magnolia on the way to the garden. Follow the shale drive where the garden can be viewed from above, or follow the path through the flowerbeds. There are over 100 species of flowers in the garden. In one part there is a thick groundcover of Myrtle, and iris and tulip are shaded by an ornamental horse chestnut tree. This is a short trail, and not the most productive for birds, but it has been known to surprise birders with hummingbirds and sparrows.

The Ridge Trail is 0.75 mile long and is distinguished by yellow trail markers. It begins across the paved road from the service vehicle entrance. The Ridge Trail passes through a dense grove of eastern hemlock, which thrives in this typically cool, rocky terrain. This area is reminiscent of the Catskill Mountains. This

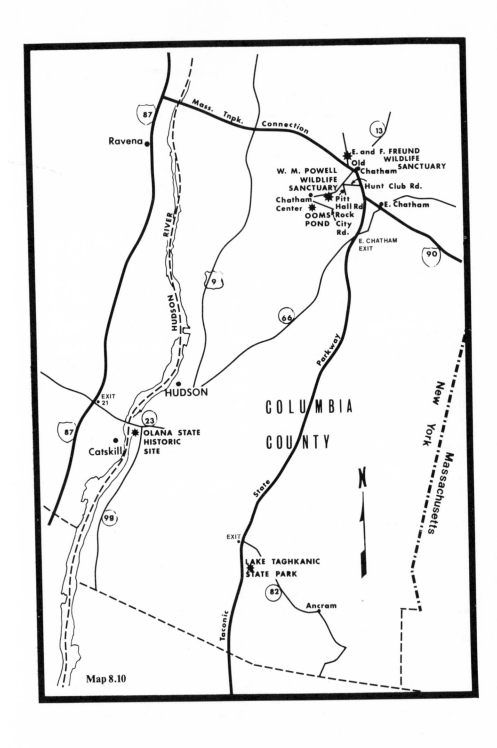

Map 8.10

trail progresses from the hemlock stand to a bright open field bordered by mature maples. These trees have wide-spread branches and a full canopy. Some of the open field has reached the later stages of plant succession, and is composed of oak and hickory hardwoods. Secondary stages of succession are evident along the field edges, where shrubs, vines and small trees are taking hold. Ruffed Grouse nest near the edge of the clearing. There are numerous white spruce, white pine, and red cedar interspersed along the trail. The path proceeds through a recently disturbed area, where only grass varieties have begun to grow. From there it passes again through a dense oak-hickory forest with an understory of black cherry, elm, hop hornbeam and black birch. At the end of the trail, turn left to return to the trailhead.

The Hemlock Trail is 1.0 mile long and is signalled by red trail markers. It also begins across the paved road from the service vehicle entrance. The Hemlock Trail proceeds along the north road of the estate, which once led to a northern entrance to the grounds. Along this trail the dominant trees are oak and maple, with some dogwood, black birch, and wild cherry. There are witch hazel, spicebush, and viburnum shrubs along it. Wild grape vines climb the trunks of many of the trees.

The trail passes through an area of dead and downed hardwoods in which Common Flickers, Pileated, Hairy and Downy woodpeckers nest. The trail then passes through a dense, cool, dark hemlock forest and beside a deep ravine, which drains moisture from the hillside above it. Hemlock Trail traverses a section of decomposing hemlocks and then an area of exposed shale and limestone rock outcroppings. Along the next stretch of the Hemlock Trail there is quite a bit of natural "forest litter," which provides protection for ground-nesting bird species. Additionally the Red-tailed Hawk reportedly nests in this section of the estate. The trail then approaches the edge of the woods and a cleared field. This change in habitats should be conducive to "edge-effect" birds. Listen for their songs. Beyond, there are farm fields and a pleasant vista of the Taconic and Berkshire Mountains to the east. The path comes to a junction where the birder can either continue toward the Victorian Garden or branch off and follow the Lake Trail to a ten-acre lake south of the mansion.

The Lake Trail is 1.25 miles long and is distinguished by blue trail markers. It begins near the barns and icehouse southeast of the mansion, at the north end of the lake. A shagbark hickory shrubby fencerow forms one of the boundaries around the lake. Proceed along the trail until at one point, it leaves the lakeside and follows a carriage drive to the top of Crown Hill. At the beginning of this spur, there are large oak trees. In the woods beyond, one can see American sycamore. Farther along the trail one reaches a swamp edge. A boardwalk has been constructed so that access can be gained to this habitat. From the boardwalk one sees duckweed floating on top of the water, and along the swamp edge are oaks, red maple, swamp white oak, gray birch, and shrubs of blueberry. Wood Ducks and Mallards nest in the area of the pond, as well as at least four frog species and several small *Microtus* species. Continue on to the top of Crown Hill. The view at this point is unobstructed, and it makes a fine hawk migration lookout. After leaving Crown Hill, walk to the south end of the lake and from there up along its west side.

Each of the trails has alternate routes to those described, and there are several points at which one trail connects with another. Birders can shorten or lengthen their walks, as they like.

COLUMBIA COUNTY – INLAND
The Emanuel and Frances Freund Wildlife Sanctuary,
The Wilson M. Powell Wildlife Sanctuary, Ooms Pond,
a Tour in North-central Columbia County (Map 8.10)

Rating: Spring***, Summer**, Autumn**, Winter*

Birding all three of these sites is rated moderately easy to easy. This area is part of the old Rensselaerwyck, the largest and longest lived of the enormous colonial estates known as patroonships. Rensselaerwyck was, in effect, a feudal manor, with members of the Van Rensselaer family acting as feudal lords. They leased all land to tenant farmers, but, never sold as much as a rock to anyone. The farmers were obligated to pay a heavy tax in crops, animals, fowl, etc., on the first day of every new year, to the manor lord. This system and Rensselaerwyck remained intact until the mid-nineteenth century. Columbia County, on the east side of the Hudson River, has some of the most peaceful and beautiful countryside in New York State.

Farming is still actively pursued in the rural areas. There are horse, cattle, poultry, and swine farms. There are apple orchards and dairy farms, as well as truck farms that provide vegetables for local markets and roadside stands. In the wilder areas, rather far up in the hills, deer, bear, Turkey, pheasant, and grouse still maintain respectable populations. The fast-running streams are rich with brook and brown trout, large mouth bass, and pickerel. The countryside is a chessboard of old, hand-hauled stone walls, and many of the houses and villages built before Revolutionary War retain their pastoral charm.

This is not to say that Columbia County is monotypic. Its bucolic aspect is only one of its facets. Much of the land has reverted to its former wild and diversified state. These three sites sample extensive pine and spruce communities, mixed deciduous-coniferous communities, woodland pond communities, and urban-farm communities.

If approaching from the south, take the Taconic State Parkway north of New York City. This roadway is an example of a product of powerful technology combined with enormous creativity. It has transformed the natural countryside, while still respecting its character. Leave the parkway at the East Chatham Exit. Immediately gain NY 295, by turning right off the parkway, and proceed northeast to the village of East Chatham. At the Post Office, turn left and cross the railroad bridge. Follow signs to Old Chatham. At Old Chatham bear right on the Old Albany Turnpike. Proceed approximately 0.5 mile, crossing the Thruway bridge to Pitts Road. Turn right on Pitts Road. Immediately on the right is the Emanuel and Frances Freund Wildlife Sanctuary. Park along Pitts Road, as far off the road as possible. Take the trail on the right into the preserve.

The preserve property was last actively farmed over fifty years ago. After that the fields became brushy and the forest began creeping over the fence lines, until today the preserve is a maturing second-growth forest. The large scattered white pines appear to be either remnants of decorative plantings or isolated natural growth. In spring they are not only attractive, but also offer refuge to several bird species. Notice also the shadbush, apple trees, viburnums, and occasional pinxters. There are healthy stands of ferns and herbaceous plants throughout the sixty acres of the Freund Sanctuary.

The bedrock exposed along the creek and on the hilltops, is Nassau shale of Cambrian age. It is obviously covered with glacial till. The soil here is mainly of coarse texture and not conducive to profitable farming.

The paths throughout the preserve are kept open but are unmarked. This, however, presents no problem of getting lost. There are a few low places along the trails, that tend to be somewhat wet in the spring.

Ruffed Grouse, woodpeckers, flycatchers, all of the thrush species (except Eastern Bluebirds), Blue-winged Warblers on the brushy slopes and Golden-winged Warblers in the more open areas, Indigo Buntings, towhees, and sparrows are some of the common birds found in the sanctuary. Additionally, northern flying squirrel has been seen here as well as long-tailed weasel, skunk, raccoon, woodchuck, and deer.

The Wilson M. Powell Wildlife Sanctuary is owned and maintained by the Alan Devoe Bird Club, Inc., of Chatham. The 140-acre sanctuary has belonged to the Devoe Bird Club for over twenty years and is open to the public year-round without admission or parking fees. There are about three miles of trails throughout the sanctuary, and a well has been drilled to provide fresh drinking water.

From the Freund Sanctuary on Pitts Road retrace your way back to the center of the village of Old Chatham. At the intersection of the Old Albany Turnpike and NY 13, turn right and proceed for about 1 mile to the first road coming in on the left. Here at Pitt Hall Road (also Powell House Road) turn left. Proceed to Hunt Club Road, which comes in on the left. Here, turn left and bear left up the hill (which can be icy in winter), to the small parking lot and entrance to the Powell Sanctuary. All visitors are asked to sign the register, which is in the small shed near the parking lot.

Wilson M. Powell, after whom the preserve is named and whose widow gave the property to the Devoe Club, was a well-known conservationist during the governorship of F. D. Roosevelt. Powell had been very active in reforestation efforts throughout the state and on this, his own property, he planted more than 180,000 trees of larch, Scotch pine, and Norway spruce. To these fine stands, nature has added a mixed woodland of white ash, chestnut oak, shagbark hickory, beech, red maple, hemlock, white pine and white oak. This sanctuary has been considered a botanists' delight, because of its great variety of shrubs and herbs. Over 200 plant species have been recorded on the preserve.

The same mammal species occurring at the Freund Preserve have also been sighted here. Additionally opossum, red and gray fox, bobcat, ermine, mink, Brewer's and star-nosed moles, deer and red-backed mice, three shrew species, and three bat species have also been observed here.

A small intermittent stream elbows its way between some of the surrounding hills and contributes its water to a six-acre shallow impoundment, where, Dusky, Alleghany and Two-lined salamanders live and breed, especially in the permanent seeps of the creek. The lowest point on the grounds is at the outlet of this pond. A country road runs close to the impounding dike, and from a small parking lot north of the pond, one can walk onto the dam behind a specially-built log blind. From there the wildlife can be observed without the observer being observed. The highest point on the preserve is Dorson's Bluff, in the southwest section of the preserve. The elevation here is 924 feet above sea level. From here one can see, on a clear day, a panorama of farms and orchards in the Hudson River Valley, the Shawangunks, the Catskills, and the Helderberg Escarpment. This is an excellent vantage point from which to observe migrating and resident Turkey Vultures and hawks soaring down the river valley and over nearby fields and glens. The bedrock here is composed of slates and shales of Cambrian-Ordovician age, and the bluffs were scraped and scoured clean by the advancing glacier, which melted back only 20,000 years ago.

More than 200 bird species have been recorded on the sanctuary property. The marsh and pond are especially attractive to Pied-billed Grebes, Great Blue and Green herons, American Bitterns, Canada Geese, Wood Ducks, Green-winged Teals, Mallards, Black Ducks, Blue-winged Teals, and American Coots.

A Goshawk resides at the sanctuary from autumn through late winter and can usually be seen in the vicinity of the stocked feeders, near the parking lot, or down the White Trail. During some winters a Cooper's Hawk takes up residence in the vicinity of the pond. Great Horned and Barred owls can be heard calling year-round. Pileated Woodpecker is a permanent resident in the woods. Flycatchers and swallows nest on the preserve.

In winter, in the area of the well-maintained feeders, there are usually fifteen to twenty bird species, including jays, chickadees, titmice, White-breasted and Red-breasted nuthatches, Cardinals, Evening Grosbeaks, Common Redpolls, Pine Siskins, American Goldfinches, and various sparrow species.

Along the White Trail, especially at the higher elevations, all of the eastern thrush species can be found in spring, summer, and autumn. Breeding Veery is especially easy to locate. Red-eyed Vireo is a common nester on the preserve, as are Blue-winged and Golden-winged warblers, Yellow and Chestnut-sided warblers, Ovenbirds, waterthrushes, Common Yellowthroats, and American Redstarts, Scarlet Tanagers, Rose-breasted Grosbeaks, Indigo Buntings, and a whole array of sparrow species.

During every season there is some interesting bird activity going on here. It is well worth taking the time out to visit the W. M. Powell Wildlife Sanctuary.

Before leaving the hamlet of Old Chatham, one might find a visit to the Shaker Museum, off NY 66, informative. The Shakers, one of America's most successful utopian religious sects, first settled in the Hudson River Valley, after fleeing England, more than 200 years ago. Since then, they have spread their communal and largely celibate mode of living and unique way of worshipping throughout New England and even into the midwest. They are, as a group, renowned for their craftmanship, ingenuity, and egalitarianism. This museum, comprising four buildings, portrays their heritage through the finest collection of

Shaker-made artifacts in the world. It includes 16,000 items of furniture, textiles, blacksmithing, medicines and other examples of Shaker inventiveness. Drive through the hamlet of Old Chatham, which was a recognized community prior to 1812, where some of the comodious houses are over 150 years old.

Ooms Pond covers approximately ten acres about three miles southeast of the hamlet of Old Chatham. It is on private land, but access through the property is not necessary, as birding can easily be done from the road. The owner of the pond and surrounding land, is a friendly farmer who may stop to chat. He has always been hospitable to birders. A scope would be extremely useful from the higher vantage point over the pond.

From the center of the hamlet of Old Chatham, turn left (west) on NY 13 and proceed about three miles to its intersection with Rock City Road. At that junction there is a large gravel bank on the right side of NY 13. Turn left and continue for about 1 mile to the pond site, which is on the right side of the road. There are two basic vantage points, which will be evident once you reach the site: the top of the Rock City Road hill and the lower elevation of the pond itself.

Ooms Pond has also been known previously as Sutherland Pond. It is an especially fine site for migrating waterfowl in both spring and autumn. Horned and Pied-billed grebes stop here; herons occasionally frequent the pond; flocks of Snow Geese (including the blue form), and Canada Geese can be viewed in the pond and grazing its slopes. The most common duck species observed here are American Wigeon, Green-winged Teal, Mallard, Black Duck, Blue-winged Teal, Ring-necked Duck, scaup, Common and Hooded mergansers, and Ruddy Duck; American Coot, Killdeer, Greater Yellowlegs, and Spotted Sandpiper. Black Tern has been observed here. Flycatchers breed in the area. Look for Tree and Barn swallows in season, as well as warblers along the road, and blackbirds and sparrows in the more marshy sections surrounding the pond.

Lake Taghkanic State Park (Map 8.10)

Rating: Spring****, Summer*, Autumn***, Winter*

This state park is about fifteen miles southeast of the village of Hudson. There is an admission and parking charge of $2.00 per day during the summer season. The park provides a beautiful natural setting among forests, shrubs, wildflowers, rock outcroppings, lakeside, and generally rolling terrain. Within the park there are paved paths to hike. The park is highly recommended during the off-season, when all of its areas can be explored with relatively little traffic and noise from campers, swimmers, boaters, and picnickers. This park has proven a fine place to see both spring and autumn migrants.

Leave the Taconic State Parkway at the Lake Taghkanic Exit. Gain NY 82, headed southeast. Proceed to the park.

Go to the West Beach parking area, where one can leave the car. Walk along the west shore of Lake Taghkanic. During migration seasons there are usually some resting waterfowl on the lake. Along the road are abundant thickets which seem to act as magnets for migrating passerines, especially warblers, in spring. Walk all along the road, exploring any of the short turn-off paths and roads. Behind the parking area is a camping site. There are paved roads here also and walking along in this section is often very profitable, especially in the height of the spring migration season. Cars of noncampers are not allowed in this area but, the best birding here is done on foot.

GREENE COUNTY
The Coxsackie and Athens Flats, a Tour (Map 8.11)

Rating: Spring*, Summer***, Autumn***, Winter****

Appropriately, the Indian name, "Coxsackie," is said to mean the "hoot of an owl." This tour is geared to find as many wintering owls and other raptors as possible, and samples some of the most pristine habitat for raptors in the entire Hudson Valley.

The Coxsackie and Athens Flats extend some three miles north and six miles south and west of the village of Coxsackie. The area encompasses about ten square miles; about 90 percent of the land is privately held, while the other 10 percent is publicly owned. In spite of the high proportion of private lands, permission to gain access to those areas is not necessary, as in general, birding can be done from the car or from public roadways. The Hudson-Mohawk Bird Club keeps abreast of the birdlife in this area and instructs that the best time to bird the flats is mid-December through March, or late-July through August. In general, the area is easily birded; mostly from the car with some meandering down hedgerows and into deep conifer stands, or walking through recently cut hay fields.

The following tour encompasses several roads and sites at which stops should be made in order to do justice to the area. Allow plenty of time for the tour and it will yield the best birds.

Leave I-87 (the New York State Thruway) at Exit 21B. Turn left on Schoolhouse Lane and proceed west on it only a short distance, to its intersection with US 9W. At US 9W, turn left (south), and work your way south to the Schoharie Turnpike (NY 28), or Athens Road (NY 74). Along the way the following stops should be made.

Proceed for about 1 mile on US 9W to some open fields on both the east and west sides of the road. Stop. Scan the fields and hedgerows around them carefully for wintering raptors. Open fields with scattered cedars invariably harbor harriers and owls. The entire area of the Coxsackie Flats has been termed by the Hudson-Mohawk Bird Club members who bird it most, "Winter Raptor Wonderland." In winter, at any of the prescribed stops look for Northern Harrier, Red-tailed and Rough-legged hawks, American Kestrel, Merlin, Peregrine Falcon, Barn Owl, Screech, Great Horned, Snowy, Short-eared, and Saw-whet owls. Be alert also for wintering Northern and Loggerhead shrikes, along with owls in dense conifer stands.

In all of the open fields, but especially in the fields spread with manure in winter, search out Horned Lark, Lapland Longspur, and Snow Bunting.

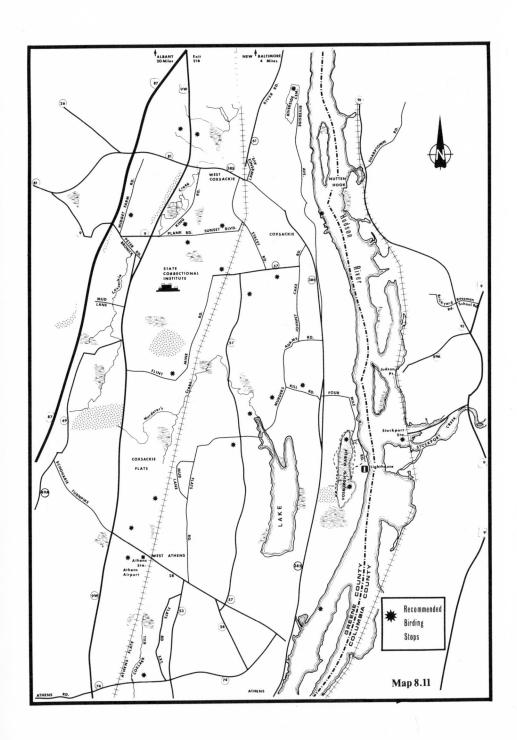

Map 8.11

In late summer and autumn search the same fields, and at the same open sites for American Golden Plover, possibly breeding Upland Sandpiper, Pectoral or Buff-breasted sandpipers. In these seasons seek out recently cut hay fields, hedgerows and fencelines. This often requires leaving the car to find the birds. A scope would be very useful.

This is a large and underworked area, and there will be some interesting surprises here, including nesters and winter rarities. It may be possible some day to turn up a Boreal Owl in one of the more remote areas or dense conifer stands. It is likely that in late spring or in summer one might be able to turn up breeding grass-associated sparrows or those elusive and unpredictable nesting or migrating "grass-pipers."

Continue south on US 9W. Soon, just before intersecting NY 81, it forks; stay to the right here, cross NY 81 and continue south on US 9W for 1.1 miles. This is the intersection of US 9W and NY 9 (Plank Road). Turn right (west) and go as far as the next intersection. Turn right again on Midget Farm Road (Bronck Mill Road). Approximately 0.5 mile north on Midget Farm Road there will be large open fields on the right (east) side. Search this stretch carefully. The west side of this road is very steep and trees here should also be birded. Proceed up the road birding its right side to a convenient turnaround. Then reverse directions and come back to the intersection. Here, instead of regaining NY 9 by making a full left turn, cross NY 9 (Plank Road) and veer slightly left to gain Peter Bronck Road. Proceed approximately 0.3 mile to the junction of NY 42.

Here, on the corner of Peter Bronck Road and NY 42 is part of the historic homestead of the early Dutch Bronck family. Among the buildings is one of stone (1663–1685), brick (1738), a Dutch Victorian, and a thirteen-sided "freedom" barn, family cemetery, artifacts, farm implements, etc. This museum is maintained by the Greene County Historical Society and is open to visitors. The entrance is ahead on US 9W.

Turn right on either NY 42 or US 9W and proceed south 0.7 mile to Mud Lane, which comes in on the right. Turn right and check this road out and then return to US 9W. Turn right and proceed south another 0.9 mile (just past a drive-in theater on the right), to Flint Mine Road. Turn left (west) here and proceed 0.7 mile. North of Flint Mine Road here, the Coxsackie Correctional Institute (state prison) maintains an active farm and orchard. These fields, in the southeast quadrant of the institute lands, offer good birding, especially from Flint Mine Road. Keep to the roads here. The institute has a horse patrol, whose job it is to keep people out (or in).

After covering these fields and orchards, reverse directions and go back (west) on Flint Mine Road to its junction with US 9W. Turn right (north) and drive only 0.4 mile to NY 49. Turn left (west) and follow it west and south, stopping to bird both sides of the road as you go. At 0.6 mile, NY 49 turns decidedly south and follows a ridge line approximately 1.3 miles to NY 28 (the Schoharie Turnpike). All along this drive check out the fields and flats on the left side of the road. These should be productive so spend time and effort on it. It might pay off.

At the junction of NY 49 and NY 28, turn left and follow the Schoharie Turnpike east. Cross US 9W and proceed approximately 0.4 mile. Park off the road and bird the fields on both sides of the road. South of NY 28 is the Athens Airport and its surrounding open fields and flats. North of the road is a marvelous section of flats through which Murderer's Creek flows; it should be very productive. Drive another 0.2 mile and turn right. Park in the Athens train station parking lot. You are now at the hamlet of West Athens. Walk north on the railroad tracks from there for about 1 mile. Bird both sides of the tracks, but pay special attention to the flats on the left (west) side. Be alert for trains and walk back to the parked car after thoroughly covering this avian-rich area.

From the station parking lot, turn left (west) on NY 28 and proceed to the junction of US 9W. Turn left and proceed south on US 9W, stopping to bird any areas, especially the flats on the left (east) side of the road. At NY 74 (Athens Road), turn left (east). Proceed approximately 0.8 mile and stop to bird the rich area on the north side of Athens Road. The Corlaer Kill runs through these Athens Flats, and this area deserves some special investigation.

Continue east. Pass the next road on the left (Flats Road extension) and go just a bit farther to the next road: NY 53. Turn left (north) here and proceed north on it to NY 28. Turn right and then immediately turn left onto Flats Road. Drive approximately 1.9 miles where Murders Kill Road comes in on the right. Turn right and proceed east almost as far as the junction of NY 57. Just west of that junction and south of Murders Kill Road is a good place to check.

Continue east driving and birding all the while. Cross NY 57 and proceed north and east on Murders Kill Road. This area is extremely vulnerable to development. A large second-home community has been built on the right side of the road. The long lake you will have passed on the right has been impounded by the developers of this community, and it has encroached on what used to be prime raptor habitat. This community is private and posted. As you travel north past the lake, bird the area on the left (west) side of the road, but only from the road. Turn right where Murders Kill Road turns right, and just past the point where Murderer's Creek crosses the road, stop to check the north side of the road.

Continue east to the junction of NY 385. Turn left and proceed north 0.8 mile to Adams Road. Here turn left again, and proceed approximately 0.3 mile to Johnny Cake Road. Turn right and bird along its 1-mile length to its intersection with NY 57. Here turn left (west). At 0.4 mile stop and bird the area on the left (south) side of the road. Do the same another 0.4 mile west on NY 57. One may even wish to walk south on NY 57, which turns left at this point, in order to better bird the area east of it. Park the car off the road where NY 57 turns left (south) and walk due west for 0.3 mile, across the railroad tracks, to bird the area just northwest of the tracks.

Walk back to the car. Turn it around to head due east. Drive approximately 0.3 mile to Stacey Road, which comes in on the left (north). Turn left here and proceed north on Stacey Road. Immediately after crossing the railroad tracks, make a sharp left turn. After paralleling the tracks for a short distance, Sunset Boulevard turns right, away from the tracks, and proceeds west. Follow it through these meanderings, and approximately 0.4 mile after the sharp turn, stop to bird the area on the north side of Sunset Boulevard. You will see the telephone line right-of-way, which will identify this spot. Continue west as Sunset Boulevard becomes Plank

Road. Note where Coxsackie Creek crosses Plank Road and scan the area just west and south of this point. Birding here must be done from the road as these are now the grounds of the Coxsackie Correctional Institution. Continue west a short distance. King Road comes in at a sharp angle from the right. Make a sharp right turn and proceed northeast up King Road. Approximately 0.4 mile after the sharp turn onto King Road, check out the east side of the road just beyond a small cemetery.

Continue driving up King Road to its intersection with NY 385. Turn right and follow NY 385 through the hamlet of West Coxsackie. The first road on the left after crossing the railroad tracks, is Lawrence Avenue. Turn left here and follow it to NY 61. Turn right, thereby gaining NY 61. Continue north on NY 61 (River Road). From the last right turn, stop at 0.3 mile and check the fields on the right (east), and at 0.6 mile to check the fields on the left (west) side of NY 61.

In addition to the above-outlined tour, there are two sites worth mentioning here. In the northeast section of the village of Coxsackie, east of NY 385 and on the shore of the Hudson River, is a New York State boat launching site. From here one has fine views of the river and can search out waterfowl, gulls, etc., if one has tired of the lengthy raptor reconnoiter.

The Riverside Cemetery, north of the village of Coxsackie, is a tranquil and worthwhile place to bird. It can be reached by heading due north out of Coxsackie on Riverside Avenue. From the center of town, the entrance to the cemetery is on the left (west) side of the road, at about 1.6 miles.

Vosburgh's Marsh (Map 8.11)

Rating: Spring****, Summer**, Autumn***, Winter*

Vosburgh's Marsh lies about 5.0 miles south of the village of Coxsackie, entirely on private land. The area is in multiple ownership. It is not posted, but there is really no public access to the area. There are no admission or parking charges for birding the area; however, birders should signify their good-will and acknowledge that of their host, by "checking-in" with either Mr. Vosburgh, at Vosburgh's Greenhouse on NY 385, or Mr. Elmer at River Farm, at the end of Four Mile Point Road. Both usually give the go-ahead to birders to wander freely within the area.

The Vosburgh's Marsh is an especially good place to bird in spring, when one can see and hear Great Blue and Green herons, Black-crowned Night Herons, Least and American bitterns, a large variety of duck species, several rail species, marsh wrens, and marsh-breeding sparrow species.

Birding the marsh is not physically taxing at all and one need only prepare oneself with waterproofed footwear and perhaps a little bit of insect repellent to make life in the marsh a birding treat.

Leave I-87 (the New York State Thruway) at Exit 21. Follow feeder roads and signs south to NY 23. Drive east on NY 23 to the junction of it and NY 385. This will be just west of the Rip Van Winkle Bridge. Turn left to gain NY 385 north. Continue north through the village of Athens.

Athens is, in fact, an interesting old town. It was settled in the late 1680s by the Van Loon family. Two of the family's old stone houses still stand. In downtown Athens there are several Federal-period houses. These exemplify Greek Revival and bracketed Victorian styles. There is also a marvelous prototype of a board-and-batten church. Along the river look to the southern end of a midriver island for a lighthouse more than 100 years old. If you have time, inspect the sleepy hamlet at leisure. It's full of history.

Proceed north on NY 385 about 2.4 miles from the Athens village northern limits. Look for Vosburgh's Nursery and Greenhouse on the right (east) side of the road. Park and go to the greenhouse to advise Mr. Vosburgh that you would like to bird his marsh. Proceed east and enter as the map trails indicate.

If it is early morning, the east entrance to the marsh is not the best place to begin birding, for the glare from the early sun poses a real obstacle. At that time of day, continue north on NY 385, past Vosburgh's Nursery, to Four Mile Road. Turn right and follow Four Mile Road east and south to its terminus at the River Farm. Park, making sure to not block road access and go to the farm to inform Mr. Elmer of your plans to bird the Vosburgh's Marsh. Then enter by walking west as the map trails indicate.

After birding the marsh, exit on the east side of the swamp and walk to the Four Mile Point Lighthouse. This stone structure with a clapboard wing, was built in the early nineteenth century. The river bank here is a good place to picnic. Four Mile Point Road affords an excellent overlook of the Hudson River; along it one might look for waterfowl, Osprey, an occasional Bald Eagle, and various gull species.

If you did not pack a picnic lunch to eat overlooking the river, regain NY 385 headed south. Drive about 1.7 miles, then turn left (east). Advance to Hagar's Harbor restaurant and bar. There one can get a bite to eat and still have fine river views.

ACKNOWLEDGMENTS

The author expresses thanks to the following people for the information they supplied on the Hudson-Mohawk region: Edgar M. Reilly; Kate Dunham; members of the Alan Devoe Bird Club; Peter Feinberg; Richard P. Guthrie; Bill Cook; Paul F. Connor; Kristine A. Kelly, former Executive Director of the Eastern New York Chapter of the Nature Conservancy; Louis F. Ismay, Chairman of the Board of Trustees of the Eastern New York Chapter of the Nature Conservancy; Alan Mapes; Robert L. Marx; Anne Williams, Executive Director of the Eastern New York Chapter of the Nature Conservancy; and especially to Edgar M. Reilly, of the New York State Museum, for his thorough and helpful reading of the Hudson-Mohawk chapter.

Hudson-Delaware—Region 9

THE HUDSON-DELAWARE REGION includes all of Ulster, Sullivan, Orange, and Rockland counties on the west side of the Hudson River, and all of Dutchess and Putnam counties, and that part of Westchester County north of NY 117 and NY 120 (North Tarrytown to Chappaqua to the Connecticut state line), on the east side of the river. This region amounts to about 4500 square miles (Map 9.1).

The lowest-lying areas in the region are those on the Hudson River, at sea-level and the highest elevation is that of Slide Mountain at more than 4200 feet above sea-level. In between there are more than 100 Catskill peaks with elevations of more than 3000 feet, the cliff-riddled Shawangunk Range with an average elevation of 1600 feet, the Hudson Highlands and Ramapo Mountains, which rise to more than 1400 feet in some places, and the outlying ridges of the Taconic Range in Dutchess and Putnam counties, on the east side of the region, which run to 1300 feet. Finally there is the western rampart of the Hudson River, the Palisades, whose cliffs rise dramatically to 450 feet.

Throughout, the terrain could be described as rolling-to-rugged with the Hudson, Delaware and Wallkill river valleys presenting wide variations in the cliff-ledge-summit theme. Up to approximately the 3000-foot elevation, the vegetation is that of beech-birch-maple-hemlock association and above 3000 feet it is mostly of the northern spruce-fir association. There are a few relatively small islands of boreal association.

The water resources of this region are extensive, with approximately 20,000 surface acres or, in the aggregate, 32 square miles of inland reservoirs helping to supply the water needs of New York City, and countless miles of mountain streams and brooks famous for their outstanding quality.

The historic, scenic, and cultural resources of the "Rhineland of America" in combination with the birding opportunities afforded by such enormous forest preserves as Catskill Park, Bear Mountain/Harriman State Parks, Harvard Black Rock Forest, Clarence Fahnstock State Park, and Ward Pound Ridge Reservation, as well as smaller preserves, are seemingly limitless. The onion-growing district known as the Orange County Mucklands—a 25-square mile rich, black-earth area interlaced with a labyrinth of drainage ditches—presents another unique area to bird.

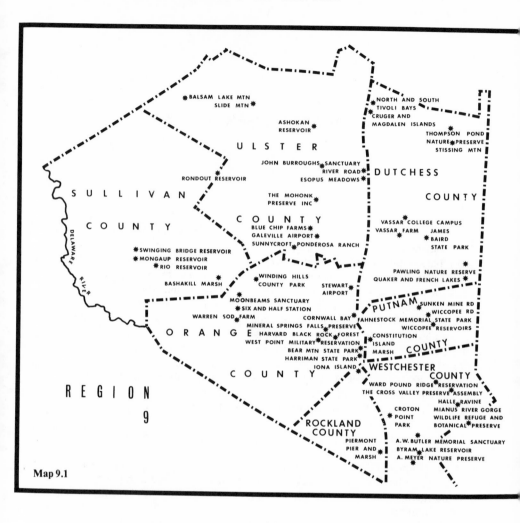

Map 9.1

Bird species with northern affinities can be found breeding and on migration throughout the higher reaches of the region, while birds with southern affinities can be found in appropriate seasons at lower elevations.

In spite of the rapid suburbanization throughout the Hudson-Delaware there are areas remaining whose avian life has still barely been explored. Some of these areas are described below, with the hope that the adventurous birders will be moved to widen their horizons—and will widen ours.

ULSTER COUNTY
Balsam Lake Mountain (Map 9.2)

Rating: Spring***, Summer***, Autumn**, Winter*

The primary reason for including Balsam Lake Mountain in this book is to give the birder accustomed to the strident shrieking of ambulances, fire engines, and burglar alarms, the grinding and crunching of garbage trucks, the threatening chorus of squealing car brakes, and the other various and sundry sounds of big-city discord heard at typical fortissimo levels, an opportunity to listen to some of the area's most infrequently heard bird songs — and this without having to tolerate the swelling, noisy masses winding up and down the trail to Slide Mountain, not to mention the boisterous recreationalists at its summit.

The best time to take this hike is mid-to-late June. At the mountain's summit, near the fire tower on the summit, and along the dirt road that leads to it, Saw-whet Owls, Yellow-bellied Sapsuckers, Bicknell's race of the Gray-cheeked Thrush, and Mourning Warblers nest. However, these are only a few of the species encountered along the way.

To reach Balsam Lake Mountain: leave I-81 (coming from the north or west), or I-84, or I-87 and then I-84 (coming from the east or southeast), and gain NY 17. Proceed on NY 17 to the Livingston Manor Exit. It will not be necessary to go into the village of Livingston Manor; however, if you are without something to eat or drink, this will be your last opportunity to avail yourself of eateries and grocery stores. From the exit or from the village gain NY (county road) 151, heading north. Proceed north through the hamlet of Deckertown, staying on NY 151 as it becomes Johnson Hill Road and then as it gradually curves eastward and becomes NY (county road) 152 and Beaver Kill Road. Continue northeast, passing through the hamlet of Lew Beach, a distance of a little more than ten miles from NY 17.

Continue on Beaver Kill Road to its first right turn after leaving Lew Beach. The sign may indicate the hamlets of Turnwood and Hardenburg. Follow Beaver Kill Road through both of these tiny crossroad hamlets. Beaver Kill Road will have been closely paralleling the Beaver Kill and you will have seen any number of places that deserve pulling off to bird for a few minutes. Be sure to do so. There is wonderful birding habitat all along this route.

From Hardenburg, for which you will have to be keenly looking for fear of missing it, clock 3.5 miles on the odometer. You will be at Quaker Clearing. From here one can park and begin the 2.7 miles hike up to the summit of Balsam Lake Mountain, or one can turn left and proceed for 1 mile north on this old truck-supply road, park off of it near Balsam Lake, and then climb the remaining 1.7 miles to the summit of Balsam Lake Mountain.

An alternative way to reach Balsam Lake Mountain is to leave I-87 (the New York State Thruway) at Exit 19 (Kingston exit), and immediately gain NY 28 headed west. Follow NY 28 west to the hamlet of Pine Hill. There take a left (south) on the road indicated to Bellayre and Highmount. Proceed to a T in the road and there turn left. Drive one mile and turn right at the Department of Environmental Conservation sign indicating the way to the Dry Brook Ridge. Drive for four or five miles along this road looking for signs on the left which show the trail to the summit of Balsam Lake Mountain. Park on the right side of the road in the vicinity of these signs. The trail is long but not rigorous and can be found on the left side of the road.

Along the way you might have seen Turkey Vulture, Broad-winged and Red-tailed hawks, Ruffed Grouse, Yellow-bellied Sapsucker, Downy Woodpecker, Great Crested Flycatcher, Eastern Wood Pewee, Black-capped Chickadee, White-breasted Nuthatch, Winter Wren, and Wood, Hermit and Swainson's thrushes and Veery, Cedar Waxwing, vireo species, Black-throated Blue, Black-throated Green, Chestnut-sided and Blackpoll warblers, Ovenbird, Mourning Warbler, Common Yellowthroat, Canada Warbler and American Redstart, Scarlet Tanager, American Goldfinch, Dark-eyed Junco, and White-throated Sparrow. All of these species, and more, nest within sight or earshot of the trail.

Once you have located the general area of the territory of the singing male Mourning Warbler, locate it more specifically by "spishing" in areas with thick understory of blackberry brambles and ferns, and a thin upperstory of birch. Be alert for a male which will probably leave the thick undergrowth for a few minutes and alight in the small trees above the brush. Only infrequently does this secretive species repeatedly show itself, so be alert the first time it reveals itself.

Walk to the fire tower at the summit (elevation 3723 feet), which was built in 1930. While you are at the summit, do not miss investigating the fifteen-acre sphagnum bog, located approximately 200 yards off the trail through dense balsams. This habitat adds another dimension to the balsam fir habitat through which the birder has been roaming, and introduces the possibility of an unexpected breeding species or two. Near the summit there are two lean-tos; one more primitive than the other and containing a dirt, not wooden, floor. Either of these provide a marvelous place to lunch and both provide comprehensive views of the valley to the south and Balsam Lake. Except in years of severe drought there is clean drinking water available at a spring located underneath a ledge to the left of the trail in the vicinity of the lean-tos.

Slide Mountain (Map 9.2)

Rating: Spring***, Summer**, Autumn***, Winter*

The Guinness Book of World Records is filled with scores of achievements, some awesome, some absurd, but all attesting to the insatiable need of our peculiar species for trophies tangible and intangible. To people who hike in the Catskills, Slide Mountain stands as a sort of trophy peak at 4204 feet in elevation. To obtain its summit has been the objective of most hikers in the Hudson Valley since 1885, when John Burroughs himself added to its summit statistic a special enticement, describing it variously as "hedged about so completely by other peaks—the greatest mountain of them all, and apparently the least willing to be seen"; "we came plump upon the almost perpendicular battlements of Slide. The mountain rose like a huge, rock-bound fortress"; "solitude of mountaintop is peculiarly impressive"; "it has a thrush peculiar to itself. The song is in a minor key, finer, more attenuated, and more under the breath than that of any other thrush."

This birdsong of course, was that of the thrush that Eugene P. Bicknell had observed and collected on the summit of Slide Mountain in 1881. For some years it enjoyed notoriety as a separate species (Bicknell's Thrush) from The Gray-cheeked Thrush, but it was finally reduced to subspecies status by taxonomists. In addition to breeding atop Slide Mountain, it breeds on neighboring mountaintops (usually above 3000 feet) in the Catskills, as well as in selected sites in the Adirondacks. It also breeds in other states east and north of New York, always on remote mountaintops.

Slide Mountain, in my mind, can only be compared to Grand Central Station at rush hour, although admittedly, I've hiked it with the greatest pleasure and reaped great rewards during some seasons; namely, mid-April through mid-May and the last ten days of October. The last part of October will usually not yield more than ten species of birds, but the mountain is peaceful, lovely, colorful and mysterious at that time. The late-April through mid-May period should yield many bird species, and, I suppose if you are willing to be a part of the madding crowd, you can hike it toward the end of May—which, in fact, ought to be even more rewarding in terms of birds.

Whatever you do, do not - *repeat,* do not stay away from Slide Mountain on the basis of my curmudgeonish treatment of it. It is genuinely impressive, especially without the transistor-radio and cassette-recorder carriers and the more than occasional gridlock of hikers.

As a last word here on the dos and don'ts of Slide Mountain: whether the "layered look" is in or out in circles of fashion, be sure you have it. It gets sometimes foggy and wet at the peak, while it can be quite warm at the base of the mountain. Bring something light (in weight) to eat and drink. There is a spring a short distance from the summit, which is usually reliable, but be safe not sorry.

The following directions guide one to the least arduous and least lengthy route to the summit of Slide Mountain. There are other longer, more dramatic and less traveled routes. For those I refer the birder to a wonderful compendium of Catskill hikes by Bennet and Masia (1974).

Leave I-87 (the New York State Thruway) at Exit 19 (Kingston exit) and immediately gain NY 28 heading west. Pass by the northeastern corner of the Ashokan Reservoir on NY 28 and remain on it as it parallels the north shore of the Ashokan Reservoir. This offers the birder the opportunity of stopping to bird some of the extensive buffer zone of natural and planted trees along the impoundment's edge for migrants and/or breeders. Continue following NY 28 northwest as you pass through Phoenicia, and Shandaken.

The village after Shandaken is Big Indian, which is 32 miles from Exit 19. In the center of Big Indian you must take a left turn onto Oliverea Road leading south to the village of Oliverea. From the turn to the village is 3 miles. This is a marvelous road on which to bird. It passes right down through Big Indian Hollow and on its right (west) side, the Esopus Creek chases along, originating in the artificial impoundment of Winnisook Lake, some miles south. Proceed south through the hamlet of Oliverea, continuing on to Winnisook Lodge, which you will see on the left. Approximately 1.5 mile south of the Winnisook Lodge, on the Big Indian-West Branch Road, the state-marked trail can be found on the left (east) side of the road. From Oliverea to the state-marked trail to Slide Mountain is six miles. Winnisook Lodge is a private retreat owned by members of the Winnisook Club. This is strictly off-bounds, so do not insist upon trespassing.

The 3-mile trail to the summit of Slide Mountain is well-marked and is unquestionably the easiest and shortest ascent. At the top there are stunted balsams. The elevation at which the climb begins is approximately 2665 feet above sea-level, making it approximately 1540 feet below the summit. Along the walk one should be able to note the vegetational transition as the beech, oak, and sugar maple give way to spruce and fir. Ferns and mosses are abundant along the trail.

Breeding species one should expect to encounter are: Ruffed Grouse, Pileated Woodpecker, Yellow-bellied Sapsucker, Yellow-bellied Flycatcher, Redbreasted and White-breasted nuthatches, Brown Creeper, Winter Wren, and Hermit, Swainson's, and Gray-cheeked thrushes (including the *bicknelli* race of the Gray-cheeked Thrush). Warblers one should watch and listen for are: Blackthroated Blue, Yellow-rumped, Black-throated Green, Blackburnian, Blackpoll, and Canada. Of course, close to the summit the Dark-eyed Junco and the Whitethroated Sparrow nest.

Ashokan Reservoir (Map 9.2)

Rating: Spring**, Summer*, Autumn****, Winter***

Ashokan Reservoir, in the northeastern section of the county, is approximately 7 miles west of the city of Kingston, and was the first of New York City's reservoirs built in the Catskills. It was constructed in 1912, is 12 miles long, and averages one mile wide. It covers more than 8000 acres with a maximum depth of 190 feet and an average depth 50 feet.

To reach Ashokan Reservoir: leave I-87 (the New York State Thruway) at Exit 19 (Kingston Exit). Immediately gain NY 28 heading west. Proceed west 3 miles to the fork of NY 28A and NY 28. Turn left and gain NY 28A (County Road 50) which is also called New York City Road. Follow it west around the southern end of the reservoir, stopping along the east basin, at safe pull-offs and especially stopping to bird well the Dividing Weir area and the spillway area, at 9.3 miles from the turn onto NY 28A. Here there is limited parking on the right. On the left are buildings of New York City Water Supply. Walk along Dividing Weir road and right and left on the small roads intersecting it at the southern end of the Dividing Weir. Another very productive spot to bird is the northeastern end of the reservoir in the area of the village of West Hurley. There are nesting Cliff Swallows on the buildings of the Dividing Weir.

The prime times to visit this enormous reservoir, surrounded by the rolling, tree-covered slopes of the eastern peaks within the Catskill Park, is from mid-October through mid-December. In October and early November one may see southward-bound migrant waterbirds, including Common Loons, Red-necked, Horned and Pied-billed grebes, Canada Geese, Wood Ducks, American Wigeons, Green-winged Teals, Mallards, Black Ducks, possibly Pintails, Blue-winged Teals, Canvasbacks, possibly Redheads, and certainly Ring-necked Ducks, both scaup species, possibly all three scoter species (although White-winged and Black are the more frequently observed species). As winter progresses one should be able to see

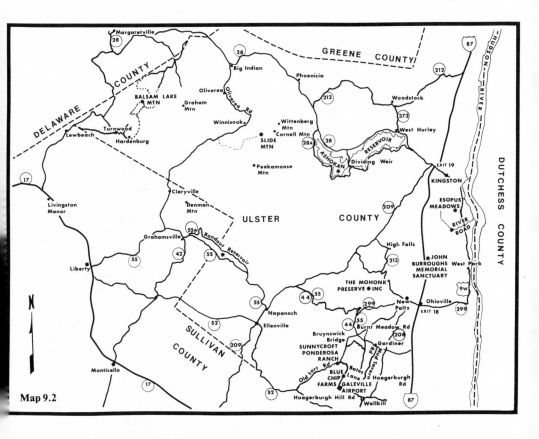

Map 9.2

Buffleheads, Common Goldeneyes, and Hooded and Common mergansers in considerable numbers. Look for migrating Sharp-shinned and/or Cooper's hawks flying over the reservoir and be observant for resident Red-tailed and wintering Rough-legged hawks in the vicinity of the reservoir or on the drive to it.

When the water in the reservoir has dropped to below-normal levels, sizable expanses of mud and rock are exposed, and it is during those years that species normally uncommon in Ulster County, *e.g.,* Black-bellied Plover, both Greater and Lesser yellowlegs, Solitary and Pectoral sandpipers, can be observed here. At least Great Black-backed, Herring, and Ring-billed gulls are to be expected and the relatively infrequent but annually occurring Bonaparte's Gull should be watched for. Search the woods for owls, woodpeckers, Boreal Chickadee and winter finches in flight years, titmice, nuthatches, Brown Creepers, and Golden-crowned and Ruby-crowned kinglets.

At the Dividing Weir, along the dike road, and near the spillway look for Horned Larks, Water Pipits, juncos, and Tree, White-crowned, White-throated, Fox, and Song sparrows. These are excellent spots at which to see also Snow Buntings.

John Burroughs Sanctuary, "Slabsides", River Road and Esopus Meadows, a Tour (Map 9.2)

Rating: Spring****, Summer**, Autumn**, Winter

The John Burroughs Memorial Association owns and maintains the 175-acre John Burroughs Sanctuary, on which is located "Slabsides," the house built by the famed naturalist, John Burroughs, in 1895. It was in this house that the writer spent many productive days at this retreat during the last twenty-five years of his life. Because of the fame of Burroughs and the natural beauty surrounding it, "Slabsides" became a mecca for such naturalists as John Muir and Theodore Roosevelt. Today it has been declared a National Historic Landmark. The area can be characterized as a plot of second-growth forested ridges and rock outcrops containing a long man-made pond, celery swamp, and a section drained by Black Creek, with various low-lying marshes maintained by surface water runoff. Approximately 90 percent of the area is covered with second-growth forest and brush communities consisting of an intermix of oak-hickory and hemlock-northern hardwoods. Generally oak and hickory species are found on the warmer and drier parts of the sanctuary, while hemlock, maple, and birch occupy the moist and cooler sites.

To reach the Burroughs Sanctuary: leave I-87 (the New York State Thruway) at Exit 18 (New Paltz exit) and immediately gain NY 299 heading east. Proceed 5 miles east to the intersection of NY 299 and US 9W. Turn left and proceed north on US 9W for 3.5 miles to Floyd Ackert Road. Turn left (west) here (just past Marcel's Restaurant) and continue 1 mile, crossing the railroad tracks and continuing up the hill to the foot of Burroughs Drive, on the left. Burroughs Drive may or may not be chained off; cars may not proceed on it. However, there is room to park off the road at the entrance, which is well marked. From there one can walk the 0.5 mile up Burroughs Drive to Slabsides Lane and "Slabsides." From there one is free to roam throughout the sanctuary trails looking for the birds it is bound to harbor during migration and listening to the songs of species breeding within the sanctuary confines, early in the breeding season.

A total of 142 species of birds has been recorded on the sanctuary grounds with just over one-half of these being species that nest there. Some of the more notable breeders are: Red-shouldered, Broad-winged and Red-tailed hawks, Ruffed Grouse, Yellow-billed and Black-billed cuckoos, Great Horned and Barred owls, Whip-poor-wills, Pileated, Hairy, and Downy woodpeckers, Brown Creepers, Winter and Carolina wrens, Wood Thrushes, Veeries, Eastern Bluebirds, Blue-gray Gnatcatchers, Yellow-throated Vireos, Worm-eating, Golden-winged, and Blue-winged warblers, Black-throated Blue, Yellow-rumped, Black-throated Green, Cerulean and Chestnut-sided warblers, Ovenbirds, and Northern and Louisiana waterthrushes.

After leaving the John Burroughs Sanctuary, follow Floyd Ackert Road back out to the junction of US 9W. Turn left and continue north 4.5 miles, where River Road comes in on the right. Turn right and follow along the almost semicircular River Road for 1.4 miles, where there is a place to stop with a fine vantage of the Hudson River. Along River Road one can see a shallow section of the river, with some flats exposed at low tide. This area is commonly referred to as Esopus Meadows. Fish production is high at these meadows and hence it is an excellent feeding place for duck species, especially in migration. Proceeding 0.7 mile farther on River Road there is another excellent view of the river in the area of a small beach. It is easy to stop, park, and scope the Hudson from this vantage point. Continue to the end of River Road.

The Mohonk Preserve, Inc. (Map 9.2)

Rating: Spring***, Summer****, Autumn***, Winter*

The long, narrow ridge known as the Shawangunk Mountains extends from the town of Rosendale, in Ulster County, southwest approximately fifty miles to the New Jersey state line. There the name of this ridge becomes the Kittatinny Mountains, although geologically each of these two is simply an extension of the other. The Shawangunks and the Kittatinnys are both components of the Appalachian Mountain Chain. The Mohonk Preserve is a 5500-acre tract lying on the ridge of the northern Shawangunk Mountains, in Ulster County.

The preserve is of interest from several viewpoints, as well as that of the breeding and migrant birds recorded on the property.

The Shawangunks are a meeting place for several diverse vegetational communities: southern hardwoods, characterized by oak, hickory and pitch pine are dominant in the Hudson Valley and throughout large areas on the mountain ridge; relict stands of northern coniferous spruce forest (not fir), can be seen in several scattered patches. Northern hardwoods, characterized by beech, yellow birch, sugar maple, white pine, and hemlock are a large component of the vegetation on the mountain; there are additionally a few remnant alpine plants which have, except for these, been replaced by forest.

Below the high ridges within the preserve is a long, narrow, poorly drained strip of woods. Here one finds ash, elm, hemlock, black birch, red maple, and infrequently yellow birch. The understory contains highbush blueberry, alders, willows, and silky dogwood. Here also ferns, sedges, and bryophytes are found in abundance.

Dramatic slab cliffs and scarps provide marvelous long vistas of the Rondout Valley, the Catskills, and Wallkill River. Ice and compacted snow does not leave the deeply creviced rocks until nearly July. This provides a unique habitat for plant species not usually found in southeastern New York State. The upper expanse of rock is visible for some distance and resembles a slightly tilted table, thus the local name of "Table Rock." The cliff face and talus slopes have not been disturbed since the last Ice Age, thus providing a fascinating area in which to study those plant species that are capable of tolerating the most difficult conditions. In this habitat, scrub oak, chestnut oak, pitch pine, and white pine grow, as well as lowbush blueberries, huckleberries, and sheep laurel in the understory. Lichens and bryophytes are common throughout this, the southeastern portion of the preserve. Within the preserve there are 300-year-old trees, growing slowly as they cling to these rocks.

There is an extensive network of old but substantial wood roads, trails, and even some stone walls that point up infrequent sites previously suitable for some sort of human habitation or use. Along the slopes here the vegetation is mixed conifer-hardwoods, including red oak, black oak, red maple, hemlock, and black and yellow birch in the cool ravines and on the north-facing slopes. The understory contains flowering dogwood and an abundance of mountain laurel.

Owing to the diversity of terrain and microclimate and the near-absence of human habitation and interference, the preserve is home to nineteen species of mosses and ferns, two tree species, and six wildflower species found on the *Rare and Endangered Vascular Plant Species in New York State*. These species, of course, should not be picked or removed from their natural habitats.

To reach the Mohonk Preserve, Inc.: leave I-87 at Exit 18 (New Paltz exit) and head west on NY 299. Pass through the village of New Paltz. Pass over the bridge spanning the Wallkill River, and stay on NY 299, heading west. Proceed to its junction with NY 55 and US 44. Here turn right and proceed up the hill to the entrance of the Mohonk Preserve lands, which will be well marked. There is a nominal daily-use and parking fee. Frequent users may purchase an annual pass.

Turkey Vultures nest on these lands. The birder might expect to see nesting Goshawks; this is one of the most southern locations for a nest of that species. Red-tailed Hawks and American Kestrels nest in appropriate habitat here. Black-billed Cuckoos can be found nesting in June and July as can Great Horned and Barred owls. Among picids, Pileated, Hairy, and Downy woodpeckers commonly nest. Great Crested Flycatchers, White-breasted and Red-breasted nuthatches, and Brown Creepers nest on the grounds. One of the most vociferous and delightful birds to be found breeding on the talus slopes and in the cool ravines is the Winter Wren. Wood Thrushes, Golden-crowned Kinglets, and Cedar Waxwings can all be found nesting here. Among the many species of breeding warblers to be found here are: Worm-eating, Cape May, Yellow-rumped, Black-throated Green, Ovenbird, and Canada Warbler.

Blue Chip Farms, the Galeville Airport, and Sunnycroft Ponderosa Ranch, a Tour
(Map 9.2)

Rating: Spring****, Summer***, Autumn**, Winter****

The Shawangunk Kill originates in a number of small streams in the towns of Greenville, Wallkill and Mt. Hope, in southwestern Orange County. It is one of the few rivers flowing north in the eastern United States. It proceeds northeast from Orange County, and for approximately ten miles forms the border between Orange and Sullivan Counties. After that, it forms the border, for approximately eight miles, between Orange and Ulster Counties. Then it continues north and east, joining the Wallkill River. The two creeks are a part of the Rondout River system; they flow together northeasterly, joining the Rondout Creek near Rosendale. The Rondout flows into the Hudson River at Kingston.

The Shawangunk Mountains dominate the landscape along the Shawangunk Kill and the Wallkill Rivers. Their eastern ridge forms quite a dramatic escarpment. The vistas in the valleys are very broad, owing to the level-to-slightly-rolling character of the land. Large portions of this land have been cleared for agriculture.

The road patterns in the Shawangunk and Wallkill valleys generally follow the stream patterns, and human settlement patterns parallel the roads. There are actually no major population centers in these valleys but there are a number of rural corners which survive as local centers, offering limited retail business or community services to local residents. Most of the buildings in the valleys are private dwellings; farms, summer homes, religious communities, etc. Most of the land is in agriculture; but, as the land rises toward the mountains, the woodland becomes heavier, approaching almost totally forested mountain slopes.

These three named sites lie quite near one another in the south-central part of Ulster County. They lie northwest of the village of Wallkill and east of Rutsonville and Dwaarkill.

The Blue Chip Farms is a very large horse-breeding operation, with holdings on both sides of Hoagerburgh Road, east of the village of Dwaarkill. By far the most significant tourist business in this section of Ulster County is the Sunnycroft Ponderosa Ranch, which owns and leases approximately 300 acres; it serves as a resort ranch and is located north of the Blue Chip Farms. The United States Government owns the approximately 550-acre Galeville Airport site on the Hoagburgh Ridge near the old Bruynswick Center, 3000 feet east of the Shawangunk Kill. It was developed as an airport during World War II and is now used as a parachute jumping site for West Point cadets. The Galeville Airport is located very close to the farm and resort ranch, so the three places make an ideal birding tour, taking the better part of an early spring morning or most of a mid-winter birding day.

Leave I-87 at Exit 17, west of Newburgh. Turn right and then quickly right again, onto NY 17K, headed west. Proceed west on NY 17K to its intersection with NY 208. Here, turn right and head north, passing through the village of Walden, in Orange County. Continue north, passing over the border between Orange and Ulster Counties to the village of Wallkill. You will have been paralleling the Wallkill River from just north of Walden and should have stopped at habitats along it that looked productive for spring migrants in that season. From Wallkill, turn left on Hoagerburgh Hill Road. Cross over the Wallkill River and over the Dwaar Kill. Proceed west to the intersection of Hoagerburgh Hill Road and Hoagerburgh Road. Here, turn right and proceed north on Hoagerburgh Road past two or three dwellings on the left (west) side of the road. Within sight of the next intersecting road, Bates Lane on the east and Old Fort Road on the west, you will see the Blue Chip Farms and its extensive acreage beginning on the left side of the road. Bird all along this stretch, on both sides of the road, from this point to the next road intersecting Hoagerburgh Road. Here, turn left (west) on Old Fort Road.

The agricultural fields on both sides of the road here are within the holdings of Blue Chip Farms. Bird along this road to a historical marker located on the right (north) side of the road. This marker indicates the traditionally accepted site of a palisaded Esopus Indian fort, attacked by Dutch troops in 1663. Just beyond these flats, a bit north of and at the end of Old Fort Road, are perhaps the finest woodlands along the Shawangunk Kill in Ulster County. They are located on steep slopes and are composed primarily of mature beech, sugar maple, hickory, white oak, red oak, American sycamore and tulip poplar. There is a spring in the woods just above the flats. Bird along it and throughout the woods, after having parked the car safely off Old Fort Road.

After birding in the forested sections return (east) along Old Fort Road to its intersection with Bates Lane and Hoagerburgh Road. Here, turn left (north) and note a row of thick red cedars on the right (east) side of the road. These cedars are just south of the Galeville Airport site, the entrance to which can be seen a few hundred feet north of the intersection, on the right (east) side of Hoagerburgh Road. The clapboard shingled dwelling, south of the airport, is one of the holdings of the Blue Chip Farms. Its surrounding open land should be carefully birded. Bird the old airport grounds as thoroughly as possible. In some years, driving east along Bates Lane to Mud Tavern Road has proven a good bird-finding strategy.

After thoroughly covering that site, continue north on Hoagerburgh Road, following it as it turns west. Just past the intersection of Burnt Meadow Road, which comes in on the right, is the Shawangunk Reformed Church, parsonage, and burial grounds. This group of buildings is now on the left (south) side of Hoagerburgh Road. Proceed west and downhill to the Bruynswick Bridge. Here the Shawangunk Kill is approximately 150 feet wide but only about 1.5 feet deep.

Just west of the Bruynswick Bridge are the holdings of the Sunnycroft Ponderosa Ranch resort. It owns the land south of Hoagerburgh Road, east to Bruynswick Road, on both the east and west sides of the Shawangunk Kill. It is basically a dude ranch whose business flourishes in summer but slacks off in the colder weather.

North of the Bruynswick Bridge the land rises rather steeply. Most of it is cleared and is presently used as pasture or cropland. Approximately 2500 feet north of the bridge, a substantial floodplain opens up on the east and west sides of the Shawangunk Kill. Uphill and east from the floodplain is a rich, mixed-deciduous woodland, encompassing nearly 140 acres. To bird this forest one must return east on Hoagerburgh road to its junction with Burnt Meadow Road. There, turn left (north) and drive north a short distance, just past the Shawangunk-Gardiner town line. Park the car off the road and walk west and slightly north into the woods.

Blue Chip Farms, a large manicured farm specializing in breeding registered Morgan horses, and Sunnycroft Ponderosa Ranch are excellent sites at which to see nesting upland meadow birds during late spring and early summer as well as sometimes during spring and autumn migrations. During the winter, the fields, meadows, uplands and surrounding farms, and especially the Galeville Airport, attract winter resident hawks, owls, picids, and half-hardy larks, finches, sparrows, longspurs, and buntings.

Blue Chip Farms and Sunnycroft Ponderosa Ranch are two of the ever-decreasing strongholds of nesting Upland Sandpipers. The number of sites from which that species has disappeared as a breeder in New York State is astonishing. Other nesting species to be expected in this vicinity are: Broad-winged and Red-tailed hawks, American Kestrel, Ruffed Grouse, Killdeer, American Woodcock, Common Nighthawk, Chimney Swift, Ruby-throated Hummingbird, Common Flicker, Pileated Woodpecker, Yellow-bellied Sapsucker, Hairy and Downy woodpeckers, several flycatcher species, several swallow species, jay, crow, chickadee, titmouse, nuthatches and creeper, as well as wrens of several species, mimids and several thrush species, including Eastern Bluebird. The Blue-gray Gnatcatcher has expanded its range within Ulster County and one should be able to find it in this area. If one birds the area in late May and early June, it is no trick at all to turn up breeding Red-eyed, and Warbling vireos. At the same time one should be able to find Blue-winged, Golden-winged, Nashville, and Chestnut-sided warblers all acting territorially. Among uplands-nesting species Bobolink, Eastern Meadowlark, Grasshopper, Field, and Song sparrows can all be found at these sites. On the periphery of Sunnycroft Ponderosa Ranch there are extremely productive upland meadows and grasslands. Here, there is a colony of the locally rare Vesper Sparrow nesting.

From late November through the end of February these three areas are birded especially for the numbers and variety of raptor species that usually winter throughout the Wallkill Valley from approximately New Paltz to the Galeville Airport. The majority of the trips are held from the end of the first week in January through the middle of February. Times outside of this period have been also productive, but, most birders have had luck during this prescribed time span. The entire Wallkill Valley with its open fields and well-sodded wet meadows, provide ideal raptor habitat; but, the area of Galeville Airport and Blue Chip Farms are among the most gratifying winter birding areas in *Kingbird* Region 9.

If all of the conditions are just right one can expect to see numbers of wintering Marsh, Red-tailed, and Rough-legged hawks of both light, dark and intermediate color phases, American Kestrels, pheasants, and possibly a Killdeer or two. In the low red cedars at the Galeville Airport and abutting Blue Chip Farms look for Short-eared Owls roosting, sometimes only eight to ten feet up. If one times the trip in order to arrive in late afternoon, the chances are quite likely that the birder will see owls as they begin their crepuscular hunting. With even more luck, one will be able to approach within fifty to seventy-five feet of one of the relatively tame owls as it perches atop a fencepost. A few winters ago Jeheber *et al.* witnessed ten to twelve Short-eared Owls roosting in one red cedar tree at the airport site. Brayton *et al.* have effectively used the car as a blind, approaching the owls to a distance within which, by using a 20X scope, they could get views "of every feather."

Other species to be expected wintering in the Wallkill Valley are Horned Lark, an occasional Northern Shrike, Evening Grosbeak, Purple Finch, Pine Grosbeak, Common Redpoll, Pine Siskin, possibly Red and White-winged crossbills, and Tree Sparrow.

In August and September, occasionally, a late-migrant shorebird, such as American Golden and Black-bellied plovers, can be seen in the manicured fields of the farm and ranch.

ULSTER AND SULLIVAN COUNTIES
Rondout Reservoir and Swinging Bridge, Mongaup, and Rio Reservoirs, a Tour
(Maps 9.2 and 9.3)

Rating: Spring**, Summer**, Autumn***, Winter****

Rondout Reservoir straddles the boundary line between northeastern Sullivan and southwestern Ulster counties. It is about seven miles northwest of Ellenville and encompasses more than 2000 acres. It is about six miles long and averages half a mile wide. Its maximum water depth is nearly 180 feet, and 50 percent of the reservoir averages 75 feet deep. This reservoir was built in 1951 and is owned, maintained, and services New York City. Approximately 2.25 miles of its total length lies in Sullivan County and 3.75 miles stretches east into Ulster County.

The northern end of a string of impoundments — Swinging Bridge, Mongaup, and Rio Reservoirs — is located approximately eighteen miles directly southwest of the western end of Rondout Reservoir. This distance translates into nearly 25 miles by automobile. The northern end of Swinging Bridge Reservoir lies six miles west of Monticello. These reservoirs stretch south from the hamlet of Mongaup Valley for nearly eleven miles in Sullivan County, and the last mile or so of Rio Reservoir extends into Orange County.

On all of these reserves the birder can usually find various waterfowl species in winter, but their main attraction is the concentration of Bald Eagles, of various ages, that winter at these sites. The New York State Department of Environmental Conservation is carefully monitoring the winter movement of Bald Eagles in this part of the state, and a recent estimate places the maximum number of eagles in the Ulster-Sullivan-Orange counties area at 35 individuals.

This wintering concentration is certainly the largest in the entire northeastern United States. The eagles appear to fly from one reservoir to another, in search of the most abundant food sources. Some water at each of the sites is usually open, even at the coldest temperatures, owing to streams feeding into the reservoirs and the existence of power plants on the reservoirs. Fish, after being caught in the turbines of the hydroelectric generators, come up dead or severely traumatized, and provide a fairly reliable and stable food source for the eagles and several waterfowl species. The Alewife (*Pomolobus pseudoharengus*) was probably introduced into the reservoir complex in southern Sullivan County in the early 1960s. In the early 1970s there was an Alewife die-off in this complex which may have been responsible, in part, for the high eagle densities observed there in the winters of 1972–1973 and 1973–1974. Conservative estimates place the number of individuals present in those winters between twenty-one and twenty-eight.

The time to view the eagles can be anytime from mid-December through the first week in March. The birds appear to remain dispersed at various open-water sites until ice has closed them. After a period of below-freezing temperatures, the eagles tend to concentrate at the open-water feeding sites at one or all of the reservoirs. Peak numbers of eagles seem to be correlated with long periods of very cold weather in January and February.

Nota Bene: The Bald Eagle unquestionably typifies the ultimate in sensa-

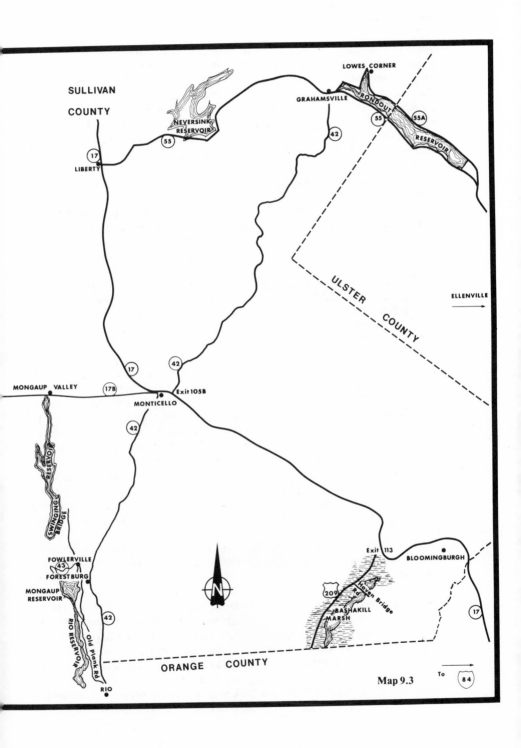

SULLIVAN
COUNTY

LOWES CORNER

GRAHAMSVILLE

RONDOUT

NEVERSINK
RESERVOIR

55A

RESERVOIR

55

42

17
LIBERTY

55

ULSTER

COUNTY

ELLENVILLE

17

42

MONGAUP VALLEY

17B

Exit 105B
MONTICELLO

42

SWINGING BRIDGE RESERVOIR

Exit 113

BLOOMINGBURGH

FOWLERVILLE

43

FORESTBURG

MONGAUP
RESERVOIR

209

Haven Bridge Rd.

BASHAKILL
MARSH

17

42

RIO RESERVOIR

Old Plank Rd.

N

ORANGE   COUNTY

Map 9.3

To
84

RIO

tional and impressive sights. To see more than one simultaneously, or in close succession, affords the birder a rare and treasured treat. However, this is still an endangered species, and long-term studies confirm that it is especially sensitive to close human disturbance. It is required therefore, that no birder attempt to approach these birds closely. In addition to constituting unethical behavior, there are very severe federal penalties for acting in ways that could be construed as molesting or disturbing these birds; these include approaching them on foot, approaching roosting sites too closely, etc. At each of the above-named reservoirs there are ample DEC designated sites from which the birds can be viewed.

To reach Rondout Reservoir, leave I-87 (the New York State Thruway), at Exit 18 (New Paltz). Turn left onto NY 299 headed west. Continue west for about 8 miles where NY 299 becomes US 44 and NY 55. Proceed west approximately ten miles to the intersection of US 44 (NY 55) and US 209. Turn left and proceed south on US 209 (NY 55) approximately 3.8 miles, where NY 55 turns north, on the right, in the community of Napanoch. Turn right and proceed northwest approximately 5 miles to the junction of NY 55 and NY 55A. Stay right, thereby gaining NY 55A, and proceed along the north side of the reservoir. This road parallels the reservoir shoreline, and one should look for eagles perched in the trees close to the road or fishing in open-water patches. Proceed to the northwest corner of the reservoir at the hamlet of Lowes Corner. Here Rondout Creek enters the reservoir and the power plant is located here. Look for eagles again in the trees lining the shore, or fishing in the open water near the power facility.

If there is no eagle within 0.25 mile, park the car and scan the open water and scour the woods close to the road. At this site and at the more southwesterly reservoir complex one should easily see Canada Geese, Mallards, Black Ducks, possibly Pintails or Buffleheads, goldeneyes, certainly Hooded and Common mergansers, one or two Red-tailed Hawks and an equal number of American Kestrels, and an occasional Ring-necked Pheasant. One of the specialties of the Rondout Reservoir and the other reservoirs as well, is Turkey. Sometimes as many as five to ten have been sighted at each place. Look for the very rare Iceland Gull among the infrequently occurring black-backeds, and the regularly occurring Herring and Ring-billed gulls. Land birds to look for in the surrounding mix of deciduous and evergreen trees are Barred Owls, Downy Woodpeckers, jays, crows, chickadees, and titmice, Red-breasted Nuthatches, Brown Creepers and especially wintering Evening and Pine grosbeaks, and Pine Siskins, and American Goldfinches. The sparrow species one might expect in groups of two or three are Tree and Song.

To reach the Swinging Bridge, Mongaup, Rio reservoirs complex from Rondout Reservoir, turn left at Lowes Corner and drive southwest on NY 153 to the hamlet of Grahamsville. There, turn left on NY 42 and proceed south on it for about 16 miles to its junction with NY 17. Cross NY 17 into the village of Monticello and follow NY 42 as it turns right and then left and then continues south. Proceed south on NY 42 for approximately 10 miles to the hamlet of Forestburg.

To reach the Swinging Bridge, Mongaup, Rio reservoirs complex from Newburgh (Exit 17 on I-87: the New York State Thruway), proceed southwest on I-84 for a distance of 18 miles to the intersection of NY 17 and I-84. Here, turn right onto NY 17. Proceed northwest for about 26 miles to Exit 105B. Here exit NY 17 and follow signs for NY 42 heading south. This will take you into the village of Monticello. Follow NY 42 as it turns right and then left in Monticello and then proceeds south. Follow it south for approximately 10 miles to the hamlet of Forestburg.

Turn right at Forestburg and follow NY (country route) 43 northwest, through the hamlet of Fowlerville, and then south and west to the bridge crossing the north end of Mongaup Reservoir. Swinging Bridge Reservoir can be reached by turning right (north) on the small road on the east side of the bridge and driving approximately 1.2 miles. It is highly inadvisable to try this road if it is not cleared of snow, and even if it is, one takes some risk if his vehicle is not equipped with four-wheel drive. From the bridge itself and the pull-offs on both sides of it one can get marvelous views both north and south. All birders can probably get excellent views of the eagles without leaving the car.

To reach Rio Reservoir, turn around and drive east, back to Fowlerville. Here, turn right on the only road intersecting NY (county road) 43. Proceed south on Old Plank Road (Section A), as it parallels Black Brook and then closely parallels the east side of Rio Reservoir.

All along these routes the birder should have been scanning the trees lining the shores of Swinging Bridge, Mongaup, and Rio reservoirs for eagles, as well as the other various species mentioned as occurring at Rondout Reservoir. Make a point of stopping to scour any open water. These are likely places to stop, exercise patience, and wait for an eagle to swoop in and lift a fat fish out of the water. To the person who has never witnessed that feat, all I can say is: it is well worth waiting for, and once seen, is memorable in the extreme.

SULLIVAN COUNTY
Bashakill Marsh (Map 9.3)

Rating: Spring****, Summer**, Autumn***, Winter**

Freshwater wetlands—marshes, swamps, bogs, pools, sloughs, bayous, etc.— assume various forms and differ from one another in a number of fundamental ways. Some are entirely independent of above-ground freshwater ways. Some are closely interconnected with nearby rivers, which overflow periodically, and with lakes. Primarily, freshwater wetlands are created by unique characteristics of the surrounding landscape. Essentially, they are above-ground extensions of the groundwater in an area. Although, there is no daily tidal influence, as is the case of coastal marine environments, the water level fluctuates, on annual or seasonal schedules, in response to variations in rainfall. Some wetlands are relatively small or only intermittent and seasonal, while others are quite large and are permanent features of the landscape. All wetlands, large or small, however, are ephemeral on a geological time scale.

Some freshwater wetlands tend to be infertile, such as bogs; others are naturally fertile. Because organic nutrients and organic matter tend to accumulate in wetlands, there is a great temptation for humankind to drain them and use the drained land for agriculture. This, of course, displaces the myriad plant species and wildlife forms that are entirely dependent upon a wetlands habitat for life.

Wetlands nourish many species of mosses, grasses, sedges, rushes, reeds, and bulrushes. Most of these wetlands additionally have an enormous capacity for absorbing water during wet periods, for filtering out silt, and for gradually releasing clear water during any dry period. Rivers or creeks that originate in, or pass through, marshes and other wetlands are filtered and cleared and are more constant in flow throughout the year, than are those that depend altogether upon surface runoff, especially upon runoff from cultivated landscapes. Nevertheless, despite their contributions to other ecosystems, most wetlands are dependent upon the specific features of the surrounding landscape from which they developed. Any significant alteration or disruption, therefore, of the system that maintains a wetland may lead to its destruction.

Whether through design, accident, or ignorance, we have polluted our wetlands and desecrated our landscapes. We have, to a large extent, destroyed nonreplaceable natural wetlands and damaged others irreversibly, and all of this in the name of progress.

Within the last decade there has been an increasing awareness of the need for making conscious, informed choices regarding further modification of the natural wetlands remaining in the country. Inland wetlands, formerly quick to be filled and built upon or drained and planted, are now seen as valuable and endangered natural resources. Most wetlands are readily identifiable by the presence of typical emergent vegetation or by various amounts of submerged and floating plant life.

The Bashakill Marsh is a freshwater marsh. As such, it is covered with shallow water, the level of which rises in periods of heavy rainfall and heavy river runoff, and recedes during more dry periods. The Bashakill is fed by ground water, by surface springs, by streams, by runoff from the surrounding terrain, and by rainwater.

The Bashakill's vegetation is characterized by soft-stemmed plants — especially grasses, sedges, and rushes — which may emerge above or float on the surface. This vegetation includes waterlilies, cattails, reeds, arrowheads, pickerelweed, and smartweed.

The species that feed in such freshwater marshes as the Bashakill Marsh inhabit one of the most extraordinarily fertile of all environments. the 2200-acre Bashakill wetland is the largest freshwater marsh southeast of Montezuma National Wildlife Refuge, in New York State. It harbors a vast array of flora and fauna, some of which species are listed on rare and endangered lists.

The Bashakill literally teems with wildlife, no small proportion of which is birds. It is spoken of as a "birder's paradise" during spring and autumn, for upwards of 220 bird species have been recorded as breeders or migrants at the Bashakill Marsh.

More than thirty mammalian, twenty-five reptilian and amphibian, and thirty fish species inhabit the expansive marsh.

On the eastern edge of the marsh on the South Road, which parallels NY

209, there lie Sullivan County's only limestone formations. Below these outcroppings are extensive underground caves, which provide roosting habitat for several bat species. Over these limestone formations tumble millions of gallons of water in early spring, resulting in some very aesthetically pleasing waterfalls.

To reach the Bashakill, leave NY 17 (the Quickway) at Exit 113 and head south on NY 209. Proceed 2 miles south and take the first possible left turn (just past Moose Lodge). This is Haven Bridge Road, which dissects the marsh; the bridge road is often just called the causeway. The old, now abandoned, Delaware & Hudson (D&H), canal lies on the western edge of the marsh, east of NY 209, and provides a natural footpath on which to walk to observe the marsh wildlife. The now defunct Ontario and Western Railway bed lies on the eastern edge of the Bashakill and forms an easy trail from which to view the birds and beasts of the marsh on the east side of the marsh. NY 17 forms the northern boundary of the Bashakill Marsh, and the southernmost tip of the marsh juts slightly into Orange County.

Pied-billed Grebes and Great Blue and Green herons are commonly seen here. Watch for Least Bitterns clinging to the stalks of purple loosestrife, and American Bitterns hunting along the marsh edges. These latter two species nest within the marsh. From the third week in March through mid-April, there is an extensive waterfowl migration to be seen here. Ospreys are frequently seen during both spring and autumn migrations. Bald Eagles put in an appearance or two nearly every spring, autumn, and winter and do not be surprised to see Northern Harrier flying low over a playful otter. Sharp-shinned, Cooper's, Broad-winged, and Red-tailed hawks can all be expected in mid-September through October. During nesting season, especially early in the morning, listen for the gobbling mating call of the tom Turkey. Virginia Rails, Soras, and Common Gallinules nest in the marsh.

In autumn look for migrating Killdeers, American Golden Plovers, American Woodcocks, and Common Snipe in numbers. Infrequently, Greater and Lesser yellowlegs, and Solitary and Least sandpipers occur during autumn migrations.

Birding the marsh and its environs around Memorial Day weekend should certainly yield both Yellow-billed and Black-billed cuckoos. At this time, examine the beautiful scarlet cardinal flowers which have such a strong attraction for the helicopter-like Ruby-throated Hummingbird. The rattle of Belted Kingfishers overhead will be unmistakable.

Picids of five or six species occur during spring and autumn. From the end of April through May tremendous waves of migrating flycatchers, vireos, and warblers can be seen poised on the buttonbush, viburnums, and red maples. From mid- to late May it is not unusual for birders out for a "big day" to tally over ninety species at Bashakill Marsh by early afternoon. The northbound passerine migration usually peaks here during the second and third weeks in May. During this period listen for Long-billed Marsh Wrens defending one or more territories in the marsh. Winter birding at the Bashakill has its own rewards: in addition to spotting the very possible Bald Eagle or Rough-legged Hawk, look also for Northern

Shrike, Evening or Pine grosbeaks, Common Redpoll, Pine Siskin, and American Goldfinch. At this time Tree, Fox, and Song sparrows can usually be found in the brushy hedgerows lining the paths.

ORANGE COUNTY
Winding Hills County Park (Map 9.4)

Rating: Spring***, Summer**, Autumn**, Winter*

This relatively new birding site is monitored mainly by the members of the Goudy Wildlife Club and the Edgar A. Mearns Bird Club. It lies about five miles north of the village of Montgomery. There is neither an admission nor a parking fee. Some of the park can be birded by car but more thorough coverage is guaranteed by traveling the trails and paths on foot. Birding here is rated as "moderately easy." There are blackflies and wood ticks present in season, and one should note and avoid poison ivy patches. At Winding Hills Park there is a Nature Center where a naturalist is on duty during the day. Although there is not a central bird list kept at this center, it is advisable to check with the naturalist, as this person will know of any notable current bird sightings.

To reach Winding Hills County Park, leave I-87 (the New York State Thruway) at Exit 17, west of the city of Newburgh. Gain I-84, headed west, and proceed to its junction with NY 17K. At this point, leave I-84, gain NY 17K, and proceed west on it for about 6.4 miles to the village of Montgomery. Stay on NY 17K as it passes through the village and crosses the Wallkill River. Continue west for approximately 5 miles, at which point a sign will indicate the park. Turn right here and proceed to the parking area off Valley Road.

The following is an alternate route from NY 17 (the Quickway): leave NY 17 at Exit 120, which is just northeast of Middletown. Gain NY 211 (Montgomery State Road) heading north. From Exit 120, proceed north about 10 miles on Montgomery State Road, passing through the village of Montgomery. At the intersection of NY 211 and NY 17K (here, also called Ward Street), turn left onto NY 17K and follow the above directions to the entrance of the park.

Once inside the park, check in at the Nature Center, which is reached by walking about 350 feet north on a small dirt path leading from the parking area. From there follow the yellow-blazed trail, which will pass through several different habitat types and will eventually bring the birder to the Upper Pond. In the trees and bushes fringing the pond one might be lucky enough to witness a fine passerine migrational flight in mid-to-late May. Those who actively bird the park have seen and heard resident Barred Owls and their young, nesting and migrant picids, several flycatcher species in passage, mimids, nesting thrush species, many warbler species, including Mourning Warbler, and Northern Oriole, Summer Tanager, and at least five sparrow species, all of which nest within the park.

There is a more primitive, less manicured, trail leading through another large tract, which is also part of this park. This trail can be found on the opposite (west) side of Valley Road. One should hike it for some distance, if time allows, as it is in this section that some of the more noteworthy avian sightings have been made, especially in the height of spring migration.

Moonbeams Sanctuary (Map 9.4)

Rating: Spring****, Summer***, Autumn**, Winter*

This preserve encompasses 150 acres in the extreme northwestern section of the county. The loveliness of this sanctuary lies in its diversity and size. From the grounds one sees the Shawangunk Mountains in the distance and the trout-filled Shawangunk Kill forms the preserve's western boundary. In addition to fine old-growth forests in which majestic hemlock and white pine are mixed with northern hardwoods, there are several distinct swamps, each adding its own unique flora and fauna. Finally, eight large, open upland meadows (approximately 35 acres total) are being kept open by a program of mowing, thereby maintaining a variety of habitats.

This Nature Conservancy–owned sanctuary is one of the gems in the Orange County crown, and it should be a visiting "must" on any New York birder's list, if simply to experience its marvelous botanically and zoologically complex ecosystems.

Moonbeams Sanctuary is an isolated, little known or frequented, scantly advertised (even locally) preserve, that intentionally keeps a low profile in order to avoid some of the more common problems—trail erosion, littering, and vandalism—of heavily-used areas. It is however, open during daylight hours, seven days per week, and the Moon family, which donated the property to the Nature Conservancy, welcomes all birders, natural history students, and just observers, to the preserve. There are informal trails, unmarked but frequently cleared of brush, running through the property as well as a network of fairly recognizable roads and cow paths connecting different fields and old pastures. These pastures and fields are well delineated by stone walls and hedgerows. The stout, well-laid stone walls cause one to speculate on the amount of labor that was once required to make these fields suitable for cultivation. The trails are placed in areas of high interest and possible study.

Quite practically, the birder leaving the trails will all too often find himself in impassable swamps or thickets. In this way, pedestrian traffic is somewhat controlled. I hasten to add, however, that this network of trails allows access to all of the diverse plant and animal communities represented on the grounds and the birder will be well rewarded for following as many of them as possible.

To reach Moonbeams Sanctuary, leave NY 17 (the Quickway) at Exit 118, the Fair Oaks Exit, northwest of Middletown. Follow the exit ramp down and under NY 17 to NY 17M, which is on the west side of NY 17. Turn right (north) on 17M (Bloomingburg Road) and proceed approximately 2 miles to Prosperous Valley Road. (The road on the right is Shawangunk Road.) Here turn left and continue 0.9 mile where Prosperous Valley Road tees with Howells Turnpike. Bear right and stay on Prosperous Valley Road. Continue for approximately 1.5 miles,

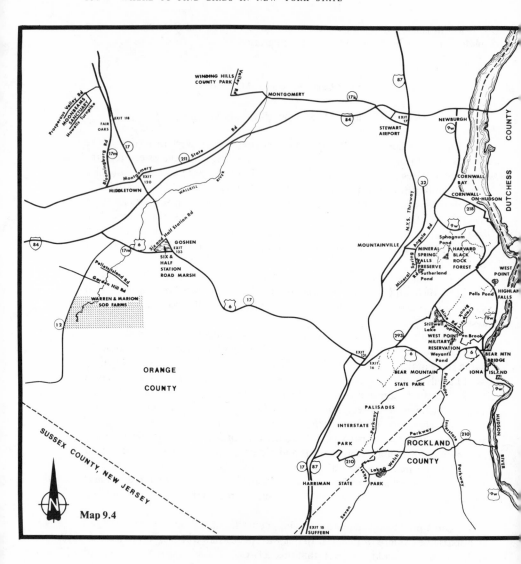

Map 9.4

where there will be a sign designating Moonbeams Sanctuary, on the left side of the road. Park along the roadside. Visitors are asked to register at the small booth near the entrance to the main trail near Prosperous Valley Road, in order to provide information on public use of the preserve.

The preserve spans both sides of Prosperous Valley Road, and there are trails through both sections. The property on the east side of Prosperous Valley Road and lying south of it reaches the highest elevations (approximately 650 feet above sea level). On this side there are two ample swamps: Bear Swamp and The

Downs, and a smaller swamp known as Lily Swamp. There are several rivulets, some of which are ephemeral, and also the largest and only year-round open body of water. Crayfish Crawl, the longest stream on the property, was impounded in the 1960s to form this pond to attract migrant waterbirds. It covers approximately 1.5 acres and has a maximum depth of four feet. This pond drains to the southwest and empties into Crayfish Crawl and eventually into the Shawangunk Kill. It is in the area of this pond that Horned and Pied-billed grebes, Great Blue and Green herons, breeding Least Bitterns, migrant wigeons, Gadwalls, teals, scoters, and Hooded and Common mergansers can be found. As you walk this trail, notice the marshy area across the trail from the pond. Here, spiked or purple loosestrife and black gum grow abundantly, as well as common cattail around the pond, and two or three of the eight clover species, and fifteen to eighteen of the sixty wildflower and herb species occurring throughout the sanctuary. Continue past the pond and note the hemlock, chestnut, northern red oak and eastern white pine crowding the high rocky ground. Follow the ridge down to The Downs, which is at an elevation of approximately 620 feet. This area is shaded by both speckled and smooth alders and numerous willows. Although some of this area is fairly rocky, it is on this side of the property that four of the open fields are located and are in part separated from pastures by enclosing rock walls.

As you continue toward Bear Swamp you will pass through large stands of mixed deciduous tree species, including black oak, American elm, and hornbeam in the wetter area; trembling aspen, red cedar, and gray birch in the openings; sugar maple and shagbark hickory on the drier sites; and white pine and hemlock on the slopes. Look for the few scattered black cherry, white ash, and red maples.

The sanctuary has some notable geological vestiges of the last glacial retreat, in the form of eskers, ice-block kettles, and boulder fields, most of which are found on this (the south) side of Prosperous Valley Road.

Walk down through The Downs, noting that marsh fern occurs here. Royal, interrupted, and cinnamon ferns are perhaps the most abundant ferns on the preserve, but look also for sensitive fern, New York, Spinulose, and wood ferns in the moderate-to-well-shaded areas on poorly drained soils. Blue-eyed grass is found in all of the fields. Common green or bullbrier is found in heavy concentrations at the northeast end of The Downs in well-shaded areas.

Proceed farther along the trail turning left (east) from an open field into mixed deciduous woods. You will be approaching Bear Swamp, which is on your left. All along this walk the birder should have been looking and listening for breeding Turkey Vultures, possibly breeding Goshawks, and definitely but locally rare breeding Red-shouldered Hawks. Broad-winged and Red-tailed hawks as well as American Kestrels have bred on the property. Ospreys, Marsh Hawks, and Merlins are frequent migrants spotted on the preserve. Ruffed Grouse, Killdeer, American Woodcock, and Common Snipe breed on the property in appropriate habitats. Keep an eye out for breeding or migrant Yellow-billed Cuckoos. Screech, Great Horned, and Barred owls are all year-round residents at Moonbeams Sanctuary. Be alert for the distraction display of the Common Nighthawk, indicating that its nest is relatively close-by.

In the area of Bear Swamp look for breeding Ruby-throated Hummingbirds. Flicker, Pileated, Red-headed, Hairy, and Downy woodpeckers are year-round residents and Yellow-bellied Sapsuckers can be found here during autumn migration. Bear Swamp has a thick canopy of red maple and black gum, and winged sumac and sweet fern occupy different parts of the ice channel fill-ridge which nearly encircles it. Crested woodfern is abundant in the swampy area. Indian Cucumber is found to the north of Bear Swamp underneath the mixed hardwoods. Wild blue iris grows in profusion here and great rhododendron is the main component of the understory. Mountain laurel and azaleas are well supported here in this easternmost swamp, and nearly twenty species of wildflowers can be found in its immediate vicinity. It is in the general vicinity of Bear Swamp that fresh black bear tracks have been recently found.

Tree, Bank, Rough-winged, and Barn swallows feed or nest on this side of the preserve. White-breasted Nuthatch and Brown Creeper can be found here year-round. Winter Wren has recently been added as a breeder to the sanctuary list. The three eastern Mimids nest here and Wood and Hermit thrushes and Veeries can be found in some numbers in spring, summer, and autumn. Eastern Bluebirds nest in the edge-area of the open fields. Be sure to listen for the distinctive song of the Blue-gray Gnatcatcher, and both kinglets.

Behind the Moon residence, the property slopes considerably at an accelerated rate until the last 0.1 mile, the land drops sixty feet to an elevation of 500 feet above sea level at the Shawangunk Kill itself. This sloping terrain, on the north (west) side of Prosperous Valley Road holds open hay fields, and rocky wet pastures being claimed by successional forest.

The end product of succession, climax, will be different in each of these diverse areas. These pastures in various stages of succession host bracken, hay-scented, and marginal shield ferns on the drier sites, Christmas fern in areas of deep shade and well-drained soil, and the lovely lady fern along stream beds. Six species of goldenrod are to be found here; five of which can be located in open, dry, sunny sites and the sixth, large-leaved goldenrod, is only to be found along the Shawangunk Kill in fairly deep shade and under less than well-drained soil conditions.

Here on the west side of Prosperous Valley Road, ebony spleenwort is found on the rock wall in front of the Moon residence. This is the only site on the preserve where this spleenwort occurs. On this side there are also paths leading by and over intermittent streams and low-profile swamps that differ in form and flora from those on the upper property. A stand of "cathedral," mature hemlock and a white pine grove also await exploration. There is a stand of small white pine and hemlock approximately sixty to sixty-five feet tall along the Kill, with a needle-covered forest floor along with wintergreen and partridgeberry and pink lady's slipper growing there. The banks of the Kill, of course, are quite rich.

Farther upslope, there are more deciduous and coniferous trees as well as ferns and a small stand of American beech that overlooks the Kill. There are yellow and black birch, hop hornbeam, white ash, and sugar maple in moist rocky coves. Here, alder and red maple are a large component as well as a red maple/white ash association on the moist land sloping down to the north end of the alder swamp (located in the northeast corner of the preserve). In the well-drained sections in a few of the successional pastures, there are red maple and gray birch along with trembling aspen, and scattered tulip trees, hemlocks, and white pine.

Running clubmosses in company with Canada mayflower and Solomon's seal, ground pine, and ground cedar are rather widespread over the preserve, preferring the drier, shaded sites. Look for wild yellow or Canada lily and the trout lily throughout. Along the northern trail, look for Queen Anne's lace in the fields and swamp dogwood near the marshy areas. Indian pipe occurs abundantly in the northeastern corner of the sanctuary as well as along the Shawangunk Kill.

Walk everywhere within the preserve and be sure not to neglect the abandoned railroad right-of-way. From there the birder can usually spot waxwings, Solitary, Red-eyed, and Warbling vireos. Look for breeding warblers throughout the preserve: Black-and-white, Blue-winged, and Nashville. Northern Parula, and Yellow, Magnolia, Cape May, Yellow-throated and Chestnut-sided warblers have been recorded as nesters here. Look for Blackpoll and Pine warblers in migration. Ovenbirds and Louisiana Waterthrushes both nest on the property, while Connecticut and Wilson's warblers have been frequently recorded during migrations. Naturally, American Redstarts also nest here. In the more open habitat Bobolinks, Eastern Meadowlarks, and Grasshopper, Vesper, and Field sparrows nest.

Visit Moonbeams Sanctuary and see what a valuable outdoor living museum the Nature Conservancy has saved for posterity.

Six and Half Station Road Marsh (Map 9.4)

Rating: Spring****, Summer*, Autumn*, Winter*

Six and Half Station Road is located just west of Goshen, and is therefore easily birded going to or coming from the Warren Sod Farm, Moonbeams Sanctuary, or Winding Hills County Park. Although the road itself is little more than a mile long, it skirts an extremely productive marsh. Most of the marsh is in private ownership but birding can be done either from the road, which overlooks the marsh or from the public railroad right-of-way, which cuts straight through the marsh.

In spring and early summer, this area can virtually teem with water-associated birds: migrant herons, Least Bitterns setting up breeding territories, migrant waterfowl including geese, teal, Pintails, and Ring-necked Ducks. Look overhead for locally breeding Turkey Vultures, Red-shouldered, Broad-winged, and Red-tailed hawks. Various rail species have nested in the marsh and possibly American Coots. The Pileated Woodpecker is often sighted here. On a good spring morning flycatchers and swallows are "all over the place." Long-billed Marsh Wrens nest in the marsh as do Red-winged Blackbirds in great numbers, and Field, Swamp and Song sparrows.

To reach Six and Half Station Road marsh: leave NY 17 (the Quickway) at Exit 123, which is just west of the village of Goshen. Head west on NY 17M (US 6). Proceed west 1 mile where Six and Half Station Road meets 17M on the right. Turn right and follow the road, birding the marsh on your right and the open fields on the left. Pull off wherever it is safe and convenient, and walk the railway right-of-way to bird the marsh from its interior.

Warren Sod Farm (Map 9.4)

Rating: Spring**, Summer*, Autumn****, Winter**

The Warren Sod Farm covers hundreds of acres of large fields of alternately mowed rich turf grass, dirt tracks where sod has been recently harvested, and tracts of seeded fields where the sod grass is in an intermediate stage. This sod farm and others close by is located in an area of Orange County whose parent soil materials are glacial in origin. The surface soils are rich, well-drained where irrigated, well-soaked where not, silty/gravelly loams and are extremely fertile. On these sod farms, the upper stratum of the humus is filled with vegetable molds, and roots of grasses, herbs, and other small plants, so that it forms a thick mat. Pieces are pared from this upper stratum, with its green growth preserved, and sold as ready-grown lawn. While travelling to the sod farm, note the onion, lettuce, and celery farms all thriving throughout this rich "black dirt" area.

The manager of Warren Sod Farm informed us that they have been sodding this area for close to twenty years. He commented on the recent upsurge of birders frequenting the farm fields in early autumn right through early winter. The management of Warren Sod Farm is sympathetic to birders and allows roaming the sod farm roads. It is advisable to check in at the sod farm office, which is located directly across the road from the fields.

To reach the Warren Sod Farm, from the intersection of I-84 and NY 17 (the Quickway), continue south on NY 17 to Exit 123, just west of the village of Goshen. Head west on NY 17M (US 6). (Incidentally, this intersection is surrounded on all four sides by relatively large fields. In late autumn, as many as four Rough-legged Hawks have been seen simultaneously here, coursing over the fields in pursuit of prey. They unquestionably overwinter in the area, in those years when the species comes south in high numbers.) Proceed west on NY 17M for 3.6 miles. Here, turn left, thereby gaining NY (county road) 12. Continue south on NY 12 through the rolling terrain, being sure to note the "black dirt" district off on the left. At 6.1 miles after gaining NY 12, just after Garden Hill Road, which comes in on the right, you will see the large warehouse and offices of the Warren Sod Farm. This will be on the right, so driving up to the office involves only a simple right-hand turn.

After leaving the office, turn right on NY 12 and take the first possible left turn, which will put you on one of the main roads of the sod farm. Drive east on it as far as it will take you, turning right or left, where necessary, to explore one or another of the farm roads. Looking eastward, one sees a long, approximately north-south, row of trees, obviously lining a coursing body of water. This is the Wallkill River and several of the farm roads lead east to it and some cross it, giving access to more sod fields on the east side of the Wallkill River. This row of trees, in addition to the fields, can sometimes be productive, so do not neglect to search there for birds also.

The Warren Sod Farm is most productive from late August through September. The ideal time seems to be centered around the Labor Day weekend. During this part of the fall migration the fields attract upland meadow and tundra bird species not usually found away from the barrier beaches and ocean shoreline.

Look for Killdeer. American Golden Plovers begin arriving about September 10 to 15 with as many as 200 individuals, then build up to 300 to 350 by the last week in September, and seem to dwindle to groups of 10 to 15 by the last week of October; they appear to be completely gone by November 1. Black-bellied Plovers appear in small groups of up to ten individuals. Expect the "grasspipers"; Upland and Solitary sandpipers, Pectoral Sandpipers (which reach a maximum of 8 to 12 in late September and is last seen near the last week in October); Baird's Sandpipers, in small groups of 2 to 5 appearing in mid-to-late September; Short-billed Dowitchers appearing approximately September 20 to 25 with a maximum of 25 to 30. Occasionally 2 to 5 Stilt Sandpipers occur on the turf farm September 20 to 25. Buff-breasted Sandpipers can appear as early as mid-September in small groups of 2 to 5, but, they are usually completely gone by the end of the first week in October.

Upland meadow species should be present in autumn; Horned Larks, Bobolinks, large numbers of Water Pipits reach a maximum of 225 to 275 in mid-October, with some still present by mid-November; and Eastern Meadowlark in varying numbers. From mid-November on one should be able to locate Lapland Longspurs and Snow Buntings throughout the farm fields.

Another sod farm in the neighborhood is the Marion Sod Farm, which can be reached by turning east off NY (county road) 12 at Pelletts Island Road. This farm will be indicated by a large sign on NY 12 at its junction with Pelletts Island Road.

Stewart Airport (Map 9.4)

Rating: Spring****, Summer*, Autumn****, Winter*

Stewart Airport, located just west of the city of Newburgh, can, during autumn and sometimes spring, be a good-to-excellent site for viewing transient shorebirds.

Although NY 207 runs for some miles along the airport's southern boundary, do not approach it from that side. Leave I-87 (the New York State Thruway), at Exit 17, west of Newburgh. Gain NY 17K heading west, and proceed west approximately 1.75 miles, to a drive road on the left, which gives access to the airport grounds. Turn left and follow the roads down and around in those areas that are legal and unobstructed. If, for some reason, that drive road is closed, turn around and proceed east for 0.75 miles to an entrance on the right. Here enter the airport grounds and proceed to the open, grassy areas that may harbor shorebirds in transit.

In spring, Upland Sandpiper has appeared at the airport as early as April 29 and has stayed right on through the remainder of the spring. This may be construed by some as some sort of nesting evidence; but much more support would be needed before we could claim this area as a nesting site.

In autumn look for Killdeers from early September through November. American Golden Plovers usually arrive approximately September 1, and build to a maximum of 150 to 175 near October 1; it has usually departed the area entirely by November 10 to 12. A few Upland Sandpipers, with a maximum of 10 to 15 can be seen September 1 to 15. Pectoral Sandpipers may begin to appear about the third week in September and then build to a maximum of 30 to 35 by the first week in October, before departing entirely by mid-October. Small numbers of Baird's Sandpipers, usually 2 to 5, may occur at the very beginning of October. Also, in the first week of October, Dunlins start to arrive and shortly thereafter reach peak numbers of 40 to 60. Look for Buff-breasted Sandpipers in groups of 2 or 3 the last few days of September, and expect numbers to quickly build to 15 to 20 by the end of the first week of October; after that they almost immediately depart. Horned Larks occur on the large grassy fields and Water Pipits can be expected in numbers approaching 50 to 75 in mid-October.

The real virtue of Stewart Airport is its proximity to a large city and the relative ease in locating migrant "grasspipers" on its grounds.

Cornwall Bay (Map 9.4)

Rating: Spring****, Summer**, Autumn***, Winter****

Cornwall Bay is certainly one of the finest places in Orange County to view an extensive variety and sometimes very high numbers of a whole spectrum of bird species throughout the year.

To reach Cornwall Bay proceed north or south on US 9W: from the city of Newburgh (on the north), or the Bear Mountain Bridge (on the south), to the junction of NY 218. Coming from the north, turn left (east) off US 9W and follow NY 218 into the village of Cornwall-on-Hudson. Coming from the south turn right (northeast), off US 9W, approximately 1.5 miles north of the village of Highland Falls. Follow NY 218 into the village of Cornwall-on-Hudson. At the junction of NY 218 and River Road, inside the village, turn south on River Road, heading toward the Hudson River. Proceed less than a mile to Shore Road, which parallels the river and also the railroad tracks on the river side of Shore Road. Turn left and park along the right side of the road. Walk to and across the railroad tracks looking for a suitable place to set up your scope.

If you are without a scope, walk north (left) up the tracks where you will have a good vantage of Cornwall Bay on the right and Moodna Creek on the left. Approximately 250 yards up the tracks from the foot of Shore Road there is an extremely productive freshwater pond on the right (river side) which never fails to provide avian results. Upwards of 135 species have been recorded within one calender year (1980), at Cornwall Bay, by Jeheber and Lorch. Throughout the various seasons it appears to be a consistently gratifying birding site with each season offering several specialties.

In spring, Green Heron, Least and American bitterns, Wood Duck, Red-tailed Hawk, Virginia Rail, Sora, American Woodcock, Common Snipe, Great Horned Owl, Belted Kingfisher, Pileated Woodpecker, Eastern Kingbird, Eastern Phoebe, Tree and Barn swallows, and Long-billed Marsh Wren, to name just a few, arrive to set up territories and complete their breeding cycles here. During the same season Blue-winged Teal, Turkey Vulture, Osprey, Broad-winged Hawk, Spotted and Solitary sandpipers, Forster's and Black terns, Great Crested Fly-catcher, Bank Swallow, Wood Thrush, several warbler species, and Rose-breasted Grosbeak are moving through on migration.

In late summer heron species on their post-breeding wanderings drop in at the bay, and it is possible to then see Great Blue and Little Blue herons, Great and Snowy egrets, in some years Louisiana Herons, and Black-crowned Night Herons annually. Additionally, in late summer, one should look for the occasional Bald Eagle and Northern Harrier that stop by the bay. Cornwall Bay is of course, tidal, and so at low tide, in late summer and early autumn one is likely to see Semipalm-ated Plovers, Greater and Lesser yellowlegs, Pectoral, Least and Semipalmated sandpipers, as well as Sanderlings. Carolina Wrens and Cedar Waxwings are seen in late summer here.

In autumn, the bird activity seems to pick up momentum in about mid-September, with the arrival of Double-crested Cormorants, Pintails, American Wigeons, and possibly Cooper's Hawks, American Kestrels, and Peregrine Fal-cons. Laughing Gulls can be seen from about the middle of September on. White-eyed Vireo and Black-throated Blue, Yellow-rumped, Black-throated Green, Blackpoll, and Palm warblers move through. In late autumn one should look for Common Loons, rafts of Brants, Snow Geese, Redheads, Buffleheads, Black and White-winged scoters, and Ruddy Ducks. Late autumn is also the time to look for the unexpected and infrequently occurring Golden Eagle.

In winter, although the breezes blowing off the river are sometimes cold and strong, it is still worth visiting the bay, if it has not reached a stage of being com-pletely frozen-over. During the winter months Horned and Pied-billed grebes, Green-winged Teals, rafts of Ring-necked Ducks, Canvasbacks, and Greater Scaups are impressive indeed. Common Goldeneyes and Oldsquaws, as well as Hooded and Common mergansers are quite abundant at this point in the Hudson River. There are usually small flocks of American Coots feeding near the shoreline and Great Black-backed, Herring, Ring-billed, and small numbers of Bonaparte's gulls can be seen flying overhead. On one visit to Cornwall Bay, in December, we had four Pileated Woodpeckers in view simultaneously. Both Golden-crowned and Ruby-crowned kinglets can be expected from the last week of October through mid-December. In very late autumn through the winter months, one can "spish up" House Finch, goldfinches, and Savannah, White-throated, Swamp and Song sparrows.

Mineral Spring Falls Preserve and Harvard Black Rock Forest (Map 9.4)

Rating: Spring****, Summer***, Autumn***, Winter*

Mineral Spring Falls Preserve is owned and maintained by the Nature Conservancy. It is a 120-acre tract abutting restricted use, wooded lands of the United States Mil-itary Academy of West Point on two sides, and is adjacent to the 3600-acre Harvard Black Rock Forest. Obviously Mineral Spring Falls Preserve is an important component of the complex of protected lands in the Hudson Highlands.

Black Rock Forest takes its name from a prominent, tree covered summit within the preserve, from which there are marvelous views to the north, east, and west. This tract was donated to Harvard University by Dr. Ernest Stillman in 1949, and since then it has been used for ongoing scientific research, which status ought to ensure its permanent protection. East of the forest's boundary lies lands of the Palisades Interstate Park as well as lands of the U.S. Military Academy, which also adjoin the forest on the southeast, south and southwest.

The ecological components of both sites are typical of the Hudson Highlands, with the exception of a 100-foot cascading waterfall located at the northwestern part of Mineral Spring Falls Preserve. Both sites are located in the Moodna Creek drainage system. The waterfall is part of Mineral Spring Brook, which flows through the northern part of Mineral Spring Falls Preserve. The brook feeds into Woodbury Creek, which feeds into Moodna Creek, which in turn, feeds into the Hudson River. Mineral Spring Brook is intermittent and ranges in width from eight to twenty feet. It forms pools and cascades of varying depths for more than 100 feet, as it flows through a steep-sided hemlock-lined ravine. The deepest pool, below the main cascade, is over four feet deep during the late spring and early summer.

If both tracts are thought of as one ecological unit, the elevation ranges from 760 feet above sea level at the base of Mineral Spring Falls to more than 1460 feet at the summit of Spy Rock in Black Rock Forest. Small pockets of massive rock outcroppings are distributed throughout both tracts. Most of these are unforested, rugged granite of Precambrian age. The fauna one might expect to encounter is characteristic of wooded areas of the Hudson Highlands and includes copperhead and timber rattlesnakes, which although poisonous, are rarely seen. Black bear occasionally wander through the area.

The vegetation throughout the preserves includes four distinct communities: hemlock-hardwood; mixed hardwood; red oak; and beech-maple. The hemlock-hardwood association is found almost exclusively in the vicinity of Mineral Spring Falls gorge and forms an almost pure stand. Eastern hemlocks line the ravine and are surrounded by yellow birch and sugar maple. In this area the shrubbery consists of flowering dogwood and maple leaf viburnum and the main ferns are Christmas and royal. In some places, the banks of the stream are covered with dense mats of mosses. There is low light penetration through the dense hemlock canopy here, and therefore groundcover is quite sparse.

Farther north and east throughout Harvard Black Rock Forest, there are large areas almost completely forested with northern mixed deciduous hardwoods. In this association the tree species are mainly red oak, sugar maple, white ash and yellow birch. The mixed hardwood association displays the greatest diversity of all of the associations represented, and the transition zone between this and the others is often extensive. One might expect the greatest avian densities and diversities here. Witch hazel and mountain laurel are abundant in these areas and marginal wood fern and interrupted fern will be evident.

Within the red oak association the primary tree species is, of course, red oak. There is also a high percentage of chestnut oak present in these areas and again mountain laurel and witch hazel grow abundantly. This association is usually found on the upper, gentler slopes throughout the combined tracks.

American beech and sugar maple are the main components in the association. In these areas the canopy is quite open and light penetration picks up considerably. The shrub and herbaceous layer is not at all dense and the groundcover is primarily leaf litter.

Nineteen plant species found on these tracts appear on the *New York State Protected Native Plant List.* These include numerous species of ferns and at least two orchid species.

To reach Mineral Spring Falls Preserve and Harvard Black Rock Forest, follow the traffic circle on the west side of the Bear Mountain Bridge to the signs indicating US 9W, heading north. Gain US 9W north, and proceed approximately ten miles to the Angola Road Exit. Here, leave US 9W and gain Angola Road by turning left onto it. Proceed southwest on Angola Road for almost two miles, where it forks. Stay to the left, thereby gaining Mineral Spring Road. Continue south on Mineral Spring Road for approximately 1.3 miles to the entrance to Mineral Spring Falls Preserve, which is on the left on Old Mineral Spring Road. *Nota Bene:* One may *not* park on this dirt entrance road. However, one may legally park on the east side of Mineral Spring Road on its shoulder, in a fifty-foot stretch south of the entrance to the dirt road. This area is large enough to accommodate about four cars.

From the parking site it is a walk of approximately 0.5 mile to the waterfall. Walk in on the dirt Old Mineral Spring Road and turn left where an old wagon road leads one into the forested preserve. The path follows the remains of that old wagon road, which was constructed so that bottled sulphur water from the mineral spring could be carried out easily. Note the old cellar foundations and stone walls along the trail, remains of a long-forgotten farming homestead.

The main trail through this preserve is white-blazed; lengthened and maintained by the New York – New Jersey Trail Conference. The white trail originates at NY 32, in the hamlet of Mountainville, heads east crossing Mineral Spring Road and then follows the Old Mineral Spring Road southeast to the base of the falls. From here, follow the trail up the slope north of the falls. Cross Merrill Road and continue to "Fifth of July" rock and from there into the holdings of Harvard Black Rock Forest. From here one can follow the same trail north to Sutherland Pond and then connect with a yellow-blazed trail which leads to Sphagnum Pond. Any one of a number of intersecting trails can be taken, and in any direction from there.

While still in the vicinity of Mineral Spring Falls, the birder may wish to connect with an orange-blazed trail, which is south of Merrill Road and just above the falls. This is a shorter trail than those continuing into Black Rock Forest. The blazing of the orange trail was accomplished within the last few years and much of it follows old logging roads twisting through the preserve. The orange trail follows along the north ridge of Mineral Spring Brook and then crosses over it and continues toward the southeastern corner of the preserve. It ends at the approximate boundary line of the U.S. Military Academy.

The avian life of the two tracts combined includes the following species: Great Blue and Green herons, Least and American bitterns, Canada Goose, Wood duck, teals, mergansers, Turkey Vulture, Osprey, Marsh Hawk, Goshawk, Cooper's, Red-shouldered, Broad-winged, and Red-tailed hawks, and American Kestrel; also breeding Ruffed Grouse, Killdeer, American Woodcock, and Common Snipe. Mourning Dove can be heard cooing and both Yellow-billed and Black-billed cuckoos nest on this large diversified tract. Screech, Great Horned, and Barred owls nest here. Some birders suspect that Saw-whet Owl nests here but to date there is no conclusive evidence that it does. Several picids can be found here throughout the year. Spring sees the advent of wave after wave of flycatchers, Corvids, nesting House and Winter wrens, mimids, and nesting thrushes.

The following warblers have been recorded as nesting on these tracts: Black-and-white, Blue-winged, Golden-winged, Northern Parula, Yellow, Magnolia, Black-throated Blue, Black-throated Green, Cerulean, Blackburnian, Chestnut-sided, and Bay-breasted. Look also for the nesting Ovenbird, Northern Waterthrush, Common Yellowthroat and American Redstart. The Red-winged Blackbird, Northern Oriole, Scarlet Tanager and Rose-breasted Grosbeak can be found throughout the breeding season. Seven sparrow species have been recorded throughout these grounds.

In addition to its obvious virtues, these preserves are very near the Whitehorse Mountain Hawkwatch (see chapter on Hawkwatches), which is itself something of an amazement.

West Point Military Reservation (Map 9.4)

Rating: Spring****, Summer***, Autumn*, Winter* (specialties)

The United States Military Academy was founded by an act of Congress in 1802. It was constructed on a 3500-acre tract at West Point, a site that was then considered the most strategic point on the Hudson River.

Today birders in the Hudson-Delaware Region sing the praises of the West Point Military Reservation loud and long—especially in the height of spring migration and in the early summer when birds are actively setting up territories.

Mine Road cuts a circuitous course through the southern end of the military reservation, affording the birder access to a multitude of habitats: old farm fields, woodlands in various stages of succession, extensive natural hardwood forests, spring-fed freshwater lakes, alder swampy areas and low-lying wetlands. The most productive areas along Mine Road are surrounding Stillwell Lake, Popolopen Brook, Weyants Pond and the wide mouth of the Popolopen Brook. The mouth of the brook was once flanked by the twin outposts of Fort Montgomery and Fort Clinton. Today the Popolopen Creek is crossed by a bridge bearing its name and over which US 9W passes. From it, the Bear Mountain Bridge can be seen to the southwest, spanning the narrow gorge of the Hudson River between Bear Mountain on the river's west side and Anthony's Nose on the river's east side.

From the west side of the Hudson River, leave I-87 (the New York State

Thruway) at Exit 16. Follow signs indicating US 6, heading east. Gain US 6, and continue east to the junction of it and NY 293. Here turn left (north) and follow NY 293 to the entrance of Mine Road, which is approximately 3.5 miles. Turn right (east) and proceed along Mine Road.

Drive slowly along Mine Road, stopping first to bird both sides of the road as it passes Stillwell Lake. Immediately east of Stillwell Lake there is a hiking trail meeting Mine Road on its right (south) side. There is a fine place to bird on foot. Continuing east, notice that Popolopen Brook parallels Mine Road. At approximately one mile past the east end of Stillwell Lake, the creek widens into a broad, marshy area. This is often productive and should be birded in the spring when the concentration of birds is most dense. Less than one-half mile farther, Mine Road makes a sharp right turn. Just before the right turn, Cranberry Brook crosses the road. On the north side of the road here, there is a path that can be hiked to good advantage. It lead northeast for approximately one mile and then branches both east and west. Continuing southeast on Mine Road you will see that the Timp-Torne Trail crosses the road. This is easily identifiable as it is marked with blue blazes. The Timp-Torne Trail is maintained by the New York–New Jersey Trail Conference. Pull off Mine Road and walk west on the blue-blazed trail. This trail passes through the most rewarding area for both density and abundance of birds in this southeastern section of West Point Military Reservation.

Walk west following the blue markers, making sure to bird the lower end of the Popolopen Brook marsh, which will now be on the right. Cross the bridge over Popolopen Brook. This bridge is a superb location from which to see marsh birds. Just west of the bridge there is an unmarked footpath one can follow north along the inside of the marsh, or south to the outlet of Weyant's Pond. Both walks are easy and quite short. After birding Weyant's Pond, continue east on Mine Road, stopping to bird wherever the habitat look luxuriant and where pulling off the road is safe. Continue along it headed east to its junction with US 9W. Here one can turn left and continue north for two miles to Pells Pond, which is on the southeastern edge of the village of Highland Falls and on the east (right) side of US 9W, or one can turn right and proceed south to the Bear Mountain Circle and Bridge or even a little bit farther south to Bear Mountain and Harriman state parks or the Iona Island Bird Sanctuary in Rockland County.

The entire length of Mine Road, from its entrance off NY 293, to its exit at US 9W, is five miles and can probably be covered in about three hours. This assumes stopping at all of the above-cited areas and also some birding on foot. It is a bird-rich area, containing ample varied habitats and a wide range of species. Birding Mine Road should be on the spring agenda of every birder within easy reach of the West Point Military Reservation.

If one follows the reports of the long-respected editor of the Delaware-Hudson *Kingbird* Region 9, such species as the following appear consistently in spring reports: Brant, more than 3000 flying over Mine Road in Mid-May; Bald Eagle over Stillwell Lake as late as the end of April; Red-shouldered, Broad-winged, and Red-tailed hawks nesting in the wooded slopes north of Mine Road; Ruffed Grouse heard all along the road throughout spring; American Woodcock seen displaying in late March and in April; Pileated Woodpeckers nesting in the hardwood forests lining the road; along with more commonly encountered fly-catcher species, Olive-sided Flycatchers as early as the first week of May at Wey-ant's Pond; Bank Swallows nesting along Popolopen Brook sandbanks; the Win-ter Wren heard singing Mine Road on May 10; Brown thrashers returning to set up nesting territory at Weyant's Pond by mid-April; Blue-gray Gnatcatchers nesting along the Timp-Torne Trail; White-eyed Vireo, "always a good find," Along Mine Road May 7 to 24.

Many warbler species nest along and off Mine Road. Among the more nota-ble finds are: Black-and-white, which is rather uncommon, off Timp-Torne Trail; Golden-winged nesting in the area near the Timp-Torne trailhead; "Lawrence's" hybrid male singing off Mine Road in mid-May; several Ceruleans off the road it-self and a colony of approximately six pairs nesting at Weyant's Pond; Chestnut-sided setting up territory in mid- to late May; Prairie is usually recorded in good numbers singing by the first week in June; Kentucky observed by mid-May; Mourning at several places along Mine Road by mid-May; Yellow-breasted Chat by mid-May.

Orchard Orioles recorded along Mine Road in the last week of May, also a good find. Indigo Buntings may be fairly scarce throughout some of the region ex-cept locally but they are found "every few yards along Mine Road" in June. There are other noteworthy records of flycatchers, marsh wrens, vireo and warbler spe-cies throughout the southern part of West Point Military Reservation. The trick is to find the rare transient or breeder yourself.

In winter Stillwell Lake seems to attract eagles overwintering in the region. In winter 1979–1980, dead deer on the surface ice of the lake provided food for two Golden Eagles throughout the winter. In other winters similar attractions have acted as magnets for eagles and Rough-legged Hawks. It is certainly worth check-ing out the area in seasons other than spring.

ORANGE AND ROCKLAND COUNTIES
Bear Mountain and Harriman State Parks and Iona Island, a Tour (Map 9.4)

Rating: Spring****, Summer**, Autumn**, Winter**

Far more bird species, with the total standing at 250 species (in 1981), have been re-corded on this huge tract of land than any other class of vertebrate (25 reptile spe-cies, and 40 mammal species). Ninety-five to one hundred of those 250 were re-corded as breeders by 1981. There is a wide variety of habitats represented here. More than 51,000 acres of the park are located in the Hudson Highlands and their southern extension, the Ramapo Mountains. In fact these holdings constitute ap-proximately 75 percent of the Palisades Interstate Park acreage. Harriman State Park alone encompasses 80 square miles, much of which straddles the border be-tween Orange and Rockland Counties and includes most of the New York State portion of the Ramapo Mountain Range. Both Bear Mountain and Harriman are laced with hundreds of miles of hiking trails. The Appalachian Trail passes through the park on its way from Maine to Georgia. Its many offshoots and spurs offer the birder access to all of the varied habitat types.

In 1965, the Palisades Interstate Park Commission acquired the approximately 120-acre, bouldery headland, Iona Island, and the adjacent, brackish Salisbury Marsh. These lie almost 2 miles south of the main gateway to Bear Mountain State Park and furnish yet another habitat type included in the jointly operated state parks. Iona Island, at sea level in the Hudson River, and the summit of Mount Aramah, at 1470 feet above sea level, bracket the altitudinal range of the parks. There are numerous lakes, swamps, bogs, open fields, hardwood ridges, and hemlock forests harboring animal life. Deciduous tree species within the park include oak, maple, hickory, ash, tulip, beech, tamarack, and sweet and sour gums. Evergreen tree species include cedar, spruce, hemlock, balsam, and pine. Mountain laurel, rhododendron, witch hazel, spice bush, wild azalea, and wild blueberry can be found in profusion throughout the tracts. Numerous wildflowers, ferns, mosses and lichens can be found interspersed throughout the complex mosaic of forest communities.

Bear Mountain and Harriman state parks can most easily be reached from the following major highways: from the west *via* NY 17 (the Quickway) at Exit 131. This gives access to US 6, which crosses the northern section of the park and leads to the Newburgh connection to I-87. From the north and south the parks can be approached *via* I-87 (the New York State Thruway) using Exit 16 (Harriman Exit) to connect with US 6, or Exit 15 (Suffern Exit) to gain access to the southern end of the parks and NY 17 south. The Palisades Interstate Parkway (Exit 13 on I-87/287) from approximately Letchworth Village on the south to the Queensboro Circle on the north, winds through the northeastern section of the parks and from it many excellent hiking trails and pull-offs are available. At the Queensboro Circle one can get onto US 6 headed east and proceed to the Bear Mountain Bridge. Finally, US 9W runs along the western shore of the Hudson River east of the park. They can be reached from north or south by way of US 9W, which connects with NY 210 near the south end or US 6 at the Bear Mountain Bridge on the northeast.

NY 210 cuts east-west through the southern half of the parks and can be reached by US 9W, the Palisades Interstate Parkway, or NY 17 south (on the west side of the parks). This is a superior drive and will give the birder access to extensive tracts of various habitat types.

From the Bear Mountain traffic circle proceed south on US 9W for one-half mile to the main gateway to Bear Mountain State Park and the site of the sprawling, rustically baronial, Bear Mountain Inn. At this hub of activity there is a ski jump, a swimming pool, a skating rink, a row-boat lake, an athletic field and a picnic grounds. "Anything here for the birder?" you ask. Well, yes. Park and then walk along the white-blazed Appalachian Trail behind and east of the inn; this is a nature trail at this point. After a short distance one reaches the Trailside Museum, which consists of four buildings. Here you can pick up several hiking-trail maps and also obtain the most current information on notable bird sightings throughout the parks.

From the inn, follow signs indicating Seven Lakes Parkway to Perkins Memorial Drive. In mid-May listen for nesting Prairie Warblers along the drive. From Perkins Memorial Tower atop Bear Mountain, one has views of three states as well as views of migrating hawks during autumn.

Approximately 1.5 miles south of Bear Mountain Inn US 9W descends to water level. On the left you will notice a causeway leading out to Iona Island. Park just north of the causeway on the east side of US 9W. Walk out the causeway looking for Double-crested Cormorants in mid-May, nesting herons, Least and American bitterns, and duck species. King Rails have been found there in numbers by mid-May and Virginia Rails can be heard in the marsh as early as the end of the first week in April. Listen for Soras, and watch for gull and tern species. At the end of the first week in May watch for numbers of Common Nighthawks migrating northward. Notice fishing Belted Kingfishers, which nest in the area. Swallows will be commonly seen overhead as the weather grows warmer. Listen for Long-billed Marsh Wrens nesting in the marsh.

Iona Island has earned a reputation as a fine place to view overwintering Rough-legged Hawks and Bald and Golden eagles. Be sure to check out the areas then.

In Bear Mountain and Harriman state parks Wood Ducks nest in high numbers on many of the Bear Mountain Lakes. Turkey Vultures also nest. Goshawks nest in Harriman along with an occasional pair of Red-shouldered Hawks and good numbers of Broad-winged Hawks after mid-April. There are many active nests of Red-tailed Hawks throughout the parks. This species can be seen acting territorially by the end of March and the beginning of April. Ruffed Grouse, cuckoos, owls, Common Nighthawks and Ruby-throated Hummingbirds can be seen in their preferred habitats. Several woodpecker species including Pileated, and several flycatcher species nest within the grounds. Nuthatches, Brown Creepers, and wrens can be found throughout. Look for nesting Carolina Wrens along the trail up Dunderberg Mountain. Thrush species can be heard actively singing by mid-to-late May. Blue-gray Gnatcatchers, Solitary Vireo, and upwards of fifteen warbler species nest on the park grounds.

Golden and Bald eagles have been spotted in Bear Mountain and Harriman state parks in the late autumn and throughout the winter.

A few of the most productive sites in the parks are Dunderberg Mountain, where male Kentucky Warblers have been heard singing in mid-to-late May; Island Pond, the site of nesting Solitary Vireos; Dismal Swamp, where singing Nashville Warblers and Northern Waterthrushes defend territories; Pine Swamp, where Cerulean and Pine warblers can be found nesting; and the White Cross hiking trail, which extends from just south of Parker Cabin Mountain south to the Tuxedo–Mount Ivy trail. This trail is 2 miles in length and surpasses most others in terms of the avian life found from it.

ROCKLAND COUNTY
Piermont Marsh and Pier (Map 9.1)

Rating: Spring****, Summer**, Autumn***, Winter***

Piermont Marsh is located south of the Tappan Zee Bridge on the west side of the Hudson River, in the village of Piermont. The village, once called Tappan Landing, takes its present name from an earthen pier constructed in 1839 as the eastern terminus of the Erie Railroad. The length of the pier was one mile. From here at one time, both freight and passengers were transferred from trains to boats in order to complete the trip to New York City. Now of course, the railroad extends south to Jersey City, New Jersey, and the pier has long since ceased serving its original function. While the pier was in use, the village of Piermont flourished as one of a string of riverfront villages along the western shore of the Hudson River. Today it is a quiet outlier of Manhattan, retaining much of its handsome old character.

The Piermont Pier is presently occupied by several factories, but its eastern end, owing to many years of disuse, has turned into a fine place from which to watch birds. The shallows south of the pier have reverted to rich marsh that provides the required breeding habitat for many bird species and food and shelter for numerous others on migration.

Birding the pier and marsh is rated as "easy." Some birders cover the extensive marsh by canoe, which is highly recommended, although the area can be well-covered partially by car and the remainder on foot.

From the western end of the Tappan Zee Bridge and I-87 (I-287) proceed south on US 9W (South Broadway Avenue) approximately 2.8 miles to the hamlet of Piermont. Follow signs for the business district and obtain Piermont Avenue. Turn east (left) off Piermont Avenue onto Paradise Avenue and then Ferry Road. Park in the parking lot of the large paper factory that dominates the western end of the pier.

Walk out onto the pier. There is a baseball playing field here and along its edges Killdeer, Horned Lark, and sparrow species nest. Proceed to the dump area, which is an excellent birding site and probably the only inland location for nesting Seaside Sparrow. Continue along the road and check the streams it crosses for King, Clapper, and Virginia rails, as well as for Sora. All have been recorded here. Continue on to the end of the pier, checking the mudflats on its south side for duck species and shorebirds. Check the north side of the pier for herons, ducks, and gulls.

South of the pier there is an excellent mix of woodland and marsh habitat. Walk the fire road along this south edge of the marsh, looking for the following species: nesting Green Heron, Black-crowned Night Heron, Least and American bitterns, American Kestrel, rails, Belted Kingfisher, Pileated Woodpecker, flycatchers, swallows, Carolina Wren and Long-billed Marsh Wren, possibly nesting Blue-gray Gnatcatcher, breeding Yellow-throated, Red-eyed and Warbling vireos, Yellow and Prairie warblers, Ovenbird, Common Yellowthroat, American Redstart, Eastern Meadowlark, Red-winged Blackbird, Northern Oriole, American Goldfinch, possibly Sharp-tailed Sparrow, and Swamp and Song sparrows.

In spring the marsh and adjacent woodlands have yielded such species as Double-crested Cormorant, Glossy Ibis, Osprey, Common Gallinule, Common Tern, Yellow-billed and Black-billed cuckoos, Whip-poor-will and Common Nighthawk, Purple Martin, Fish Crow, Prothonotary, Tennessee, Orange-crowned, Magnolia, Cape May, Black-throated Green, Cerulean, Blackburnian, Bay-breasted, and Pine warblers. Some of these, of course, are only occasional visitors, but in spring the woods and marsh are alive with bird activity.

In autumn Piermont Marsh and Pier harbor waterfowl, several raptor species, sporadically Black-bellied Plovers and Ruddy Turnstones, commonly American Woodcocks, Common Snipe, yellowlegs, Spotted, Solitary and Semipalmated sandpipers, Bonaparte's Gulls, Common and Least terns, *Empidonax* and *Myiarchus* flycatchers, vireos, and a full complement of warbler and sparrow species.

In winter one can observe loons, grebes, various waterfowl, possibly Bald and Golden eagles, gull species, and small half-hardy landbirds.

DUTCHESS COUNTY
North and South Tivoli Bays and Cruger and Magdalen Islands (Map 9.5)

Rating: Spring****, Summer***, Autumn***, Winter*

The Hudson River is characterized as fresh-tidal water from approximately the city of Poughkeepsie to the city of Troy, nearly 75 miles north. Cruger and Magdalen Islands and the Tivoli Bays are located on the eastern bank of the river in northwestern Dutchess County. North Tivoli Bay was created more than 100 years ago by the partial diking of a shallow edge of the Hudson estuary. North Tivoli Bay a freshwater tidal marsh and its varied peripheral habitat has a higher diversity of bird and fish life than South Tivoli Bay, which is essentially shallow tidewater and mudflats. However, South Tivoli Bay is rapidly becoming a marsh. The tidal swamp between the two bays contains red maple and ash. Deciduous woods grow along Cruger Island Road and from there one also has access to non-tidal woodland ponds, a tidal purple loosestrife marsh, and an open estuary.

Tivoli Bays with their extensive marshes, associated shallows, and tidal swamps, and Cruger and Magdalen islands with their rocky, wooded bluffs and ridges, form a three-square-mile complex of wildlands possessing a high natural diversity. The Tivoli Bays are the only large fresh-tidal cat-tail marshes on the Hudson River entirely surrounded by undeveloped forest. The uniqueness of these natural areas is further enhanced by their proximity to undisturbed uplands. In total, the area provides an important migration stopover for birds as well as post-breeding habitats. It is also important as a breeding habitat for true marsh birds. By 1981, a cumulative total of upwards of 150 bird species had been recorded here in all seasons.

Early spring and late summer are the best times to visit the area. If one wishes to bird Cruger Island, remember that driving Cruger Island Road out to it is not allowed. Walking this road/causeway is not difficult and the birding along it is excellent. Remember also that one must take the tidal cycle into account when planning to walk out to the island. During high tide the causeway is impassable and if one is on the island he will have to remain there until the tide recedes. The local weather bureau, newspapers and/or yacht clubs can be consulted for tide tables. Permission to bird the island and bays is not needed and to thoroughly cover the area plan on five to six hours. To bird the interior marshes and Magdalen Island off the northwest section of North Tivoli Bay completely, a canoe is required. Poison Ivy flourishes along Cruger Island Road, on Cruger Island, and in most upland areas.

From the city of Poughkeepsie and the Mid-Hudson Bridge, proceed north on US 9 (the Albany Post Road) for approximately 9.2 miles, passing through the

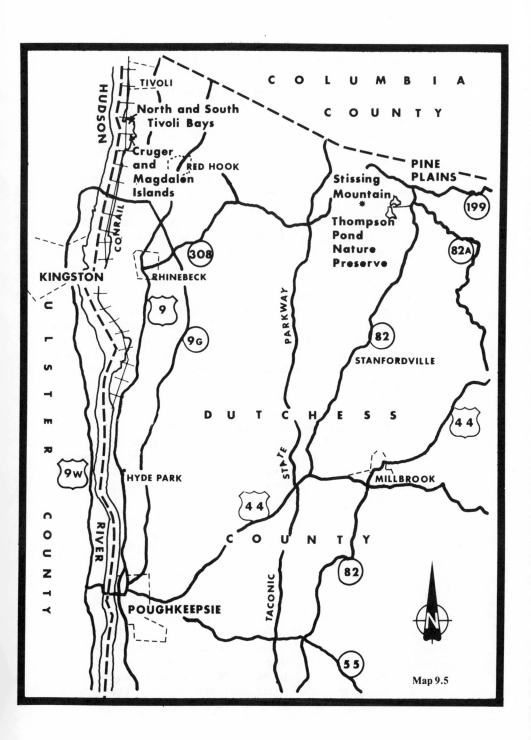

Map 9.5

villages of Hyde Park and Rhinebeck. At the intersection of US 9 and NY 9G at Weys Corners, just northeast of Rhinebeck, turn left (west) onto NY 9G. Proceed northwest and then due north on NY 9G for approximately 2.6 miles, to the junction of NY 9G and Cruger Island Road. Here turn left (west) and park in a large parking area at the top of Cruger Island Road at the Bard College Cruger Village dormitories. Walk west on the dirt Cruger Island Road to the island.

On the island there are a number of old roads and trails that when cleared, give easy access to the best birding spots. However, both North and South Tivoli Bays can be best viewed from the Conrail railroad right-of-way.

Species that breed along the railroad right-of-way, and in the cat-tail and purple loosestrife here are: Least Bittern, Virginia Rail, Sora, Common Gallinule, Eastern Kingbird, Willow Flycatcher, Long-billed Marsh Wren, Gray Catbird, Yellow Warbler, Common Yellowthroat, American Redstart, Red-winged Black-bird, Northern Oriole, American Goldfinch, and Swamp and Song sparrows. The breeding population of Long-billed Marsh Wrens is especially dense here and this is a good place to watch their territorial behavior. Pileated Woodpeckers are resident in the area, and with patience they can be seen and more often heard in the vicinity of lower Cruger Island Road.

During spring and autumn migrations Great Blue and Green herons and Black-crowned Night Herons and American Bitterns can be seen in the marsh. High numbers of Wood Ducks, and hundreds of Green-winged Teals, Mallards, Black Ducks, and Blue-winged Teals can be seen in migration from the duck hunters' blinds (all of which are accessible only by boat), or by scoping from the railroad bridges or river shores. Ospreys are usually present throughout the area in April, May, and September. In late August and early September Ruby-throated Hummingbird can be seen in the jewelweed.

The time of ice-out and soon thereafter (February or March), is the best time to see the Bald Eagle, although, of course, it is not present every year. Look for migrant shorebirds in spring and autumn as well as gulls, swallows, and vireo species. Birding the trails of Cruger Island can be especially productive for northbound warblers in late April and through May. Note the hordes of breeding Red-winged Blackbirds.

Thompson Pond Nature Preserve and Stissing Mountain (Map 9.5)

Rating: Spring****, Summer**, Autumn***, Winter**

Thompson Pond is a 100-acre glacial lake fringed by upland farm fields, mature deciduous woodlands, and a freshwater marsh in northeastern Dutchess County. It lies at the foot of Stissing Mountain, whose elevation rises to 1400 feet above sea-level. The Thompson Pond Project Committee together with the Nature Conservancy finished buying the pond and more than 200 acres surrounding it in 1972, and in 1975 the preserve was dedicated as a National Natural Landmark.

The preserve is open daily year-round, and permission is not required to bird it. There is neither an admission nor a parking fee.

From the city of Poughkeepsie and the Mid-Hudson Bridge; proceed north-east out of the center of the city on US 44. Continue approximately 5.5 miles to the hamlet of Washington Hollow and the junction of NY 82 and US 44. Here turn left (north) and continue driving north on NY 82 for 7.7 miles to the village of Pine Plains. At Lake Road Causeway, which is south of the junction of NY 82 and NY 199, there is a firehouse on the west side of NY 82. Turn left (west) onto Lake Road Causeway. Follow it west for approximately 1.6 miles to the main entrance of the preserve, at the foot of Stissing Mountain. The preserve is located on the left (south) side of Lake Road Causeway. Park along the road near the preserve's entrance.

Follow the preserve's main trail, the entrance to which is located on the south side of Lake Road Causeway. This trail around Thompson Pond passes first through dense stands of cat-tails and bulrushes, next through deciduous woods, and then to the open field at the south end of the preserve. It then passes through a wooded swamp and up the east side of Thompson Pond to a point where there is a boardwalk allowing passage over the headwaters of Wappingers Creek. Continue up the east side of the pond where the trail passes through wooded hills and farm-land bordering the preserve on its east side, and then to its terminus at Lake Road Causeway. The time required to walk the main trail is about three hours. All along the walk the birder should have passed profuse growths of mountain laurel, tril-lium, patches of lily-of-the-valley and evening primrose, and clumps of various fern species.

By 1981 more than 170 species of birds had been recorded within the Thomp-son Pond Preserve, including waterfowl, nesting Turkey Vultures, buteos and ac-cipiters, nesting Ruffed Grouse, Virginia Rails and Soras, migrant shorebirds and gulls, woodpeckers both resident and migrants, various flycatcher species, several thrush species including Eastern Bluebirds in considerable numbers, five vireo spe-cies, twenty-seven warbler species, including Worm-eating, Blue-winged and Golden-winged, and Black-throated Green warblers, blackbirds, tanagers, finches and a variety of sparrow species including Savannah, Grasshopper, and Vesper sparrows.

Walking throughout the preserve is considered quite easy as the trails are well-maintained and marked. In 1981 the total trail mileage was four. On the west side of the preserve Stissing Mountain rises, casting its shadow over the ground be-low. There is a trail to the top of the mountain, where the view is panoramic. This trail begins approximately 0.2 mile north of the main entrance to the preserve. The walk up to the summit of Stissing Mountain can take an hour or more. In recent winters one or more Golden Eagles have been seen coursing effortlessly over the mountain.

Vassar College Campus, Vassar Farm, and James Baird State Park (Map 9.6)

Rating: Spring****, Summer**, Autumn**, Winter***

The city of Poughkeepsie had, in 1980, a population of approximately 40,000 peo-ple. Growth pattern census projections indicate that Poughkeepsie and municipal-ities nearby will reflect modest increases within cities and dramatic increases in suburban towns by 1990. This, of course, means the reduction of open space areas and forests as they are converted to housing developments. Vassar College Campus, Vassar Farm and James Baird State Park are therefore valuable assets as some of the last remaining open spaces in the vicinity of Poughkeepsie. Within these three sites one will find a variety of habitats that provide excellent birding.

Vassar College Campus and Vassar Farm lie along the southeastern edge of the city of Poughkeepsie. The area of the campus is approximately 1000 acres and of the farm approximately 450 acres. Habitats on the campus include Sunset Lake at the southern end of the campus, a large marshy area rimmed with dense growths of rushes and bamboo grass, the landscaped campus proper containing large, old trees of many varieties, a low-lying area with a hemlock grove of considerable size south of the buildings, and a golf course extending east of the campus. The Vassar Farm, inactive since the early 1960s, is located just south of the campus and consists of fallow cropland interspersed with woodlands and shrubby fields, several ponds in the northern portion of the tract, and creeks of intermittent and permanent flow. The fields and woodlands of the southern section of the farm are excellent representations of vegetative successional trends, and up to 200 acres are kept in permanent grassland through a program of periodic mowing. Casper's Kill Creek is the largest water course common to these two sites. It flows through the campus behind the buildings area and through the southern half of the farm tract.

James Baird State Park is located approximately 4.7 miles due east of Vassar College campus and farm, and the Taconic State Parkway borders it on its east side. Its area is approximately 600 acres and it contains two ponds on a golf course, extensive woodlands, a few dense evergreen plantations, open fields, and a waterfall in a secluded dell. This small ravine is lined with a variety of ferns and wildflowers.

Permission is not required to bird Vassar campus and farm. However, inform the security service at the college that you are there and for what purpose. James Baird State Park is open to the public year-round and no permission is required to bird it. An admission and parking fee is charged from mid-April through October. Heavy recreational use of the park may conflict with birding in summer. This is not a problem, however, from approximately 6:00 a.m. to 8:30 a.m., which is ample time to cover the area. At all three sites the birding is rated as "easy." There are roads and well-maintained trails giving access to virtually all of the habitat types.

From the eastern terminus of the Mid-Hudson Bridge and the intersection of US 9 and Church Street in downtown Poughkeepsie, proceed east for approximately 0.7 mile (8 or 9 blocks) to South Hamilton Street. Turn right (south) and advance one block to a fork where South Hamilton Street continues due south and Hooker Avenue heads southeast. Turn left onto Hooker Avenue and drive southeast for approximately 2.1 miles to Raymond Avenue. Here turn left (north) and proceed a short distance almost to the entrance drive to the college on the east (right) side of Raymond Avenue. Park just south of the entrance drive or turn right and drive in the main campus drive and park on either side. Vassar Farm is situated on the south side of Hooker Avenue (NY 376) between Cedar Avenue on the west and IBM Road on the east. Its northern boundary is Hooker Avenue and

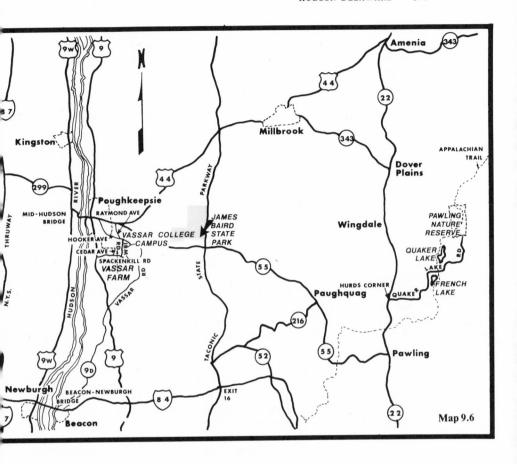

Map 9.6

its southern boundary is Spackenkill Road. There is a farm road off Hooker Avenue opposite Raymond Road. One can drive down this road and park near a grandly conspicuous old rambling barn.

James Baird State Park is located northeast of the village of Freedom Plains between NY 55 on the south and US 44 on the north. The main entrance is from the Taconic State Parkway, 1 mile north of the intersection of the parkway and NY 55. It is very well marked.

Birds to be expected at these three sites in addition to all of the more common woodland and urban dwelling species are: Great Blue and Green herons, nesting Wood Duck, Turkey Vulture, breeding Red-tailed Hawk, American Kestrel, Virginia Rail, Sora, American Woodcock, Yellow-billed and Black-billed cuckoos, Screech and Barred owls, Common Nighthawk, Pileated, Hairy and Downy woodpeckers, Yellow-bellied Sapsucker in early spring and late fall, nesting Eastern Phoebe, Barn Swallow, Carolina Wren, Wood Thrush, Eastern Bluebird, Blue-gray Gnatcatcher, Warbling Vireo, upwards of twenty warbler species during migration with Worm-eating, Golden-winged, Blue-winged, Cerulean and Blackburnian warblers nesting. In the open fields look for nesting Bobolink, Eastern Meadowlark, Savannah, Grasshopper, and Vesper sparrows.

Pawling Nature Reserve, Quaker and French Lakes (Map 9.6)

Rating: Spring****, Summer**, Autumn***, Winter*

The Pawling Nature Reserve of the Nature Conservancy is located in the southeast corner of Dutchess County. Quaker Lake abuts the southeastern section of the reserve. It is however, not in ownership by the Nature Conservancy. French Lake is south and slightly east of Quaker Lake.

The Pawling Nature Reserve is approximately 1000 acres, nearly 80 percent of which is mature second-growth, mixed-hardwood, upland forest. The remaining acreage consists of birch groves and swamps, an old apple orchard, laurel fields, swamp alder, a rare natural, open meadow and an easily accessible, hemlock-lined and shaded gorge. The reserve is surrounded by approximately 2000 acres of heavily forested, privately held land. The upland terrain is traversed by Hammersly Ridge, a north-south ridge whose elevation is 1055 feet above sea-level.

The Appalachian Trail (marked with white blazes) passes through the reserve on its route between Maine and Georgia and three miles of other trails have been developed here. They are marked by yellow blazes and red blazes. The red-blazed trail serves also as a bridle path. By hiking these trails one has access to stands of maple, tulip, poplar, oak, elm, hickory, ash, red cedar, birch of four species, and groves of eastern hemlock. Wildflowers are abundant here, and be sure to note the several-acre beaver complex near the northern trailhead.

The preserve is open to the public year-round with the exception of the four-week period from the third week of November through the second week of December, when it is closed. There is neither an admission nor a parking fee. Birding the preserve is rated as moderately difficult, owing to the steep trails winding to the top of the ridge. Completing any of the marked trails would take about three hours, birding at a fairly leisurely pace. There are two parking lots, each accommodating three or four cars, one at the north end of the reserve and another 1.5 miles farther south. The Nature Conservancy sign is located at the more southern parking area.

From the eastern end of the Beacon-Newburgh Bridge and the city of Beacon, proceed east on I-84 to its junction with the Taconic State Parkway. Here head north on the parkway and proceed to the first exit. Here leave the Taconic State Parkway and head east on NY 52. Proceed east a short distance (approximately 1 mile) where NY 52 forks with NY 216. Turn left onto NY 216 and continue on it northeast to the hamlet of Paughquag. Here turn right (east) onto NY 55. Take NY 55 southeast into the village of Pawling. At the junction of NY 55 and NY 22 turn left (north) and proceed north on NY 22 for about three miles to the hamlet of Hurds Corners. Watch for Hurds Corners Road coming in on the right (east) side of NY 22. Turn right (east) at Hurds Corners Road and follow it east to its junction with Quaker Lake Road. Turn left (north) onto Quaker Lake Road and follow it north (stopping to bird French Lake on the right and Quaker Lake on the left) for approximately 1.2 miles, where you will see the south parking area of Pawling Nature Reserve and the Nature Conservancy sign on the left. Park here and walk west to Hammersly Ridge or north to the hemlock gorge. From here one can walk either north or south along any of the above-mentioned trails. The reserve's north parking lot can be reached by continuing north on Quaker Lake Road past the south parking area for 1.4 miles, where it is located on the left (west) side of the road. From here one can walk south on the Appalachian Trail and meet the yellow-blazed or red-blazed side trails.

This area is most productive during spring and autumn (in that order) migrations, when the woods host waves of migrating passerines. Hawks nest in the dense woodlands as do grouse and owls.

PUTNAM COUNTY

The total land area of Putnam County is a mere 321 square miles, and its human population in 1980 was estimated at fewer than 60,000 people. It lies north of Westchester County, south of Dutchess County, east of the Hudson River and west of Connecticut. It is the least intensively birded county within easy reach of metropolitan New York City—a fact quite incomprehensible, for the county is rich in scenic beauty and diversified habitats. These habitats include extensive open deciduous woodlands, marshy swales, weedy fields and thickets, a wealth of reservoirs and lakes, old coniferous plantations, and upland fields, meadows, and farms. There are true wildernesses in Putnam County. The vegetative profile or structural diversity here offers the opportunity for fulfilling the nesting and food needs of a wide variety of bird species. A rich birdlife is found in some of the more complex vegetative communities. The following places are included in this book because they are accessible, little-known sites that have proven repeatedly productive, especially in spring and autumn. Any observations of breeding birds, rare birds, or common birds out-of-season, should be submitted with supporting details to the *Kingbird* Region 9 editor and the New York State Avian Records Committee (page 4).

Sunken Mine Road, Wiccopee Road, Clarence Fahnestock Memorial State Park, and the Wiccopee Reservoirs, a Tour (Map 9.7)

Rating: Spring****, Summer***, Autumn***, Winter*

This tour should take about five or six hours to bird at a leisurely pace, and it includes some of the finest birding habitats in western Putnam County.

From north or south of the county approach it on US 9. At the intersection of US 9 and NY 301 at McKeel Corners, east of the village of Cold Spring, gain NY

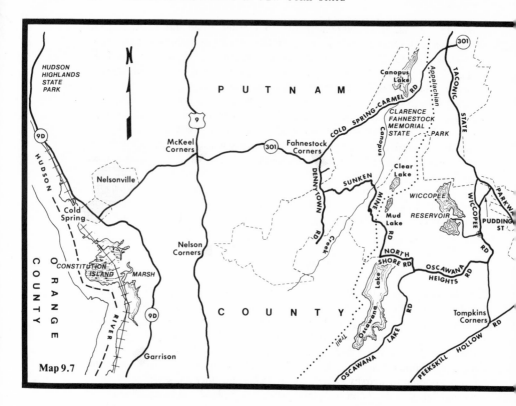

Map 9.7

301 (the Cold Spring-Carmel Road) headed east and proceed east to the almost indistinguishable hamlet of Fahnestock Corners, a distance of approximately 2.6 miles. Look for Dennytown Road coming in on the right. Turn right (south) on Dennytown Road and proceed south a short distance (approximately 0.6 mile) to Sunken Mine Road (the second road), which comes in from the east (left). Turn left on Sunken Mine Road and proceed along it, stopping to bird at every safe pull-off. Continue birding along the entire distance (approximately 3 miles) of Sunken Mine Road, being sure to especially bird well the area north of the road approximately 1.5 miles from the turn, which is the site of the old Sunk Mine, and where Canopus Creek crosses the road, to its junction with North Shore Road at the north end of Oscawana Lake. This is a good place to stop to bird the north end of the lake.

Afterward, proceed east and south (after a sharp right turn) on North Shore Road to its intersection with Oscawana Heights Road. Turn left (east) and follow this road to its junction with Wiccopee Road. All along the way you should have been stopping to bird the various open fields and mixed deciduous woodlands through which these roads traverse. Along Sunken Mine Road there are beaver pond meadows and deep, water-filled old mines obscured by lakes and marshes to investigate and cold clear brooks along which Winter Wrens and waterthrushes nest. At Wiccopee Road turn left (north) and proceed approximately 0.6 mile, where you will see a gate on the left (west) side of the road bearing a "posted" sign of the City of Peekskill Water Supply Department, banning vehicles but not walkers. Pull over off the road here and park. Walk around the gate and down the road and between the two parts of the Wiccopee Reservoir, a distance of 1.3 miles. This is an excellent birding road. Walk out the same way you came in and drive approximately 1.2 miles north on Wiccopee Road as it becomes Pudding Street and then to the junction of Pudding Street and the Taconic State Parkway. At the parkway, turn left (north) and continue north to the junction of the Taconic State Parkway and NY 301. From Pudding Street north you will have been in Clarence Fahnestock Memorial State Park, which comprises 6200 acres. Here leave the parkway and gain NY 301 west, and proceed west to Canopus Lake on the north side of NY 301. Bird the periphery of the lake at its southern end, and, at its southern tip, look for the Appalachian Trail. Obtain the trail and bird south on it to Clear or Mud lakes (nearly 2.2 miles). Return to the parking lot and NY 301 via the same trail in.

Along this tour, if taken in spring or early summer, the birder might have encountered all of the following species: breeding Great Blue Heron, Least and American bitterns, Wood Duck, Mallard and Black Duck, nesting Turkey Vulture, Sharp-shinned, Red-shouldered, Broad-winged, and Red-tailed hawks, American Kestrel, Ruffed Grouse, Virginia Rail and Sora, Spotted Sandpiper, two or three gull species, breeding Screech, Great Horned and Barred owls, Ruby-throated Hummingbird, kingfisher, Common Flicker and Pileated Woodpecker, Yellow-bellied Sapsucker on migration, flycatcher species, Tree, Rough-winged and Barn swallows, Brown Creeper, wrens, mimids, thrushes on migration and breeding, Blue-gray Gnatcatcher, two or three vireo species, upwards of twenty warbler species, including Black-and-white, Blue-winged, Golden-winged, Yellow, Magnolia, Cape May, Black-throated Blue, Black-throated Green, Cerulean, and Prairie warblers, Ovenbird, Northern Oriole, Scarlet Tanager, grosbeak, Indigo Bunting, finches, and five or six sparrow species. It is no trick at all to log over 100 species per day on this tour, if it is done in the peak of spring migration.

Constitution Island Marsh (Map 9.7)

Rating: Spring***, Summer**, Autumn***, Winter**

This tidal marsh of more than 330 acres is administered by the National Audubon Society. It is bounded on the north and south by state-owned tidal wetlands; to the west by the New York Central Railroad, Constitution Island which is owned by the United States Military Academy and the Hudson River; and by Boscobel Restorations on the east. The marsh protrudes into the Hudson River just 1.5 miles south of the village of Cold Spring. Permission to bird the island can be obtained from the National Audubon Society's warden/biologist who resides directly across from the entrance to Boscobel Restoration. The habitat here consists of almost

270 acres of low-lying marshy swales, more than 60 acres of heavily wooded uplands and a beautiful ravine and waterfall. A multitude of northbound migrants stop here in early and late spring and approximately 25 breeding species can be found in late spring and through summer.

To reach Constitution Island from either the north or the south, proceed toward the village of Cold Spring on US 9. At the junction of US 9 and NY 301 leave US 9 and gain NY 301 headed west. Proceed west to the village of Cold Spring. Watch for NY 9D coming in on the left (south) and turn left on it. It may also be called the Bear Mountain-Beacon Highway. Continue south on NY 9D for approximately 1.5 miles to Indian Brook Road. Turn right and proceed one-half mile to an iron fence and gate. Proceed on foot down this road which terminates at the cove south of Constitution Island Marsh.

Walk north and south along the railroad tracks, which afford a fine vantage over the marsh. Wood Ducks nest in the marsh and Ospreys can be seen in autumn. Look and listen especially for Clapper and Virginia rails, which not only breed here but remain well into winter. Pileated Woodpeckers and Blue-gray Gnatcatchers are specialties of this area, and a keen birder should be able to find nesting Long-billed Marsh Wrens in the narrow-leaved cattail, and Cerulean Warblers in the area as well as Savannah, Vesper, Swamp, and Song sparrows.

NORTHERN WESTCHESTER COUNTY
Croton Point Park (Map 9.8)

Rating: Spring****, Summer*, Autumn**, Winter***

Croton Point Park was once part of a huge semi-feudal estate established in the Hudson River Valley during the colonial period. The first families to arrive in the valley were Dutch and to them belonged the first fortunes to be made in the valley. The Van Cortlandt family owned the manor, which originally comprised 86,000 acres, from 1688 to 1945, when the manor house, adjoining formal gardens, and several outbuildings were finally bought and eventually restored. The Van Cortlandt Manor, located just east of the entrance to Croton Point Park, today houses an extensive collection of authentic antique furnishings.

Croton Point Park has traditionally proven one of the better birding sites along the east side of the Hudson River in northern Westchester County and it is at this location that Weissman and Howe *et al.* have substantiated such records as an adult Yellow-nosed Albatross in August 1976, a Golden Eagle in October 1979, an American Avocet in August 1979, and a Royal Tern in August 1976.

The most productive seasons to bird Croton Point Park are spring and winter. In winter, concentrations of loons, grebes, waterfowl, sometimes Bald Eagle, and gulls (including "white-winged" species) can be found on ice-floes in open water in the Hudson River or at the dump, and owls and passerine species can be found elsewhere in the park, especially in the pine grove. In spring, landbirds using the Hudson River flyway on their way north can be found throughout.

To reach Croton Point Park from north or south of it, take US 9 to the vil-

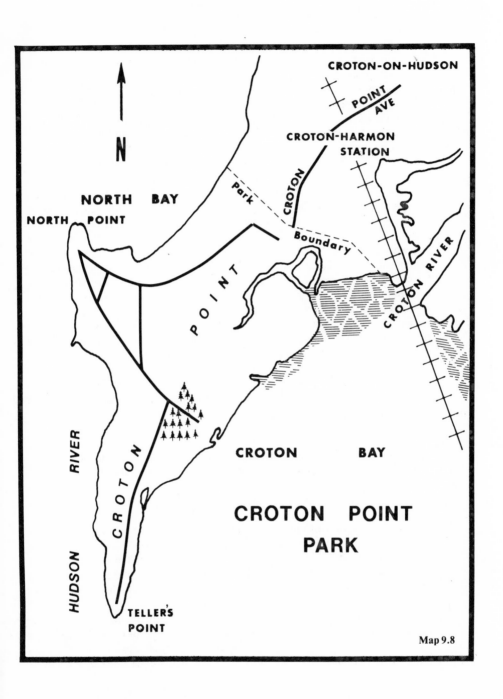

CROTON POINT
PARK

Map 9.8

lage of Croton-on-Hudson. Leave US 9 at the Croton Point Avenue interchange just north of the point where the Croton River widens and feeds into Croton Bay. Turn west on Croton Point Avenue and proceed west past the Croton-Harmon train station on the left and continue into the park.

Bird the entire park, searching out the fields, marshes, pine groves, picnic and parade grounds, dump area, and tidal flats in Croton Bay at low tide. Be sure to bird Teller's Point, the thirty-foot high bluff extending southwest into the river at the park's southern tip. This is especially rewarding when the winds are from the northwest. Also investigate Enoch's Neck and the North Point at the park's northern tip.

Bird species to especially watch for are: late-lingering Horned Grebes, Double-crested Cormorant, Cattle and Snowy egrets, Mute Swan, Least Sandpiper, over-wintering Iceland and Bonaparte's gulls, first-arrival Hermit Thrushes, and possibly the casual "Lawrence's" Warbler hybrid and Indigo Bunting, along with various sparrow species in the early spring; Glossy Ibis has been reported in summer along with Least, Semipalmated, and Western sandpipers in late August. Look also for breeding Long-billed Marsh Wren and early-migrating Bobolink in late summer. In autumn one should be able to find Red-throated Loon, post-breeding herons and egrets, migrating waterfowl and raptors, Virginia Rail, Black-bellied Plover, kinglets, warblers, possibly the uncommon Blue Grosbeak and hordes of sparrows in migration. In winter examine any peculiar looking gull in the dump southeast of the railroad tracks for Glaucous and Iceland gulls and look for sparrows, Lapland Longspur, and Snow Bunting in the open fields. The pine grove on the map has a dirt road running between it and the marsh below it. This is an excellent birding area between two productive habitats, especially for owls and passerines.

The Arthur W. Butler Memorial Sanctuary, Byram Lake Reservoir, and The Eugene and Agnes Meyer Nature Preserve (Map 9.9)

Rating: Spring***, Summer**, Autumn***, Winter*

The Butler Sanctuary of the Nature Conservancy comprises more than 360 acres of predominantly ridge and swale topography, approximately 1.2 miles in length and varying in width from 0.7 mile at its northern end to 0.3 mile at its southern extreme. It is bordered by Byram Lake Road and Byram Lake Reservoir on the south, I-684 and Chestnut Ridge Road on the east, and several large private land holdings on the west and north.

The Meyer Nature Preserve, donated to the Nature Conservancy in 1973, comprises approximately 350 acres and lies west and south of the southern end of Byram Lake Reservoir, which is its eastern boundary. It extends west as far as Sarles Street, although the main portion of the preserve has Oregon Road as its western boundary.

Although Meyer Preserve is somewhat topographically similar to Butler Sanctuary, it has some natural features that Butler lacks — including a rocky ravine through which flows a fast-running brook, a number of high, open meadows, and many small temporary woodland ponds.

There is an extensive, well-marked, and well-maintained network of trails at both sanctuaries, and interconnecting trails between the two make all parts of every habitat type easily accessible. Detailed trail maps as well as checklists of flora and fauna are available from the resident caretaker-naturalist, who lives in a house at the entrance to the Butler Sanctuary.

The largest portions of both preserves are covered by upland deciduous forest, with oaks accounting for 65 percent of the trees and another over forty-five tree species and sixty shrub species also represented. Wildflowers, including pink and yellow lady's slippers, showy and green orchids, and rattlesnake plaintain, can be found along the paths. There are several small swamps, a large swamp dominated by red maple an also containing pure stands of red maple as well as American and black ash, black gum, and yellow birch. There are plantations of white pine and Norway spruce with some conifer specimens at least 100 years old. Both preserves are largely rugged and wild with boulders, cliffs, and rock outcrops found throughout. The two ridges that make up the main body of the Arthur W. Butler Memorial Sanctuary trend in a north-south direction and are broken into a series of ridges which are typically more than 700 feet in elevation. The highest point on the easternmost ridge is 775 feet. This prominent overlook is the top of the road-cut for I-684. From here one can see Long Island Sound, ten miles distant. At this point the Robert J. Hamershlag Hawk Watch, complete with observation platform, is located. In the fall thousands of hawks migrate past this point (see the chapter on hawk watches).

Of the land in Westchester County not developed, approximately 80 percent is forested and 20 percent is in fields. Obviously, old fields are a premium resource in this country. The Butler Sanctuary contains three old fields and the Meyer Preserve contains one large old meadow, all of which are maintained through a program of mowing and new-tree removal. In this way the present state of succession is maintained for numerous species that best thrive in open space.

To reach the Butler Sanctuary from the north, west or northeast: approach the village of Mt. Kisco by way of I-84 and I-684; from the south approach Mt. Kisco on I-684. Leave I-684 at Exit 4 (east of the village of Mt. Kisco). Turn west on NY 172 (South Bedford Road) and proceed approximately 200 yards up a hill where Chestnut Ridge Road comes in on the left (south). Turn left (south) here and continue south on Chestnut Ridge Road for approximately 1.5 miles. You will have passed over I-684. Take the first right. Cross back over to the west side of I-684 and at the end of the bridge turn right onto a small dirt road. Follow it only a short distance into Butler Sanctuary.

If one decides to drive rather than walk *via* the interconnected trails system from Butler Sanctuary to Meyer Preserve, exit Butler by the same route in. At Chestnut Ridge Road, on the east side of I-684, turn right (south) and proceed south to a fork where Lake Ridge Road heads west and Chestnut Ridge Road continues south. Turn right (west) onto Lake Ridge Road and proceed west, passing underneath I-684, to Byram Lake Road. Turn left and proceed as slowly as possible down the east side of Byram Lake Reservoir, stopping to bird it wherever safe.

Proceed south to the end of the lake and turn right at the first road meeting Byram Lake Road. Continue 0.1 mile to the first trailhead on the right, and park in the designated area.

The combined bird lists for the two sanctuaries, plus those recorded on and around the reservoir, stand near 180 species, with approximately 75 of those breeding. The more notable nesting species include Red-shouldered and Red-tailed hawks, Ruffed Grouse, American Woodcock, Yellow-billed and Black-billed cuckoos, Screech and Great Horned owls, Cedar Waxwing, and Worm-eating, Yellow-throated, Chestnut-sided, Hooded, and Canada warblers.

Please remember to sign the register at both places so that the approximate visitor usage can be determined and to also report any notable rare bird or common bird out-of-season sightings to the *Kingbird* Region 9 editor and/or the resident naturalist at Butler.

Mianus River Gorge Wildlife Refuge and Botanical Preserve (Map 9.9)

Rating: Spring****, Summer***, Autumn***, Winter*

This 360-acre preserve of the Nature Conservancy stands as a priceless remnant of what was once virgin America. The Mianus River courses through noisy rapids and through a deep gorge lined by virgin stands of hemlock. It is surrounded by a buffer zone of woods in various stages of succession, including mature hardwood forest. In the mature forest the birder can find oak, hemlock, beech, and black birch. The successional woodlands contain maple, birch, beech, white pine, spruce, aspen, ash, elm, and American chestnut. Wildflowers of 45 families and 245 species abound on the preserve. Thirty-five species of grasses, rushes and sedges have been identified here. The tree species mentioned above are only a few of the 52 found on the preserve; and additional 55 species of shrubs and vines have also been identified here. Ferns and their allies are well-represented with more than 30 species and a preliminary survey of the area has yielded more than 50 species of lichens, 70 of mosses, 40 liverwort species and more than 100 species of mushrooms (fungi) of fluorescent white, sulphur yellow, orange, vermilion, purple, green, and black coloration.

More than 180 bird species have been recorded in the Mianus Gorge area, and it is likely that at least twenty to twenty-five more could be added to the list if the preserve were well-birded year-round.

More than five miles of well-marked and well-maintained trails are layed out at Mianus River Gorge, and birders, hikers, and naturalists are welcomed to investigate them from the beginning of April through early December. There is a visitors' parking lot and a trail shelter, which includes a guest register, trail maps and various other information. There is no admission or parking fee to roam the preserve (rustic restrooms are located near the trail shelter).

To reach the gorge from east, north, or west, proceed toward the village of Bedford on I-84 and I-684; from south of Bedford Village proceed north toward it on I-684. Leave I-684 at Exit 4 and proceed east on south Bedford Road to its junction with NY 22. Turn left (northeast) and follow NY 22 through Bedford Village to the intersection at the triangular green of Poundridge road. Turn right (east) and take Poundridge Road approximately 1 mile. Here (the second possible right turn) turn right (south) onto Stamford (Long Ridge) Road. Proceed a short distance to Miller's Mill Road. Here turn right (west) and advance 0.1 mile over a bridge and then turn left onto Mianus River Road. Continue 0.7 mile to the entrance to the preserve indicated by a sign on the left.

Most of the preserve is forested, so it naturally tends to attract woodland birds with a preference for tall trees, deep shade, shrubby undergrowth, cool temperatures, abundant insect life, and the vegetable or animal food produced by such an environment. Of the total number of birds recorded at the preserve, upwards of 95 species nest, including Sharp-shinned, Cooper's, Red-shouldered, Broad-winged, and Red-tailed hawks, as well as American Kestrels. Ruffed Grouse and Ring-necked Pheasants are permanent residents. Killdeer, American Woodcocks, and Spotted Sandpipers nest here, as do both Yellow-billed and Black-billed cuckoos. Although the shade conceals and distorts bird coloration in the forest, one might be able to locate nesting Barn, Screech, Great Horned, and Barred owls. Common Flickers, Pileated, Hairy, and Downy woodpeckers can be found year-round. Eastern Kingbirds, Great Crested Flycatchers, Eastern Phoebes, Alder and Least flycatchers, and Eastern Wood Pewees can be located in spring and summer. Approximately thirty warbler species have been recorded in the preserve, with some of the more notable nesters being: Worm-eating, Magnolia, Black-throated Green, Blackburnian, Chestnut-sided, Prairie, Palm, both waterthrushes, Hooded and Canada warblers. At least nine sparrow species have been documented here with Vesper, Field, Swamp, and Song sparrows as breeders.

This wildlife refuge has been relatively little explored by the modern army of bird watchers in and around the New York City area. I wholeheartedly suggest that more of us spend more time in these primeval surroundings.

Halle Ravine and Ward Pound Ridge Reservation, a Tour (Map 9.9)

Rating: Spring***, Summer****, Autumn**, Winter*

The Halle Ravine, locally known as "Slawson's Meadow," is a superb ecological rarity of approximately forty acres dominated by a commanding gorge through which water from wetlands to the north plummets and then runs for nearly 0.5 mile through a hemlock forest of high, huge evergreens, including mature trees 125 feet or more in height. The water flows from the ravine into a substantial alder swamp supporting its own unique biotic community. Where the ravine opens into the alder swamp, the terrain levels out and becomes a wetlands, which serves as the drainage basin for the ravine and the surrounding area. This preserve, owned and maintained by the Nature Conservancy, displays contrasts of different seral stages of succession. In the ravine, the mature hemlock forest exemplifies vegetation typical of the Canadian biothermal zone. There is an open field adjacent to the preserve property on its southwest side, which was once used for grazing; it is a fine

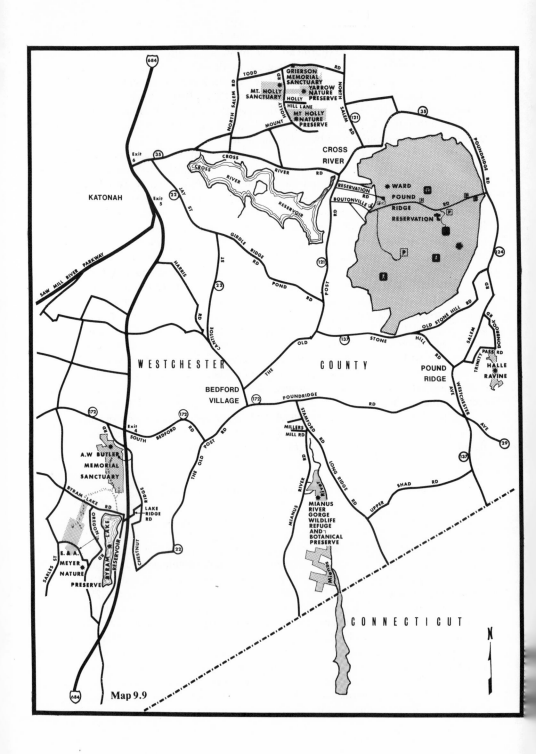

Map 9.9

example of the primary stages of succession. The majority of the preserve consists of mixed hardwood forest with a wetland complex surrounding two ponds, and an area of rough horsetail, doubly interesting because of its dense growth. Rough horsetail's latin name (*Equisetum hiemale*) loosely means "of winter," which refers to the fact that this species of horsetail is evergreen. It grows on the hillside southeast of the ravine. Horsetails, taxonomically, are ranked as the most primitive of vascular plants and usually grow in sandy soils and in damp semi-shaded areas. The size of this stand (approximately fifty feet by twelve feet) is unusual in Westchester County at all.

The dominant trees in the hardwood forest are sugar maple, tulip, poplar, and white ash; the primary constituents of the shrub layer are viburnum and witch hazel; and the main herbaceous plants are poison ivy, garlic mustard, Christmas fern, and hay-scented fern. Throughout the preserve there are abundant rock outcroppings covered with mosses and lichens.

The northern part of the preserve has a moderately sloping terrain with hills of approximately 575 feet sloping gradually down to an old streambed, which has been enlarged into two 0.25-acre ponds. The ravine drains into the alder swamp, which is the lowest elevation in the area — approximately 400 feet.

Yellow lady slipper, purple trillium, Jack-in-the-pulpit, rattlesnake plantain, partridgeberry, Indian pipe, white beanberry, bloodroot, hepatica and spotted wintergreen are a few of the wildflowers growing throughout the Halle Ravine.

None of the successional communities has well-defined boundaries, so there is additionally some transitional zone consisting of conifers intermingled with hardwoods. Here the understory is more substantial than in the hemlock forest, but not as dense as in the mixed hardwood forest.

Halle Ravine is open to the public for "passive" recreational use — birding, hiking — that will not disturb the preserve's ecosystem. There are maintained paths incorporating several foot bridges over the stream, throughout. At the south end of the preserve there is a trail system that connects it with the Pound Ridge town park, southwest of it.

Ward Pound Ridge Reservation located north of Halle Ravine and southeast of the village of Cross River is the largest park in Westchester County and certainly, in that it comprises an area of six square miles, it is larger than many of the hamlets and villages surrounding it. In the wildest sections of the park the biotic community is that of more northern communities, and the birdlife reflects that. Here the shrub and herbaceous plants are limited to those preferring a cool shady environment. Christmas fern and polypody are common in these areas as well as occasional wildflowers and ground pine, shining and tree clubmosses. In addition, there are relatively open upland fields whose edges are dominated by red cedar, shrubs of gray dogwood, smooth sumac, and highbush blueberry, and whose dominant herbaceous plants are goldenrod species, sweet vernal grass, and broomsage. A variety of colorful wildflowers grow here. Deep hollows, especially in the eastern portion of the park, are surrounded by hills that rise sharply to more than 850 feet. Climbing to the top of the fire tower, near the southeastern park boundary, places one at approximately 940 feet above sea level. The Cross River runs east-west through the northern section of the park and, along the southeastern park boundary, the Stone Hill River flows. The mixed mesophytic hardwood forest has a series of intermittent and permanent streams coursing through it. The predominant rock outcroppings found here are of Precambrian age and many Indian artifacts have been found throughout. There is abundant wildlife living in the reservation, including deer, red fox, raccoon, opossum, as well as a wealth of birds of prey, game, and songbirds.

To reach Halle Ravine and Ward Pound Ridge Reservation from east, west, or north: take I-87 and I-84 to I-684 and proceed toward the village of Bedford Village; from the south approach Bedford Village on I-684. Leave I-684 at Exit 4 and proceed east on South Bedford Road. At the junction of it and NY 22 turn left (east) onto NY 22. Continue northeast to the intersection of NY 22 and Poundridge Road at the Bedford Village triangle. Turn right on Poundridge Road and continue 7.2 miles to the junction of Westchester Avenue (NY 29). Turn left and proceed north to the junction of Westchester Avenue and Salem Road (NY 124). From here go north on Salem Road approximately 250 yards to Trinity Pass Road, which comes in on the right (east). Turn right and follow Trinity Pass Road approximately 0.7 mile. Just west of the intersection of Trinity Pass Road and Donbrook Road you will see the preserve on the right. Park along the road shoulder and walk into the Halle Ravine.

After birding the Halle Ravine exit *via* the same route in; turn left at Salem Road (NY 124) and then immediately right; thereby gaining Stone Hill Road (NY 137). Continue west on Stone Hill Road to its junction with the Old Post Road (NY 121). Here turn right (north) and continue driving north to either Boutonville Road or to Reservation Road, where there are signs indicating Ward Pound Ridge Reservation County Park. Turn right (east) on either of these entrance roads and continue to one of several parking areas.

In 1981 Halle Ravine had a bird list of approximately 130 species and Ward Pound Ridge Reservation one of upwards of 160 species. Neither of these areas had been consistently and intensively birded, so with more thorough coverage both area lists will undoubtedly grow. Many of the species found at these sites duplicate those recorded at the Mianus River Gorge Preserve (above) with some specialties and additions being Green Heron and Black-crowned Night Heron recorded at both sites during migration, Wood Duck nesting at Ward Pound Ridge Reservation and rarely occurring at Halle Ravine, Turkey Vulture nesting at both sites, Goshawk nesting in Halle Ravine, Sharp-shinned Hawk commonly nesting at Halle Ravine.

Halle Ravine is one of a few natural areas in this region where all three New York nesting species of buteo (Red-shouldered, Red-tailed and Broad-winged hawks) can be expected as breeding birds. Northern Harrier (Marsh Hawk) is rarely encountered at both sites during autumn migration.

Look in low wetlands of Pound Ridge Reservation and especially the alder swales of Halle Ravine for nesting American Woodcock. Solitary Sandpiper has been seen in transit at both locations. Herring and Ring-billed gulls are commonly seen flying over.

The Screech Owl nests at the reservation and it has been known to be widely distributed as a breeder within the ravine; it has been consistently breeding in hollow trees at the entrance to the ravine and also at the foot of Winterbottom Lane on the west side of the ravine. Great Horned and Barred owls breed with remarkable consistency at both sites. Long-eared Owl has been recorded as a winter visitant at Pound Ridge Reservation. Whip-poor-will nests at the reservation, and Common Nighthawk nests at the ravine. Pileated Woodpecker is a year-round resident at both places, while Red-bellied Woodpecker has been recorded as a breeder at Halle Ravine. Bank Swallow is ranked as a common-to-abundant breeder at the ravine. Boreal Chickadee has been recorded at both sites in irruptive years. Red-breasted Nuthatch definitely and Brown Creeper possibly, nest at the reservation. Look for nesting Winter Wren in the root systems of hemlock groves at both places. Gray Catbird is a common breeder in the densest thickets and in swampy areas of both sites. Wood Thrush and Veery are regular breeders in moist open woodlands with plenty of undergrowth at both places. Blue-gray Gnatcatcher nests in the southern section of Halle Ravine, while Cedar Waxwing is found as a breeder at Ward Pound Ridge Reservation. Of warblers, Worm-eating breeds in small numbers within the confines of the ravine, Blue-winged breeds exclusively along the ravine's outer edges and surrounding woodlands as well as at Pound Ridge Reservation. Black-throated Green and Blackburnian warblers can be found breeding in the hemlock stands of both places, especially those of Halle Ravine. Chestnut-sided Warbler is a rare breeder at the ravine and is restricted to the dry slashings of the woodlands. Listen for the loud song of the Ovenbird, which can be found breeding at both places. Northern Waterthrush breeds in the alder swale of the ravine and Louisiana Waterthrush regularly nests in crevices along the length of the ravine as well as along the ravines of the reservation. Its penetrating song is one of the special sounds to be heard in spring and early summer. Hooded Warbler nests at both sites and Canada Warbler is a confirmed breeder at Halle Ravine.

During some winters highly erratic and irruptive species *e.g.,* Pine Grosbeak, Common and Hoary redpolls, Pine Siskin, and Red and White-winged crossbills (with Red Crossbill predominating ten to one), can be found throughout both places.

The Cross River Valley Preserve Assembly (including Mt. Holly Nature Preserve, the Marian Yarrow Nature Preserve, the Mildred E. Grierson Memorial Sanctuary, and the Mt. Holly Sanctuary), a Tour (Map 9.9)

Rating: Spring****, Summer***, Autumn**, Winter*

The total 365 acres of these four preserves, owned and maintained by the Nature Conservancy, are included in this book owing to their diversified habitat and potential for birding. They are located in the northeastern rolling uplands section of the county, north of NY 35 and east of I-684. The topography is quite rugged in some places with a total relief of 270 feet and a maximum elevation of 620 feet. Higher acres are dominated by chestnut oaks, black oaks, and red oaks. Farther downslope, white oaks, sugar maples and black birches predominate. In the lower areas red maple, tulip, and poplar frequently occur. In the swamps and wetlands, hummock grass forms small islands that rise above the water level and fern species are common. They are situated within easy walking distance of one another and comprise approximately 215 acres of rich hardwood forest of predominantly oak-maple association (with several outstanding oak and maple specimens more than a hundred years old), and various other stands of mixed second-growth containing a wide variety of flora: approximately eighty acres of red maple and speckled alder swamp; approximately forty acres of freshwater ponds, swift-flowing brooks and marshy wetlands that feed into the Cross River Reservoir; and nearly twenty-five acres of open meadows that are "brushed out" or mowed annually to preserve the ecology of the meadow. There is also a thirty-foot high waterfall, and a stretch of spectacular eighty-foot cliffs and steep rocky terrain.

These sites are open to the public for hiking and birding and there are limited bridle paths intersecting some sections of one or two of the preserves. There is an extensive system of maintained trails and old roads giving access to all habitats. The variety of habitats in these areas has encouraged a diverse wildlife community and, on the basis of preliminary surveys, it appears to harbor an extensive avian community both in and out of breeding season.

To reach the preserves from east, west, or north: take I-87 and I-84 to I-684 and proceed toward the village of Katonah; from the south approach Katonah on I-684. Leave I-684 at Exit 6. Gain NY 35 (Cross River Road) and proceed east approximately 1 mile to North Salem Road. Turn left (north). Continue 0.25 mile on North Salem Road to its junction with Mt. Holly Road. Turn right (east) and proceed east for 1.5 miles, at which point Mt. Holly Road makes a right-angle turn north. Turn left (north) and you will see a sign 75 feet farther on the right side of Mt. Holly Road indicating the Mt. Holly Nature Preserve. Park the car on the side of the road and follow a red-blazed footpath east to the preserve entrance.

To reach Mt. Holly Sanctuary and the Marian Yarrow Nature Preserve continue north from the Mt. Holly Nature Preserve on Mt. Holly Road just past its intersection with Holly Hill Lane (on the east). Park the car on either Holly Hill Lane or on Mt. Holly Road. Mt. Holly Sanctuary and its entrance are located on the west side of Mt. Holly Road and the Marian Yarrow Nature Preserve and its entrance are on the east side of the road. The entrance to each is clearly marked.

To reach the Mildred E. Grierson Memorial Sanctuary proceed north on Mt. Holly Road approximately 0.5 mile to its intersection with Todd Road. It is advisable to leave your car parked on Mt. Holly Road, as there are parking restrictions on Todd Road. Turn right (east) on Todd Road and walk approximately 100 yards to the Nature Conservancy sanctuary signs on the right. A slightly overgrown fire trail leads from Todd Road into the heart of the sanctuary. Extreme caution is urged near the edges of the swamp on the eastern part of the preserve as some quicksand is known to be present along its rim.

ACKNOWLEDGMENTS

The author would like to express her most sincere thanks to the following people for the information supplied by them for the Hudson-Delaware region: Paul

Jeheber, James P. Stapleton, Robert S. Arbib, Jr., Martin Borko, Thomas L. Carrolan, members of the Dutchess County Bird Club, Robert F. Deed, Thelma Haight, William Howe, Alice Jones, Erik Kiviat, Bud Lorch, Helen Manson, Gertrude A. Pope, Eric Single, the late Robert W. Smart, Edward D. Treacy, Hans Weber, Berna Weissman, David Wells, Executive Director of the Lower Hudson Chapter of the Nature Conservancy. Special thanks to Robert S. Arbib, Jr., Jim Stapleton, and Al Brayton, whose reading of the Hudson-Delaware chapter and comments on it improved it.

Marine — Region 10

THE MARINE REGION encompasses that part of Westchester County lying south of NY 117 and NY 120 (North Tarrytown to Chappaqua to the Connecticut state line) and all of Bronx, New York, Richmond, Kings, Queens, Nassau, and Suffolk counties. The islands lying east of Long Island — Plum, Great Gull, Fisher's, and Gardiner's — are included in this region. With its approximately 1660 square miles, it is the smallest of the ten *Kingbird* reporting regions. With its estimated population (based on 1980 U.S. Census data) of ten million and its population density of 6,024 people per square mile, it is unquestionably the most populous of the ten *Kingbird* regions. In fact, there is no other area of comparable size in North America with as high a population density as *Kingbird* reporting region 10 (Map 10.1).

Nassau and Suffolk counties encompass more than two-thirds of the region's total area. They are less densely populated than the five boroughs of New York City and lower Westchester but their growth rates in the last 35 years, since the end of World War II, have been much higher than those of lower Westchester and New York City. They too have been classified by the United States Census Bureau as metropolitan (as opposed to rural) counties.

The unique aspect of this region can be summed up in two words: salt water. Salt water surrounds most of the land mass of the region or touches at least one boundary. The surface water movements and the ocean currents, have a considerable influence on this region's climate and are largely responsible for the mild winters in the Marine Region. The fact that Long Island lies in the prevailing westerly wind belt also means that its weather is strongly affected by the mainland lying to its west. The ocean acts as an efficient modifying agent, heating and cooling air masses that pass over it to Long Island. Its physiographic make-up as part of the northern terminus of the Atlantic Coastal Plain and its formation by two separate glacial moraines, the Ronkonkoma Moraine and the Harbor Hill Moraine, have determined its vegetation, the shape and composition of its shorelines, and its flora and fauna (see Physiographic Regions of New York State).

These factors are crucial also in the determination of the avifauna in the region. One might expect that with the dense population that there would be a dearth of birds and areas in which to bird in the region. Of course, quite the oppo-

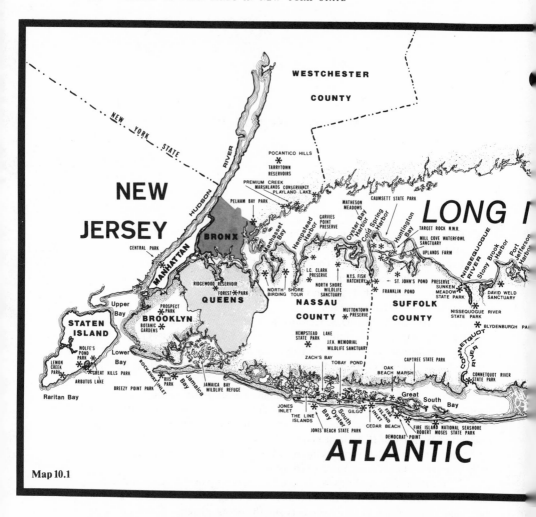

Map 10.1

site is true. The sea provides unique habitats and biomes with altogether different ornithological aspects than are found elsewhere in the state (see also the following chapter, Seabirds and Pelagic Birding). The sea replenishes the bays, estuaries, and tidal flats with a mixture of mineralized salts and microscopic plants and animals whose complexity is such that sea water itself has yet to be reproduced in the laboratory. This then is the stage on which the lives of many of the birds of the Marine Region are played out.

To the birder some of the most interesting aspects of the Marine Region are: (a) those breeding species whose range is limited in New York State to this reporting region. Among those species are Little Blue Heron, Great and Snowy egrets, Louisiana Heron, Yellow-crowned Night Heron, Glossy Ibis, Clapper and Black rails, American Oystercatcher, Piping Plover, Willet, Laughing Gull, Gull-billed,

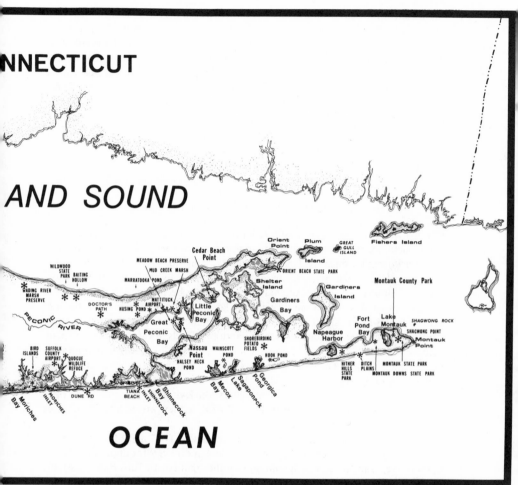

Roseate and Least terns, Black Skimmer, Chuck-will's-widow, Boat-tailed Grackle, and Sharp-tailed, and Seaside sparrows; and (b) those species that are recorded as regular coastal migrants or seasonal visitors, that are either unrecorded or infrequently reported from other *Kingbird* regions, some of which are: Eared Grebe; Cory's, Greater, and Sooty shearwaters; Leach's and Wilson's storm-petrels; Gannet; Great Cormorant; White Ibis; European Wigeon; Common and King eiders; Harlequin Duck; Oldsquaw; Surf and White-winged scoters; Merlin; Purple and Curlew sandpipers; Marbled Godwit; Red Phalarope; Pomarine Jaeger; Black-legged Kittiwake; Royal Tern; Dovekie; Western Kingbird; Blue Grosbeak; Dick-cissel; Lark Bunting; and Savannah (Ipswich), Lark, and Clay-colored sparrows all qualify.

With a line-up like that, almost any birder in the Marine Region can thoroughly enjoy himself birding at any season. However, the above named birds constitute only the tip of the iceberg—the specialities of the region. Even average spring and autumn migrations are sensational at any of several concentration points throughout the region. Summer and winter birding is normally quite exciting owing to the abundance and variety of species present. Here detailed are some of the region's excellent places at which to find birds and in which seasons. Those sites described have been selected for their obvious merits and their accessibility. This does not include *all* of the rewarding sites in the Marine Region but, rather, the cream of the crop.

SOUTHERN WESTCHESTER
The Pocantico Hills and Tarrytown Reservoirs (Map 10.2)

Rating: Spring***, Summer**, Autumn***, Winter**

In northwestern lower Westchester County there is a gneiss ridge, which, as dynamic metamorphic rock, is made up of granite, sandstone, and clay with mineral bands primarily of mica, feldspar, and quartz. This ridge is locally known as the Pocantico Hills. It lies west of the long valley eroded by the Saw Mill River. Because the majority of the Pocantico Hills acreage has been owned by the Rockefeller families for more than ten decades, its large oak-hickory stands on the drier sites, ash-elm-maple stands on the moist to wet sites, active open crop and haylands, dense conifer plantations, long-abandoned fields in varying stages of plant succession, marshes, ponds, and reservoir form a complex network of communities in which to bird. The handsome Rockefeller estate house on the southwestern part of the property sits atop Kykuit Hill. Although the families have opened much of the estate lands to birders and hikers, the Kykuit Hill portion is not open to the public. Northeast and northwest of the estate there are many miles of open trails on which one is free to bird.

From the eastern end of the Tappan Zee Bridge (I-87 and I-287) turn north on US 9 (South-North Broadway). Proceed north on US 9 for approximately 1.1 miles to Neperan Road, which comes in on the right (east). Turn right (east) and follow it as it passes Marymount College and then runs between the upper and lower parts of the Tarrytown Reservoir. Pull off here to bird these two water bodies. Paths skirting the reservoir and its peripheral habitats make birding very easy.

At the intersection of Neperan Road (as it becomes County House Road) and Lake Road turn left (north) and proceed north to NY 448 (Bedford Road). Turn right (east) and proceed a short distance to an old church on the right. The Union Church of Pocantico Hills contains several stained glass windows by Chagall and one by Matisse commissioned by the Rockefeller family.

Turn right and then left, into the parking area behind the church. From here one can bird east on an obvious trail for approximately 0.5 mile and then follow the trail northeast to the highest point (650 feet) on the property, Buttermilk Hill, which is approximately 2.1 miles from the parking lot behind the church. From here follow the path west and across NY 448 and continue west and south to an

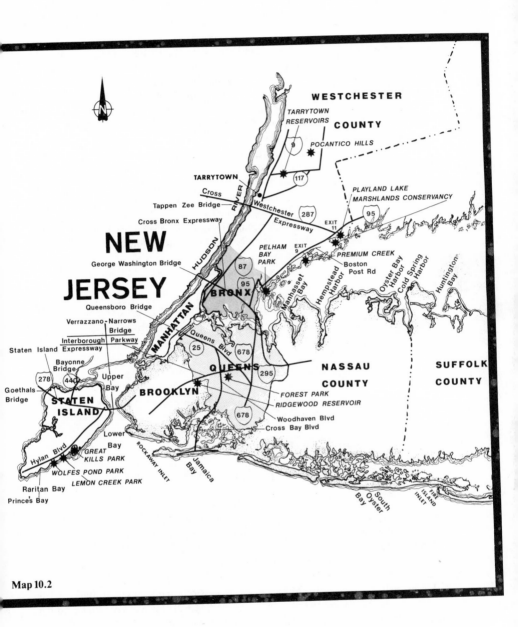

Map 10.2

area where a magnificent stable, built in the style of a French country house, is located. Bird behind the stable and west of it down to a secluded, shaded lake. At the north end of this lake there is a fairly marshy section worth investigating. Follow the path on the eastern side of the lake south for approximately 0.8 mile and then east for 0.2 mile, which will return you to the church and NY 448 or east back up to the stable, from which you can walk out to NY 448 on a dirt road. From there turn right and walk southwest to the church and parking lot.

This walk takes the birder through a variety of habitats, which should be most productive during spring and autumn. The route described is not, however, the only one to bird in the hills. It is simply the one I know the best. There are several other obvious side spurs that should also be interesting.

Some of the birds in this area are: herons, waterfowl, rails and gulls on the reservoirs and ponds. Raptors, owls, nightjars, woodpeckers, flycatchers, thrushes, and warblers nest in appropriate habitats. Half-hardy landbirds can be found throughout the woodlands in winter.

Playland Lake (Map 10.2)

Rating: Spring***, Summer*, Autumn***, Winter***

Marshlands Conservancy (Maps 10.2 and 10.3)

Rating: Spring****, Summer**, Autumn**, Winter*

Premium Creek (Map 10.2)

Rating: Spring***, Summer*, Autumn**, Winter*

These three sites are consistently the most productive places for birding in Westchester County on Long Island Sound. Playland is a 273-acre amusement park in Rye, with athletic fields, picnic area, amusement park, beach, and wildlife reservation. Most of its concessions are open from late spring through Labor Day. This includes rental boats to be used on Playland Lake. Once the boats are put in mothballs, following Labor Day, waterfowl can be found on the lake through December or early January. Some of the park's concessions are open year-round. The deciduous and coniferous woodlands south and east of the lake hold high numbers and an interesting variety of migrants in spring and autumn, and the tidal flats and bays abutting the Playland should be checked.

Marshlands Conservancy, a county-owned oasis for wildlife of nearly 140 acres, is located in Rye. Its diversity of habitats includes mowed fields, forests and an extensive saltmarsh. It is a sanctuary and not a recreational park, where dogs, bicycles, motorbikes, swimming, radios, etc., are not allowed. There is a 1.5 mile trail leading through the woodlands, meadows and across a causeway to a small is-

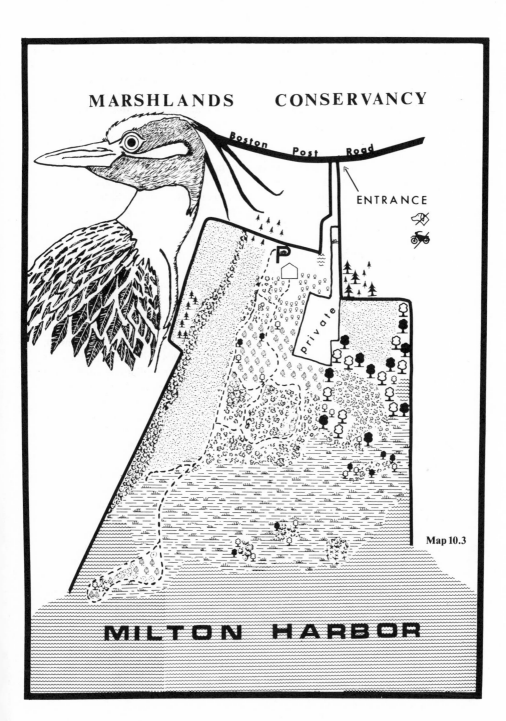

MARSHLANDS CONSERVANCY

Boston Post Road

ENTRANCE

private

Map 10.3

MILTON HARBOR

land in the saltmarsh. The Friends of Marshlands is a membership-supported volunteer organization that tangibly assists the Conservancy through a program of annual educational lectures, membership drives, fund-raising events, direct purchases of equipment and volunteer assistance at the sanctuary. Marshlands Conservancy is a natural area yet undiminished by humans, where more than 230 species of birds have been recorded. All of the habitat types are within easy walking distance of the parking lot and visitor's shelter.

Premium Creek, in Larchmont, is a vest-pocket birding site at which many of the county's most notable avian records have been established. It should especially be checked out in spring and autumn migrations.

To reach Playland Lake, leave I-95 (the New England Thruway) at Exit 11, and travel east on the Playland Parkway approximately 1.6 miles. Enter Playland. There is no admission for birders in the off-season.

To reach the Marshlands Conservancy, leave I-95 at Exit 11, and drive east on the Playland Parkway to its junction with US 1 (East Boston Post Road). Gain US 1 headed south and proceed approximately 1.6 miles, where the entrance to Marshlands is on the left (east) side of the road.

To reach Premium Creek, leave I-95 at Exit 9. Turn right (east) on Chatsworth Avenue and proceed east on it to US 1. Turn right (south) and proceed south on US 1 approximately 0.9 mile to Deane Place, which comes in on the left (east) side of US 1. Here turn left and park. Walk south to a path skirting the creek.

The prime times to visit Marshlands are generally mid-May, late summer, and in September and October. Walk through the fields and forests looking for passerines. Landbird migration can usually be seen from the end of the first through the third weeks in May and from late August through mid-October. Migrant raptors pass through from mid-September through the third week in October. Transient shorebird species come through mid-July through the last week in August. Herons not breeding on the sanctuary visit it on their post-breeding wanderings from mid-July through the last week in August. Walk south along the trail, down a hill to an old causeway. Continue walking around the island to view the shore of Milton Harbor. During early winter, especially mid-November through January, and late winter, particularly mid-February through the end of March, waterfowl species can be viewed on Long Island Sound, which borders the east end of the conservancy.

In 1981, the list of breeding birds at Marshlands totalled about fifty species, with the pride of the preserve a colony of Yellow-crowned Night Herons that have shown strong site fidelity. The nesting territory of these birds is admirably and zealously protected by the full-time naturalist. Additionally, Clapper Rail, American Woodcock, Long-billed Marsh Wren, Blue-winged Warbler, Yellow-breasted Chat, Bobolink, and Sharp-tailed, Chipping, and Field sparrows nest here. Among rarities recorded between 1976 and 1981 are Louisiana Heron, White Ibis, Redhead, Lesser Scaup, Common Merganser, Bald Eagle, Rough-legged Hawk, Golden Eagle, Peregrine Falcon, Sora, Little Gull, Royal Tern, Black Skimmer, Snowy Owl, Red-bellied and Red-headed woodpeckers, Acadian and Olive-sided flycatchers, Loggerhead Shrike, Yellow-throated Vireo, Cerulean Warbler, Louisiana Waterthrush, and Kentucky, Mourning, and Hooded warblers. More than thirty species of warbler have been recorded in spring on the grounds. Twenty shorebird species, six gull and seven owl species round out the list.

Premium Creek is the site where Westchester County's first Louisiana Heron (S. Bahrt) occurred. In November 1970, Sid Bahrt color-photographed an Ash-throated Flycatcher here for the first New York State record. In February 1973, the state's first Fieldfare (P. Lehman) was discovered at Premium Creek. In spring and autumn this tiny natural wetland area comes alive with migrants and is certainly worth checking out if one is birding the Westchester Long Island Sound.

Red-throated Loons, Pied-billed Grebes, and Great Blue Herons have been recorded on the lake at Playland in autumn and winter. Cattle and Snowy egrets, Black-crowned and Yellow-crowned night herons appear in spring and summer. However, the autumn and winter are the prime times to visit the lake to see American Wigeon, Canvasback, sometimes thousands of Greater Scaup, Common Goldeneye, Bufflehead, and all three merganser species, plus Red-tailed and Rough-legged hawks. There is further the possibility of finding wintering Black-bellied Plover, Greater Yellowlegs, and Purple Sandpiper. In the grass and borders of fields look for wintering sparrows and an occasional Snow Bunting.

BRONX COUNTY
Pelham Bay Park (Maps 10.2 and 10.4)

Rating: Spring**, Summer*, Autumn***, Winter****

Although the over 2000 acres of this county park, lying chiefly between the Hutchinson River and Long Island Sound, are a boon for recreationalists, there are three main areas of interest to birders. These are (a) Hunters and Twin Islands and their surrounding waters located north of Orchard Beach, (b) the age-old oak, beech, willow, maple, hickory, and poplar groves lining the road into and east of the Greek Revival Bartow mansion, and the evergreen plantations and pines on both sides and especially in back of the mansion, and (c) the Thomas Pell Wildlife Refuge. Although one can bird the park year-round, the most profitable time of year is winter when geese, ducks, raptors, owls, and over-wintering landbirds are present.

To reach Pelham Bay Park: from north or south of the park, approach it on the Hutchinson River Parkway. Leave the Hutchinson River Parkway at the Pelham Bay Park/City Island/Orchard Beach Exit. Continue east farther into the park past the traffic circle to the parking area on Hunters Island. From the west approach the park by driving east on the Pelham Parkway, which enters the park on the southwest. Continue north to the traffic circle. Turn east and proceed to the Orchard Beach parking area on Hunters Island. Pelham Bay Park can also be reached by public transportation. From New York City, take the Lexington Avenue IRT Number 6 subway train north to the end of the line at Pelham Bay Park. From there it is an easy walk of approximately 2.0 miles north and east to Hunters Island.

Over the past several years this park has become increasingly threatened by leaching from the nearby mainland sanitary landfill, oxidation from abandoned

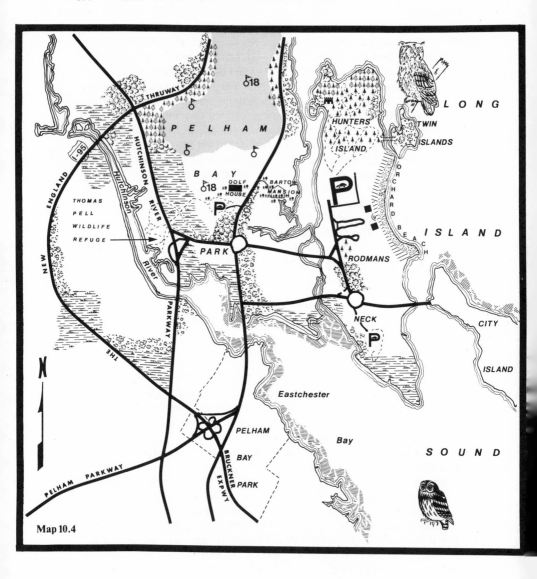

Map 10.4

cars, cadmium saturation into important swamp areas, residue from oil barges draining into the Hutchinson River, as well as the usual deterioration from lack of maintenance, and litter. It is difficult to judge just how long this will be a productive winter birding site. In late spring and summer, and early autumn the park receives heavy use by a wide variety of pleasure-seekers.

Begin therefore at Hunters and Twin Islands. Search the waters off both to the east and the rowing lagoon (basin) on the west side of Hunters Island for the occasional loon, or grebe. Look also for Great Cormorant, Canada Goose, Black Duck, American Wigeon, Canvasback, scaup, Bufflehead, Common Goldeneye, and Red-breasted Merganser. Look diligently among the high reeds in the rowing lagoon for Clapper Rail and on the rocks of Twin Island for the rare Purple Sandpiper. Before leaving Hunters Island, walk on the dirt and gravel trails north from the parking lot into the forest above. This area is heavily wooded with stands of tulip trees and chestnut oak, among others, and there is a pine-spruce grove in which to look for Great Horned and Barred owls and in some years, large roosts of Long-eared Owls.

From the traffic circle in the park proceed north on Shore Road to the clubhouse of Split Rock Golf Course on the west (left) side of the road. Pull into the parking area and search the low evergreens and shrubbery around the clubhouse building and golf course edge for wintering Saw-whet Owls. Then walk north along Shore Road a short distance, to the drive road of the Bartow mansion on the right (east) side. Walk down the road searching the trees along it and the open fields beyond.

Search the evergreens at the entrance, then walk around to the back of the house, watching attentively, especially on the bay side, for Barn, Screech, Great Horned, Long-eared, and Saw-whet owls. All of them are possible here with the most regularly occurring being Barn, Long-eared, and Saw-whet. In 1981, this was still the most reliable place around New York City to find a variety and numbers of wintering owls.

The Thomas Pell Wildlife Refuge is a natural tract within the park that includes wooded uplands, saltmarsh, and tidal estuaries. Goose Island in the Hutchinson River, which borders one side of the refuge, is a lowland through which Goose Creek winds. There are specimens of many mature deciduous trees, and resident and migrant species can often be seen on the grass hummocks in the marsh.

From the parking area of the clubhouse of Split Rock Golf Course walk west and southwest toward the Hutchinson River and Eastchester Bay on paved paths and a bridle path, which traverse much of the refuge.

MANHATTAN, NEW YORK COUNTY
Central Park (Map 10.5)

Rating: Spring****, Summer*, Autumn***, Winter*

Spring comes flying into New York City in the form of colorful, insectivorous wood warblers, which seem the embodiment of perpetual motion. They usually arrive on southwest winds and land wholesale on the hospitable greenery of Central Park. Most of these migrant warblers have wintered in Central and South America and have gone through several arduous non-stop flights before arriving in the Big Apple. The longest of these non-stop flights is probably from the Yucatan Peninsula across the Gulf of Mexico to the coasts of Louisiana or Texas. A considerable

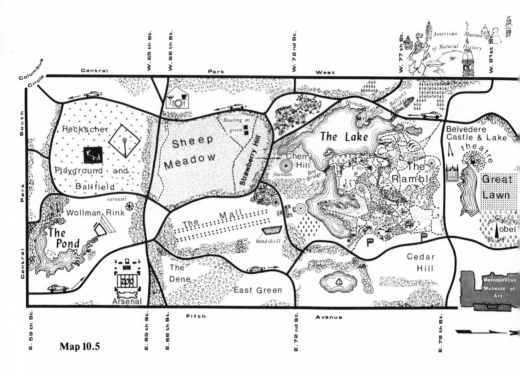

Map 10.5

feat for a creature whose body weight rarely exceeds 16 or 18 grams. Although warblers migrate primarily at night, they also undertake lengthy travels by day, stopping only to refuel and rest.

Witnessing the spring migration in Central Park is for hundreds of New
Yorkers the highlight of the birding year. It is a time for welcoming and recognizing old friends, avian and human, returning from temporal and spatial absences.
In the park, binocular-wielding zealots personally experience the reawakening that
is proceeding all around them.

Warblers typify springtime for many birders, but warblers are not the only
visitors in the more than 840 acres of Central Park. The total list of bird forms
(species and subspecies) recorded in Central Park stands, in 1981, at almost 270.
Between March first and the first week in June, usually 140 to 150 species are
sighted in various habitats throughout the park. The peak of spring bird activity
usually occurs throughout the month of May. Autumn is by no means dull in Central Park, with the first southbound migrants occurring mid-to-late July, in Au-

gust, and with greater numbers and varieties seen in September and October. Usually more water-associated species are recorded in autumn than in spring. Throughout an average autumn 115 to 120 species are observed.

Central Park is rectangular, with Central Park South (59th Street) and the Cathedral Parkway (West 110th Street) as its southern and northern boundaries, respectively. Fifth Avenue borders it on the east and Central Park West (Eighth Avenue) borders it on its west side. It is, as its name denotes, central. It can be reached by convenient public transportation; the Madison Avenue bus (north) deposits the rider one block east of the park at East 72nd or East 79th Streets. On the west side of town the Eighth Avenue IND "AA" subway train (north or south) has convenient stops at West 81st and West 72nd Streets, both on Central Park West. From the canyonesque maze of apartment and office buildings enveloping the park on all sides it is an invigorating walk to the Ramble or to the Pond on the park's southeast corner.

Parts of the park are underwater, parts undermaintained, and several parts under continual and ever-increasing surveillance by birders. It has for so long been a well-loved haunt that it is doubtful that hardly a bird of interest migrates through without being seen. Following are specific locations where migrants can most easily be seen, but feel free to investigate the rest of the park. After all, birds do. The optimal time to bird the park is in the early hours of morning. Although New York's Finest police the park, it is not advisable to bird alone. In recent years the increasing number of joggers in the park cause one to think of it as the locker room of the city, but even they run mostly in twos or threes. Migrants add pageantry and drama to spring, but dramas of altogether different dimensions also take place in the park, of which none of us would wish to be a part.

The Ramble is located between the 72nd Street and 79th Street Transverse Roads, and the East (northbound traffic) Drive and West (southbound traffic) Drive. It can be reached from either the west or east sides of the park by walking toward the park's center and between the northwest and east arms of the Lake. This favorite haunt of birders encompasses thirty-three acres and supports thick canopied groves of locust, weeping willows, swampy pin oaks, swamp oaks, tulip, and cherry trees and small stands of mulberry, sour gum, elm, and maple trees. There are large knotweed patches, dense shrubbery, azalea bushes and wisteria providing protective covering for migrants. Dead tree limbs, numerous bathing pools and a flowing stream make the area more attractive to birds.

The hilly character of the Ramble is the result of glacial rock outcroppings. The perched erratics and the intensely deformed and impressive outcroppings of schist provide excellent vantage points for observing tree-top foraging birds at nearly tree-top level. The Ramble is interlaced with macadam pathways and is bordered on the north by Belvedere Castle and hill, Belvedere Lake, meadows, and the western edge of the Great Lawn. These areas as well as Cherry Hill south of the Ramble, Cedar Hill and the Dene are sometimes productive and should not be overlooked. Also bird the area around the Pond located northwest of the entrance at Fifth Avenue and Central Park South. Also try the more treed areas north of the Reservoir but do not bird them alone.

Within the Ramble there are several spots that seem to be concentration points for spring and autumn migrants. If time is limited these should be visited, preferably in the order presented.

The Point is a peninsula with the Lake (Rowboat Lake) surrounding it on three sides, a path running from its base to its tip, a cove at the northwestern side of its base and a boathouse at the northeast side of its base. It is covered with locust trees, sloping woodlands and thick underbrush. From the high pedestrian path one can look west to rock outcroppings that slope to a muddy/marshy area at the lake edge. High weeping willows tower above the rock formations and it is here that bird feeding activity is high during a spring "wave." Along and across the point and in the cove look for American Woodcock, Spotted and Solitary sandpipers, Whip-poor-will, and Common Nighthawk. Eight flycatcher species are recorded annually in May and all ten species breeding in the eastern part of the country have been reliably reported here. Brown Creeper, Long-billed Marsh Wren, Wood, Hermit, Swainson's, and Gray-cheeked thrushes, Veery, Blue-gray Gnatcatcher, Golden-crowned and Ruby-crowned kinglets and vireos occur yearly. Hordes of warblers can be seen, the most abundant of which are Black-and-white and Tennessee warblers, Northern Parula, Yellow, Magnolia, Black-throated Blue, Yellow-rumped, Black-throated Green, Blackpoll, Prairie, and Palm warblers, Ovenbird, Northern Waterthrush, Common Yellowthroat, and American Redstart. Less numerous but regularly observed are Blue-winged, Nashville, Cape May, and Bay-breasted warblers, Louisiana Waterthrush, Mourning Warbler, Yellow-breasted Chat, Wilson's, and Canada warblers. Observed only intermittently are Prothonotary, Worm-eating, Golden-winged, Blackburnian, Chestnut-sided, Pine, and Hooded warblers. Warblers rarely occurring in the Ramble are Swainson's, Orange-crowned, Cerulean, Yellow-throated, Kentucky, and Connecticut warblers. The "Brewster's" and "Lawrence's" warbler hybrids are both seen, however not annually, in the Ramble.

The Azalea Pond is reached by walking north and slightly west of the Point. It is located in the approximate center of the Ramble and features a small pond surrounded by dense azalea bushes. There are rock outcroppings and sloping hardwoods on the north and a small lawn on the east. Ruby-throated Hummingbird, many warbler species, Orchard and Northern orioles, Scarlet Tanager, towhee, and sparrow species including Fox, Lincoln's, and Swamp are attracted to this general area.

The Swamp-High Rock-Warbler Rock area, reached by walking southwest of the west end of Azalea Pond, is excellent during both spring and autumn migrations. The predominant tree species here is pin oak and these oaks are literally alive with birds during both seasons. High Rock, the northern prominence and Warbler Rock, the western prominence are superb spots to view passerines feeding in the cherry trees and knotweed patches below.

Warbler Walk is the pathway west of and over the hill from Warbler Rock, which extends along the shore of the Lake south to Bow Bridge. Cherry trees line the slope east of the walk and willows thrive between the path and the lakeshore. A variety of migrants can be seen entering the Ramble from along the path. The vicinity of Bow Bridge is well planted and maintained and is also a productive birding spot, as well as the Cherry Hill area southwest of Bow Bridge.

Between Bow Bridge and the Point there is a slope wooded with tulip trees along which migrants feed and rest. Look for vireos, warblers, tanagers, Rose-breasted Grosbeaks, Indigo Buntings, and various sparrow species throughout this stretch.

Outside of the Ramble, the Reservoir—reached from either West or East 90th Streets and birded from the bridle path surrounding it—regularly harbors several duck species and flocks of gulls. It should be most heavily birded in autumn and winter for the occasional loon or grebe, waterfowl, and not infrequently "white-winged" gulls. The best times to view these gulls is usually on overcast days in mid-to-late February and early March.

If time permits, bird the area north of East 97th Street on the east side of the park. The Conservatory Garden and the area around the Harlem Meer and west to the Loch often harbors birds.

STATEN ISLAND—RICHMOND COUNTY
Great Kills Park of the Gateway National Recreation Area, Oakwood Beach, Arbutus Lake, Wolfe's Pond Park, and Lemon Creek Park, a Tour (Map 10.2)

Rating: Spring***, Summer*, Autumn****, Winter***

Staten Island, roughly triangular in shape, is sometimes referred to as the "forgotten" borough, although in area it is the third largest. From north to south it is approximately 13.0 miles and at its widest point is 7.8 miles. This most southern section of New York State encompasses approximately 60 square miles and enjoys nearly 60 miles of waterfront. The island is bisected by a series of hills that run from the northeast part of the island to its approximate center. Here, just south of the intersection of Todt Hill Road and Ocean Terrace, is the highest point in New York City: Todt Hill, with an elevation of 410 feet above sea level. Todt Hill further enjoys the distinction of being the highest tidewater elevation along the entire Atlantic seaboard south of Maine.

Great Kills Park, Arbutus Lake, Lemon Creek Park, and Wolfe's Pond Park lie on the island's south shore facing the usually calm waters of Lower New York and Raritan bays. Great Kills Park, encompassing approximately 1455 acres, was at one time the site of an Algonquin Indian village. Its topography has been drastically changed since that time by periodic massive landfilling. In the 1940s, Crookes Island at its southern end became Crookes Point and presently the interior section of the Point is being managed to reestablish and maintain natural conditions. A good deal of the northern section of the park is also being managed to preserve the natural landscape and these areas are being safeguarded from indiscriminant and random vehicular access and excessive use. Wolfe's Pond Park has a freshwater pond, mature woodlands, a bog and saltwater beach which afford the birder the rare combination of upland and marine ecosystems in which to search out birds. In spite of landfill operations and the ongoing building boom since the early 1960s, these parks are worth a visit at any time—but particularly in the autumn and winter after the summer season beach crowds have thinned.

To reach Great Kills Park by car, from the Verrazzano-Narrows Bridge or the Goethals Bridge, take I-278 (the Staten Island Expressway) east or west to the Hylan Boulevard Exit. From the Bayonne Bridge proceed south on NY 440 to I-278 and then proceed east on I-278 to the Hylan Boulevard Exit. Leave I-278 and continue south on Hylan Boulevard approximately 5.0 miles to the entrance to Great Kills Park.

Great Kills Park can also be reached by public transportation. Take the ferry from Battery Park on Manhattan to St. George on Staten Island, and from the bus terminal there, take the Number 103 or Number 111 bus to the Great Kills Park entrance. Alternatively, from the Port Authority Bus Terminal at 8th Avenue and West 40th Street, the Domenico Bus Line has scheduled service to Great Kills Park.

From the Administration Building adjacent to the parking lot inside the park, walk east of the park road toward the bay, which is visible from the parking area. Where the road tees turn left and walk north for approximately 0.5 mile along the beach. In the phragmites marshes one can find herons, ibises, and Common Gallinules. In the surrounding area look for various sparrow species. In the bay and Great Kills Harbor look for Common and Red-throated loons, Horned Grebes, sea ducks, mergansers, and "white-winged" and Bonaparte's gulls in late autumn and winter. Continue walking north to Oakwood Beach, which is north of the park where hawk flights, shorebirds, and numerous passerines on migration can be viewed.

If this walk is taken at low tide, shorebirds in transit and overwintering can be viewed on the exposed mud flats. Most winters Killdeer, Black-bellied Plover, Ruddy Turnstone, Purple Sandpiper, Dunlin, and Sanderling can be seen on the flats. In spring these shorebird species and many more frequent the mud flats at low tide. In the short grasses along the dunes and near the parking lot one can find Horned Lark and Eastern Meadowlark year-round, and Lapland Longspur and Snow Bunting in winter.

Walk back along the beach from Oakwood Beach and proceed south from the road tee toward Crookes Point. Watch overhead for migrating or wintering hawks. The charmed birder will turn up a Snowy Owl and the lucky birder will probably rouse one or two Short-eared Owls in winter. The interior section of Crookes Point may harbor several sparrow species, including Savannah, Sharp-tailed, Seaside, and Swamp. The large, pale, sand-colored race of the Savannah Sparrow, the Ipswich Sparrow, likely can be found wintering along this stretch from the boat marina south to Crookes Point.

The wooded areas of the park host woodpeckers, chickadees, titmouse, mimids, thrushes, warblers, and sparrows appropriate to that habitat.

In September and early October along the beach walk watch for migrant butterflies and insects. The Monarch Butterfly migration can be viewed south to the park's southern terminus.

Arbutus Lake can be reached by exiting Great Kills Park, turning left (south) on Hylan Boulevard and following it southwest approximately 3.4 miles to Arbutus Avenue. Here turn left (south) and proceed to tree-cleared overviews of the lake. This lake can be thoroughly viewed from the roadside. Double-crested

Cormorant can be seen in September and October on pilings in the lower bay at the foot of Arbutus Avenue. Wood Duck and Hooded Merganser winter in Arbutus Lake. Screech Owl can usually be found in winter on the brushy wooded slopes bordering the lake, as can Pine and Palm warblers in spring.

To reach Wolfe's Pond Park, regain Hylan Boulevard and turn left. Drive only a short distance to the entrance to the park. The small pond at the southeast end of the park harbors Green-winged Teals and Buffleheads throughout the fall and winter. Ruby-throated Hummingbirds can usually be found in September hovering above patches of jewelweed. Kingfisher is a year-round resident here, and Eastern Bluebird can most easily be found here in April and May.

Lemon Creek Park is less than a mile south of Wolfe's Pond Park. Lemon Creek empties into Prince's Bay here. Great Blue Heron can often be found along the creek in winter. Clapper Rail is a not-uncommon visitor. Bonaparte's Gull is usually visible from the bayshore. Here along the south shore of Staten Island may be the only nesting stronghold of Whip-poor-will in New York City and calling Chuck-will's-widows have been heard in recent years. In spring look for Tree, Bank, Rough-winged, and Barn swallows catching insects over the creek. The only Purple Martin colony within New York City can be seen from the Lemon Creek overpass on Hylan Boulevard from April 15 on.

BROOKLYN, KINGS COUNTY
Prospect Park and the Brooklyn Botanic Gardens (Map 10.6)

Rating: Spring***, Summer**, Autumn****, Winter*

This park and the adjacent botanic gardens comprise a tract of forested hills, landscaped gardens, ponds, brooks, ravines, and meadows of approximately 580 acres sitting atop the east-west trending glacial deposit known as the Harbor Hill Moraine. This ridge acts as a leading flight line for migrating birds in both spring and autumn. Frederick Law Olmsted and Calvert Vaux, partners and far-sighted landscape architects, had already designed Central Park in Manhattan by the mid-1860s. About that time they set out to create Prospect Park, bringing to the job an imaginative design that incorporated the land contours and existing vegetation. They took advantage of high rocky overlooks, serpentine streams, swampy areas, and the lake at the park's southern end. The end product was a stunning city park that would be a credit to any urban area. For many decades the park retained its natural and man-made beauty. In the 1950s, lacking an overall preservation plan and consequently a firm commitment for the upkeep and betterment of Prospect Park from the city's Department of Parks, the park sadly deteriorated. In the mid-1960s with a new surge of energy and the infusion of millions of dollars, the park underwent a great restoration. Since then, however, Prospect Park has fallen on hard times. Its unique geological features have been defaced, its native wildflowers trampled, its woods littered, and many splendid trees despoiled by vandals. It looks shabby and wasted. The late 1960s and the 1970s have taken a great toll on this once magnificent park. Yet, in spite of its rapid decline birders have

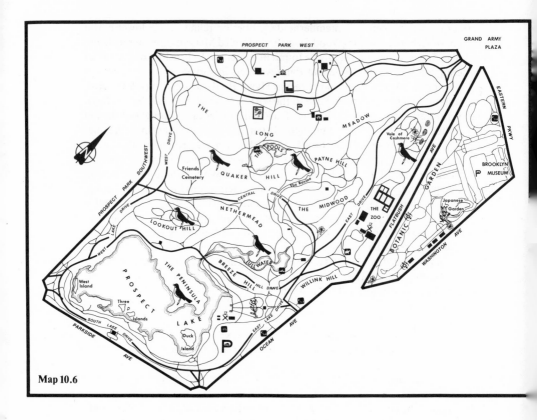

Map 10.6

steadily visited the park and recorded its avian life. More than 250 bird species have been recorded in all seasons here and of that total many are local rarities.

The outlook may not be as dim as it might first appear. In 1980 the Parks Department did come forth with a preliminary preservation plan for Prospect Park which details the restoration of the Ravine, Boathouse, Litchfield Villa and its Annex, the Nethermead Arches, Meadport Arch, Lullwater Bridge, Cleft Ridge Span and Oriental Pavilion. The citizen's organization, Friends of Prospect Park, through private funding has begun restoration of the Vale of Kashmir, a garden intimate in scale and formerly elegant of design. Perhaps with a resurgence of interest and new provisions made for its improvement, policing, and protection,

Prospect Park will enjoy its former status as one of the fine, easily accessible birding sites within the New York City limits. Do not bird here alone.

The accompanying map indicates some of the trails and better birding spots (shown by singing robins) within the park. In spring 1981 a Common Gallinule was on the Prospect Lake in late April. Red-shouldered Hawks were attempting to nest across from the zoo and the Yellow-throated Warbler was seen in the park. Spring migrants with southern affinities seen here are Blue-gray Gnatcatcher, Kentucky Warbler, and Orchard Oriole. Capper Rail and Sora have rarely been recorded here. Peregrine Falcon is seen some years in autumn.

To reach Prospect Park by automobile from lower Manhattan, cross the East River on the Manhattan Bridge at the eastern end of Canal Street. The bridge leads into Flatbush Avenue. Follow Flatbush Avenue 1.4 miles south to the entrance to Prospect Park, just south of the Grand Army Plaza.

By public transportation from the east side of Manhattan, take the Lexington Avenue IRT Number 4 subway train south (downtown) to the Grand Army Plaza stop. From there, walk south into the park. From the west side of Manhattan take the Sixth Avenue IND "D" train south (downtown) to the Prospect Park stop. From there walk east into the park.

QUEENS COUNTY
Forest Park and Ridgewood Reservoir (Maps 10.2 and 10.7)

Rating: Spring****, Summer*, Autumn**, Winter*

Queens, the largest of New York's five boroughs, is on Long Island, adjoins Brooklyn on the west, Nassau County on the east, the East River on the north, and Jamaica Bay, the Rockaway Peninsula, and the Atlantic Ocean on the south. This borough is largely residential and air travelers enter New York City here at either La Guardia or John F. Kennedy International airports.

The 540-acre Forest Park sits squarely atop Long Island's younger glacial margin; the Harbor Hill Moraine. This rough-edged, morainal ridge is marked by vast piles and blankets of glacial rock waste. When the glacier that carried the rubble readvanced north this east-west trending terminal moraine was left, marking the southern front of the ice-sheet's advance. A large section of the park is devoted to golf course, but the section lying east of the intersection of Woodhaven Boulevard and Myrtle Avenue is heavily forested and interlaced with a series of trails making birding areas accessible. It is this 330 acres of the eastern section on which we shall concentrate. From the Ridgewood Reservoir and the southwest section of Forest Park one has views of the outwash plain one hundred feet below, Jamaica Bay and the Atlantic Ocean eight to ten miles distant. In landscaping Forest Park, some care was taken to preserve some of its outstanding glacial kettles. These depressions mark the former position of ice blocks that detached from the stabilized ice sheet, were covered by glacial drift, and at some later time melted, allowing the debris to settle. These hollows have proven marvelous places to bird. The forest cover is primarily mature oak with fine stands of maple, beech and tulip trees in-

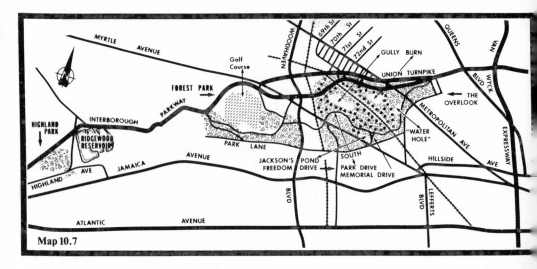

Map 10.7

terspersed throughout, and an understory of brushy low trees, and blackberry thickets. The only seasonal water in the most densely forested section of the park fills a small concavity which Tom Davis, a Forest Park devotee, calls the "water hole." The term "east end" refers to the area defined as a circle with the "water hole" as center and a radius of 300 yards. As this is the only standing water over a relatively wide area birds are attracted to it to drink and bathe. The waterhole is especially productive in mid-to-late afternoon. The area indicated as "the gully" on the accompanying map is additionally rewarding.

Early morning movement of migrants is best viewed at Ridgewood Reservoir. Here the most productive period is from dawn to about three hours later.

Since approximately 1970, Forest Park and its environs have increasingly become a metropolitan spring hot-spot. Rarities found here have been publicized on the New York Rare Bird Alert, which has resulted in more thorough coverage. Since 1960 upwards of 200 species have been sighted at the park, with more than 160 of those seen between 1975 and 1980. In spring 1976, 34 warbler species were recorded and a mid-May single-day Forest Park/Ridgewood Reservoir roundup tallied 102 species by T. H. Davis *et al.*

To reach Ridgewood Reservoir by car from midtown Manhattan, cross the East River on the Queensboro Bridge, located at East 59th Street and First Avenue. Gain Queens Boulevard heading east and follow it to Woodhaven Boulevard. Turn right (south), thereby gaining Woodhaven Boulevard, and proceed south on it to the Interboro Parkway. Exit Woodhaven Boulevard and gain the Interboro Parkway heading west. Proceed to Cypress Avenue. Exit. Turn left (south) and continue approximately 200 yards to a parking lot for Highland Park on the right.

Walk across the street to the southwest corner of Ridgewood Reservoir. Continue east along the south side of the reservoir, being sure to bird the brush and trees to the right (south) of the walk. Bird up the east side of the most eastern pool as far as the caretaker's house at its northeast corner. This pool also contains a phragmites marsh, so it should be searched for water-associated birds. South of the caretaker's house turn around and retrace the route taken to that point.

To reach the east end of Forest Park from Ridgewood Reservoir and the Highland Park parking lot, return north to the Interboro Parkway. Obtain it heading east. Proceed east to the Metropolitan Avenue exit. Turn right (southeast) onto Metropolitan Avenue. Almost immediately, Forest Park Drive comes in on the right (west). Park near this intersection and enter the park along Forest Park Drive.

To reach Forest Park from Manhattan by automobile, follow the above directions for Ridgewood Reservoir from Manhattan as far as Woodhaven Boulevard. Proceed south on Woodhaven Boulevard to Metropolitan Avenue. Turn left (east) onto Metropolitan Avenue. Almost immediately, Forest Park Drive crosses Metropolitan Avenue. Park near this intersection and enter the park along the drive.

To reach Forest Park by public transportation from the west side of Manhattan: take either the Eighth Avenue IND "E" (uptown) subway train or the Sixth Avenue IND "F" (uptown) subway train to the Union Turnpike-Queens Boulevard stop. Above ground, walk west a few blocks on 80th Road and south along Park Lane South to Forest Park and enter it along Forest Park Drive off of Park Lane South.

Some of the rarities recorded at the park in recent years are (in varying numbers): Chuck-will's-widow, Red-headed Woodpecker, Olive-sided and Acadian flycatchers, Prothonotary, Swainson's, Golden-winged, "Brewster's" hybrid, Cerulean, Yellow-throated, Kentucky, and Hooded warblers, Orchard Oriole, Summer Tanager, Blue-Grosbeak, Painted Bunting, and Clay-colored Sparrow. In Forest Park Guy Tudor *et al.* have found upwards of thirty breeding species most notable of which were Red-bellied Woodpecker and Brown Creeper. Others were Great Crested Flycatcher, Eastern Wood Pewee, Common and Fish crows, Wood Thrush, Chimney Swift, and Indigo Bunting. This park is usually not productive before mid-April or after early October.

BROOKLYN AND QUEENS
Jamaica Bay Wildlife Refuge of the Gateway National Recreation Area (Map 10.8)

Rating: Spring***, Summer****, Autumn****, Winter**

In 1980, 12.9 million overseas passengers arrived and departed from John F. Kennedy International Airport. That staggering figure, however, is rivaled only by the numbers of international avian arrivals and departures from Jamaica Bay, adjacent to the airport.

This is a golden place for birders—serious and casual, expert and amateur. Its continued and marked success as a wildlife refuge perpetuates the memory of the late Herbert W. Johnson, who brought imagination and ingenuity to the 12,000

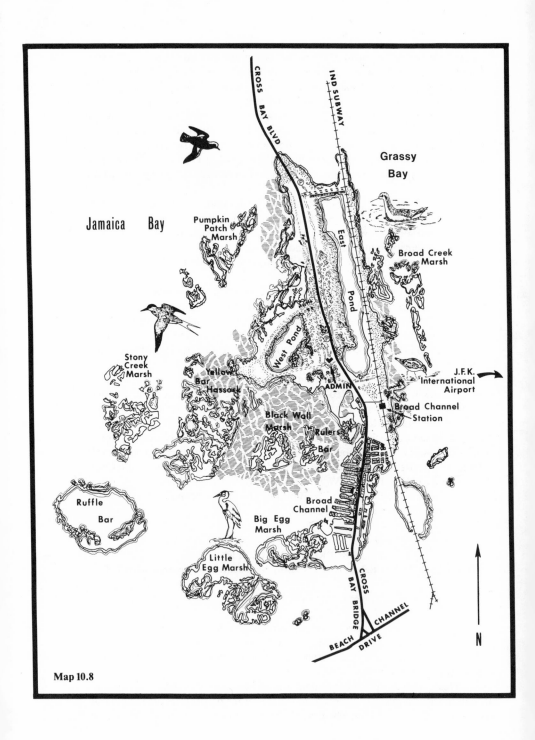

CROSS BAY BLVD

IND SUBWAY

Grassy
Bay

Jamaica   Bay

Pumpkin
Patch
Marsh

East
Pond

Broad Creek
Marsh

West Pond

Stony
Creek
Marsh

Yellow
Bar
Hassock

ADMIN.

J.F.K.
International
Airport

Broad Channel
Station

Black Wall
Marsh

Rulers
Bar

Ruffle
Bar

Broad
Channel

Big Egg
Marsh

Little
Egg Marsh

CROSS BAY BRIDGE

BEACH   DRIVE

CHANNEL

N

Map 10.8

acre reclamation project in 1953. As the reserve's horticulturist he skillfully masterminded the clearing and planting that, in tandem with its strategic location at the junction of the Hudson River and Atlantic Coast flyways, today attracts millions of water-, land-, and shorebirds on spring and autumn migrations. During breeding season the sanctuary supports populations of more than 60 species. In 1981 the cumulative list of species observed on the refuge was approximately 315. What a remarkable transformation from the depauperate, neglected site whose bird list stood at fewer than 80 species in the mid-1950s. Today the over eighteen square miles of refuge consists of islands, tidal marshes, an upland tract landscaped with important food plants attractive to migrating passerines, and two freshwater ponds rimmed with phragmites and beach grass. Dikes, stabilized with vegetation suitable for helping control erosion, provide prominent viewing points from which to observe herons, waterfowl, marsh and shorebirds, larids, and sparrows.

The period from late June through early October is the most exhilarating, for it is then that first adult and then immature shorebirds pass through, and then that the tidal creeks, ponds, and mud flats support herons, ducks, gulls, and terns shore-to-shore. Migrant passerines veneer the trees and shrubs, and varsity birders have all they can do to keep pace with the galvanizing surprises that often appear during the fall.

To reach Jamaica Bay by car from Manhattan or points north, cross the East River on the Queensboro Bridge. Gain Queens Boulevard headed east to Woodhaven Boulevard. Turn south on Woodhaven Boulevard. Alternatively, cross the river through the Queens Midtown Tunnel. Follow I-495 (the Long Island Expressway) east to Woodhaven Boulevard. There turn south. From the north, take the Triborough Bridge into Queens. Follow the Grand Central Parkway to I-278 (the Brooklyn-Queens Expressway). Gain it heading south. Exit at its intersection with I-495 (the Long Island Expressway). Proceed east on I-495 to Woodhaven Boulevard and there turn south.

From Woodhaven Boulevard and the junction of Queens Boulevard and I-495 proceed south on Woodhaven Boulevard to Cross Bay Boulevard. Continue south on it to the entrance to Jamaica Bay Wildlife Refuge on the west (right) side of Cross Bay Boulevard. In 1981 there was no fee to park in the refuge lot.

To reach Jamaica Bay by mass transit from Manhattan, take the Eighth Avenue IND "A" Far Rockaway line (downtown) subway train to the Broad Channel stop. From there walk west a few blocks and then north along Cross Bay Boulevard to the refuge.

Be sure to check the log in the visitors center for current sightings and their locations before beginning to walk the refuge trails. Access permits can also be obtained here, and they are required if you wish to walk the trails.

Birding the area is best from the two-mile, well-maintained path along the dike that impounds the forty-acre freshwater West Pond, located on the west side of Cross Bay Boulevard. Alternatively, you can walk east down a well-trodden path a few hundred feet north of the parking area, on the opposite side of Cross Bay Boulevard to the one hundred-acre freshwater East Pond, which is maintained in a largely undeveloped state. You can also bird the East Pond by hiking north about a mile along Cross Bay Boulevard and then proceeding east on the East Pond's north dike. This method of coverage affords the birder a wider overview of the pond and removes him somewhat from the crowds on the west side of the refuge — which seem to grow as the day wanes. When the water level is low, an impressive number and variety of migrating shorebirds (34 species noted between late July and mid-September 1977) gather along the edge of the East Pond. Birders can also walk the periphery, if they are prepared to end up with wet and muddy feet, but biting insects are not bothersome on the open flats.

Curlew Sandpiper is odd-man-out among Scolopacidae. Since the mid-1970s, when it began drawing down the water levels of both freshwater ponds in May (West Pond) and from July to early September (East Pond), Jamaica Bay has cornered the Curlew Sandpiper market on the Long Island south shore mud flats. Between 1975 and 1980 nine records of occurrence involving thirteen different birds were established on the two ponds. Look for Baird's and Buff-breasted sandpipers, Long-billed Dowitchers, Marbled and Hudsonian godwits, American Avocet, and Red, Wilson's and Northern phalaropes on their southbound migrations. In winter, search all the open fields, especially the area west of West Pond, on which terns nest in summer, for Snowy or Short-eared owls, Horned Larks, Savannah (Ipswich) Sparrows, Lapland Longspurs and Snow Buntings. Search the pine groves for Barn, Long-eared, and Saw-whet owls and in irruption years both crossbill species. The surrounding bays support large flocks of wintering Brant, Black Ducks, and Greater Scaup, plus smaller numbers of Horned Grebes, American Wigeons, Canvasbacks, Common Goldeneyes, Buffleheads and Red-breasted Mergansers. Rarities here have included White Pelican, White and White-faced ibises, Fulvous Whistling Duck, Cinnamon Teal, Swallow-tailed Kite, Sage Thrasher, Townsend's Warbler and Western Tanager.

Gateway National Recreation Area, Breezy Point Unit
The Rockaway Peninsula: Jacob Riis and Breezy Point Parks (Map 10.9)

Rating: Spring**, Summer*, Autumn****, Winter*

Jacob Riis and Breezy Point parks are located south and west of Jamaica Bay on the Rockaway Peninsula in Queens County. The two sites encompass some undeveloped natural resources and habitats — tern-nesting sites and dune systems, some forested uplands, and a bay shore unsuitable for high-impact public use but superb for hiking and barrier beach birding. The recreation focus of this eastern part of the peninsula is Jacob Riis-Fort Tilden, where there are public beaches and recreational complexes, including a wide variety of outdoor and indoor sports programs, and various other concessions. Obviously, one must pick and choose the best times to bird these areas, preferably, if possible, after summer and the parks' peak-use periods.

To reach the parks from Jamaica Bay Wildlife Refuge: exit its parking area and turn right (south) on Cross Bay Boulevard. Follow it south, and cross the

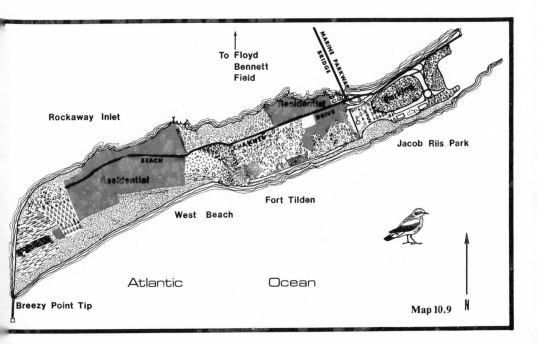

Map 10.9

Cross Bay Bridge, a toll bridge. Turn right (west) on Beach Channel Drive and proceed west into Jacob Riis Park.

All of the shrubbery in the mall area should be birded. Walk east and west looking for field birds around the park's open areas and especially the edges of the golf course and parking area fields. In autumn this is a good place to search out downed migrants. Watch overhead for migrating raptors. This area is also often productive in late autumn and early winter for Western Kingbirds, and especially in "winter finch" invasion years. Henslow's Sparrow has occurred here in recent years in October.

From Jacob Riis Park drive west as far as possible through the area occupied by the Breezy Point Cooperative buildings. A fee was being charged in 1981 for entery to this area. Drive as far west as possible and then walk out to the Breezy Point Tip. The Gateway National Recreation Area's general management plan projects establishing a buffer zone around the Co-op, a stabilized path located behind the foredune, on the ocean side of the Cooperative, and over these dunes, a boardwalk, which will bypass the Cooperative. When (and if) this is completed, birding out to Breezy Point will become considerably easier, and there will be no fee to pay. In 1981 this plan had not been effected. When it is, the area may be accessible on a use-by-permit basis only.

From the Breezy Point tip and jetty one can search the Rockaway Inlet for seabirds in winter and migrating raptors throughout autumn. The intermediate grassland–reed marsh often harbors wintering Snowy and Short-eared owls. This is a superb place to look for gulls and one should be able to find several species of field birds and sparrows.

NASSAU COUNTY, THE NORTH SHORE GOLD COAST
Little Neck Bay to Oyster Bay Harbor, a Tour (Map 10.10)

Rating: Spring****, Summer**, Autumn***, Winter**

The north shore of Nassau County is renowned for its imposing turn-of-the-century estates, opulent soirées, splendid formal gardens, fabulous yachting parties, and fashionable polo matches and fox hunts. No one who has ever read F. Scott Fitzgerald's *The Great Gatsby* can forget the luxurious indulgence he so richly depicts. Such was the splendor of this now bygone era, but the north shore still echoes its spirit and grandeur. There are still thousands of acres of mixed deciduous forest with black, white, and scarlet oaks of impressive girth and black birch, red maple, sassafras, and flowering dogwood in impressive densites. There are still mixed hardwood bottomlands, ephemeral streams and red maple swamps. Mile after mile of mountain oak and mountain laurel are found on the steep north-northwest facing slopes of the Harbor Hill Moraine, which forms the major drainage divide for the North Shore. The ruggedness of much of the area is largely due to underlying sediments and glaciation of the moraine. Areas north of the moraine drain into Long Island Sound. Little Neck, Manhasset, Hempstead, and Oyster bays are likely the result of heavy erosion from the Sound. Where the moraine is most dense, it forms the heads of bays along the North Shore. In water discharge basins, there are moist soil conditions and here red maple is the dominant tree, with arrowwood, spicebush, and winter berry the dominant shrubs and Virginia creeper, false Solomon's seal, and nightshade dominating the herbaceous layer.

Maple-leaved viburnum, lowbush blueberry, Canada mayflower, pink lady slipper, dwarf rattlesnake plaintain, indian pipe, small stands of spotted wintergreen, and at least six fern species provide diversity in the shrub and ground layers and dot the forests with color in spring and summer. Where light enters through the canopy, trailing arbutus and several species of mosses grow, and many ferns cover the forest floor. Along the North Shore there are still fine old stands of skyscraping tulip trees, open orchards, regularly-cleared fields, kettle-hole ponds, typical salt creek habitat, and of course, sheltered rocky-beach harbors.

Birding along the North Shore of Nassau County is wonderful in spring. Some of the birds one can expect and their months of arrival are listed here. Those species denoted by an asterisk (*) breed along the North Shore.

In April look for Great Blue and *Green herons and *Black and *Yellow-crowned night herons, Osprey, *Broad-winged and *Red-tailed hawks, *Northern Harrier (Marsh Hawk), Merlin, *Clapper and *Virginia rails and Sora, Semipalmated and Black-bellied plovers, *Spotted, Solitary, Least and Semipalmated sandpipers, *Screech and *Great Horned owls, *Chimney Swift, Yellow-bellied Sapsucker, *Red-bellied Woodpecker, *Bank, *Rough-winged, and *Barn swallows, Brown Creeper, Hermit Thrush, Blue-gray Gnatcatcher and Solitary Vireo. Of warblers search out *Black-and-white, Nashville, *Yellow, Yellow-rumped, Black-throated Green, *Prairie, and Palm, along with Louisiana Waterthrush. *Chipping and *Swamp sparrows should be around by mid-April.

In May look for *Common and *Least terns, *Yellow-billed and *Black-billed cuckoos, *Great Crested, Olive-sided, Yellow-bellied, Acadian, *Least, Willow and Alder flycatchers, *Long-billed Marsh Wren, *Wood, Swainson's, and Gray-cheeked thrushes and *Veery. Warblers to be expected along the North Shore during migration and breeding seasons are *Blue-winged, Tennessee, Magnolia, Cape May, Black-throated Blue, Blackburnian, Chestnut-sided, Bay-breasted, Blackpoll, *Ovenbird, Northern Waterthrush, Mourning, *Common Yellowthroat, *Yellow-breasted Chat, Hooded, Canada, and Wilson's, as well as American Redstart. Look for Bobolink, *Northern Oriole, *Scarlet Tanager, *Rose-breasted Grosbeak, *Indigo Bunting, and *Savannah, *Grasshopper, *Sharp-tailed, *Field, White-crowned, and Lincoln's sparrows in May.

The following tour should be begun in the early morning, with frequent stops to get out of the car to investigate the singing and foraging birds. I have indicated especially good birding places by the word *stop.*

The first part of the tour is approximately 33 miles in length. It begins on the southwest side of the Great Neck Peninsula and ends at the south end of Hempstead Harbor. The second part of the tour begins at the head of Hempstead Harbor and ends in Mill Neck, just west of the head of Oyster Bay Harbor.

Leave I-495 (the Long Island Expressway) at Exit 32. Proceed north on the Little Neck Parkway to NY 25A (Northern Boulevard). Turn east. Follow Northern Boulevard to Middle Neck Road. Turn north on Middle Neck Road to Old Mill Road. Turn left (west) and continue to Bayview Avenue. Turn right (north) and proceed to Grist Mill Lane on the left and Udalls Mill Pond. Stop. Proceed north over Saddle Rock Bridge and on West Shore Road for 1.3 miles to Kings Point Road. Turn right. Stop. Bird Kings Point Park on the right. At Steamboat Road turn left and then a quick right on Stepping Stone Lane. Proceed to Stepping Stone Park on the left, next to the United States Merchant Marine Academy (former Walter Chrysler estate). Stop. Walk into the Stepping Stone Park for views overlooking Long Island Sound. Follow Stepping Stone Lane around to Kings Point Road. Turn left. Proceed 1.5 miles to Kings Point and Kings Point Pond. Stop. Return south on Kings Point Road 0.7 mile to Split Rock Drive on the left. Turn left. Follow it around to Wildwood Road. Turn left. At creek stop. Continue east on Wildwood Road to Broadlawn Avenue. Turn right. Continue south to Blossom Road and south to Cow Lane. Turn left and slowly bird Cow Lane east to its junction with East Shore Road. Here turn right and bird south along the wooded, wetlands strip. There is marsh vegetation and an amazing variety of birdlife along here. Stop to bird Manhasset Inlet on the left. Continue south past Bayview Avenue to Northern Boulevard. Turn left and scan the pond on the right. Proceed east to Plandome Road. Turn left and proceed 1.5 miles to Leeds Pond Preserve. Stop. From the preserve causeway there are unobstructed views (left) of Manhasset Bay. Bird the preserve. Continue north for 1.3 miles on North Plandome Road to the town docks on the left. Stop here if traffic is light and you want better views of the bay from the dock. Continue up a steep hill to Shore Road coming in on the left. Gain it and continue north 2.8 miles to Middle Neck Road. Turn left. Go to the end of Middle Neck Road. Turn left on Lighthouse Road and return back to Middle Neck Road. Turn right. Proceed south 4.0 miles to Beacon Hill Road on the left. Turn left. Proceed east 1.0 mile to West Shore Road. This road borders Hempstead Harbor Park. Stop wherever possible and bird the harbor on the left and the park on the right. Continue south for approximately 2.2 miles. Proceed east over the Hempstead Harbor inlet on Northern Boulevard.

Continue east on Northern Boulevard past Glen Cove Road to Valentines Lane. Here, turn left (north) and proceed 0.8 mile to an overgrown lane on the right side of the road. This is a cattail, buttonbush, red maple swamp. Through this small freshwater wetland with a narrow woodland border a stream flows northward. It is a holding of the Nature Conservancy and is called the Louis C. Clark Sanctuary. Stop. Walk east on the path to bird the area. Continue north to Glen Woods Road. Turn left (west) and proceed to Glen Cove Avenue. Here turn right.

Bird north through the village of Sea Cliff and into Glen Cove on Glen Cove Avenue. Continue on the Glen Cove Bypass by keeping left at the fork, and go to the last traffic light at the Glen Cove Fire House. From here follow the green-and-white signs to Garvies Point Preserve. At the preserve stop.

The Garvies Point Preserve comprises approximately 5.0 miles of trails through an area of glacial moraine covered by thickets, forests and meadows. From the tops of the cliffs waterfowl, gulls and terns can be seen in Long Island Sound. Typical meadow birds can be seen southwest of the parking area. By walking northeast of the parking lot one passes by a pond and into a small valley forested with pin oaks and sassafras trees. This is an especially productive area in which to find warblers and small landbirds.

After leaving the preserve, continue one block north to Landing Road. Turn right (east) and follow it east as it becomes Cottage Row. At Forest Avenue turn left and proceed east past the Nassau Country Club (on the right) and stay on Forest Avenue as it becomes Buckram Road. Follow Buckram Road to Bayville Road, which comes in on the left. Turn left (north) on it and follow it north to Eden Rock Drive, which comes in on the left. Here, turn left and go a short distance to Sheep Lane. Turn right on it and follow it to its terminus, which is Fox Point. This is a great place to find herons, Clapper Rails, swallows, and singing marsh wrens.

Return on Sheep Lane and Eden Rock Drive. At Bayville Road turn left and then take the first possible right onto Factory Pond Road. Bird along this road to Feeks Lane. Turn left and cross Mill Neck Bay-Beaver Lake on Cleft Road. Stop. Continue on to Frost Mill Road, the first road coming in on the right after crossing the bridge. Turn right on Frost Mill Road. Proceed south along the east side of Beaver Lake. South of Beaver Lake, along Mill Neck Creek, Clapper Rails and Long-billed Marsh Wrens nest. Continue past the Mill Neck railroad station. Just past the railroad underpass are the entrance gates to a marvelous birding area. Turn right into the sanctuary. Please note! Parking on roads in Mill Neck is not permitted. Be sure to park inside the sanctuary or at the railroad station.

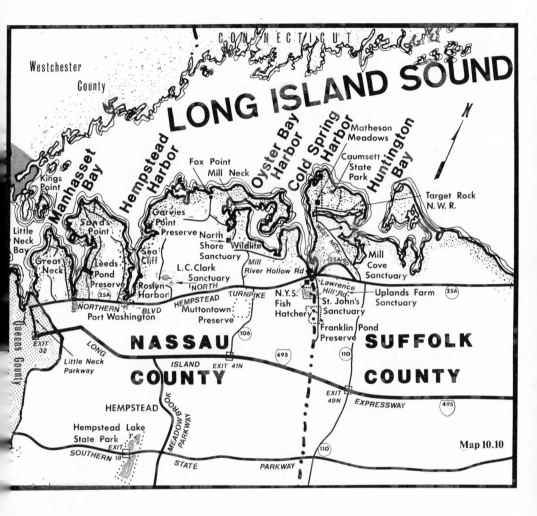

This is the North Shore Wildlife Sanctuary, whose old name was Shu Swamp. It is one of the last remaining woodland swamps on the North Shore. It has a magnificent stand of lofty tulip trees more than 100 feet high and 150 years old, some of which have trunk diameters of four feet or more. They thrive here along estuarial waters surrounded by a dense undergrowth of semi-aquatic plants with strong, fibrous, moisture-retaining root systems. This is a fragile ravine site with streams, ponds, and underground springs. Trails wind through the sanctuary, and the visitor is requested to stay on them.

This area, about forty acres, is filled with birds. Nesting species include Wood Duck, Virginia Rail, Red-bellied Woodpecker, Carolina Wren, Wood Thrush, Veery, Prothonotary Warbler, Ovenbird, Rusty Blackbird, and Swamp Sparrow. It is productive in spring migration as well as in winter, when hawks, Common Snipe, Winter Wren, Purple Finch, and other finch species can be found.

Muttontown Preserve (Map 10.10)

Rating: Spring****, Summer**, Autumn***, Winter*

This 500-acre preserve consists of moist, poorly drained woodlands, pioneer woodland with trees younger than fifty years, mature upland forest, with its typical shrub and herbaceous layers, conifer stands dominated by eastern white pine and European larch, old reverting fields, and rolling terrain with kame-and-kettle ponds.

A number of color-marked trails twist and turn through the preserve, enabling the birder to fully sample the various habitats.

Leave I-495 (the Long Island Expressway) at Exit 41N. Proceed north on NY 106–NY 107, watching for feeder roads to NY 106 north (East Norwich Road). Gain NY 106 and proceed north on it as it merges into the Jericho–Oyster Bay Road. Continue north to the intersection of NY 25A (here, the North Hempstead Turnpike). Turn left (west) and after one block turn left again onto Muttontown Lane. Drive south for three blocks and enter the preserve. Park in the lot in front of the preserve's Nature Center Building.

Birds breeding on the Muttontown Preserve include: Broad-winged and Red-tailed hawks, Bobwhite, American Woodcock, Yellow-billed and Black-billed cuckoos, Long-eared Owl, Red-bellied Woodpecker, White-eyed Vireo, Blue-winged and Chestnut-sided warblers, Bobolink, Scarlet Tanager, Rose-breasted Grosbeak, Indigo Bunting, and Savannah, Grasshopper, and Field sparrows.

In the more remote southeastern parts of the preserve hawk species nest and winter, and this is the area in which to find nesting Red-bellied Woodpeckers. Walk along the equestrian trails to reach this more southern section. While walking, notice the vigorous sprouts continuing to grow from the old American chestnut rootstocks, some of which are more than ten to fifteen feet high. In 1904 the chestnut blight hit Long Island and nearly exterminated this formerly dominant tree of the eastern hardwood forests. The present state of the woodlands here represents the natural readjustment that occurred following the demise of the large chestnuts.

SOUTHERN NASSAU COUNTY
Hempstead Lake State Park (Map 10.10)

Rating: Spring***, Summer*, Autumn***, Winter**

For many, this 870-acre park is a favorite birding area, for it can be interesting in almost any season. Summer birding, however, is not recommended, because park facilities include picnic and play areas, refreshment stands, a carousel, sports fields, and tennis courts. It receives heavy use during the warm months – which is incompatible with productive birding. However, in 1981, with the cumulative park list standing at 135 species, there had been 35 species recorded as park breeders — approximately 115 species recorded in spring, 100 in autumn and 50 in winter. Of the total, about 12 shorebird species were recorded in drought years when water levels were low and more extensive mud flats were exposed. Similarly, if there is open water during the winter, waterfowl can be found in variety and numbers.

To reach Hempstead Lake State Park, leave the Southern State Parkway at Exit 18 and then proceed into the park. Parking Area Number 3 is convenient to several of the park's better birding sites. A fee is charged May through September for the use of certain parking lots.

Bird along the edges of Hempstead Lake from the bridle path-foot trail along its eastern shore. This area is productive for waterfowl in autumn, in winter (if it isn't frozen over), and in spring. Shodack Pond, south of the parking lot number 3, and its surrounding woodlands is a good place to find migrant passerines in spring, especially on "wave days." Townsend's Warbler, a western vagrant, has been recorded here in recent years. Walk around the perimeter of the Hempstead South Pond. At its southern end, walk west and south between Peninsula Boulevard and the aquaduct, searching the trees and shrubs for migrants. Proceed south from Lakeview Avenue and continue across Main Avenue, which can be excellent when birds are moving through. The area south of Maine Avenue around Smith Lake can be equally productive in spring and autumn. Smith Lake often holds waterfowl species, and can be viewed best from its northwest corner.

THE NORTH SHORE — EASTERN NASSAU-WESTERN SUFFOLK COUNTIES
The Cold Spring Harbor Assemblage, a Tour (Map 10.10)

Rating: Spring****, Summer***, Autumn***, Winter*

This tour includes stops at three nature preserves, each with varying habitats, the New York State Fish Hatchery, and Cold Spring Harbor itself. To reach the beginning of the tour, follow the directions detailed under Muttontown Preserve from Exit 41N of the Long Island Expressway to NY 25A. At NY 25A, instead of turning left (west), turn right (east) and continue east approximately 4.2 miles to the New York State Fish Hatchery. Pull into the Fish Hatchery parking lot. This is the beginning of the tour.

St. John's Pond lies at the head of Cold Spring Harbor, adjacent to the hatchery, on the Nassau County side of the marsh, and just south of St. John's Church. Walk south from the parking lot through an entrance gate to the preserve. This area is owned by the Nature Conservancy. Owing to the steepness of the slope in some places, the preserve has continuing problems with erosion. Birders are requested to use the steps provided and the trails, when traversing the upland woods.

This is a fine place for treetop birding, because the land slopes down from the westernmost point very steeply along the north half of the preserve until it reaches the pond edge. The altitude ranges from 40 feet above sea-level to over 180 feet, creating a watershed for areas west of the preserve. The trail runs the length of the preserve and passes through woodlands dominated by oak and a heavy mountain laurel understory. A semi-circular subordinate trail east of the main trail winds through maple-dominated lowlands and a lowland red maple swamp with skunk cabbage and cinnamon fern in its mucky edges. Mosses are slowly covering fallen trees, creating nursery trees throughout the woodland swamp. The northwestern portion of the preserve supports a fine stand of chestnut oak, with a dense understory of young American chestnut, pinxter, and mountain laurel. Canada mayflower, trailing arbutus, hay-scented fern, and pink lady slipper can be found on the western side of the trail.

Great Blue and Green herons, Great and Snowy egrets, and Black-crowned Night Herons regularly occur here. Wood Ducks nest here, and great numbers have been banded. Ospreys pass through on migration, often stopping to rest and feed. Broad-winged Hawks nest, and it is suspected that both Sharp-shinned and Red-tailed hawks do also. One can find Great Black-backed, Herring, Ring-billed, and Laughing gulls as well as Common and Least terns in spring and summer and Black Tern stopping over on migration. Yellow-billed and Black-billed cuckoos and Screech and Great Horned owls nest here. At least five picids and four swallow species have been recorded here. Wood Thrushes and Veeries breed here in summer and at least twenty-three warbler species have been recorded in spring and summer.

Upon leaving the parking area, turn right (east) and almost immediately turn right (south) again, onto NY 108 (Harbor Road). Proceed approximately 0.4 mile and turn left into the right-of-way opposite Saw Mill Road, which comes in on the left, just over the Suffolk County line. This right-of-way extends into the parking area for Franklin Pond Preserve. This is also a property of the Nature Conservancy, and it includes an early nineteenth-century mill pond bordered by woodland swamp, wooded embankment, and a moist oak community edge. On the southeast section a thick carpet of periwinkle is covered by tulip and white ash. The bird species to be expected here are those found at St. John's Pond just north. In addition, look for nesting Whip-poor-wills and Yellow-throated Vireos here.

Leave the parking area. Turn left. Proceed a short distance to Lawrence Hill Road, the next road on the right. Turn right. Proceed east on Lawrence Hill Road to a sign indicating Uplands Farm Sanctuary on the right. Turn right and proceed south to the first drive road on the left. Turn left and continue to a parking area behind a renovated barn, which serves as the environmental center of the sanctuary. South of and adjacent to the center is the headquarters of the Long Island Chapter of the Nature Conservancy. Check in here, and let the staff know that you are here to bird.

The farm property comprises about fifty-five acres of flat-to-gently rolling hills and hollows with the acreage along the eastern boundary reverting from farm to woodland. Here there are young deciduous trees and bittersweet and briar tangles. More than one-half the sanctuary is hay meadow and pasture dominated by grasses and more than thirty-five wildflower species. These meadows are mowed annually in mid-August to preserve the meadow habitat and to maintain the meadow-forest interface. One trail here crosses a hay meadow, parallels hedgerows, passes through the forest, and follows a lane bordered on both sides by rows of black walnut trees. Freshwater on the property is limited to wells and precipitation.

The primary bird species to be here are those of upland fields, with more than twenty-five species nesting on the sanctuary grounds.

SUFFOLK COUNTY, THE NORTH SHORE
Lloyd Neck—Matheson Meadows, Caumsett State Park,
Target Rock National Wildlife Refuge, Mill Cove Waterfowl Sanctuary, a Tour (Map 10.10)

Rating: Spring***, Summer*, Autumn**, Winter***

To reach Lloyd Neck, leave I-495 at Exit 49N. Gain NY 110 (New York Avenue and Broad Hollow Road) and continue north to its intersection with NY 25A in Huntington. Turn left (west) and proceed west only a short distance to the intersection of NY 25A and West Neck Road. Here turn right and proceed north on West Neck Road. Cross Lloyd Harbor on the narrow isthmus connecting it with the mainland and turn right on Lloyd Harbor Road. Continue a short distance east to Lloyd Lane. Turn left here, and after a short distance Matheson Meadows will appear on the left.

Matheson Meadows is forty acres of rolling grassland surrounded by a wooded edge. It is a preserve of the Nature Conservancy and one where an effort is being made to re-establish the Bobolink as a breeding species through a program of rearing fledglings there and releasing them upon maturity. Of course, the hope is that these artificially bred birds will have imprinted upon the site and will return following migration to rear broods of their own.

Caumsett State Park is north of Lloyd Harbor Road as one continues east along it. This is a 1500-acre, undeveloped state park with marvelous birding habitats. Old roads twist through the area, giving access to a wealth of birding main and side trails. In the park and along its boundaries two Great Gray Owls wintered in 1979.

At the eastern end of Lloyd Harbor Road is a large black oak tree. This tree is famous for its 20-foot circumference, 100-foot height and 150-foot branch spread. It is said to be the largest black oak in the United States.

Past the black oak tree, up an unmarked road, is Target Rock National Wildlife Refuge at the east end of the peninsula. It sits on the eighty-acre former estate of financier Ferdinand Eberstadt. The refuge offices are now in the thirty-two-room Georgian mansion built by Eberstadt. Here there are parking facilities and miles of trails leading through hardwoods and open areas, past a brackish pond to overviews of 2,000 feet of tidal shoreline.

Upwards of 190 bird species have been recorded with 50 species nesting. Warbler migration peaks in the third week of May here and coincides with the rhododendron blooming season. From the refuge bluffs, there are excellent views of Huntington Bay. Colonies of Bank Swallows nest in the bluffs.

Follow the same roads back to Huntington. At the intersection of West Neck Road and Mill Road, turn left and proceed east to Southdown Road. Just east of this junction is the Mill Cove Waterfowl Sanctuary, which is a large brackish impoundment of tidal waters regulated by the ebb and flow of the water from Huntington Harbor. The pond also benefits from freshwater from springs and estuaries. In the late eighteenth century this was the site of a grist mill. Along the pond's periphery there are swamp maple, black cherry, black walnut, and black locust trees. It is a well protected feeding and resting stopover for migratory waterfowl, and several species nest in a marshy portion of the pond.

The Nissequogue River (Map 10.11)

Rating: Spring**, Summer***, Autumn***, Winter**

The Nissequogue River is the only river on Long Island that starts in the Ronkonkoma Moraine and flows north through the Harbor Hill Moraine to Long Island Sound. Along its route it crosses some of the most beautiful and well-preserved sections of the island. Blydenburgh County Park comprises more than 600 acres on what was formerly the Weld Estate. It lies east of Old Willets Pass and north of I-495 (the Long Island Expressway). Within Blydenburgh are several branches of the Nissequogue River, each dammed near its headwaters to form spring-fed mill ponds. The largest of these is New Mill Pond, sometimes referred to as Sump Pond. New Mill Pond is the second-largest lake on Long Island, after Lake Ronkonkoma. It is long and L-shaped and covers 120 acres. The area surrounding New Mill Pond is dominated by dry, wooded glacial hills and ravines. The forest community contains black and chestnut oak, black or sweet birch, hickory, beech, and American chestnut and extensive growths of laurel. The understory also contains blueberry, bayberry, maple-leaved viburnum, trailing arbutus, and wild azalea.

The river winds north through an extensive cat-tail marsh. Watercress and marsh marigold grow abundantly in this spring water seepage. At the river's lower end it is fresh, then brackish, and farther north it turns to saltwater. The farther north it goes the wider the river valley becomes. The mouth is wide and is filled with grass-covered islands. It finally flows through a wide estuary with high bluffs of the North Shore surrounding it and a white pebbly beach lining the Sound.

All along the course of the Nissequogue River there is abundant birdlife. Around the headwaters and in New Mill Pond look for river ducks in autumn. In the marsh, American Woodcocks and Common Snipes breed. On both sides of the river in the woodlands Brown Creepers and Winter and Carolina wrens can be found. The Wood Thrush nests here, and the Eastern Bluebird is sometimes recorded in autumn. Look for both kinglet species and Cedar Waxwing, along with migrating and wintering blackbird, finch, and sparrow species along the river's length. At the wide mouth of the river there are sweet birches, dogwoods, sassafras, locusts, and cedars growing on the eastern shore. Among them the Yellow-crowned Night Heron nests and should not be disturbed during the breeding season. Clapper and possibly Virginia rails nest in this area also. On the sandbar just off the beach, shorebirds, gulls, and terns feed.

The following are good points along the Nissequogue River to find birds,

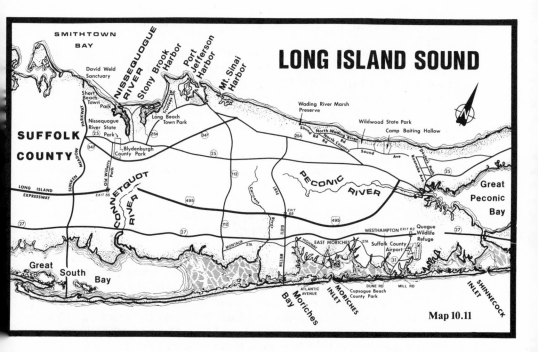

but perhaps with a bit of searching the birder can find his own secluded, most-favored sites.

Leave the Long Island Expressway (I-495) at Exit 55 and head north on Old Willets Path. Follow it to NY 25 (the Jericho Turnpike). Here turn right (east) and follow NY 25 to Brooksite Drive. Turn south. Go one block to New Mill Road. Turn west and head into the park.

Follow the trail from the parking area near the old house down to New Mill Pond. Continue southward across a bridge near the mill and along the western side of the pond to the stream. Follow the stream until the trail divides. Take the left spur across the stream and uphill. Follow it back to the high bluff of the peninsula. There you will be able to overlook the entire pond. You have (to this point) walked approximately 1.1 miles. Return to the mill. Walk east along the north shore of New Mill Pond for approximately 1.1 miles more. Cross the narrow east end, then proceed in a westerly direction. Stop and return to your starting point anywhere along here.

Leave the parking lot of Blydenburgh County Park. Turn left and go one block on Brooksite Drive to NY 25. Turn right and follow NY 25 to Nissequogue River Road. Turn left (north) and continue north about 3.0 miles to the road's end. Turn right on Moriches Road and then immediately left on Long Beach Road. After 0.5 mile Boney Lane comes in on the left. Turn left and approximately 0.4 mile farther is the entrance to David Weld Sanctuary. It is marked by an old woods road and a white gate on the right.

Walk north along the main trail, which parallels a private drive on the east. At a gate, turn west and continue walking parallel to a fence west and then north to some stunted oaks at the cliff edge. You will now be overlooking Long Island Sound.

Continue driving west on Boney Road to its end. On the west will be Short Beach Town Park. Walking through the park one has excellent views of Smithtown Bay in the Sound and the Nissequogue River estuary.

Drive back on Boney Road and turn left on Long Beach Road. Drive east to Long Beach Town Park, from which you will be able to overlook the Sound and Stony Brook Harbor. From here one can walk east out to the end of Long Beach.

Stony Brook Harbor drew quite a bit of attention in the summer of 1977 when seven immature White Ibises were found here feeding and roosting with a mixed heron-egret colony that nested on Long Beach. The area is relatively undisturbed and can be observed from Long Beach (described above) or can be observed from the entrance to the harbor on its east side, at the end of West Meadow Road. This road is located north of the village of Stony Brook, and its southward extension along the sand spit is sometimes called Trustees Road.

The best time to visit this area is probably in late afternoon in the summer and if at all possible on an ebb tide. It is then that large numbers of birds of several species gather at or near the narrow mouth of the Porpoise Channel, in the vicinity of several small islands there, to feed along the shoreline.

Species one can expect to see are: Green Heron, Great and Snowy egrets, Black-crowned and Yellow-crowned night herons, several pairs of Piping Plovers, several species of gulls with more than 100 Laughing Gulls in mid-to-late July, Common, Roseate and Least terns, and small mixed flocks of Common and Fish crows.

Wading River/Baiting Hollow, a Tour (Map 10.11)

Rating: Spring*, Summer**, Autumn***, Winter**

Wildwood State Park comprises approximately 700 acres located on Long Island Sound northwest of Riverhead. It sits on high bluffs and has a broad macadam walkway of about 0.6 mile that leads to the boulder-strewn and beautifully rounded quartz and quartzite pebble beach. This stretch of beach is more protected by the sandy cliffs stretching east and west, than those of the South Shore; it receives less buffeting by the wind and the waters freeze more quickly in winter and thaw sooner in spring. There is a rather extensive camping facility at this state park and it is particularly crowded in the summer season. Plan to bird here during the off-season, when it will be more rewarding.

Wading River Marsh is a priceless expanse of about sixty acres maintained by the Nature Conservancy. It is one of the few remaining salt marsh ecosystems of any size on the North Shore. There are distributions of characteristic salt meadow grasses — cordgrass (*Spartina alterniflora*), salt hay *(Spartina patens)*, spike grass (*Distichlis spicata*), and plume grass (*Phragmites communis*) throughout. There are freshwater streams, brooks, and ponds that add nutrients to the wetlands community. A remote freshwater pond overhung with willows and pepperidge trees, on the east side of this preserve, rimmed in part by a bamboo grove and containing a wooded island hummock, adds further dimension to the sanctuary. This environment provides migratory resting and feeding grounds for hundreds of birds. It is a source for food including small invertebrates and shellfish. Ducks, Clapper and Virginia rails, and marsh-associated sparrow species nest here. Herons, egrets, and terns can be found feeding in the marsh and along its edges. Watch for Northern Harrier (Marsh Hawk) coursing over the *Spartina.*

Several of the roads in Baiting Hollow are excellent for numbers and varieties of transient shorebirds in late summer and autumn. Notable among these are Doctor's Path and Sound Avenue.

Leave I-495 (the Long Island Expressway) at Exit 68. Gain NY 46 (the William Floyd Parkway) north and proceed to NY 25A (Sound Avenue). Turn right (east) and proceed to Hulse Landing Road (NY 54). There turn left (north) and proceed to the entrance to Wildwood State Park, which comes in on the right (east). Continue to the parking area.

There are over five miles of cleared, well-maintained walking, jeep, and fire trails throughout the wooded sections of this park as well as the trail to the beach previously mentioned. Bird them all to good advantage.

From the Wildwood State Park parking lot, gain North Wading River Road headed west and proceed approximately 2.0 miles to Sound Road. Turn right. Continue north on Sound Road approximately 0.8 mile to the entrance to Wading River Marsh Preserve. The entrance to the preserve is by way of the Kempf driveway on the left. There is parking space available on the lawn adjacent to the garage at the end of the driveway. Stay on the established trails that pass along the edge of the marsh and through much of the upland area. They are adequate to cover the terrain.

If one wishes to bird the marsh from another angle, continue north on Sound Road to Creek Road. Turn left. Continue west to the end of Creek Road and the Wading River outlet on Long Island Sound.

Return on Sound Road to its junction with North Country Road. Continue south to Sound Avenue (NY 25A). Turn left (east). Drive east to the intersection of Sound Avenue and Doctor's Path. You should have been searching the fields right and left of Sound Avenue for migrant shorebirds. Turn right (south), thereby gaining Doctor's Path and proceed slowly along it, scanning the fields in late summer and autumn for Cattle Egret, Killdeer, American Golden and Black-bellied plovers, Common Snipe, Whimbrel, Upland, Baird's and Buff-breasted sandpipers, Common Nighthawk, Horned Lark, Cliff Swallow, Water Pipit, and Grasshopper, Vesper, and Field sparrows. In mid-to-late autumn and early winter look for Sharp-shinned and Rough-legged hawks streaking by or hunting over the fields. In spring, Cliff Swallow, Orchard Oriole, and Summer Tanager have been recorded in the Wading River-Baiting Hollow area.

THE NORTH FORK (Maps 10.11 and 10.12)

Generally, the North Fork of Long Island is not so seasonally exciting as the South Shore or South Fork. However, over the whole year it has proven a most rewarding birding area. In spring nesting Ospreys can be seen fishing. At spring's end and into the summer, herons, Northern Harriers, rails, Piping Plovers, gulls and terns, and a select number of passerine species can be found breeding at this end of the island. The active heronry at Plum Island nearby is the source of the few Little Blue Herons, and numerous Great and Snowy egrets and Glossy Ibises that frequent the salt marshes from Greenport to Orient Point. Sometimes non-breeding Double-crested Cormorants, Oldsquaws, and White-winged Scoters can be found in July and August in Orient Harbor. Autumn brings another scene. Crisp winds blow off Gardiner's Bay, and this is a good time to see prickly pear cactus, beach plum, salt spray rose, horn poppy, mullein, and groundsel in winter dress. Migrants not deflected to the south shore pass along the sound side. At this time the numerous marshes overlooking Gardiner's, Southold, and Peconic bays teem with activity. Also in recent years numerous reports of Red-bellied and Red-headed woodpeckers have come from the areas around Mattituck, Cutchogue, and Southold and these two species should be looked for, especially in October and November.

Winter along the North Fork is probably the most rewarding season. From Orient Beach State Park, Cedar Beach Point on Great Hog Neck, and Nassau Point on Little Hog Neck look for the following: loons, grebes, Great Cormorant, Whistling Swan, hundreds of Mallard, Black Duck, Greater Scaup, Common Goldeneye, Bufflehead, Oldsquaw, and White-winged, Surf and Black scoters, various numbers of all three merganser species, and numbers ranging from one to twenty-five of each of the following raptor species: Northern Harrier, Sharp-shinned, Red-tailed and Rough-legged hawks, American Kestrel and Merlin. Great Black-backed, Herring, Ring-billed and Bonaparte's gulls are present. Usually Barn, Screech, Great Horned and Short-eared owls are also here, and some years it takes almost no searching at all to find one or more Snowy Owls. All of the usual half-hardy, over-wintering landbirds can be found, and diligent searching can almost always turn up an impressive list of sparrow species, including Savannah, Sharp-tailed, Tree, Field, White-throated, Fox, Swamp, and Song. Upwards of one-hundred Snow Buntings and an occasional Lapland Longspur can be found. A vigorous day of birding on the North Shore in winter can result in a list of about ninety bird species.

Take I-495 (the Long Island Expressway) to its terminus at Exit 73 and head

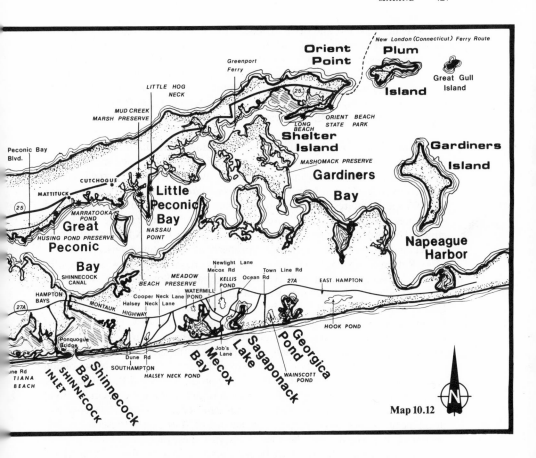

New London (Connecticut) Ferry Route

Orient
Point

Plum
Island

Great Gull
Island

Greenport
Ferry

LITTLE HOG
NECK

MUD CREEK
MARSH PRESERVE

ORIENT BEACH
STATE PARK

LONG
BEACH

Shelter
Island

Gardiners
Island

Peconic Bay
Blvd.

MASHOMACK PRESERVE

Gardiners

CUTCHOGUE

Little
Peconic
Bay

Bay

Napeague
Harbor

MATTITUCK

MARRATOOKA
POND

Great

NASSAU
POINT

HUSING POND PRESERVE

Peconic

Bay

SHINNECOCK
CANAL

MEADOW
BEACH PRESERVE

Newlight Lane
Mecox Rd
KELLIS Ocean Rd
POND

Town Line Rd

27A

EAST HAMPTON

HAMPTON
BAYS

Cooper Neck Lane
Halsey Neck Lane

WATERMILL
POND

MONTAUK
HIGHWAY

HOOK POND

Ponquogue
Bridge

Dune Rd

Job's
Lane

Mecox
Bay

Sagaponack
Lake

Georgica
Pond

une Rd
TIANA
BEACH

Shinnecock
Bay
SHINNECOCK
INLET

SOUTHAMPTON

HALSEY NECK POND

WAINSCOTT
POND

N

Map 10.12

east on NY 58 (Old Country Road). Continue east to its junction with NY 25. Take NY 25 east to Bay Avenue in Mattituck. Turn right (south) and continue to Great Peconic Bay Boulevard. Turn right (west) and drive a short distance west to Husing Pond Preserve opposite the Mattituck Yacht Club. Parking is available either on Great Peconic Bay Boulevard or at the town park across the street.

With the construction of Great Peconic Bay Boulevard, a dam was formed and the result is a freshwater woods complex that provides essential habitat for such species as Wood Duck, Mallard, Black Duck, Green-winged Teal, American Woodcock, and Common Snipe. There is a trail running up the east side of the preserve that gives access to the woods, fields and streams.

Go back to Bay Avenue and proceed north to its junction with NY 25. Turn right (east) and proceed to Marratooka Pond on the right. Check it out for any migrant or wintering waterfowl.

Continue east on NY 25 through the village of Cutchogue. Turn right on Eugene's Road and then right again on Skunk Lane. Follow Skunk Lane south to

Bay Avenue. Turn left and drive to Nassau Point Road. Turn right and cross the causeway over Broadwater Cove, thereby gaining Little Hog Neck peninsula. Over the causeway turn right on the first street, which is Broadwaters Road. Turn right almost immediately on Haywater Road and bear left to its end. Park at the waterfront and walk left at low tide to Meadow Beach Preserve. A bulkhead obstructs passage at high tide; in this case, park at Nassau Point Park south of the preserve, and walk north and west along the shore edge to Meadow Beach Preserve.

This preserve is a beautiful little salt meadow peninsula formed by the littoral drift along the west shore of Nassau Point in Little Peconic Bay. There is a small marsh pond with a stream to Horseshoe Cove, and a pebbly barrier beach on the western side. Thatch (cord) grass (*Spartina alterniflora*) covers most of this area. Of special interest are the shorebirds that use the beach during autumn, and sometimes spring, migrations. Great Blue Herons, Snowy Egrets, and Black-crowned Night Herons can frequently be found feeding here. A few pair of Piping Plovers nest along the shore. This is a property of the Nature Conservancy.

Return to NY 25. Continue driving east to Bayview Road, which is southwest of the village of Southold. Here turn right and follow Bayview Road to its end. At Cedar Beach Road turn left and continue to Cedar Beach County Park. Park and walk to Cedar Beach Point, from which there are superb views of Little Peconic Bay to the west and Shelter Island Sound to the east. You are now on Great Hog Neck peninsula. Walk southwest around the point to the salt marsh behind Cedar Pond. Birds are usually stirring here in the shallows.

Regain NY 25 and follow it east to its end at the Orient Point-New London (Connecticut) ferry pier. Scan Gardiner's Bay on the south and Plum Gut to the east. On Gardiner's, Shelter, and Plum islands and in the area of Orient Point there are clusters of active Osprey nests. This semi-social raptor can be viewed easily on the north fork and also in the vast marshes of Mashomack Preserve on Shelter Island. It is worth a trip there to closely view these marvelous birds.

All along this ride there have been attractive seascapes mingled with earthy, fertile farm fields. It is this sort of setting that makes the North Fork so attractive to some birders. At several points along the way the peninsula is very narrow, so the Sound, Orient Harbor, and Gardiner's Bay are visible from the same point. At these points, one should look for a good mix of birds in late summer and wintering loons, grebes, ducks, and gulls in winter.

Approximately 0.3 miles west of the ferry slip is a road on the right that leads to Orient Beach State Park. Turn right on it and proceed west down the narrow road about 2.0 miles to the park's parking area. Along this road there are marvelous views of Gardiner's Bay to the south and of Little Bay to the north. Little Bay is filled with marsh grasses and swarms with avian life. Be sure to investigate this area by walking back along the road once you have safely parked the car. Be alert for Ospreys from any of the loose colonies on Fisher's, Gardiner's or Shelter islands.

Walk from the parking lot southwest along the shoreline or into the interior, where there are scrubby, stunted pines and at the peninsula's widest point four smallish ponds that never fail to attract waterbirds.

At the very tip of the spit is Long Beach Point and the ruins of an old lighthouse foundation. The former lighthouse, long since gone, guided fishing vessels in and out of Greenport. The southwest end of the peninsula is a bird sanctuary, and the birder should search out its avian treasures.

NASSAU AND SUFFOLK COUNTIES, THE SOUTH SHORE
The Strip (Maps 10.13 and 10.14)

Rating: Spring**, Summer***, Autumn****, Winter***

For the restless binocular birder there may not be anywhere so seductive as the narrow sand barriers bordering the ocean. These barrier islands owe their existence and topographical formation to wind and sea erosion. The beaches are made up almost entirely of white quartz pulverized by marine action, and shaped and reshaped into ridges and mounds by the wind at the ocean edge. This protean impermanence characterizes the winds, the tides, and, more intriguingly, the birdlife along "the strip."

The tropical cyclone or revolving storm generally originates east of the West Indies and Gulf of Mexico, takes a westward course, and then recurves to the northeast. A tropical cyclone officially qualifies as a hurricane when its winds reach a mean velocity of 75 miles per hour. The months of greatest hurricane frequency are September and October, although the season extends from early June through November. Most tropical cyclones are carried away from the northeastern Atlantic coastline, but, occasionally the entire Atlantic seaboard is affected. As a tropical cyclone moves westward, before taking its northward curve, the coastline of New York becomes progressively more vulnerable. Some of these storms do not recurve north and east before wreaking havoc and devastation on those areas in its path. Most hurricanes follow this roughly parabolic curve, and their accompanying easterly winds and counter-clockwise air flows provide excellent conditions for bringing subtropical bird species north to the coast, and for taking primarily coastal species inland.

For barrier beach birders, the prospect of the avian fallout associated with tropical storms adds to the siren quality of the strip. The hope of finding a storm-displaced exotic prompts the following behavior. As the months of greatest hurricane frequency approach, birders increasingly keep abreast of any developing tropical cyclone. If one materializes they stay well-informed as to the position of the eye, as reported by the United States Weather Bureau. They realize that the number of avian residuals will be at or close to where the eye crosses the coast. Virtually all exotics are found east of the eye, where wind velocities are greatest. They listen to or watch broadcasts that revise weather information every two hours or more often, if the situation warrants. It is most important that the barrier beach birder take to the beaches at the critical time, for displaced and disoriented birds will most likely be found only on the day of and the day following the hurricane. The number of hurricane-blown exotics fall off sharply by the second day after the storm. It has been observed that the daybreak hours are crucial to the distribution

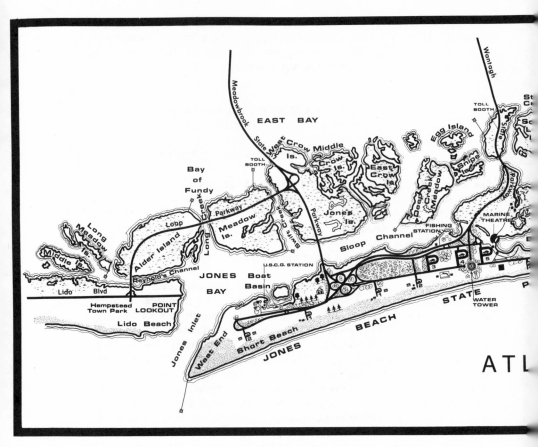

of storm-borne waifs. Few extralimital birds are found in areas crossed by storms at night, but rarities dramatically increase when storms pass just after first light. Gale-blown birds disperse quickly, so that only a scattering of individuals remains by the late afternoon of the day after the storm. So, being at the right place, although critical, is only part of the success equation. To reap the bounty of a tropical cyclone one must also be out before dawn.

What species might one see under optimal post-storm conditions? The hurricanes of this century, especially since 1917, have deposited the following birds, many are rarities, along the strip: Cory's, Greater, Sooty, Manx, and Audubon's shearwaters, Black-capped Petrel, White-faced, Leach's and Wilson's storm-petrels, White-tailed Tropicbird, Brown Pelican, Brown Booby, Magnificent Frigatebird, Pomarine and Long-tailed jaegers, Sabine's Gull, and Sooty, Bridled and Sandwich terns.

In addition to this legacy, various hurricanes have grounded numbers of

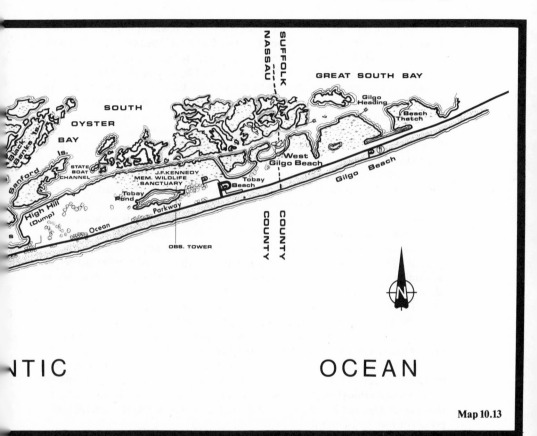

Map 10.13

over-ocean shorebird migrants—Hudsonian Godwit and Northern Phalarope, for example. In some years gulls and terns are blown far inland and they can be seen along the Hudson River and in upstate and western New York as well as along the barrier islands. Who says an ill wind blows no good?

Jones Beach State Park, John F. Kennedy Memorial Wildlife Sanctuary, Gilgo, Captree State Park, Robert Moses State Park (Fire Island), a Tour (Maps 10.13 and 10.14)

The barrier island between Jones Inlet on the west and Fire Island Inlet on the east is about seventeen miles long. More than two-thirds of the island lie within state park lands. The Ocean Parkway extends along the center of the strip to within one-half mile of either end. Large parking fields front the ocean and smaller ones are built slightly back from the ocean.

To reach this island, leave the Southern State Parkway at Exits 22, 27, or 40. Proceed south on the Meadowbrook State Parkway, the Wantagh State Parkway, or the Robert Moses Twin Causeway to the island. Robert Moses State Park, located on the western end of Fire Island is reached by leaving the Southern State Parkway at Exit 40, taking the Robert Moses Twin Causeway across the eastern end of the island (across Captree State Park) and then taking the Robert Moses Bridge across Fire Island Inlet to Fire Island. Parking is available at the terminus of all of these parkways and causeways and at points in between. At Jones Beach State Park the parking fields are numbered (see accompanying map). Parking along Ocean Parkway is strictly prohibited.

Overland southbound migrants must terminate at the barrier beaches or find themselves over water. Frequently passerines overshoot the coast, so that in the early morning (approximately daylight), on a heavy flight day, one can see hundreds of migrants heading back to land. This phenomenon is associated with bad weather and occurs most often when there are offshore northwest winds. If the winds are strong, small birds — some warblers, for instance — encounter resistance in trying to regain shore and, once there, light down to rest and refuel before continuing south. These migrants sometimes occur in the thousands along the entire strip.

In autumn scan the ocean with a telescope for loons and grebes. Occasionally, especially in strong northeasterly winds, Gannets migrate in small flocks and can be seen near the surfline. More often, they migrate farther offshore and out of sight. Double-crested Cormorants fly offshore but within sight, in long stringy lines. Herons, egrets, American Bitterns, and Glossy Ibises can be seen on their post-nuptial travels before they undertake migration. Various waterfowl species can be viewed throughout the entire autumn in the numerous small bays on the island's inner edge and in the boat channel that separates the island from the archipelago of marshy islands north of it.

A raptor flight down the outer strip provides beautiful views of Ospreys, Northern Harriers, Sharp-shinned and sometimes Cooper's hawks. Falcons include American Kestrels, Merlins and Peregrine Falcons.

Good late-summer records of shorebird species along the strip include Ruff, Baird's and Buff-breasted sandpipers, American Avocet, and Wilson's and Northern phalaropes.

On a peak flight day one can see hundreds of flickers, kingbirds (including Western Kingbirds, reported almost annually), *Empidonax* flycatchers, thousands of swallows hawking insects in flight, nuthatches, Brown Creepers, mimids, thrushes of several species, kinglets, Water Pipits, occasionally Loggerhead Shrikes along the wires and on bushes and rooftops, vireos in abundance and hundreds upon hundreds of warblers, including Nashville, Pine, Connecticut, and Canada.

Almost every year the Yellow-headed Blackbird is discovered along the strip. Western Tanager occurs some years. Blue Grosbeak, Dickcissel, Lark and Clay-colored sparrows are recorded in most years in autumn.

In winter the scene changes, and Common and King eiders, Purple Sandpipers, Dunlins, Sanderlings, and occasionally Black-headed and Little gulls, and

Black-legged Kittiwakes can be seen from the beaches and jetties. Snowy, Short-eared, and Saw-whet owls take up winter residence along the strip. Horned Larks, grosbeaks, finches, redpolls, Pine Siskins, American Goldfinches, Red and White-winged crossbills (especially in irruption years), and Snow Buntings can be found with a diligent search. Very rarely, a powerful Gyrfalcon is seen coursing over the inlets, bays, and marshy hideaways, scaring the life out of ducks and small wintering shorebirds.

In summer the barrier island on which Jones Beach State Park is located hosts the best collection of breeding colonies on the entire South Shore particularly at West End Beach and Cedar Beach and Captree State Park, 12 to 15 miles to the east. Green and Little Blue herons, Great and Snowy egrets, Louisiana Herons, Black-crowned and Yellow-crowned night herons, Glossy Ibises, Piping Plovers, Great Black-backed and Herring gulls, Common, Roseate, and Least terns, and Black Skimmers nest in considerable numbers along the strip. Chuck-will's-widow can be heard along Ocean Parkway at night in the vicinity of Oak Beach.

There are numerous places from which viewing is excellent or around which birders should concentrate their efforts.

From parking field number 2 walk west to Short Beach and the jetty to bird the Jones Inlet. On the west end of the strip poke around the pines, lawns and sand dunes surrounding and easily accessible from west end parking field number 1 in autumn and winter. Gain permission from the park police to go through the plantings around the barracks and superintendent's houses across from parking field number 1, especially in autumn. Go to parking field number 4, which is open year-round. Walk north to the fishing station where parking is also allowed, and ferret out any migrant landbirds in the Japanese black pines, Russian olives, sycamores, evergreens, and wild rose and blackberry thickets surrounding the fishing station area. Walk south and east from the fishing station, past parking field number 5 to a point overlooking Zach's Bay. Carefully bird the bay and all of the vegetation surrounding it.

Drive east along the Ocean Parkway, scrutinizing the crab grass growing on the strip and bordering the parkway for sparrows. After 1.6 miles, turn north (left) into the Tobay Beach parking lot. Make a sharp left turn just past the entrance onto the southwest corner of the parking lot and the access road to the John F. Kennedy Memorial Wildlife Sanctuary, otherwise known as "Tobay," "Tobay Pond," and (formerly) "Guggenheim Pond." Park in the area indicated north of the access road. This is one of the super special birding places on the strip. The barrier beach birder who frequents Tobay is more often than not amply rewarded for the effort of applying in writing for the permit (no fee) to bird the sanctuary. The sanctuary is owned by the town of Oyster Bay, and by applying to the Superintendent of Beaches, Town Hall, Oyster Bay, New York 11771, a permit will be issued upon request.

The sanctuary, a 500-acre expanse of sand dunes, salt marsh, bayberry, catbrier, pine groves, and freshwater and brackish ponds, is a gem. It has an observation tower and well-maintained trails, and is open dawn until dusk during the summer season and until 4:00 p.m. in the off-season.

Farther east, Gilgo is a small residential community that is well known as a

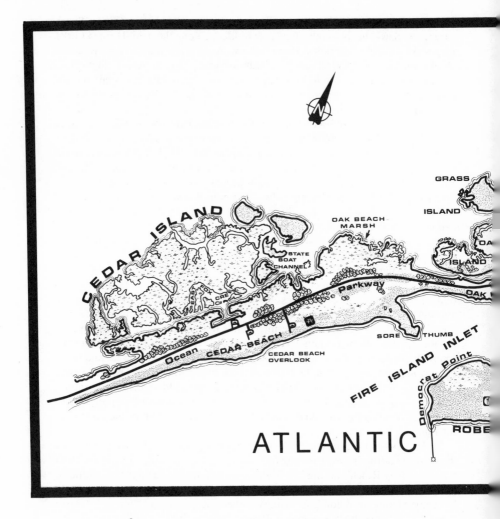

migrant trap. Park in the Gilgo village parking lot and walk east investigating the lawns and pine groves along the bay. The Gilgo village parking area is 6.7 miles east of the watertower at Jones Beach. There is a stiff fine for nonresidents parking there in any but the off-season.

East of Gilgo is Cedar Beach. Search the dense pine groves here for mi-

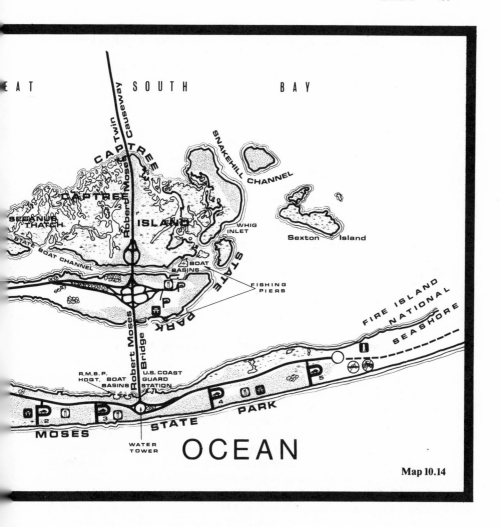

Map 10.14

grants in autumn. These and similar groves farther east at Oak Beach are the most productive winter owling sites on the strip. Cedar Beach is the site of one of the largest tern and skimmer colonies on the East Coast. Common, Roseate and Least terns and Black Skimmers have produced thousands of young annually here.

For many birders Oak Beach marsh, east of Cedar Beach and on the south side of Great South Bay, is the true gem of the strip. The marsh is primarily *Spartina alterniflora* and in it Virginia Rails breed and the rare Black Rail has bred. Chuck-will's-widow breeds in the area of Oak Beach marsh and can be heard calling on summer evenings. Bird the pine groves here in winter. Short-eared Owls breed at Oak Beach marsh and can be seen hunting over the marsh at dawn and dusk.

Captree State Park is at the eastern end of this barrier beach. The best times here are autumn, when it serves as a land trap for migrants and following tropical disturbances, when storm-carried species can sometimes be seen, and during summer when large numbers of breeding gulls and terns can be seen.

Cross south on the Robert Moses Bridge to Fire Island and the Robert Moses State Park. This is located on the western end of Fire Island and offers sensational birding during autumn. Drive to the farthest west open parking field and walk west from there to parking field number 6 and Democrat Point where one can bird Fire Island Inlet. Just one good migration day spent here watching raptors stream through and observing the frenzied activity of southbound birds farther east in the pine grove near the old lighthouse (parking field number 2), will certainly convert you also into a barrier beach devotee.

The Line Islands (Map 10.15)

Rating: Spring*, Summer****, Autumn**, Winter*

North Line and Middle and South Line islands are located at the western end of South Oyster Bay in Nassau County. These are, in part, spoil islands formed by deposition of large amounts of sand. These islands are partially inundated at high tide. They have been attractive to resting and preening large shorebirds and feeding "peeps" since sand was dredged over the islands in 1974. However, the Line Islands did not become the object of birders' affection and concentrated attention until July 4, 1975. On that day three of Long Island's crackerjack birders, without benefit of firecrackers and roman candles, discovered a Long-billed Curlew there (A.J. and B. Lauro, and S. Dempsey). Telephones started ringing, the New York Rare Bird Alert was continuously busy and business started picking up at nearby boat liveries. The potential these islands held was at last manifest. The word was out. Their promoters were claiming that the Line Islands, especially North Line, were even more productive than Moriches Inlet, long recognized as Long Island's best place to observe shorebirds.

Today, after six full summers and autumns of shorebirding there, one must admit that these islands host a regular extravaganza from the first week of July through August. They will continue to do so as long as the islands lack vegetation. Additional spoils deposits will act to maintain the islands' character.

The island's ambience includes biting mosquitoes and sand fleas, especially on windless days, wet feet, ditches, sometimes unstable footing and, of course, all of the inconvenience of hitting the wake of a large boat the wrong way. What to do? The answers are quite simple, really. Carry malodorous insect repellent, be

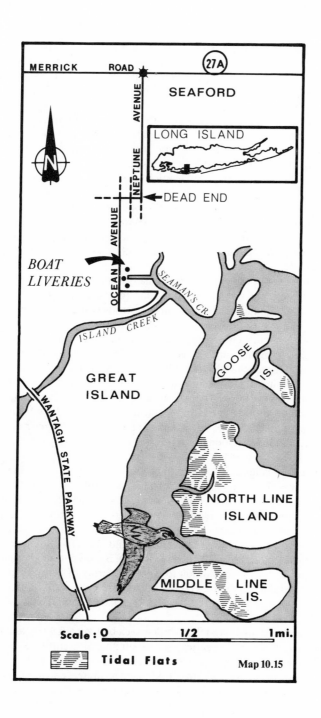

MERRICK   ROAD   27A

SEAFORD

NEPTUNE   AVENUE

LONG ISLAND

←DEAD END

N

BOAT
LIVERIES

OCEAN   AVENUE

SEAMAN'S CR.

ISLAND CREEK

GREAT
ISLAND

GOOSE   IS.

WANTAGH STATE PARKWAY

NORTH LINE
ISLAND

MIDDLE   LINE
IS.

Scale: 0   1/2   1mi.

Tidal Flats

Map 10.15

sure someone in your party has a reasonable knowledge of channel navigation and small boat handling, wear a waterproofed windbreaker on the boat ride, and be prepared to wade in knee-deep when you reach the islands. Further, I would never attempt it without a sun hat. Overlook anything else. Consider these to be "avocational hazards."

To reach the islands, rent a boat with an outboard motor, available from a number of fishing stations in nearby Seaford Harbor. Some of the boat liveries offer birders their "senior citizen's discount" (perhaps with the view that renting a boat for anything other than fishing suggests the advent of senility) which cuts approximately one-third off the regular price for a motor boat capable of carrying five people.

To reach Seaford Harbor, leave the Southern State Parkway at Exit 27 and turn south onto the Wantagh State Parkway. Proceed south and leave this parkway at Exit 6 (Merrick Road). Continue east on Merrick Road for approximately 1.1 miles. At Neptune Avenue, turn right and proceed south to Seaford Harbor.

Consult the tide tables published in many of the local newspapers or call the U.S. Coast Guard for tide information, and time your visit to coincide with a flood-tide. The optimal time to arrive at the islands is approximately three hours before high tide. As the tide recedes, shorebirds disperse to other islands and tidal flats throughout South Oyster Bay. Be careful during ebb-tide that your boat is not grounded. Being high and dry is more trouble than it is worth.

The best approach involves a thirty-minute boat ride from Seaford Harbor south to North Line Island then through a saltmarsh creek at the midwest side of the island that empties into the bay. If the tide is low but still advancing, wait in the neighborhood of Middle Line Island and proceed to North Line Island approximately three hours before the tide is full. If you are especially good at small boat navigation, continue east along the creek for a few hundred yards to some scattered pools in the saltmarsh.

The Line Islands are really exciting July and August, when, in some summers, as many as thirty shorebird species have been observed. Look for American Oystercatcher, Semipalmated, Piping, and Black-bellied plovers, Ruddy Turnstone, Whimbrel, Greater and Lesser yellowlegs, Willet, Spotted and Solitary sandpipers, Red Knot, Pectoral, White-rumped, and Least sandpipers, Dunlin, Short-billed Dowitchers, Stilt, Semipalmated and Western sandpipers, Marbled and Hudsonian godwits, Ruff (watch for Reeve), Sanderling and Wilson's Phalarope. This, of course, is in addition to a dizzying array of herons, egrets, ibises, gulls, terns, and Black Skimmers constantly in view. Look for Gull-billed Tern on South Line Island. Currently this is the only place on Long Island where it has bred.

Rest assured, the Long-billed Curlew seen by Lauro, Lauro, and Dempsey, and hundreds of birders thereafter was not an annual occurrence. In fact, that sighting broke a thirty-nine-year record absence. This species with an incredible beak has not occurred since 1975 (as of 1981) and its presence signaled the first preview of continuing attractions as far as charadriiophiles are concerned.

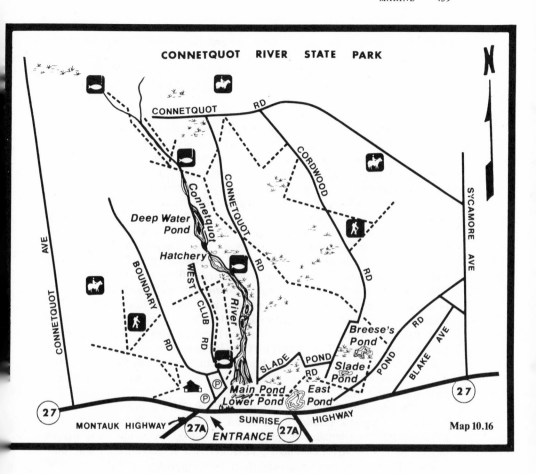

Connetquot River State Park (Map 10.16)

Rating: Spring****, Summer****, Autumn**, Winter*

Connetquot River State Park comprises approximately 3500 acres of extensive, unspoiled original woodland on Long Island. It has large tracts of pitch pine, upland hardwoods, and swampy woodland. It provides nesting habitat for large populations of many formerly common species whose numbers have been drastically reduced throughout the rest of the South Shore of Suffolk County. In addition, some six to ten species of songbirds that nest here regularly are unknown from anywhere else in Nassau and western Suffolk counties. In pristine beauty and wildness this area far outshadows other "wilderness" areas on Long Island. The Connetquot River is truly unpolluted, a statement made not lightly. Its water is drinkable from its headwaters to its estuary. Amazing! Its drainage pattern forms an ecotone, and the abundance of wildlife and flora this transition zone supports is surprising. The Connetquot sweetwater flows for about six miles, and all along it birds sing and nest and raise broods year after year. This is the gift of wildness and purity it has to offer a sometimes jaded, crowded, urbanized Long Island.

In Connetquot River State Park there is such a diversity of habitat that it supports a heterogeneous bird population. Recently recorded nesters in the park include: a large concentration of Wood Ducks, several pairs of Red-tailed Hawks, the only significant population of Ruffed Grouse on Long Island, Bobwhites, Ring-necked Pheasants, strutting Turkeys, American Woodcocks, Screech and Great Horned owls in very healthy numbers, an amazingly high population of Whip-poor-wills, Chimney Swifts, Belted Kingfishers, large numbers of Hairy Woodpeckers and exceptionally high numbers of Downy Woodpeckers, Great Crested Flycatchers, and Acadian Flycatcher (a recently arrived southern species), both Common and Fish crows, several pairs of Brown Creepers, Winter and Carolina wrens, between six and ten pairs of Wood Thrushes, Hermit Thrushes, Veeries and Eastern Bluebirds (which is indeed surprising), Blue-gray Gnatcatchers, and White-eyed Vireos. Warblers nesting at Connetquot include Black-and-white, Blue-winged, and probably the only present-day breeding Northern Parulas on Long Island. Chestnut-sided Warbler is a rare breeder on Long Island but it breeds here. Pine and Prairie warblers are common breeders. This is the only currently known breeding site on Long Island for Canada Warbler (as well as Winter Wren). Orchard Orioles possibly breed here. Red Crossbills rarely breed on the island, but it is part of the Connetquot scene. There are healthy populations of Chipping and Field sparrows here.

The administration of Connetquot River State Park is through the Long Island State Park and Recreation Commission, Belmont Lake, Babylon, New York 11702. (In 1981 the telephone number was 516-669-1000.) A permit is required to bird the park. Write to the commission's Permit Office at this address, stating how you wish to use the park (birding) and further stating the approximate number of people the permit will cover on each visit. This does not pertain to groups. The commission will issue by return mail a three-month use permit. If visiting Long Island without sufficient time to apply for a permit call the Connetquot River State Park Preserve, Sunrise Highway, Oakdale, New York 11769. (In 1981 the telephone number was 516-581-1005.) The preserve personnel is entirely sympathetic toward birders and will arrange to issue a day-use permit which can be picked up at the preserve.

To reach Connetquot State Park, leave I-495 (the Long Island Expressway) at Exit 56 and head south on NY 17 (Wheeler's Road). Continue south as Wheeler's Road merges into Carleton Avenue. Pass through the grounds of Central Islip State Hospital and proceed south to Sunrise Highway. Here turn left (east) and proceed east to the junction of NY 27A (Montauk Highway) and Sunrise Highway (NY 27). Turn left (north) into the entrance of Connetquot. Park near the old main sportsmen's clubhouse on the west side of the dam and main pond.

Suffolk County Airport and Quogue Wildlife Refuge (Map 10.11)

Rating: Spring**, Summer**, Autumn***, Winter***

The Upland Sandpiper is a characteristic bird of broad open grass fields. Acres of half-acre house lots, shopping centers, highways, along with countless adventurous real estate developers have steadily decreased the vast plains of the eastern seaboard, and with them have gone so many of the interesting grassland breeding species. Nowhere is this more true than on Long Island. The Upland Sandpiper never bred in abundance on the island, although formerly it bred at Shinnecock Hills, Montauk, Gardiner's Island, Orient, and of course, a healthy population bred in the all but unrecognizable Hempstead Plains. God bless tract housing! At present, but a few nesting sites of this species remain in the entire state. On Long Island several pairs breed at the John F. Kennedy International Airport, but access to see those birds is impossible. It is possible that one or two pairs nest at Islip MacArthur Airport in Ronkonkoma, and Grumman Peconic River Airfield in Manorville. However, Upland Sandpiper is a confirmed breeder at Suffolk County Airport (Westhampton Airport), and it is seen in numbers every year at Quogue Wildlife Refuge. This is also an important grasslands site for Killdeer, Horned Lark, possibly Bobolink, and Eastern Meadowlark, and clearly for Grasshopper and Vesper sparrows.

A wide range of waterfowl overwinter at Quogue Wildlife Refuge. Many of those breed here also. It is an excellent place to stop on route to or from Montauk Point.

To reach the Suffolk County Airport and Quogue Wildlife Refuge, leave I-495 (Long Island Expressway) at Exit 68 and drive south on NY 46 (William Floyd Parkway) to NY 27. Head east on NY 27 and continue to Exit 63. Here take NY 31 (Old Riverhead Road) south to Stewart Avenue. Turn left (east) and enter the Suffolk County Airport grounds. Permission to walk through the fields may be obtained at a firehouse near the airfield's control tower.

Regain NY 31 and continue south to Old Meeting House Road. Turn left (east) on it and follow it east and north across the railroad tracks to its intersection with Old Quogue-Riverhead Road. Turn left (north) on this road and proceed to the entrance to the refuge.

Quogue Wildlife Refuge conducts an extensive raptor and mammal rehabilitation program at its Nature Center as well as educational programs for adults and children. At times the maze of nature trails appears quite crowded but one can escape that atmosphere by walking farther north to the area of the North Pond. Migrants as well as breeding birds can be viewed in season.

Moriches Bay and Inlet (Map 10.11)

Rating: Spring**, Summer****, Autumn****, Winter***

Two islands called Bird Island and Inlet Island by those who bird them most frequently (and collectively "Bird Islands"), located near the Moriches Bay Inlet are excellent for shorebirding in August and September. These islands are accessible by boat only.

Although American Oystercatcher, Piping Plover, Great Black-backed and Herring gulls, Common, Roseate, and Least terns, Black Skimmer, and Sharp-tailed and Seaside sparrows nest on these islands, they are more noted for the great numbers of migrant shorebirds they attract.

One should arrange to go to the islands at low tide when extensive mud flats are uncovered and when, in August, birders should expect to see one or two Common Loons that have summered in the bay or are early migrants, Green Herons, which nest in the periphery of Moriches Bay, Snowy Egrets with young, Clapper and Virginia rails, American Oystercatchers, Semipalmated, Piping, and Black-bellied plovers, Ruddy Turnstones, Whimbrels, Greater and Lesser yellowlegs, Spotted Sandpipers, Willets, Red Knot, Pectoral, and Least sandpipers, Short-billed Dowitchers, Sanderlings, Hudsonian and Marbled Godwits, two or three species of gulls, Common, Roseate, and Least terns, Black Skimmers, and various marsh and wet-grass associated sparrow species, including Sharp-tailed and Sea-side sparrows.

Take lunch, something cool to drink, an old pair of sneakers for sloshing through the mud flats, a windbreaker, and insect repellent.

Leave I-495 (the Long Island Expressway) at Exit 68. Gain the William Floyd Parkway south. Proceed south to NY 27A (Montauk Highway). Turn east on NY 27A and continue to Atlantic Avenue in the village of East Moriches. Turn right (south) on Atlantic Avenue and follow it south to its terminus and the fishing station there. Boats are available here for a small fee. The boat livery tows the boats, capable of carrying three to five persons, out to the islands and then later will tow the boats back to the livery. The ride out and back takes about thirty minutes each way across the shallow waters of the bay. The livery provides life vests.

Moriches Bay is wide but relatively shallow. The flats surrounding the Bird Islands can also be reached on foot by driving farther east on NY 27A to its inter-section with Mill Road in Westhampton. Here turn right (south) on Mill Road (NY 71) and follow it to the barrier beach and Dune Road. Turn right (west) and drive to the west end of Dune Road. Walk out to Cupsogue Beach County Park. There is no parking allowed anywhere along Dune Road from May 15 to September 15. During this period there is a parking fee charged at this county park for county nonresidents. From the west end of the park the Moriches Inlet is visible and from its north projection one can walk out to the extensive mud flats at low tide. This is wet, sloshy work but well worth the effort to see the numbers and variety of shore-birds present.

These vast flats consistently remain the best place to see migrant shorebirds on Long Island in Nassau and Suffolk counties. Although the Line Islands offer a wonderful opportunity to see shorebirds of many species, if these spoil sites are al-lowed to grow over with vegetation their attraction for shorebirds will plummet.

Premium birds one may see here are Gull-billed, Royal, or Caspian terns. After storms White-tailed Tropicbird, Magnificent Frigatebird, and Sooty Tern have occurred in Moriches Bay.

Shinnecock Bay and Inlet (Maps 10.11 and 10.12)

Rating: Spring**, Summer***, Autumn***, Winter***

The late LeRoy Wilcox, who in the early stages of this book contributed generous assistance and helpful advice, operated a banding station on Tiana Beach about 0.5 mile west of Shinnecock Inlet from September 1958. In its first seven years of operation, he had banded more than 26,000 birds in autumn at that site. Wilcox expressed the view that the best flight days here were under conditions of clear skies, after passage of a cold front, and when there are either light winds from the north or no winds at all. Take his advice and bird this area under such conditions. Remember that Shinnecock Inlet is the premier Long Island site at which to see alcids and pelagic species after November storms.

When there are strong, constant southerly winds, primarily pelagic species can sometimes be seen at Shinnecock Inlet. In recent years in such winds a Greater Shearwater was seen patrolling the surf at the inlet, four Parasitic Jaegers were present for some time, and a Bridled Tern (storm-blown) occurred in 1979. Gannets are not uncommonly seen from the inlet. Herons and egrets are quite regular and abundant. The Peregrine Falcon and other falcons and Accipiters are regularly seen on their southward migrations. American Oystercatcher breeds here. Marbled Godwit and Northern Phalarope occur here nearly every autumn. Irregular summer wanderers or regular fall transients include Laughing and Bonaparte's gulls, and Forster's, Royal, Caspian, and Black terns. Some very rare occurrences of autumn passerines include Sage Thrasher and Bell's Vireo. Lark and Clay-colored sparrows have been recorded here casually.

To reach Shinnecock Inlet and Bay, turn south on Canoe Place Road from NY 27A (Montauk Highway) just east of Hampton Bays and just west of the Shinnecock Canal. Continue south as Canoe Place Road merges into Lynn Avenue. At Shinnecock Avenue turn right and then immediately left onto Light House Road. Continue south across Ponquogue Bridge to Dune Road (Beach Road).

The Shinnecock Inlet and jetties are reached by turning left (east) on Dune Road. Taking a right (west) on Dune Road will give access to Tiana and Westhampton beaches and eventually Moriches Bay.

Dune Road is especially productive in winter. Loons, grebes, and an occasional Harlequin Duck can be seen in the bay from the jetty. Purple Sandpiper is sometimes seen on the ocean side of Dune Road. Glaucous Gulls occur here many winters. Drive west along Dune Road in winter searching the sea for Iceland Gulls and the beaches for Snowy or Short-eared owls. During Snowy Owl flight years this is an almost guaranteed place to find one or two.

THE SOUTH FORK
Mecox Bay Area, a Tour (Map 10.12)

Rating: Spring**, Summer***, Autumn****, Winter**

Take NY 27A (Montauk Highway) east past Shinnecock and Great Peconic Bays to Southampton. At Halsey Neck Lane turn right (south). This road gives access to Meadow Lane, which if driven west, ends at the east jetty of Shinnecock Inlet. Proceed south on Halsey Neck Lane to Meadow Lane. Be on the lookout for shorebirds in the fields while driving south. Turn left (east). At the next road, Cooper Neck Lane, turn left. Halsey Neck Pond will now be on the left (west) side of the road. Look for migrant ducks in the pond. If it is past autumn there may be overwintering waterfowl in the pond. Proceed north to NY 27A. Continue driving east on NY 27A past Cobb and Water Mill to Newlight Lane. Turn right (south). Proceed south slowly, looking in the fields to the right and left for various shorebirds. At Halsey Lane turn right, which gives access to overviews of additional shorebird habitat. At Mecox Road turn left and then take a right (south) at Job's Lane. Drive a short distance to Dune Road and the eastern end of Mecox Bay. Much of the birdlife of Mecox Bay is the same as Shinnecock Bay, owing to its close proximity. However, this is really not a bay at all; it is more like an ocean-fronting pond and occasionally Mecox holds a surprise. In late winter look there for a Red-necked Grebe, and around its shores in spring for a Cattle Egret. In late autumn and winter the Whistling Swan may be there, as might the Snow Goose (and often the blue form). The European Wigeon has often been recorded in Mecox Bay. In autumn watch Sharp-shinned and Cooper's hawks and, perhaps, a Peregrine Falcon streak by. Baird's Sandpiper and Black-necked Stilt have been recorded here. Look along Dune Road here in winter for Snowy Owl. Large migrating flocks of Tree, Bank, Barn, and Cliff swallows can be seen from here. In autumn Loggerhead Shrikes and Yellow-headed Blackbirds, along with Savannah (Ipswich) Sparrows occur here. Mecox Bay freezes over in December or January every year, but as long as it is open bird it.

Proceed back up Job's Lane and at Mecox Road turn right (east) and then left on Ocean Road. This gives another view of fields possibly holding migrant American Golden and Black-bellied plovers, and Baird's or Buff-breasted sandpipers. East and west are the flat, fertile, silty farmlands that seem to be rapidly giving way to subdivisions. In this famous Bridgehampton loam shorebirds probe. At Halsey Lane, turn right and find a place to walk left to Kellis Pond. This is a deep kettle pond, set down in its hollow almost out of sight. Black Ducks, Green-winged Teals, and occasionally a European Wigeon can be found.

Continue north to NY 27A. Turn right (east) and proceed to Sagaponack and Town Line Road. Turn right (south) and again slowly drive down the road, looking right and left in the farm fields. Sagaponack and Wainscott are still relatively small farming communities with old houses, large fields, and too obviously newer houses. These farm fields are often exceedingly productive of shorebirds. Look for Baird's and Buff-breasted sandpipers. At Wainscott Main Road turn left (east) and after birding Wainscott Pond on the right (south), continue to the western edge of Georgica Pond. This is no more than a large low-lying area blocked, over time, by sandbars. It catches rainwater, is partly brook-fed, and sometimes in a heavy storm, the surf rides up and over into it. While it is open, various waterfowl species can be found on it and often shorebirds feed along its shores.

Turn north on Wainscott Road and once again regain NY 27A. Turn right (east) and drive as far as Hook Pond in East Hampton. A visit here in late September or early October may turn up one or two Soras, Pectoral and White-rumped sandpipers, and Dunlins. Whistling Swans can be found here in winter when Hook Pond is ice-free.

The South Fork, Montauk Peninsula (Map 10.17)

Rating: Spring**, Summer**, Autumn****, Winter****

The more southern of Long Island's glacial moraines, the Ronkonkoma Moraine, meets the sea at Montauk Point. Centuries of wave action have eroded the sandy face of the moraine and carried the debris to the west, heaping it in long strips of sand and segregating lagoons *e.g.,* Shinnecock, Moriches, and Great South bays, from the ocean.

In *Fish-Shape Paumanok,* a natural history of the island, Robert Cushman Murphy states that white cedar, white pine, oak, hickory, and isolated stands of red maple, red cedar, red spruce, and holly covered the rolling hills of the Montauk peninsula before its settlement, just as pitch pine did the barrens and grass species did the plains. However, the trees of the South Fork were so marketable that by the late 1600s, villages established there were passing municipal legislation restricting the harvesting and selling of timber. We have all heard stories of how South Shore farmers drove their cattle into glacial holes to hide them from the beef-hungry British Navy patrolling the shoreline during the American Revolution. The denuding of the South Shore and Fork did not take very long. However, today Montauk peninsula is essentially unchanged with moorland and many large forests and dense stands of wind-sheared oaks adding to the long list of features that make this peninsula one of the finest birding locations in the state. The low pines, stunted oaks, and dense brushy thickets are often found thriving in hollows, valleys and kettles and the birds they harbor are relatively easily seen. Sometimes the impenetrable matrices of bayberry, blackberry, shadbush, and wild rose shelter overwintering migrants and half-hardy landbirds. The pine groves shelter owls and siskins, crossbills, finches, and kinglets. Snowy Owls take up winter residence along the dunes and Northern Harriers and Short-eared Owls can be found coursing the low-lying marshes. The hardwood forests provide refuge for small numbers of rare birds annually, to the delight of birders. The tree-bald moors make viewing hunting raptors, *e.g.,* Red-tailed and Rough-legged hawks, American Kestrel, and Northern Shrike effortless. Horned Larks, Water Pipits, Lapland Longspurs, and Snow Buntings occur on dune and field edges.

Montauk Point is a place where the land, the sea, and the birds are closely allied. From the bluffs near the lighthouse commissioned in 1795 by President George Washington, one can scope the ocean and find tens of thousands of rafting scoters of all three species. One can also see loons, grebes, Gannets, Great and rarely Double-crested cormorants, Common Goldeneyes, Oldsquaws, irregularly Harlequin Ducks, Common and King eiders, and thousands of Red-breasted Mergansers from the lighthouse point and Shagwong Point. From here it is often possible to see any alcids present.

The famous Montauk Point Christmas Bird Count, having completed in

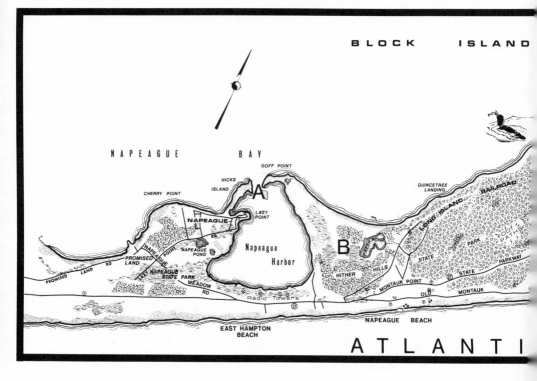

1980 its fiftieth year and having ranked top in number of species seen throughout New York State from 1977–1980, regularly tallies Glaucous, Iceland, Ring-billed, Laughing and Bonaparte's gulls, and Black-legged Kittiwake. Casually recorded are Lesser Black-backed, Black-headed, and Little gulls. Razorbills and Dovekies are often seen on the count, and Common and Thick-billed murres and Black Guillemots have rarely occurred.

From west to east places that afford prime birding in late autumn and winter on the Montauk peninsula include: (a) Napeague Harbor and Goff Point seen from Lazy Point Road, (b) the 1800-acre Hither Hills State Park, (c) the Montauk village dump, (d) Montauk village and Fort Pond, (e) Lake Montauk and its jetties at the north ends of West and East Lake Drives, (f) Ditch Plains, (g) Little Reed Pond, (h) Big Reed Pond, (i) Shagwong Point, (j) Oyster Pond, and (k) Montauk Point Lighthouse and environs. The last six areas are within Montauk County and Montauk State parks. The accompanying map shows the location and access roads to these lettered sites.

Check the areas above indicated as (a), (g), (h), and (j) for a variety of duck species and water-associated birds. Areas lettered (b), (c), (d), and (f) should be in-

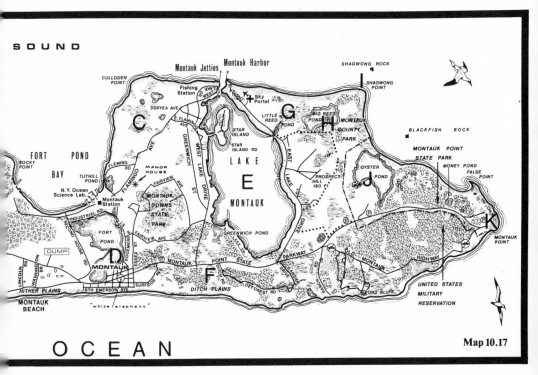

Map 10.17

tensively birded for owls, sandpipers, woodpeckers, chickadees, nuthatches, creeper, wrens, mimids, thrushes, warblers, blackbirds, finches, and sparrows. The Montauk village dump, reached on a road about two miles west of the village and on the north side of the Montauk Highway, attracts a myriad of seed-eating birds, especially in the vicinity of the sewage disposal area at the dump's lower end. Blue Grosbeaks are regular here in late September and early October. For loons, cormorants, sea ducks, gulls, alcids, and Purple Sandpiper, Dunlin and Sanderling bird areas (e), (i), and (k).

ACKNOWLEDGMENTS

Many thanks to the following people who provided information on the sites in the Marine Region: Robert S. Arbib, Jr., Sid Bahrt, Alison Beall, Thomas Burke, Thomas H. Davis, Matthew Paul and Matt Drennan, Stephen B. Dempsey, John Farrand, Jr., Jack Foehrenbach, Anthony J. Lauro, Manny Levine, Sheila Madden, Helen McClure, Roger S. Payne, Robert O. Paxton, James Romansky, Mr. &

Mrs. William Siebenheller, Richard A. Sloss, Martin Sohmer, Barbara J. Spencer, Guy A. Tudor, the late LeRoy Wilcox. I am also grateful to Robert S. Arbib, Jr., Thomas H. Davis, Anthony J. Lauro, and Roger S. Payne for their careful reading of and comments on this chapter.

Seabirds and Pelagic Birding

HISTORICALLY, SEABIRDS HAVE PROVIDED COMMERCIAL BENEFITS to human societies in many nations—from egg gathering and bird hunting. Eggs, guano, down, plumes, and meat are the major products seabirds continue to supply today. In general, the unregulated consumption of eggs and birds of decades ago is no longer the rule, and most nations prevent over-harvesting by providing at least some protection to their seabird resources, based on the management concept of maximum sustainable yield.

Birders increasingly expend considerable sums of money to reach seabird viewing areas off the Northeastern coast of the United States, as well as off the Southern, Gulf, and West coasts. This nonconsumptive exploitation of seabirds contributes significantly to regional economies and provides substantial benefits to local businesses. The esthetic appreciation of the birder and the scientific study of the ornithologist demonstrate the broad spectrum of growing awareness and interest in seabirds. More importantly, they indicate a critical need for a strong and effective governmental seabird conservation program, whose innovations and enforcement will ensure long-term social and economic benefits for future generations.

In the same way that the diversity (or richness) of terrestrial birdlife can be used as a measure of the environmental constraints on a particular community, the stability of the complex structure of various ocean communities (habitats) can be measured by the variety of their avifauna.

Birders can be of incalculable value in helping to determine the quality of oceanic habitats by identifying regions where there tend to be greater-than-average congregations of seabirds and by keeping several kinds of records. Accurate numbers (or even close estimates), of species identified—with notations regarding adult or immature status are useful. So are accurate notes on location, either the latitude and longitude or just distance, if within sight of land; time and date, weather conditions, or behaviors. Also of fundamental value are careful notes on the presence of other animals—small fish, whales, porpoises, etc.—and the physical conditions of the surface water (if possible)—temperature, clarity, color, etc. Ascertaining these parameters may seem too ambitious for a birder simply out to enjoy a pelagic jaunt; but, only when there are enough observations at various locations, at different times of the year, will it be possible to determine ecological data and migration routes with any confidence. Certainly the general major paths taken by the majority of seabird species in the northwestern Atlantic are known; but a great deal remains to be learned regarding the timing of those movements and the detailed routes of particular species.

Oceanographic properties affect the distribution and local abundance of birds at sea. The two major water conditions governing seabird distribution are the mixing of surface water layers, and the density of available food. Where the proper conditions exist, birds are usually present.

The mixing of surface water layers refers to the fact that open ocean surface layers vary in temperature with depth. These variations result in large, discrete water masses, which have a relatively uniform internal temperature and chemical composition. The area where two such discrete masses meet is referred to as a "convergence." The locations of convergences are not fixed, but shift in a north-south direction seasonally. In addition, currents bearing warm or cold water alter the positions of surface-water layers, and like the layers, currents may vary in position or velocity seasonally. For example, the Gulf Stream, a current caused by prevailing winds, carries subtropical warm water north, essentially following the Continental Shelf along the east coast of the United States. Off Newfoundland, the Gulf Stream converges with the cold Labrador current from the north, and from there the Gulf Stream generally veers east and continues to Europe. This current and its convergence have considerable influence on climate and are, for instance, responsible for the dense fogs off Newfoundland. The eddies spinning off from the Gulf Stream often dominate the waters between the stream and inshore southward-flowing coastal currents. In this eddy zone the climate is often cold, humid, and continental, with most precipitation in the form of rain. Snowfall is common only in the more northerly areas. In general, precipitation is moderately abundant east of Long Island, abundant around Long Island, and moderately abundant south of the New York bight. Tropical storms of hurricane force strike this zone occasionally, but the intensity of rainfall is only moderate. The effect of all current patterns in the area is to drift water offshore, bringing rising, deeper water inshore to replace it. Because nitrogen and other minerals have no natural cycles, except where currents return them to the surface, this sort of vertical mixing, in large part, regulates the nutrient level of the waters, and consequently sets the stage for abundant plant and animal resources.

Each seabird species associates with one or more characteristic water masses. Some seabirds may breed in the area of one surface water mass and then, following the breeding season, undertake a migration to another water mass to feed.

Food seems to set the upper limit on animal populations; however, one must keep in mind that it is not *the* most important one or the only one, in the natural world. The density of available food is related to the concentration of nutrient salts dissolved in the water. These concentrations, like water masses and convergences, are not evenly distributed throughout the Atlantic, but are dependent on seawater circulation.

Deep, cold-water layers are richer in nutrients than those closer to the surface. Through the vertical mixing already referred to, those deep waters can be brought to the surface, and thereby supply rich nutrients utilized by the plants on which the prey of the birds depend. The transport to the surface of nutrient-rich, deep water is called "upwelling." Upwelling of nutrients occurs at convergences, where winds and currents drift surface waters away from the coast, and where cold, deep currents travel upward along the inclined slopes of coasts or islands. Some of the richest plant and bird communities occur in areas of vertically mixed upwelled water. Where cold, nutrient-rich, less saline upwelling water travels along the Continental Slope and converges with the relatively warmer, more highly saline waters from the Gulf Stream, a mix is produced that is highly attractive to seabirds as well as to marine mammals. This is why pelagic birding vessels head for known areas rich in upwellings. In suitable weather their captains try to make it to the edge of the Continental Shelf. Vertical mixing of rich water layers also occurs where there is a good deal of turbulence, especially near sea banks and small islands. Lack of turbulence results in minimal nutrient renewal.

Nutrients also feed into the ocean by land runoff from river systems and their tributaries. Peak runoff in the New York bight occurs during March, April, and May, and usually ranges between ten and twenty inches yearly. In this runoff, sediment loads are usually low but dissolved solids are higher around Long Island and the New York bight. This land drainage enriches estuaries, and the freshwater that brought it causes a variable salinity level in the sea, extending far beyond the limits of the estuarine zones along the East Coast.

Owing to competition between species for available food, similar bird species rarely occupy similar niches, but rather displace each other in such a manner that each takes possession of particular kinds of foods and employs its own special modes of feeding, giving it an advantage over its competitors. One interesting niche is easy to see on a pelagic birding trip. Some seabirds, especially terns and shearwaters, are very often associated with schools of fish. The birds feed on small fish that are chased upward by schooling predatory fish, which drive the small fish against the surface to catch them. Steady streams of birds at sea, all rapidly flying low over the water's surface, and all flying in the same direction (not quartering back and forth), usually signal the direction of an otherwise invisible feeding school of fish over which large flocks will be hovering and diving. Commercial fishermen have known of this association for countless generations, and often exploit it to catch the schooling predatory fish.

Another major consideration governing seabird distribution is species habitat preference. Often the most important consideration is the fact that most pelagic birds are dependent upon a land base for reproduction. These birds subsist entirely upon animal organisms taken from saltwater, and require only the ocean as a resting surface. These species commonly have no occasion to approach within sight of land during the greater part of their lives. There are some seabirds which are not quite so independent of land, as they either feed in shallow coastal waters or roost on islands or coasts. Thus the proximity of suitable breeding sites often determines the concentration of seabirds, even when local water conditions or available food are minimal.

The accompanying map should orient pelagic birders to their oceanic sur-

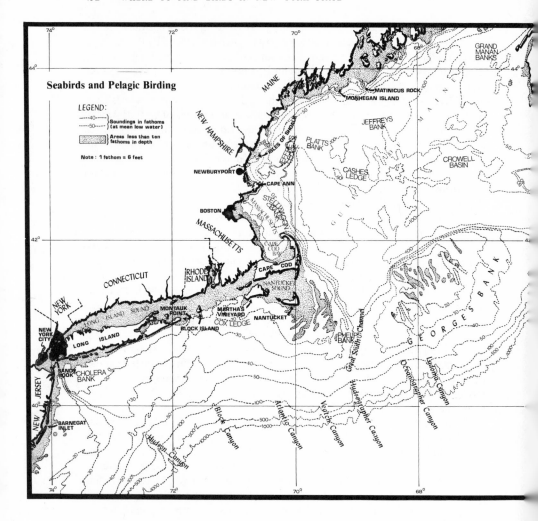

Seabirds and Pelagic Birding

LEGEND:

----40---- } Soundings in fathoms
----50---- } (at mean low water)

▨ } Areas less than ten
   } fathoms in depth

Note : 1 fathom = 6 feet

roundings. Consideration of the following terms may help in understanding the map.

Littoral Zone — the biogeographic area, including the intertidal regions, on a seacoast between high and low waterlines, or the marginal part of a body of water affected by waves and coastal currents.

Littoral Current — those flowing water masses moving along and roughly parallel to a shoreline.

Neritic Zone — that belt or region of shallow water adjoining the seacoast and usually considered to extend from the low-tide mark to a depth of 100 fathoms.

Pelagic (Oceanic) Zone — the biogeographic area consisting of the open sea, especially beyond the outer limits of the littoral zone, but within the bathypelagic zone, to which light penetrates.

Bathypelagic Zone — the deeper parts of the ocean, specifically, from the slope of the Continental Shelf at 100 fathoms to the upper limit of the abyssal zone at 1000 fathoms.

Abyssal Zone — the biogeographic area of very deep sea beginning at a depth of 1000 fathoms and continuing downward; this zone lacks higher plant life owing to the absence of light, and is chiefly occupied by animals (only a few of which are blind), many having special luminous organs, that are especially adapted to the great pressures of this level.

The seabirds commonly encountered from pelagic boats and New York coastal shores can be roughly categorized (with some overlap and exceptions, of course) according to their degree of dependence on proximity to land.

The Littoral Zone: cormorants, many gull species, many tern species, skimmers.

The Neritic Zone: shearwaters, storm-petrels, Gannets, alcids, frigatebirds, most black-capped terns, some phalaropes, some gulls.

The Pelagic Zone: albatrosses, fulmars, storm-petrels, tropicbirds, skuas, jaegers, kittiwakes, and other pelagic gulls, noddies, pelagic phalaropes.

The seabirds that pelagic enthusiasts might expect to see on those trips slated to travel to ninety miles or more southwest of Montauk Point to Block Canyon are varied and always exciting. The deeper waters off the Continental Shelf are also attractive to a host of marine mammals, including, Finback, Pilot, Humpback, and Minke whales, Harbor Porpoise, White-sided, Saddleback, and Bottlenose dolphins, or even a Right Whale or several Harbor or Gray seals. There is also the possibility of sighting a sea turtle, usually a leatherback, probably feeding on the large and abundant pelagic jellyfish.

New York state birders had been arranging passage on boats, on an irregular basis, long before the 1970s. The Federation of New York State Bird Clubs began sponsoring chartered pelagic trips to Cox's Ledge and points south and east of Montauk in 1974. Cox's Ledge, as can be seen on the map, lies about twenty-five miles east-southeast of Block Island, Rhode Island, or nearly forty miles east of Montauk Point. Its fame has been earned as a productive sport-fishing site. Birders came to know of it as a relatively close pelagic birding spot in the mid-1960s and, as of 1981, upwards of seventy trips had been conducted there during the months of May through October from Montauk Point. A cumulative total of more than 120 species had been recorded on these excursions. About 85 percent of those species seen, were disoriented migrant passerine species, apparently lost at sea, judging from their flight direction. They occur more frequently in the fall than in spring. A few more constituting the total were raptors, especially falcons, obviously not disoriented, but using the island-hopping and shoreline route in their migration southward.

Of the cumulative total of birds viewed from vessels headed offshore let us consider, in the first place, what are typical New York pelagic species. Such a species is customarily distinguished by being seen out of sight of land and, as a rule, on or near offshore fishing grounds, or close to the edge of the Continental Shelf. Using this loose definition, we automatically exclude any wind-drifted landbirds and those gull and tern species whose primary feeding and nesting niche falls within the littoral zone. The principal New York pelagic species are thus the Diomedeidae (albatrosses), Procellariidae (fulmars and shearwaters), Hydrobatidae (storm-petrels), Sulidae (gannets and boobies), Phalaropodidae (phalaropes), Stercorariidae (skuas and jaegers), and Laridae (gulls and terns). The majority of these families has at least one or two representatives that can be observed on pelagic trips out of New York harbors.

I never cease to be dumbfounded by the impossible things birds do. It is hard to imagine that in the area of Cox's Ledge, one might see a Yellow-nosed Albatross (*Diomedea chlororhynchos*), coasting, dipping, banking, and effortlessly gliding. However, that very mollymauk has been recorded at the ledge at least twice; and to further indicate the vagaries of ocean wandering, it has additionally been well-seen and photographed about two miles off Jones Beach and reliably identified at the unlikely site of Croton Point Park, located nearly thirty miles up the Hudson River! Its occurrence in New York waters is one of the marvelous unpredictables sea-going birders pray for. To attempt a generalization concerning its anomalous whereabouts would be absurd.

Thomas H. Davis, Jr., kept records of those birds sighted on each of the trips leaving Montauk Point for Cox's Ledge between 1966 and 1980 and, in summary, they indicate the following trends:

Numbers of Northern Fulmars (*Fulmarus glacialis*) seem to appear briefly in spring from late May through the end of the first week of June. They do not appear again during the summer, but peak in mid-to-late September and decline shortly thereafter. Recent seabird research, on Georges Bank, shows that Northern Fulmars are present from January to June in the thousands. This species breeds in the northern reaches of the Northern Hemisphere and has traditionally been considered an accidental straggler to southern New England waters. Perhaps the surveys in the Georges Bank area evidence a recent and substantial range extension or population increase. The advent of Memorial Day usually signals the summer arrival at Cox's Ledge of Sooty Shearwater (*Puffinus griseus*), Greater Shearwater (*Puffinus gravis*), and Wilson's Storm-Petrel (*Oceanites oceanicus*), from the Southern Hemisphere. Sooty Shearwaters peak in June and the other two species appear to reach their peak numbers from mid-July to late August. During July and all of August, Cory's Shearwater (*Puffinus diomedea*), commonly occurs in the warmer offshore waters east-southeast of Montauk. Their numbers decline considerably in September and October.

From late May through September, Manx Shearwater (*Puffinus puffinus*), occurs in consistently low numbers, flying low over the water's surface, alternately rising and then gliding, and banking low between waves, especially in offshore areas of cold-water upwelling. This species breeds in the northwestern North Atlantic, with a considerable colony discovered in Newfoundland as recently as 1977.

The very rarely seen Audubon's Shearwater (*Puffinus lherminieri*), has been recorded twice from Cox's Ledge and both occurrences were in mid-August. Trips going out at that time strive for this small black-and-white shearwater; although correctness of identification of this species depends upon thorough familiarity with all of the morphological and behavioral differences between it and two similar species, Manx and Little shearwaters. Calling an Audubon's Shearwater with assurance is directly related to the amount of time one has spent in the field. Mastering the most subtle distinctions between pelagic species closely resembling each other is one of the special pleasures of taking repeated pelagic trips.

Storm-petrels are the smallest of the tube-nosed ocean rovers, and although there are well-established, island-breeding colonies of Leach's Storm-Petrel (*Oceanodroma leucorhoa*) as close as off the coast of Maine, this species has been recorded by Tom Davis on only two trips to Cox's Ledge up to 1981. Late August and mid-September were the periods of occurrence. There are, however, numerous specimen records of this species found inland and upstate after severe storms, some of these as late in the season as the end of November. Again, separating Leach's from Wilson's Storm-Petrel calls for repeated observation, extensive field experience and actual comparative study of the two species in the field. After Wilson's Storm-Petrel peaks in late August, its numbers dramatically decline, and it has only rarely been recorded in New York waters after September. Wilson's Storm-Petrel is also apt to be deflected toward the shore by equinoctial storms, and there are several specimen records of this species from the coast, and even a few from inland New York.

The most likely time to see Gannet (*Morus bassanus*) is during the month of May or in mid-to-late October. It is seldom encountered at Cox's Ledge between the end of the first week in June and early October. This is a species about whose migratory pathways much remains to be learned. It has breeding colonies in Newfoundland and the Gulf of St. Lawrence. Gannets leave the northwestern Atlantic to winter along the Continental Shelf from Georges Bank to Cape Hatteras, but their precise migration routes in both spring and late autumn, remain undelineated.

Northern (*Lobipes lobatus*) and Red (*Phalaropus fulicarius*) phalaropes, the only pelagic phalaropes, can be nearly guaranteed in small numbers during the last ten days of May on their northbound migrations. Both species are scarce throughout the rest of the summer. Their numbers start to build again toward the end of August. Both Northerns and Reds are encountered singly or in small groups of from two to about twenty throughout September and October. They appear to have a relatively protracted autumn migration period. In the productive areas of upwelling, north and east of Cox's Ledge, prodigious concentrations of Northern Phalaropes, numbering in the millions, have been recorded in late August and early September. Unfortunately, boats leaving New York harbors have not witnessed this spectacle.

Great Skua (*Catharacta skua*) and South Polar Skua (*Catharacta maccormicki*) have both been recorded, by extremely reliable observers, in the waters of the northwestern Atlantic. There are also specimen records of each of the species taken from the same waters. This account makes no attempt whatsoever to provide identification notes on these or other pelagic birds. Skuas were recorded on only eight of the Federation sponsored trips to Cox's Ledge by 1981, and five of these trips were in either late May or June. The skuas studied on these trips had plumage characters consistent with published field descriptions of South Polar Skua, and therefore were so specified. For definitive identification characters for skuas (and other seabirds), refer to an excellent article by the seafaring Davis Finch *et al.,* "Pelagic Birds in the Gulf of Maine," *American Birds* 32:140–55, 281–94, and to another superior article by Pierre Devillers, "The Skuas of the North American Pacific Coast," *Auk* 94:417–29.

Pomarine (*Stercorarius pomarinus*), Parasitic (*Stercorarius parasiticus*), and Long-tailed (*Stercorarius longicaudus*) jaegers have all been recorded on the Cox's Ledge trips. The Pomarine is the largest, the Parasitic intermediate, and the Long-tailed the smallest of these three jaegers. All of them breed in the high arctic and winter in areas exclusively oceanic, usually in rich tropical waters of the Southern Hemisphere where upwellings attract pelagic birds of many species. Jaegers (and skuas) are famous for their spectacular aerial pursuits of gulls and terns, parasitizing them by forcing them to disgorge the food that they are carrying in their bills. The jaeger then swoops down on the forfeited booty, catching it for itself. These commandeering flights of extortion often afford the birder some of the most wonderful aerobatic displays to be seen in the bird world. The Davis data contain only one record of a single Long-tailed Jaeger, occurring September 1, 1968. However, pelagic vessels chartered by birders from Rhode Island have recorded upwards of ten Long-taileds in the area of Cox's Lodge, all occurring in the first two weeks of June. Both Pomarine and Parasitic jaegers have been recorded at Cox's Ledge in late May and early June. At this time, Pomarines are more abundant than Parasitics. Jaegers encountered from mid-June through the first week of August may possibly be non-breeding, sub-adult birds, summering in areas of upwelling south of the breeding grounds. By mid-August both species are actively migrating southward. The migration of Pomarine Jaegers appears to be more protracted than that of Parasitic, and at this season, as in spring, Pomarines are considerably more plentiful than Parasitics. Pomarine Jaegers continue to be recorded throughout October (and probably occur into November), while Parasitics are sparsely seen until the beginning of October.

Many gull species, in flocks of varying size, have been observed in the littoral and neritic zones. Only two of these, however, fit our loose definition, thereby qualifying as pelagic gulls: Black-legged Kittiwake (*Rissa tridactyla*), and Sabine's Gull (*Xema sabini*). Neither species can be expected on spring pelagic trips to Cox's Ledge.

Black-legged Kittiwakes breed commonly on suitable cliffs, in amazingly dense colonies, on the coast of Newfoundland and in the northern Gulf of St. Lawrence, as well as other high and low arctic locations. After the breeding season its dispersal is so wide that the limits of its wintering range(s) are presently unmapped. It does, however, appear to winter well within the pelagic zone. Great concentrations of kittiwakes have been surveyed wintering off southeast Labrador, and there is some evidence indicating that Georges Bank may possibly be one limit for significant concentrations of wintering kittiwakes. The species appears at Cox's Ledge singly or in pairs sporadically from the last week of August through the first weeks of October. From mid-October onward this species' status is regular and common at the ledge.

Sabine's Gull only (if that word has any meaning whatsoever when discussing birds) occurs at Cox's Ledge in the fall, and then rarely. The most likely time to see it would be within the final twenty days of September, although it is clearly a species whose appearance is erratic at best. This small gull breeds in small numbers in northern Siberia, and in greater abundance in Alaska, Greenland, and arctic Canada. Its wintering grounds remain somewhat mysterious, but there is evidence that the Greenland and Canadian arctic populations winter off the coast of southwest Africa. Presumably, its winter distribution is well out to sea. The interesting question is whether this birds' migratory route is direct and oceanic to southwest Africa or overland from the Arctic to the Atlantic and then south.

The Arctic Tern (*Sterna paradisaea*) is truly the only pelagic tern. Its globe-trotting travels take it from its arctic and north temperate coastal and offshore island nesting sites, to its wintering grounds off the Antarctic pack-ice. By flying almost the entire length of the world twice each year, it spends more of its life in daylight than any other species. Arctic Tern is seldom seen south of Long Island along the Atlantic seaboard, but records of Tom Davis indicate that three individuals of this species were seen together at Cox's Ledge on one occasion in late August.

Pelagic trips sponsored by the Federation of New York State Bird Clubs have been scheduled at the height of the projected spring and fall offshore movements, as well as several dates that would ideally overlap late summer and early winter migrations. In addition to those headed for Cox's Ledge, the Federation has arranged trips to Block Canyon, leaving from Montauk at approximately midnight and arriving near the Continental Shelf shortly after dawn, returning to Montauk that night. The Federation has also arranged pelagic trips to Hudson Canyon, leaving from Freeport, Long Island, at about 2:00 a.m. and returning that evening. Hudson Canyon is directly south of Long Island, and about one-hundred miles offshore.

The power boats chartered for these trips are trim, sturdy, and seaworthy. They can comfortably accommodate from forty to sixty participants, and are equipped with numerous cushioned bunks and chairs. The captains of these vessels are experienced navigators and exercise prudence and good judgment in questionable weather. The cost of each trip in 1981 ranged from $30 to $40 for Federation members and $35 to $45 for nonmembers.

For additional information concerning Federation-sponsored pelagic trips, contact either the *Kingbird* editor or any of the first five institutions listed on pages 15-17.

In addition to scheduled pelagic trips, some birders and bird clubs arrange passage on sports-fishing boats departing from Sheepshead Bay; Brooklyn; Freeport, Long Island; and Montauk, Long Island. The disadvantages of such trips are that the birder is never assured of going out to the Continental Shelf, these boats do not "chum" to attract birds nor slow down to observe them, and the birder is rarely charged a cut-rate but rather pays the higher rate charged fishermen.

Hawk Migration

THE TWICE-YEARLY SPECTACLES OF THE BIRDS-OF-PREY MIGRATIONS are so aesthetically pleasing and so fascinating in the unsolved migration problems they present that they attract field observers from all along the bird watcher continuum. Raptors are unquestionably complex and certainly impressive as flying machines.

Once again the New York State birder is in an enviable position, for the state has excellent places at which to watch hawk movements in both spring and autumn. Raptors concentrate along certain topographical features, which also afford easy viewing during both migrations. In general, most sites are not of equal interest in both seasons, because raptors follow different flight paths in spring and autumn. The observation points listed here are organized according to the season in which they are of greatest interest. This is not to say that no raptors will be seen at a site in other seasons. However, assuming that birders enjoy seeing large numbers of birds, each site is listed only once, under the season in which largest concentrations of birds pass.

The tables accompanying this chapter are designed to aid observers in quickly evaluating specific sites and in comparing one site with another. For each site the following information is provided: *Kingbird* reporting region and county in which the site is located, degree of ease of access, best or peak periods and in some cases also active periods, and optimal or best wind directions for seeing large hawk flights at each site. The term "best periods" means periods during which the greatest volume of birds can be seen. For the primary spring and interior autumn lookouts count ranges are given for all non-accidentally occurring raptors. The species count ranges have been included to indicate the magnitude of the seasonal flights. Count ranges do not tell the reader what he may expect to see in terms of numbers of a given species on a given day.

Primary spring and interior autumn lookouts are those that are regularly manned during all or the majority of the migration season. Other raptor migration observation sites are those that to date (1981) have not been regularly manned but at which good raptor passage has been noted. These include sites that have traditionally been used, possibly for many years in both seasons, but which lack dedicated observers throughout the majority of the season. This category also includes

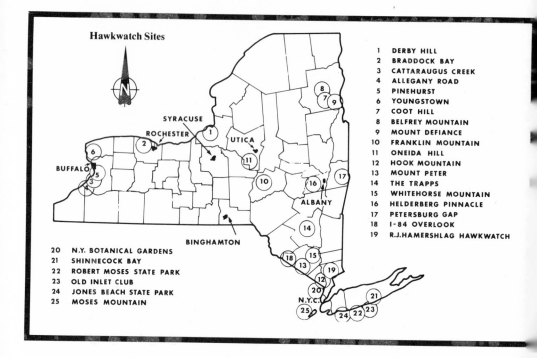

Hawkwatch Sites

1	DERBY HILL
2	BRADDOCK BAY
3	CATTARAUGUS CREEK
4	ALLEGANY ROAD
5	PINEHURST
6	YOUNGSTOWN
7	COOT HILL
8	BELFREY MOUNTAIN
9	MOUNT DEFIANCE
10	FRANKLIN MOUNTAIN
11	ONEIDA HILL
12	HOOK MOUNTAIN
13	MOUNT PETER
14	THE TRAPPS
15	WHITEHORSE MOUNTAIN
16	HELDERBERG PINNACLE
17	PETERSBURG GAP
18	I-84 OVERLOOK
19	R.J.HAMERSHLAG HAWKWATCH

20	N.Y. BOTANICAL GARDENS
21	SHINNECOCK BAY
22	ROBERT MOSES STATE PARK
23	OLD INLET CLUB
24	JONES BEACH STATE PARK
25	MOSES MOUNTAIN

sites that have only recently been discovered as hawk watches and that, after more seasons of coverage, may be upgraded to primary lookouts.

As used here and elsewhere in this guide, "early" includes the first through the tenth of the month, "mid-" the eleventh through the twentieth of the month, and "late" the twenty-first through the end of the month.

SPRING RAPTOR MOVEMENTS

Hawks tend to congregate along topographical features — coast lines, mountain ridges, and river valleys — during their migrations. These features are often referred to as leading lines. There is also ample evidence that, for one or another reasons, some raptors sacrifice directness of route in favor of over-land migration; that is, many raptor species avoid traversing large open bodies of water. If by veering in another direction a shore-hugging route is possible, many raptors will take that apparently more circuitous route.

The probable explanation for this migratory strategy is that meteorological events cause updrafts at the boundary between the shore and the water. These updrafts provide lift upon which hawks can glide. Imagine a raptor energy budget. Such a budget would undoubtedly allow for maximum energy expenditure on the breeding grounds so that any given raptor could get its genes into the population pool in general, and the next generation in particular. If this budget is to work, raptors cannot expend all of their energy getting to the breeding grounds. In fact, they should be conserving energy en route to them. It is undoubtedly more energy efficient to take advantage of updrafts and thermals instead of reaching the breeding grounds by means of powered flight. The strategy here is "Soar, don't oar." If these uplifts occur along the water edge, raptors will deviate from a direct over-water route to a less direct shoreline route.

Raptors moving north during migration tend to travel in a wide flight path, not usually following the route taken south.

If hawks are to follow a shoreline route, they will clearly have to narrow the wide front on which they have been moving to that section along the shore where updrafts are occurring. That path is restricted to a short distance either side of the shoreline. Hence, the shores of Lakes Erie and Ontario provide many sites from which hawk watching is superb. When northbound raptors come to Lakes Erie and Ontario, they deviate from a direct northbound route and travel up the eastern and western shores of the lakes in relatively concentrated narrow bands.

If ambient atmospheric conditions are right the raptor flights in spring can be simply spectacular along the lakeshore. These conditions include southerly winds, approach of a low-pressure area and cold front, and rising temperatures. Soaring on southerly winds in spring is obviously more efficient and therefore more advantageous to raptors than attempting to advance against head winds. The "Soar, don't oar" strategy calls for utilization of tail winds.

Atmospheric instability and updrafts are conditions that accompany low pressure systems and storms. Intense and concentrated northward hawk migration precedes storms and low pressure areas. The late William T. Keeton, his associates and students brilliantly demonstrated that birds are capable of detecting even slight changes in atmospheric pressure, and they suggested that hawks may be reacting to subtle atmospheric clues that enable them to sense an approaching low. It is also possible that hawks do not anticipate the low but are responding more immediately to locally advantageous atmospheric conditions associated with a low.

Table 2 summarizes the migration picture at the two most important spring hawk migration observation sites in New York State: Derby Hill and Braddock Bay. Table 3 contains several additional spring hawk observation sites that receive sporadic coverage and about which limited data were accessible. The information on both of these tables as well as others in this chapter was gleaned from numerous issues of the *Kingbird*, all of the newsletters of the Hawk Migration Association of North America published to date (1981), the relevant regional reports in recent issues of *American Birds*, as well as questionnaires completed by active hawk watchers about their favorite sites.

AUTUMN INTERIOR RAPTOR MOVEMENTS

During the fall, birds in the northeast move generally south and southwest, and they gather along any physical feature that provides thermal and deflective uplift.

Differential heating of land, air, and water causes the buoyant air currents on which southbound migrants ride. The strength of these air currents varies considerably, due primarily to topography. Thermal air currents may also be reinforced by reflective heat from nearby sparsely vegetated areas along mountain ridges. Deflective lift, on the other hand, is produced when air moving in a horizontal direction strikes the vertical surface of hills and mountain ridges. Air is forced upward, creating strong updrafts. This deflective air provides a perfect medium upon which hawks can effortlessly ride. Maximum deflective air currents are produced on windy, turbulent days. On extremely windy days, with north and northwest winds, hawks have been recorded moving at approximately forty miles per hour. If a raptor were to move at this speed for five to six hours it could cover a distance of 200 to 240 miles on an ideal migration day. Under such conditions hawks move farther away from the center of the deflective air currents along the ridge. They can easily get a free ride down the updraft some distance from the ridges where they can hold a steady balance and control the lifting power of the rising wind. Soaring hawks, particularly buteos, wait until relatively late in the morning (nine or ten o'clock) to begin a migrating day, unlike some landbirds, which often begin just before dawn. This provides time in which to hunt for food while thermals are building. After getting underway buteos, especially Broad-winged Hawks, spread out over a wide front. By proceeding south or southwest over widely spaced parallel streams there is more opportunity for some birds in flight to fortuitously hit one or another rising thermals. As soon as one or more birds finds thermal lift, others quickly join. Clearly, the migratory strategy of the majority of birds is to locate thermals visually by watching for hawks in the lead characteristically circling with rigid wings upward into a kettle. A good-sized kettle could probably be seen for several miles by the birds behind it. In this way, not all birds would have to locate rising thermals on their own but would simply take a visual cue from hawks migrating in front of them.

The optimal conditions for seeing a steady stream of migrating raptors in autumn from a mountain peak or hill summit in New York State occur immediately following the passage of a continental polar air mass (or cold front) out of Canada when there is a low pressure system to the north.

A counter-clockwise movement of air around the low causes winds from the northwest to push migrating raptors toward any north-south oriented mountain ridge. Winds from this direction also create the best updrafts along the slopes during cool dry periods following the passage of the cold front. This combination should provide good viewing from any outlook along a north-facing ridge. Foggy, rainy weather seems to ground hawks. Often viewing conditions are perfect after an extended period of moderate-to-heavy rains and southerly winds.

Table 4 shows the approximate timing of autumn hawk flights by species in New York State and includes flight range and the peak flight period of each species. By consulting the table the hawk watcher can determine the best periods in which to watch various species. Tables 5 and 6 list all of the relevant information on primary interior hawk lookouts and other secondary interior hawk migration observation sites in the state. Directions to every site may be found following page 463.

AUTUMN COASTAL RAPTOR MOVEMENTS

The coast of the Atlantic Ocean apparently serves as a visual leading line for several species of diurnal migrating raptors. The species in this group following the coast do not include buteos, which need strong updrafts for soaring flight, but includes Ospreys, Northern Harriers, Sharp-shinned Hawks, American Kestrels, Merlins, and Peregrine Falcons. Sharp-shinned Hawks, Merlins, and Peregrine Falcons are highly maneuverable, swift fliers, not built for soaring. They ignore strong updrafts and move along the coast, following shorebird and small landbird migrations, thereby assuring themselves a constant food supply of birds. The American Kestrel, although also recorded in large numbers along interior ridges, migrates primarily along the sea coast. It is not uncommon to see several hundred in a single day streaking past Robert Moses State Park and Jones Beach. The beaches and dunes serve as a major migrational route not only for birds but also for various insect families. A visit to the barrier beaches in late September and early October will be well repaid with a wealth of migrating falcons and often large assemblages of monarchs, skippers, pygmy blues, and hordes of humming dragonflies and grasshoppers on the sand and in dense growths of goldenrod, marsh fleabane, bayberry, and fleshy glassworts. Watch the diminutive kestrel falcon as it passes, often at eye-level, grabbing migrating painted ladies, buckeyes, big dragonflies and monarch butterflies in mid-air and eating them on the wing.

The migration system of these coast-following species may have evolved in response to food-gathering pressures rather than the selective pressures of navigation mechanisms.

Prerequisites for an excellent coastal flight are northwest or north winds at velocities of five to fifteen miles per hour, especially following an advancing cold front or following several days of rain and southerly winds. Birds also move occasionally on northeast winds but these movements are usually so high that they are not within binocular limit. The general tendency for hawks to move behind low-pressure areas in fall, in the same way that they move in front of lows in spring, is probably explained again by the advantage of tail winds. The lift of the low in combination with the northerly circulation of air behind the low provides advantageous tail winds to assist the birds in their southbound flight.

In autumn 1979, Cornelius (Neil) Ward, a thoroughly committed monitor of raptor movements along the strip for at least twenty years, reported that more than forty Peregrines were seen during the season. With fifteen miles per hour northwest winds on October 14, 1979, following three days of ten miles per hour southwest winds and one full day of rain, the flight picked up in a rather dramatic way. At Democrat Point, on Fire Island, Ward, along with H. Darrow, A. Dignan, E. Levine *et al.,* recorded three Ospreys, seventeen Northern Harriers, one Goshawk, more than a thousand Sharp-shinned Hawks, four Cooper's Hawks, over 525 American Kestrels, seventy-eight Merlins, and five Peregrine Falcons — a grand total of over 1630 birds in one stellar hawking day. By anyone's standards this was high-quality birding.

Table 7 designates several of the prime coastal autumn hawk-watching sites in Nassau and Suffolk counties on Long Island and on the east side of Staten Island. Every birder who can reach one of these sites in autumn should go at least once or twice. Watching raptors on a barrier beach is exhilarating — and quite different from watching raptors atop a mountain.

PRIMARY SPRING HAWK LOOKOUTS IN NEW YORK STATE
Derby Hill Hawkwatch

Derby Hill Hawkwatch is the premier spring hawk watching site in the northeast. It is also a wonderful place to see the spring landbird migration. In autumn, waterfowl, jaegers, gulls, and terns migrate past this drumlin-promontory on the southeast shore of Lake Ontario. The crest of the hill is the highest point for miles around, and the treeless hill drops off as a perpendicular cliff at the lake's edge.

The combined southern shorelines of Lakes Ontario and Erie comprise more than four-hundred miles. Hawks migrating north from southerly wintering grounds meet the lakes, which act as a natural barrier, causing raptors to veer westward or eastward and to then proceed north by way of a shore-hugging route. As one might guess, this shore-skirting maneuver tends to concentrate the numbers of hawks in a narrow flight path up the lakeshore. Flights are most spectacular here when winds are from a southerly direction. When such winds blow, hawks migrate earlier in the day and continue flying later than in north or west winds. Generally, the best flights occur when ground wind speed is ten to thirty miles per hour. The best flights also occur on days when temperature exceeds that of several previous days and when barometric pressure is low or falling. Rising barometric pressure usually results in a weak raptor flight. Study the accompanying table for trends and peaks of species flights.

To reach Derby Hill, leave I-90 (the New York State Thruway) at Exit 36 just north of Syracuse and head north on I-81. Proceed north to Exit 34, then head west on NY 104. Continue west approximately 5.5 miles to the village of Mexico. There pick up NY 3 headed north and proceed north 4.3 miles to NY 104B. Turn left (west) and proceed west about one-half mile to Sage Creek Drive, which comes in on the right (north). Turn right (north) onto Sage Creek Drive and proceed north to its end. Park along the road or turn right (east) on an unpaved road and pull off. Walk to the top of the hill on your right. This is the best place from which to see migrating hawks and passerines in spring.

Braddock Bay Hawkwatch (Map 2.9)

During the raptor migration period (see Table 2), any day with temperatures near 50°F and south-southwest winds will bring hawk flights ranging from moderate to marvelous, with the peak passage in the last ten days of April. Most days, especially weekends, will find the area manned by hawk watchers who can also direct the visiting birder to other interesting birds or areas within the park (see Braddock Bay write-up on page 88). Braddock Bay Hawkwatch is located on an embayment on Lake Ontario. Large low pressure areas moving west-to-east create strong

TABLE 2

Primary Spring Hawk Lookouts in New York State

	Derby Hill				Braddock Bay			
Region	5				2			
County	Oswego				Monroe			
Access by car	Yes				Yes			
Best period(s)	Late Mar. and Late Apr.				Mid and Late Apr.			
Best winds	SW				SW			
Species	Flight Begins	Peak Flight	Flight Ends	Count Range	Flight Begins	Peak Flight	Flight Ends	Count Range
Turkey Vulture	Early Mar.	Mid-Apr.	Late May	500–1100	Mid-Apr.	Mid-Apr.	Late Jun.	800–1800
Osprey	Mid-Apr.	Late Apr.	Mid-May	100–800	Mid-Apr.	Late Apr.	Late Jun.	40–300
Bald Eagle	Mid-Mar.	Late Apr.	Late May	5–20	Mid-Mar.	Mid-Apr.	Late Jun.	5–30
Golden Eagle	Early Mar.	Late Mar.	Mid-May	5–30	Early Apr.	Late Apr.	Early May	5–20
Northern Harrier	Early Mar.	Mid-Apr.	Late May	300–700	Early Mar.	Mid-Apr.	Late Jun.	100–600
Goshawk	Early Mar.	Early Mar.	Mid-May	30–60	Early Mar.	Late Mar.	Early May	10–50
Sharp-shinned Hawk	Early Mar.	Late Apr.	Late May	3000–10,000	Mid-Mar.	Late Apr.	Late Jun.	1000–8000
Cooper's Hawk	Early Mar.	Late Mar.	Mid-May	200–700	Early Mar.	Mid-Apr.	Mid-May	50–400
Red-shouldered Hawk	Early Mar.	Late Mar.	Mid-May	500–1300	Early Mar.	Late Mar.	Early May	100–500
Broad-winged Hawk	Mid-Apr.	Late Apr.	Late May	2000–35,000	Mid-Apr.	Late Apr.	Late Jun.	8000–25,000
Red-tailed Hawk	Late Feb.	Late Mar.	Late May	3000–8000	Early Mar.	Mid-Apr.	Late Jun.	1000–4000
Rough-legged Hawk	Late Feb.	Late Apr.	Mid-May	200–500	Early Mar.	Late Apr.	Early Jun.	100–500
American Kestrel	Early Mar.	Mid-Apr.	Mid-May	200–700	Early Mar.	Mid-Apr.	Late Jun.	100–800
Merlin	Mid-Apr.	Late Apr.	Early May	5–20	Mid-Mar.	Late Mar.	Early May	0–10
Peregrine Falcon	Early May	Mid-May	Mid-May	0–5	Late Mar.	Late Mar.	Early Jun.	0–10

Note: 'Early' designates 1st through 10th of month
'Mid' designates 11th through 20th of month
'Late' designates 21st through end of month

southerly air flows east of the low pressure. Raptors appear to move eastward, pushed along the Lake Ontario shore by southerly winds. Like Derby Hill, flights are best when the wind velocity is no more than about thirty miles per hour and no less than ten miles per hour. Although there are normal yearly fluctuations in species counts, owing to amount and quality of coverage, weather patterns, and actual status changes in abundance and frequency of raptors, substantial flights occur in most years, and large numbers of birds and excellent views of them combine to make this an outstanding migration lookout within the state. Consult the accompanying table for approximate flight periods and peaks within each species' flight duration.

To reach Braddock Bay, leave the New York State Thruway (I-90) at Exit 47 (LeRoy). Take I-490 east to NY 47 north (I-590 east). Follow it to US 104 (Ridge Road). Exit west and go 1.1 miles to Long Pond Road. Turn right (north) and proceed 4.9 miles to the Lake Ontario State Parkway entrance, labelled Hamlin Beach Park. Proceed 1.8 miles west to the East Manitou exit. The entrance to Braddock Bay State Park is just to the right of the exit ramp with the parking lot 0.5 mile into the park. For purposes of orientation: looking across the bay one sees a large white hotel and dance hall; this is north. The woods behind the observer then are southeast. The hawk lookout is conducted from the parking lot.

OTHER SPRING HAWK MIGRATION OBSERVATION SITES
Cattaraugus Creek (Maps 1.1 and 1.2)

Most raptor species recorded in the region have been observed at Cattaraugus Creek. Occurrence times and abundance vary, but usually Turkey Vultures, Red-tailed, and Red-shouldered hawks occur early in the period and Broad-winged and Sharp-shinned hawks are most abundant in the latter half of the season. Total raptors recorded are generally fewer than those at Pinehurst and Allegany Road lookouts. Migration of passerines and some non-passerines does occur here but is not as concentrated nor do as high numbers occur as at nearby Pinehurst lookout.

To reach Cattaraugus Creek, leave I-90 (the New York State Thruway) at Exit 58. Turn east onto NY 5 north. Cross Cattaraugus Creek and turn left (east). Continue to an underpass, on the other side of which you should turn left again. Follow the road to the northeast side of the mouth of the creek, park, and watch the raptor flight along the shore of Lake Erie.

Allegany Road (Maps 1.1 and 1.2)

Allegany Road is 690 feet above sea-level and has for many years been a reliable point from which to watch the spring passage of hawks — particularly when there is no precipitation, when winds are between five and fifteen miles per hour, and when there is complete or partial cloud cover for easier detection. The distance from shore and width of the raptor flight path will vary with wind direction. The more easterly the wind, the closer to shore and the more narrow; the more westerly, the farther from shore and usually the broader the path.

Observers must sometimes shift the lookout points along Allegany Road to intercept the flight from the slope toward the lake plain, or inland to beyond the bridge over I-90. All along this road there is generally unobstructed viewing to considerable distances southwest and northeast as well as parallel with the road to the north-northwest and the south-southeast.

Wear a windbreaker, especially early in the season, when temperatures are lower and winds off the lake are sometimes chilling.

To reach Allegany Road observation point, leave I-90 (the New York State Thruway) at Exit 58. Turn left onto NY 5 and NY 20. The first traffic light (0.8 mile) is Allegany Road. Turn left onto Allegany Road and proceed 0.4 mile to the first crest above the lake plain. Park on the road shoulder, and watch for passing raptors.

Pinehurst (Maps 1.1 and 1.2)

This is one of the best observation points in *Kingbird* Reporting Region 1 for watching spring raptor movements, although visibility is somewhat limited by buildings and trees to the south. At the peak of the flight, several thousand may pass in a day. Loons, Ciconiiformes, waterfowl, gulls, and passerines migrate past this point in spring. Little Blue Herons and Snowy Egrets have been recorded here. Blue Jays, American Robins, and icterids sometimes occur here by the thousands.

Wear a windbreaker and warm clothing early in the season, for lake ice can lower the temperature at the lookout. Caution is advised near the cliff, which has been known to give way on occasion, threatening serious injury to those at the brink.

To reach Pinehurst observation point, leave I-90 (the New York State Thruway) at Exit 57. Turn right (north) on NY 75. Almost immediately turn left onto NY 5 at Athol Springs. Proceed to Wanakah. Turn right on Old Lakeshore Road at Wanakah and proceed southwest about 0.8 mile to the observation point. Park on the road shoulder opposite the Bavarian Chalet Restaurant and observe the hawk flight from near the cliff edge. Look to the southwest.

Youngstown (Creek Road) (Maps 1.1 and 1.5)

Here the east-west location of hawk flights depends on area winds giving rise to air currents. If there are no lifts, the hawks proceed westward near the Lake Ontario shore and the flight is low. With mild southwest, south, and southeast winds and north, northwest or northeast offshore breezes, the flight moves inland, depending on where the hawks find the most assistance. Such flights will vary in altitude because hawks are looking for and finding rising currents where they change from moderate altitude to binocular limit. Moderate east or west winds of fifteen to twenty miles per hour usually mean flights near the lake. On north winds, there are either no flights, or the hawks avoid this part of the lakeshore, turning westward

TABLE 3

Other Spring Hawk Places in New York State

	Cattaraugus Creek	Allegany Road	Pinehurst	Youngstown (Creek Road)	Coot Hill	Belfrey Mountain	Mt. Defiance
Region	1	1	1	1	7	7	7
County	Chautauqua	Chautauqua	Chautauqua	Niagara	Essex	Essex	Essex
Access by car	Yes	Yes	Yes	Yes	No†	No‡	Yes
Active periods	Mar. thru Mid-May	Late Feb. thru Late May	Late Feb. thru Late May	Mid-Mar. thru Mid-May	Apr. thru Early May	Apr. thru Early May	Apr. thru Early May
Peak periods	Mid-Mar. and Late Apr.	Late Apr.	Late Apr.	April	Mid-Apr.	Mid-Apr.	Late Apr.
Best winds	SW	SW-S	S-SE	W-S-E	S	S	S

†Possibly several-mile hike (sometimes in snow).
‡Ten-minute uphill walk.
Note: 'Early' designates 1st through 10th of month.
    'Mid' designates 11th through 20th of month.
    'Late' designates 21st through end of month.

closer to Buffalo and Fort Erie, Ontario. On most days hawk flights begin around nine o'clock with numbers increasing to maximum rates in the hour before noon. Until mid-April, the flights drop off rapidly at noon. But with warmer weather and Broad-winged Hawk flights, the migration may continue until two or three in the afternoon. On good flight days counts reach 150 birds per hour.

To reach Youngstown (Creek Road), from downtown Niagara Falls proceed north on the Robert Moses Parkway to its terminus at NY 18. Turn right (west), thereby gaining NY 18. Look at your odometer. This point is the original Lake Road Lookout. All mileages will be from this point. Proceed west and north about 0.3 mile to the Woodcliff Lookout A, off Lake Road. The remaining points lie south on Creek Road. At 0.5 mile south on Creek Road is Lookout B. At about one mile is Lookout C. At about 1.5 miles is Lookout D. At 2.1 miles is the Hardings Restaurant parking lot, Lookout E. At about 2.6 miles is Lookout F. At about 3.5 miles is Lookout G. At 3.8 miles south is Lewiston-Porter North parking lot, Lookout H. Try any and all of these. Depending upon the winds one can easily be much better than another.

Coot Hill (Maps 7.11 and 7.12)

Here at an altitude of 1100 feet the dedicated raptor watcher often trudges through snow carrying clipboard, wind gauge, owl decoy, field guides, hawk migration reporting forms, and other assorted paraphernalia in answer to the vernal craze inspired by hawks. He or she thrills to the sight of migrating hawks drifting by on southerly April or May breezes warmed by Bermuda highs. Below lies Bulwagga Bay and Crown Point Peninsula and Lake Champlain, stretching north and south, visible for at least thirty miles. The distant summits of Camel's Hump and

Mount Mansfield (in Vermont) and the snow-covered High Peaks of the Adirondacks can be seen from atop the sun-warm rocky outcroppings.

Maximum flights recorded here since regular coverage was begun in 1976 exceed 160 birds per day (Carleton, Peterson *et al.*).

To reach Coot Hill, leave I-87 (the Adirondack Northway) at Exit 29. Proceed east across the Schroon River to North Hudson and US 9. Proceed north on US 9 to NY 4C and NY 4. Turn right (east) and follow NY 4C and then NY 4 and then NY 4A (West Moriah Road) east and to the junction of NY 7 (South Moriah Road) and NY 4A (about two miles west of the village of Port Henry). Turn right (south) onto NY 7 (South Moriah Road) and drive south for 2.8 miles to an unpaved road, Lang Road, which comes in on the left (east). Turn left (east) and proceed past the third farmhouse. There the road dips south, then goes uphill, and then east again at a clearing on the left (east) past a cemetery on the west side of the road. Proceed east past the clearing to the summit. In early spring observers may have to park near the first or second farmhouse to avoid road washouts. They would then have a hike of several miles to the summit. This climb is moderately steep but long.

Belfrey Mountain (Maps 7.1 and 7.11 and 7.12)

April brings hawks floating steadily northward on warm southerly breezes that melt snows off budding ridges. From Belfrey Mountain wedges of waterfowl and waves of landbirds signal winter's end. This peak is reached more easily than Coot Hill, which lies eight miles to the southeast, but hawks are more difficult to detect as they fly below against a background of trees. The view of Lake Champlain and the Adirondacks is lovely but not as spectacular as the view from Coot Hill. Wear good hiking shoes to cover the short trail to the 1940-foot summit. Buteos, mostly Broad-winged and Red-tailed hawks, occur most frequently with small numbers of Ospreys, Northern Harriers, Red-shouldered Hawks, and American Kestrels also seen. Maximum counts per day exceed 50 birds.

To reach Belfrey Mountain Hawkwatch, leave I-87 (the Adirondack Northway) at Exit 30. Proceed east on NY 6 (Tracy Road) to the hamlet of Witherbee, which is 5.0 miles northwest of Port Henry. Turn left (north) at the junction of NY 6 and Dalton Hill Road (NY 7C). Watch for the Belfrey Mountain fire tower on the hill on the left (west). When you sight it, look for the trailhead on the west side of Dalton Hill Road (Lincoln Pond Road). Park on the east (right) side of the road, on the shoulder opposite the trailhead, and walk the trail to the summit. This walk should take around ten minutes and is not strenuous.

Mount Defiance (Maps 7.11 and 7.12)

Of the three spring hawkwatch sites in *Kingbird* Region 7, Mount Defiance is the one most easily reached. This is a lookout you can drive up to. I have driven on the Mount Defiance toll road (actually, no toll is charged) when it has been in rugged, post-winter shape, but it was drivable. Views at the top are unobstructed and often birds fly close to the ridge, affording excellent observations. Sometimes in late April there are "sundog" circles around the sun and high numbers of geese and ducks migrating north. Maximum counts of birds per day exceed 150.

To reach Mount Defiance, leave I-87 (the Adirondack Northway) at Exit 28. Proceed east on NY 74 through the hamlets of Severance and Paradox and past Eagle Lake, traveling east toward the village of Ticonderoga. At the junction of NY 74 and NY 9N (NY 22) turn right (south) and proceed to Montcalm Street. Turn left (east) and proceed east, crossing the Ticonderoga Creek. At Champlain Avenue turn right (south) and drive south only a short distance. A street called The Portage will be the first street on the left. Take The Portage to the second street coming in on the left, Mount Defiance Street. Turn left here and cross the railroad tracks. Continue east a short distance to the toll road to Mount Defiance, which comes in on the right. Turn right and continue up the road to the summit of Mount Defiance.

PRIMARY INTERIOR FALL HAWK LOOKOUTS IN NEW YORK STATE
Franklin Mountain (Oneonta)

The summit of Franklin Mountain is 2150 feet above sea-level. Autumn hawk flights are best observed in north winds and warm, partly cloudy days appear to be the most productive. Winds from the west tend to push the leading line of hawks relatively farther east of the lookout and often only within the detectable but not identifiable range. As is clear from the accompanying table, this is an excellent place at which to see autumn flights of large numbers of Red-tailed Hawks, and in nearly every year large numbers of Golden Eagles are reported (up to ten). In autumn 1976, more than 4500 hawks of thirteen species were recorded here. In autumn, 1977, a flight consistent with that of the previous autumn was again monitored; since that time Franklin Mountain (Oneonta Hawk Lookout) has been manned regularly.

It is well known that raptors migrate along very wide fronts. The limits and routes taken by southbound migrants in autumn through central New York State are not as yet well delineated. As more time is spent observing and recording at such central summits as this lookout, the migration puzzle will be pieced together.

To reach Franklin Mountain (Oneonta Hawk Lookout), from the I-88 interchange in downtown Oneonta (in Otsego County) proceed south and west on NY 28 (NY 23). Just south of the interchange NY 23 and NY 28 diverge. Follow NY 28 right (west). At the first road (Southside Drive) turn east and follow Southside Drive about 1.1 miles to Swarthollow Road. Turn right (south) and climb Swarthollow Road up Franklin Mountain approximately 1.7 miles to the first road coming in on the right, which is Grange Hill Road. Turn right here and continue up this road one-half mile to the summit. Stop and park at an iron gate on the right. Walk beyond the gate north about one-hundred yards, toward the large screen of a billboard-like microwave reflector. Hawks are best viewed from the base of the reflector.

TABLE 4
Timing of Autumn Hawk Flights in New York State by Species

Species	Range	Peak Period
Turkey Vulture	Mid-Oct.–Nov.	Late Oct.–Early Nov.
Osprey	Early Sept.–Early Nov.	Mid-Sept.–Late Sept.
Bald Eagle	Aug.–Oct.	Late Aug.–Early Sept.
Golden Eagle	Late Oct.–Nov.	Early Nov.–Mid-Nov.
Northern Harrier	Sept.–Early Nov.	Sept.–Mid-Oct.
Goshawk	Oct.–Nov.	No peak
Sharp-shinned Hawk	Late Sept.–Late Nov.	Early and Mid-Oct.
Cooper's Hawk	Oct.–Late Nov.	Mid-Oct.–Early Nov.
Red-shouldered Hawk	Mid-Sept.–Mid-Nov.	Mid-Oct.–Early Nov.
Broad-winged Hawk	Sept.–Mid-Oct.	Mid-Sept.–Late Sept.
Red-tailed Hawk	Sept.–Nov.	Mid-Oct.–Early Nov.
Rough-legged Hawk	Late Oct.–Nov.	No peak
American Kestrel	Sept.–Early Nov.	Late Sept.–Mid-Oct.
Merlin	Early Sept.–Late Oct.	Mid-Oct.
Peregrine Falcon	Sept.–Oct.	No peak

Note: 'Early' designates 1st through 10th of month.
'Mid' designates 11th through 20th of month.
'Late' designates 21st through end of month.

Oneida Lookout (Oneida Hill)

This is a promising lookout for those within easy reach of it who might want to observe a fine Broad-winged Hawk migration in mid-September. The second most numerous species recorded here is Sharp-shinned Hawk, with smaller numbers of other eastern raptor species following.

The optimal conditions for viewing good flights are when northwest winds blow on slightly overcast days in mid-to-late September. This site is 1420 feet above sea-level, and during autumn one should dress warmly as strong winds can be very chilling at this height, with as much exposure as the summit offers.

In addition to raptors in passage one can view long skeins of migrating Canada Geese, large flocks of southwardbound Blue Jays, and concentrations of Common Crows moving south in October.

To reach the Oneida Lookout (Oneida Hill), leave the New York State Thruway (I-90) at Exit 33 (Verona). Proceed south on NY 365 to NY 5. Turn east and continue on NY 5 to NY 26. Turn south on NY 26 and continue for 6.7 miles to the junction of Knoxboro Road. Turn right (west) on Knoxboro Road and proceed 2.2 miles up the hill to Hatalla Road. Here make a right turn and proceed on Hatalla Road past the Oneida County dump, where the road curves sharply left (west). Continue to the crest of the hill. The hawk watching area is along this paved public road, but there is very little traffic on Hatalla Road and plenty of room to park cars off the road.

TABLE 5

Primary Interior Fall Hawk Lookouts in New York State

	Franklin Mountain (Oneonta)	Oneida Hill (Oneida Lookout)	Hook Mountain	Mount Peter	The Trapps (Mohonk)	Whitehorse Mountain
Region	4	5	9	9	9	9
County	Delaware	Oneida	Rockland	Orange	Ulster	Orange
Access by car	Yes	Yes	No†	Yes	No†	No†
Best period(s)	Mid-Sept. and Late Oct.	Mid thru Late Sept.	Mid-Sept. and Early Oct.	Mid-Sept. and Early Oct.	Mid-Sept. and Early Oct.	Mid-Sept. and Early Oct.
Best winds	N	W-N-E	NW-N-NE	NW-N-NE	NW	SW
Species	Count Range					
Turkey Vulture	10–50	0–10	0–30	NA	NA	NA
Osprey	20–70	20–50	200–400	60–200	20–100	70–200
Bald Eagle	0–5	0	0–10	0–5	0–5	0–5
Golden Eagle	0–10	0–5	0–5	0–5	0–5	0–10
Northern Harrier	5–20	50–70	200–300	20–60	5–100	40–80
Goshawk	5–30	0–10	5–40	0–10	0–10	0–20
Sharp-shinned Hawk	400–900	600–800	3000–8000	500–1900	100–2000	500–1900
Cooper's Hawk	0–5	20–50	20–90	10–70	20–90	20–140
Red-shouldered Hawk	0–10	50–70	100–200	20–80	0–40	10–200
Broad-winged Hawk	1000–1400	600–3200	6000–16,000	3000–6000	600–2600	7000–9000
Red-tailed Hawk	1300–2000	200–600	200–600	60–300	30–500	300–1200
Rough-legged Hawk	0–5	5–10	0–5	0–5	0–5	0–5
American Kestrel	80–120	100–150	500–800	100–500	100–200	100–300
Merlin	0	0–5	5–30	0–15	0–5	5–30
Peregrine Falcon	0	0	0–15	0–5	0–5	0–15

NA – Not available.

†Short uphill walk.

Note: 'Early' designates 1st through 10th of month.

'Mid' designates 11th through 20th of month.

'Late' designates 21st through end of month.

Hook Mountain

This 1000-acre summit lookout sits 730 feet above sea level. It is manned daily during the autumn migration, and flights have proven best on the second or third day following passage of a cold front when winds are from the northeast, north, or northwest. In 1981, this site had been actively manned for ten years. Huge numbers of Broad-winged Hawks are recorded here annually as well as the largest numbers of Sharp-shinned Hawks recorded at any interior autumn hawk lookout in the state.

One of the best aspects of the lookout is that not only can large flights be observed here but also birds can be seen at close range, often at eye-level or below. Birds fly close to the mountain ridge to take advantage of updrafts created by the steep slope of the mountain. In addition to hawks, loons, ducks in considerable variety and numbers, geese, shorebirds, woodpeckers, and warblers can be seen. Totals in excess of 25,000 hawks and a few thousand landbirds are tallied each autumn from this mountaintop. Hook Mountain is a real winner!

To reach Hook Mountain, proceed west on I-87 (I-287) from the western terminus of the Tappan Zee Bridge, a short distance to Exit 11. Leave I-87 and turn north onto US 9W (Highland Avenue). Proceed north about 2.0 miles, pass Christian Herald Road and then park in a low-lying area on the west side of US 9W. Some refer to this area as a "dump" as it is usually quite littered. Walk south and across US 9W approximately 200 feet. On this, the east side of the road, look for a stencilled arrow and some blue paint on several rocks, which mark the trailhead to Hook Mountain. Follow the blue blazes marking the trail uphill to the exposed clearing at the summit of the mountain, and watch hawks.

Mount Peter

This hawk lookout has the advantage of being easy to reach without a taxing climb. It has been used as a site for observing southbound raptors nearly twenty-five years, having first been manned in 1957.

Large numbers of Broad-winged Hawks — upwards of 7000 to 7500 per season — pass Mount Peter in mid-to-late September. The second most numerous species seen from here is the Sharp-shinned Hawk, with the American Kestrel holding third place. Throughout an entire autumn season 11,000 to 12,000 hawks of about twelve species can be seen from this watch.

To reach Mount Peter, from the intersection of NY 17A and NY 210 in the village of Greenwood Lake, at the north end of Greenwood Lake in southeastern Orange County, proceed northwest on NY 17A for about 2.0 miles. Pull into the parking lot of the Valley View Restaurant, which is on the north side of the road. Walk approximately 200 feet to the crest of the ridge behind the parking lot. This is the place from which to watch migrating hawks. Look to the northwest.

The Trapps (Mohonk) (Map 9.2)

Part of the Shawangunk Mountain range, this hawk-watching site is approximately 1200 feet above sea level on the crest of The Trapps. From here unobstructed vistas to the west and northeast afford fine views of migrating raptors. The majority of raptors seen are usually close to the ridge and ride on updrafts created by winds off the ridge or off the ridge's northwest slope. The optimal time to visit this watch is during brisk autumn weather after passage of a low pressure system, when there are moderate winds from the northwest. Around ten in the morning thermals of warm air begin to provide the necessary lift, and soaring hawks can be seen kettling upward on these invisible warm air currents. When there are sufficient and continued updrafts, the migration steadily continues through the early afternoon.

During the month of September the most abundant species passing The Trapps is the Broad-winged Hawk. During October the most abundant species recorded is the Sharp-shinned Hawk, and during November the Red-tailed Hawk is the most abundant raptor.

Mohonk Mountain House, on whose property The Trapps sit, has in the past charged a nominal fee for a day-use permit.

To reach the Trapps Hawkwatch, leave I-87 (the New York State Thruway) at Exit 18. Turn left (west) on NY 299 and proceed 6.0 miles west to US 44 (NY 55). Turn right (west) and follow US 44 uphill past a hairpin turn. Continue west and pass under a footbridge over US 44. Park on the right (north) side of the road near the overhead bridge. Climb up to the bridge and walk south on the roadway approximately 150 yards to a blue blazed trail that angles up to the exposed northeast facing brow of the ridge. This climb should take fifteen minutes or less.

Whitehorse Mountain (Map 9.4)

This is unquestionably one of the finest, if not the finest, interior hawk watching sites in southeastern New York State. The setting is the summit of Whitehorse Mountain, where the Hudson River flows through a gorge between Breakneck Mountain on the east and Storm King Mountain on the west. Storm King Mountain is just northeast of Whitehorse Mountain, and there are spectacular views of the river and gorge from the Whitehorse peak. Paul Jeheber, proprietor of this mountain, has long ago responded to the special lure of raptor watching and in order that others can enjoy it with him, he has built not only towers high enough over the trees to afford 360-degree views but he has also constructed rustic benches at the summit and along the path to the summit, where birders can sit and enjoy the active migration of passerines moving through the property. There is a feeling of camaraderie, easiness, and natural warmth at Whitehorse Mountain; and the figures in the accompanying table give a hint of the wealth of bird life that soars in review for those lucky enough to be there at the right time. The walk to the top of the mountain is easy and there are so many interesting things to look at along the way — a fox den, for instance — that the walk seems remarkably short. Upwards of thirty species nest on the property, and on the walk up to and back from the tower many of them appear early in the season. In addition, geese, ducks, shorebirds, and landbirds other than local breeders can be seen here in good numbers. I heartily recommend Whitehorse Mountain.

To reach Whitehorse Mountain, from the western terminus of the Bear Mountain Bridge and the Bear Mountain traffic circle, proceed north on US 9W for 8.6 miles to the first break in the center island divider, which is approximately 200 yards below the crest of a hill. Cross over to the south side of US 9W through this break in the highway. Just after the turnaround, there is a driveroad on the south side marked with the name of Paul Jeheber. Turn right onto the driveroad and park. Follow the wood road (marked) past the house and up to the summit of Whitehorse Mountain and the observation towers.

TABLE 6
Other Interior Fall Hawk Places in New York State

	Helderberg Pinnacle	Petersburg Gap	I-84 Overlook (Port Jervis)	R.J. Hamershlag (Butler Sanctuary)	NY Botanical Gardens
Region	8	8	9	9	10
County	Albany	Rensselaer	Orange	Westchester	Bronx
Access by car	Yes	Yes	Yes	Yes	Yes
Active periods	Sept. thru Mid-Nov.	Sept. thru Late Oct.	Sept. thru Nov.	Sept. thru Early Nov.	Sept. thru Early Nov.
Peak periods	Mid thru Late Sept.	Mid-Sept.	Mid-Sept.	Mid-Sept.	Early thru Mid-Oct.
Best winds	W-NW-N	NW	NW-N-NE	NW	NW

Note: 'Early' designates 1st through 10th of month.
'Mid' designates 11th through 20th of month.
'Late' designates 21st through end of month.

OTHER FALL HAWK MIGRATION OBSERVATION SITES
Helderberg Pinnacle (Maps 8.6 and 8.7)

Those in the Albany area can view migrating raptors from Helderberg Pinnacle, whose summit is 1800 feet above sea-level.

Those who frequent this watch theorize that raptors following prominent ridges leave the Green Mountains of Vermont, follow the Helderbergs south and west, and then funnel along the ridges of the Catskill Mountains. Observers began coming to this site in 1975 and by 1977 it was manned on a regular basis.

Relative to most sites, this pinnacle is fortunate in that high numbers of Goshawks are annually observed.

To reach Helderberg Pinnacle, from the west side of the city of Albany, take NY 85 south and then west about 10.5 miles to Thacher Park Road, which is about 3.0 miles west of the village of New Scotland. Turn right (north) onto Thacher Park Road and proceed 0.7 mile to Beaver Dam Road, which comes in on the left (west). Turn left (west) and follow Beaver Dam Road 1.3 miles to Pinnacle Road. Turn left (south) and continue on Pinnacle Road to signs indicating Camp Pinnacle. Turn right (west) on a dirt road at the signs. Enter Camp Pinnacle and proceed to the back of the camp. Park without obstructing access. Walk to the top of the hill. This is the place from which to watch hawks in migration.

Petersburg Gap

Although not an organized and manned hawkwatch, Petersburg Gap, a notch in the Taconic/Berkshire Range at the crest of Petersburg Mountain, traditionally has been one of the most productive raptor watching locations in the capitol district. Area birders frequent this peak informally. Ten raptor species have been sighted from here. This is an area of considerable scenic beauty, and it is easily accessible; but enormous hawk flights should not be expected here. Highland migrating warblers can be seen here during autumn. The Appalachian Trail passes over the crest.

To reach Petersburg Gap, from the city of Troy, proceed east on NY 2 to the junction of NY 22 and NY 2 in the hamlet of Petersburg. Continue east on NY 2. This is a long, climbing, winding road. At the New York-Massachusetts border there is a ski lodge. Park in the lodge's parking lot. Cross NY 2, where you will find a trailhead. Follow this trail up from the road along the crest of the mountain. On the crest is the site from which to watch hawks. The Appalachian Trail crosses this mountain, and you will see blazes for it at the crest.

I-84 Overlook (Port Jervis)

This site holds promise of being a productive site at which to watch the autumn raptor migration but it has not, as of 1981, been systematically manned. As a result, firm trends have not been established, and no numbers have been tallied. It deserves mention here as a potentially fine site and the hope is that ambitious hawk counters will pursue quantitative data here so that others will eventually learn of it and enjoy its flights.

To reach the I-84 Overlook (Port Jervis Hawkwatch), at Port Jervis gain I-84 headed east and proceed about two miles to the top of a mountain, where there are pull-offs offering scenic overviews on both the east and west sides of the highway. Continue east past the overviews area to the next exit. There, leave I-84, and reverse directions so that you can re-enter I-84 westbound. Drive on I-84 west to the vicinity of the overviews again. Park in the west side overview. This parking area and 200 to 300 feet west of the end of the overlook are the points from which hawks are observed. Some people who have watched hawks from here feel that the site west of the end of the overview offers greater visibility and less distraction from passing traffic.

Robert J. Hamershlag Hawkwatch (Butler Sanctuary) (Map 9.9)

This hawkwatch was initiated in 1979, and after only two years of coverage the Hamershlag site can boast a platform that affords excellent overviews in all directions, an ever-increasing number of competent observers, and more full-days coverage. The Broad-winged Hawk is the most numerous species recorded here, and there are also smaller numbers (between 600 and 800) of Sharp-shinned Hawks. The American Kestrel, as one might suspect, is the third most numerous species observed. In 1980, the site proudly added the Bald Eagle and the Peregrine Falcon to its list. The Robert J. Hamershlag Hawkwatch actively seeks observer participation and urges birders to join its efforts.

To reach the Robert J. Hamershlag Hawkwatch, from north, west, or northeast, approach the village of Mt. Kisco on I-84 and I-684; from the south approach Mt. Kisco on I-684. Leave I-684 at Exit 4 (east of the village of Mt. Kisco). Turn west on NY 172 (South Bedford Road) and proceed approximately 200 yards up a hill where Chestnut Ridge Road comes in on the left (south). Turn left (south) here and continue south on Chestnut Ridge Road for about 1.5 miles. You will have passed over I-684. Take the first right. Cross back over to the west side of I-684 and at the end of the bridge turn right on a small dirt road. Follow it only a short distance into Butler Sanctuary. The Hamershlag Hawkwatch platform is located within sight of the parking area to the sanctuary.

New York Botanical Gardens (Map 10.1)

Observers have only recently (within the past three years) come to learn that these lovely grounds are in fact a fine place from which to view autumn southbound hawks. The advantages of the Botanical Gardens over several other sites are: one can get there on public transportation, there is no arduous hike up to a summit or along lengthy woods trails, there are restroom facilities and eateries nearby, and, there are park benches at the watch on which to sit. The numbers of birds seen are not spectacular but they are certainly respectable and it cannot be beat for convenience for New York City birders without cars.

To reach the New York Botanical Gardens by automobile, from the Triborough Bridge proceed north into the Bronx and immediately get onto the Major Deegan Expressway. Continue north to the Fordham Road Exit. Turn right (east) onto Fordham Road. Continue east to the entrance to the New York Botanical Gardens on the left (north). This entrance is opposite the New York Zoological Park on the south. To reach the gardens by public transportation from Manhattan, on the east side of town take the Lexington Avenue IRT Number 5 Express subway train (uptown) to the Pelham Parkway/White Plains Road stop; on the west side of town take the Seventh Avenue IRT Number 2 Express (uptown) subway train to the Pelham Parkway/White Plains Road stop. From the subway stop walk southeast on Fordham Road to the entrance to the park.

COASTAL FALL HAWK LOOKOUTS IN NEW YORK STATE
Shinnecock Bay (Tiana Beach) (Maps 10.11 and 10.12)

Shinnecock Bay-Tiana Beach is famous, of course, for numbers and species of migrating shorebirds passing through southbound in late summer and early autumn (see page 443). This outer beach is every bit as rich for its numbers and variety of migrating raptors. The most notable flights are all seen on northwest and north winds. Almost no flight is recorded when the wind is from any other direction, except occasionally, when the surface wind is northeast. On those winds birds fly at great heights. The most numerous species seen here in descending order are: Sharp-shinned Hawk, American Kestrel, Merlin, Marsh Hawk, Osprey, Cooper's Hawk, and Peregrine Falcon. Notice that buteos are lacking from the list. To see buteos one must definitely move inland and find a decent mountaintop situated

TABLE 7

Coastal Fall Hawk Lookouts in New York State

	Shinnecock Bay (Tiana Beach)	Robt. Moses State Park (Democrat Point)	Old Inlet Club (Fire Island)	Jones Beach State Park	Moses Mountain
Region	10	10	10	10	10
County	Suffolk	Suffolk	Suffolk	Nassau and Suffolk	Richmond
Access by car	Yes	Yes	No†	Yes	Yes
Active periods	Mid-Sept. thru Mid-Oct.	Early Sept. thru Late Nov.	Early Sept. thru Mid-Oct.	Early Sept. thru Late Oct.	Mid-Sept. thru Early Nov.
Peak periods	Late Sept. and Early Oct.	Late Sept. and Early Oct.	Late Sept. and Early Oct.	Mid-Sept. thru Mid-Oct.	Late Sept. and Early Oct.
Best winds	NW-N	NW	NW	NW	NW

†Two to three mile hike on beach.

Note: 'Early' designates 1st through 10th of month.
'Mid' designates 11th through 20th of month.
'Late' designates 21st through end of month.

on a flight path. Out here on the barrier beach the observer can watch light-bodied, long-tailed accipiters and Falconidae flapping and gliding over the beaches and darting through the trees and bayberry patches until often they emerge with a kill. They are all strong fliers and their grace and agility is strikingly beautiful. This is a superb site, and by viewing the passage of these powerful raptors one gets a real sense of the nervous energy involved in the annual push south.

To reach Shinnecock Bay and Tiana Beach, turn south on Canoe Place Road from NY 27A (Montauk Highway) just east of Hampton Bays and just west of the Shinnecock Canal. Continue south as Canoe Place Road merges into Lynn Avenue. At Shinnecock Avenue turn right and then immediately left onto Light House Road. Continue south across Ponquogue Bridge to Dune Road (Beach Road). Turn right (west). All hawks can be counted by just sitting along Dune Road, about one-half mile to one mile west of the lighthouse at the Shinnecock Inlet. Parking along Dune Road is strictly prohibited, however, from May 15 to September 15. In order to count hawks before September 15, park (fee charged) at the county park farther west and walk east to a comfortable site.

Robert Moses State Park (Democrat Point, Fire Island) (Maps 10.13 and 10.14)

Hawks are no less opportunistic than other creatures, and that is quite clear here at Democrat Point, where Peregrine Falcons have consistently and over long periods been recorded as moving west down the outer beach and then at Democrat Point veering out over the ocean and assuming a southwesterly course. By holding such a course to the southwest, birds can shorten the migration over-land distance. Darrow (1963) reported that, based on twelve years experience at Democrat Point, approximately 75 percent of the Peregrines seen here altered their direction by 30 to 45 degrees in order to fly out over the ocean. These birds could easily by-pass New Jersey altogether and not come ashore until reaching Delaware, or even more likely, the Delmarva Peninsula. Ospreys occasionally take this over-ocean course from Democrat Point also, but almost 95 percent of the other raptors seen here leave Democrat Point and cross over to Jones Beach. This, however, is an excellent funneling point as it juts out on the western end of Fire Island. It can be an exciting site at which to view southbound falcons following along the dune-lines on days of northwest winds.

This too is a marvelous place to see wave after wave of migrating passerines in the height of the autumn migration (see page 436). The view is unobstructed in every direction, and in the sparse vegetation birds light to rest — to the delight of birders.

To reach Robert Moses State Park, leave the Southern State Parkway at Exit 40. Take the Robert Moses Twin Causeway across the eastern end of the island on which Jones Beach State Park is located (across Captree State Park). Take the Robert Moses Bridge across Fire Island Inlet to Fire Island. Turn right (west) and proceed to the farthest west open parking field. From it walk west toward Democrat Point. The best place to watch the hawk flight is primarily from parking field Number 6 or secondarily, from parking field Number 2 farther east toward the old lighthouse.

Old Inlet Club (Fire Island East) (Maps 10.13 and 10.14)

Not enough long-term counting has been done at this site to make comparisons with other barrier beach sites. However, the species composition of the Old Inlet Club counts approximate the flights seen along the rest of the barrier island in autumn. The coastal flavor of Sharp-shinned Hawk, American Kestrel, Northern Harrier, Osprey, Merlin, and Peregrine flights is here consistent with those of Shinnecock Bay-Tiana Beach and Democrat Point. For those who favor long beach walks, the Old Inlet Club is a bonus site. The walk is long and the outer beach ceaselessly wild. The anticipation of seeing a Peregrine Falcon always makes a tingling autumn afternoon more stimulating.

To reach the Old Inlet Club, leave I-495 (the Long Island Expressway) at Exit 68. Proceed south on the William Floyd Parkway (NY 46) to the Smith Point Bridge. Cross the Smith Point Bridge and park the car in Smith Point Park. Walk about two miles west to the hawk-watching site.

Jones Beach State Park (Maps 10.13 and 10.14)

Parking Field Number 1, west of Jones Beach Tower is often productive. When winds are from a westerly direction and at a velocity of eight to ten miles per hour, birds can be seen coming down the beach in a southwesterly direction. Parking Field Number 4 and the area northwest of it (between it and the fishing station) is also a good place to watch. Here on "hawk days," so termed for the continuous river of hawks passing by, an observer has the opportunity of familiarizing himself with the soaring and hunting habits of accipiters and falcons. The Jones Beach Strip does not offer the all-out hawkorama days of Hawk Mountain highs, but birders experience another sort of high, by studying various hunting methods of hawks and observing the fascinating and fancy flying antics of Merlins and Peregrines. The strip also offers the autumn beach birder the opportunity to see all sorts of land- and waterbirds in southbound migrations, and that is quite impossible atop an inland mountain. Get here early in the day and explore the strip for advantageous watching places. Be alert for an Osprey or two passing overhead, sometimes carrying a fat flounder. Watch for the flapping and sailing flight pattern of the Sharpie. Strain to find the powerful oaring of the Peregrine Falcon.

To reach Jones Beach State Park, leave the Southern State Parkway at Exits 22 or 27. Proceed south on the Meadowbrook State Parkway or the Wantaugh State Parkway. Parking is available at a number of parking fields once one reaches the barrier island. Almost any open parking field along the Jones Beach strip will afford fine views of the raptors passing. To view the raptor flight, a parking field on the ocean side of the strip is more advantageous than one farther back.

Moses Mountain (Staten Island) (Map 10.2)

Although birders have used Moses Mountain for many years as a place from which to observe the southward migration of raptors, only recently (in 1979) did its most regular supporters decide to officially man it and complete the reporting forms of the Hawk Migration Association of North America. With each new September and October, there seems to be an explosive increase in the numbers of mature and immature, male and female hawk watchers throughout the state. A similar trend can be documented through the entire northeast. With more watchers than ever before, manning more lookouts than ever before, a body of impressive sightings is rapidly building from which significant new research will be gleaned. Through better and more consistent coverage at a given site every hawk watcher can contribute to the pool of scientific data on raptors and thereby tangibly benefit the welfare of birds of prey. Anyone interested in assisting at this or any of the other hawk migration sites mentioned, should contact the editors of the *Kingbird* for more information.

To reach Moses Mountain, from the western end of the Verrazano Narrows Bridge, proceed west on the Staten Island Expressway (I-278); from the western end of the Goethals Bridge, proceed east on the Staten Island Expressway (I-278). Leave I-278 at the Bradley Avenue Exit and continue south on Bradley Avenue for approximately 0.9 mile to Brielle Avenue. Turn left (east) onto Brielle Avenue and proceed 0.4 mile to Manor Road. Turn right (south) and go 1.3 miles on Manor Road to its intersection with Rockland Avenue. Park the car well off the road as close to the intersection of Manor Road and Rockland Avenue as possible. Parking is not always easy in the immediate vicinity of this intersection. Manor Road is winding and narrow and it is sometimes difficult to find places to leave your car.

One must look along side streets and, after parking the car, walk back to the intersection, where one watches hawks. Moses Mountain summit lies at the junction of these two roads.

ACKNOWLEDGMENTS

I should like to thank the following people who contributed information on the sites covered in this chapter: Robert F. Andrle, Michael Bennett, members of the Alan Devoe Bird Club, Thomas L. Carrolan, Neil W. Currie, Harry Darrow, Bruno DeSimone, James Gibson, William L. Graves, Paul Jeheber, Paul Kerlinger, James W. Key, Walter Klabunde, Walter C. Listman, Robert O. Paxton, Belle Peebles, John M. C. Peterson, James Romansky, Norma and Bill Siebenheller, Eric Single, Jim Stapleton, Cornelius (Neil) Ward, and the late LeRoy Wilcox. Special thanks to Matthew Drennan who helped immeasurably in the immense task of data gathering for the presentation of the tables in this chapter.

Bibliography

THE MOST IMPORTANT SOURCE FOR THIS BOOK was *The Kingbird,* the quarterly ornithological journal of the New York State Federation of Bird Clubs, Incorporated. It has
been published since 1950. Articles and regional reports in Volumes 1 through 32 have been
extensively used throughout the researching and writing of this book.

American Birds, a bi-monthly ornithological field journal published by the National
Audubon Society, has been an invaluable research source during the writing of this book.
Especially important were the quarterly regional reports from the Hudson-Delaware Region
(Robert O. Paxton *et al.*) from 1975 through 1980, those of the Niagara-Champlain Region
(Douglas P. Kibbe) from 1975 through 1980, and the annual Blue Lists (Robert S. Arbib)
from 1975 through 1980.

Adirondack Mountain Club, Inc. *Guide to Adirondack Trails.* Glens Falls, N.Y., 1979.

American Ornithologists' Union Check-list Committee. *Check-list of North American
Birds,* 5th ed. Baltimore: American Ornithologists' Union, 1957.

_____. "Thirty-second Supplement to the American Ornithologists' Union Check-list of
North American Birds." *Auk* 90 (1973): 411–19.

_____. "Thirty-third Supplement to the American Ornithologists' Union Check-list of
North American Birds." *Auk* 93 (1976):875–79.

Arbib, R. S., O. S. Pettingill, and S. H. Spofford. *Enjoying Birds around New York City.*
Boston: Houghton-Mifflin, 1966.

Bagg, E. "Annotated List of the Birds of Oneida County, N.Y., and of the West Canada
Creek Valley." *Trans. Oneida Hist. Soc.* 12 (1911):17–86.

Bart, J. "Winter Distribution of Red-tailed Hawks in Central New York State." *Wilson Bulletin* 89 (1977):623–25.

Beardslee, C. S., and H. D. Mitchell. "Birds of the Niagara Frontier Region." *Bull. Buffalo
Soc. Nat. Sci.* 22 (1965):478 pp.

Beehler, B. McP. *Birdlife of the Adirondack Park.* Glens Falls, N.Y.: Adirondack Mountain Club, 1978.

Buckley, P. A. "Recent Specimens of Western Vagrants at Fire Island National Seashore,
Long Island, N.Y." *Auk* 91 (1974):181–85.

Bull, J. *Birds of the New York Area.* New York: Harper and Row, 1964.

_____. "Supplement to Birds of the New York Area." *Proc. Linn. Soc. N.Y.* 71 (1970):1–54.

_____. *Birds of New York State.* New York: Doubleday/Natural History Press, 1974.

_____. *Supplement to Birds of New York State.* New York: Federation of New York State Bird Clubs, 1976.

Carleton, G. "The Birds of Central and Prospect Parks." *Proc. Linn. Soc. N.Y. Nos. 66–70* (1962).

_____. "Supplement to the Birds of Central and Prospect Parks." *Proc. Linn. Soc. N.Y.* 71 (1970):132–54.

_____. *Birds of Essex County, New York.* Essex, N.Y.: High Peaks Aud. Soc., 1975.

_____. *Supplement to Carleton's Birds of Essex County.* Essex, N.Y.: High Peaks Aud. Soc., 1976.

Comar, M. C., D. P. Kibbe, and D. W. McIlroy. *Birding in the Cayuga Lake Basin.* Ithaca: Cornell Univ. Lab. Ornith., 1974.

Cruickshank, A. D. *Birds Around New York City.* New York: Amer. Mus. Nat. Hist., 1942.

Crumb, D. W. *Bird Finding in Onondaga County.* Syracuse: Onondaga Aud. Soc., 1971.

Eaton, E. H. *Birds of New York.* Parts 1&2. Albany: University of the State of New York, 1910.

Elliott, J. J., and R. S. Arbib. "Origin and Status of the House Finch in the Eastern United States." *Auk* 70 (1953):31–37.

Ewert, D. "First Long Island, New York, Nesting Record of the Kentucky Warbler." *Proc. Linn. Soc. N.Y.* 72 (1975):77–79.

Gochfield, M., and D. B. Ford. "Reproductive Success in Common Tern Colonies near Jones Beach, Long Island, New York, in 1972: A Hurricane Year." *Proc. Linn. Soc. N.Y.* 72 (1975):63–76.

Harper, R. M. "The Hempstead Plains: A Natural Prairie on Long Island." *Bull. Amer. Geogr. Soc.* 43 (1911):351–60.

Hays, H. "Polyandry in the Spotted Sandpiper." *The Living Bird* Eleventh Annual. Ithaca: Cornell Lab. Ornith. (1972):43–57.

_____. "Probable Common X Roseate Tern Hybrids. *Auk* 92 (1975):219–34.

Holroyd, G. L., and J. G. Woods. "Migration of the Saw-whet Owl in Eastern North America." *Bird-Banding* 46 (1975):101–105.

Kendleigh, S. C. "Community Selection by Birds on the Helderberg Plateau of New York." *Auk* 62 (1945):418–36.

Kennard, J. H. "Longevity Records of North American Birds." *Bird-Banding* 46 (1975): 55–73.

Klein, H. G., and C. W. Mitchell. *Birds of Clinton County.* Plattsburgh, N.Y.: Northern Adirondack Chapter of Natl. Aud. Soc., 1979.

LaBastille, A. "The Endangered Loon." *Andirondack Life* 8, no. 3 (1977):34–38.

Levine, M. "Some Observations at Fire Island, September 23, 1974." *Linnaean News-Letter* 28, no. 8 (1975):2–3.

MacFarlane, A. E. "Roof-Nesting by Common Terns (Suffolk Co.)." *Wilson Bulletin* 89 (1977):475–76.

Madden, S. "Evening Grosbeak Summers in Stuyvesant Town, Manhattan." *Linnaean News-Letter* 30, no. 5 (1976):1–2.

Meade, G. "Sixty Years of Rochester Birding." *Goshawk* 33 (1977):67.

Merriam, C. H. "A Preliminary List of Birds Ascertained to Occur in the Adirondack Region, Northeastern New York." *Nuttall Ornithological Club Bulletin* 6 (1881):225–35.

Mitchell, H. D., and R. F. Andrle. "Supplement to Birds of the Niagara Frontier Region." *Bull. Buffalo Soc. Nat. Sci.* 22 (1970):10 pp.

Palmer, R. S., ed. *Handbook of North American Birds — Loons to Flamingos,* vol. 1. New Haven: Yale University Press, 1962.

Parkes, K. C. "Notes on Some Birds of the Adirondack and Catskill Mountains, New York." *Annals Carnegie Mus.* 33: (1954):149–78.

———. "Gulls and Terns of New York." *New York State Conservation Dept. Booklet,* 1959.

Pasquier, R. "The 1974 Spring Migration in Central Park." *Linnaean News-Letter* 28, no. 5 (1974):1.

———. "Recent Additions to the Birds of Central Park." *Proc. Linnaean Soc. N.Y.* 72 (1976):80–81.

Peterson, R. T. *A Field Guide to the Birds,* 2nd ed. Boston: Houghton Mifflin, 1947.

———. *A Field Guide to the Birds,* 4th ed. Boston: Houghton Mifflin, 1980.

Pettingill, O. S., and S. F. Hoyt. *Enjoying Birds in Upstate New York.* Ithaca: Cornell Univ. Lab. Ornith., 1963.

Pough, R. H. *Audubon Bird Guide: Eastern Land Birds.* New York: Doubleday, 1949.

———. *Audubon Water Bird Guide.* New York: Doubleday, 1951.

Radke, E. L. *Birds of the Columbia County Area.* Chatham, N.Y.: Alan Devoe Bird Club, 1972.

Robbins, C. S., B. Bruun, and H. Zim. *Birds of North America.* New York: Golden Press, 1966.

Saunders, A. A. "The Summer Birds of the Allegany State Park." *Roosevelt Wildlife Bull.* 22 (1923):239–354.

———. "The Summer Birds of the Northern Adirondacks Mountains." *Roosevelt Wildlife Bull.* 5 (1929):327–499.

———. *Ecology of the Birds of Quaker Run Valley, Allegany State Park.* Albany: New York State Mus. Handbook 16, 1936.

———. "Studies of Breeding Birds in the Allegany State Park." *New York State Mus. Bull.* 318 (1938).

———. *Summer Birds of the Allegany State Park.* Albany: New York State Mus. Handbook 18, 1942.

Spofford, W. R. "The Breeding Status of the Golden Eagle in the Appalachians." *American Birds* 25 (1971):3–7.

Tate, J., Jr. "The Cornell Laboratory of Ornithology." *American Birds* 25 (1971):10–12.

Temple, S. A., and B. L. Temple. "Avian Population Trends in Central New York State, 1935–1972." *Bird-Banding* 47 (1976):238–57.

The Nature Conservancy. *The Nature Conservancy Guide to the Natural Areas of the Eastern New York Chapter,* 5th ed. New York: Nature Conservancy, 1976.

Tudor, G. "Captree June Count." *Linnaean News-Letter* 28, no. 1 (1974):1–3; 28, no.2 (1974): 1–3.

Webb, W. L., D. F. Behrend, and B. Saisorn. "Effect of Logging on Songbird Populations in a Northern Hardwood Forest." *Wildlife Monographs* 55 (1977):35pp.

Weld, P. W., and A. Klonick. *Getting Acquainted with Birds in Genesee Country.* Rochester: Rochester Academy of Science, 1966.

Wilcox, L. "A Twenty-Year-Banding Study of the Piping Plover." *Auk* 76 (1959):129–52.

INDEX

To avoid searching an overly long species index, the following matrix, taxonomically ordered, was devised. This will assist the reader in locating birds of particular interest, by region, cited in the text. "B" indicates that a species is noted as breeding at one or more sites within a regional write-up. "o" indicates that a species is noted as occurring or having occurred, but not as breeding, at one or more sites within a regional write-up. For the likelihood of seeing any of these special species, consult the text. For the status of birds in the families Procellariidae, Hydrobatidae, Sulidae, Phalaropodidae, Stercorariidae, and Alcidae, see the chapter on Seabirds and Pelagic Birding.

SPECIES					REGION					
	1	2	3	4	5	6	7	8	9	10
Loon, Common	o	o	o	o	B	B	B		o	o
Grebe, Pied-billed	o	o	B	B	B		B	o	B	o
Heron, Little Blue										B
Heron, Louisiana									o	B
Night Heron, Yellow-crowned	o									B
Bittern, Least	o	o	o		B	B	B	B	B	
Ibis, Glossy	o	o							o	B
Osprey	o	o	o	B	o	B	B	o	o	B
Eagle, Bald	o	B	o	o	o	o	o	o	o	o
Eagle, Golden	o	o					o		o	o
Grouse, Spruce						B	B			
Partridge, Gray						B	B			
Turkey	B	B	B		B			B	B	B
Rail, King	B				B				B	o
Rail, Clapper									B	B
Rail, Virginia	B	o	B	B	B		B	B	B	B
Sora	B	B			B		B	B	B	B
Rail, Black										B
Oystercatcher, American										B
Plover, Piping										B
Willet		o					o			B
Knot, Red		o			o	o	o			o
Sandpiper, Baird's		o			o	o	o		o	o
Sandpiper, White-rumped					o		o			o
Sandpiper, Pectoral	o				o	o		o	o	o
Sandpiper, Curlew										o
Sandpiper, Western	o	o			o	o			o	o
Sandpiper, Buff-breasted	o	o	o		o	o	o	o	o	o
Sandpiper, Upland	B	B		o	B	o	o	o	B	B
Whimbrel		o			o	o				o
Godwit, Hudsonian	o	o			o	o				o
Godwit, Marbled	o	o	o		o	o				o
Avocet, American	o	o	o						o	o
Gull, Bonaparte's	o	o		o	o			o	o	o
Gull, Little	o	o								o
Tern, Gull-billed										B
Tern, Common	o	o		o	o	B			o	B
Tern, Least										B
Tern, Royal									o	o

SPECIES (Continued)	REGION									
	1	2	3	4	5	6	7	8	9	10
Tern, Sandwich										o
Tern, Caspian	o	o			o		o			o
Tern, Black	B	o	o	o	B	B	B	o	o	o
Tern, Roseate										B
Skimmer, Black										B
Owl, Barn	B							B	B	B
Owl, Snowy	o	o	o	o	o	o	o	o		o
Owl, Long-eared	B	B					o		B	B
Owl, Short-eared	B	o	o		B	o	o	o	o	B
Owl, Saw-whet	B	B			o	B	B	o	B	o
Chuck-will's-widow										B
Nighthawk, Common				B	B				B	o
Woodpecker, Red-headed	B	B	B		B	B	o			o
Sapsucker, Yellow-bellied			B	B	B	B	B	B	B	o
Woodpecker, Black-backed Three-toed						B	B			
Woodpecker, Northern Three-toed						B	B			
Kingbird, Western										o
Flycatcher, Yellow-bellied		o	B		B	B	B		B	o
Flycatcher, Least	o		B		B	B	B		B	B
Flycatcher, Acadian	o	o							B	B
Flycatcher, Willow			o		B			B	B	o
Flycatcher, Alder	B		o	B	B	B	B	B	B	o
Flycatcher, Olive-sided		o			B	B	B		o	o
Swallow, Rough-winged		o	B	o		B	B		o	B
Swallow, Cliff	B	B	o	B		B	B			o
Jay, Gray						B	B			
Raven, Common						B	B			
Crow, Fish									B	B
Chickadee, Boreal		o			o	B	B	o	o	
Nuthatch, Red-breasted		B	B	o	B	B	B	o	B	o
Creeper, Brown	B	B	B	B	B		B	B	B	B
Wren, Winter	B	B	B	o	B	B	B	B	B	B
Wren, Carolina		B		B					B	B
Marsh Wren, Long-billed		B	o	o	B		B	B	B	B
Marsh Wren, Short-billed	B			B	B			B		
Thrush, Hermit	B	B	B	B	B	B	B	B	B	B
Thrush, Swainson's	B		B	B		B	B		B	o
Thrush, Gray-cheeked		o	B				B	B	B	o
Gnatcatcher, Blue-gray	o	B	o	B	B		B	B	B	B

SPECIES (*Continued*)	1	2	3	4	REGION 5	6	7	8	9	10
Kinglet, Golden-crowned	B	B	B	B	B	B	B		B	o
Kinglet, Ruby-crowned	B				o	B	B		o	o
Pipit, Water	o		o	o	o				o	o
Waxwing, Bohemian		o				o	o			
Shrike, Northern		o	o		o	o		o	o	
Shrike, Loggerhead	o	B		o	o	B		o		o
Vireo, Solitary	B	B	B	B		B	B		B	
Vireo, Philadelphia	o	o				B	B			
Warbler, Prothonotary	B		B						o	B
Warbler, Worm-eating	o	o	B		o			B	B	
Warbler, Blue-winged	o	B	B	B	B			B	B	B
Warbler, Golden-winged	o	B	B	B	B	B		B	B	o
Warbler, Tennessee				B	o	B	B		o	o
Warbler, Nashville	o	B	B	B	B	B	B		B	o
Parula, Northern	B	B				B	B		B	B
Warbler, Magnolia	B		B	B	B	B	B	B	B	o
Warbler, Cape May	o			B		B	B		B	o
Warbler, Blackburnian	B	B	B	B	B	B	B	B	B	o
Warbler, Yellow-throated	o		o						o	o
Warbler, Chestnut-sided	B		B	B	B	B	B	B	B	B
Warbler, Bay-breasted	o			B		B	B		o	o
Warbler, Pine		B	B		B		B	B	B	B
Warbler, Prairie	o		B	B			o	B	B	B
Warbler, Cerulean	B	B	B		B	o			B	o
Waterthrush, Louisiana		B	B		B			B	B	o
Warbler, Kentucky		o	o		o				B	o
Warbler, Mourning	B	B	B	B	B	B	B		B	o
Chat, Yellow-breasted		B	B	o	o				B	B
Warbler, Hooded	B	B	B		B				B	o
Warbler, Wilson's		o				o	B		o	o
Bobolink	B	o	o	B	B		B	B	B	B
Oriole, Orchard	o	B	o						B	o
Blackbird, Rusty	o				o	B	B	o		B
Grackle, Boat-tailed										B
Grosbeak, Blue									o	o
Finch, Purple			o	o	o	B	B		B	o
Siskin, Pine		o	o	B	o	B	B	o	o	o
Crossbill, Red		o	o	o	o	B	B		o	B
Crossbill, White-winged		o	o		o		B		o	o

SPECIES (*Continued*)	REGION									
	1	2	3	4	5	6	7	8	9	10
Sparrow, Savannah	B	o	B	B	B		B		B	B
Sparrow, Savannah (Ipswich)										o
Sparrow, Grasshopper	B	B	B	B	B				B	B
Sparrow, Henslow's	o	B	B		B	B				o
Sparrow, Sharp-tailed									o	B
Sparrow, Seaside									B	B
Sparrow, Vesper	B	o	o	B	B		B	B	B	o
Sparrow, Field	B		B	B	B		B	B	B	B
Sparrow, White-throated		B	o	B	B	B	B	B	B	o
Sparrow, Lincoln's	o		o		o	B	B			o

SITES INDEX

Numbers in bold face type indicate the page number on which the map for the named site appears.

NOTES

NOTES

NOTES

WHERE TO FIND BIRDS IN NEW YORK STATE

was composed in 9-point Compugraphic Times Roman and leaded two points,
with display type also in Compugraphic Times Roman,
by Metricomp Studios;
printed by sheet-fed offset on 50-pound acid-free Glatfelter Smooth Offset,
Smythe-sewn, and bound over boards in Columbia Bayside Chambray,
also adhesive bound with laminated covers,
by Maple-Vail Book Manufacturing Group, Inc.;
and published by

SYRACUSE UNIVERSITY PRESS
SYRACUSE, NEW YORK 13210

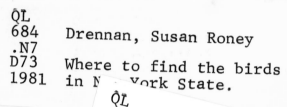

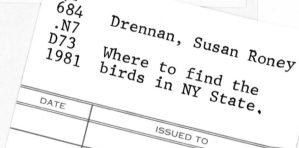

DATE	ISSUED TO